Direct Copper Interconnection for Advanced Semiconductor Technology

In the "More than Moore" era, performance requirements for leading edge semiconductor devices are demanding extremely fine pitch interconnection in semiconductor packaging. Direct copper interconnection has emerged as the technology of choice in the semiconductor industry for fine pitch interconnection, with significant benefits for interconnect density and device performance. Low-temperature direct copper bonding, in particular, will become widely adopted for a broad range of high-performance semiconductor devices in the years to come.

This book offers a comprehensive review and in-depth discussions of the key topics in this critical new technology. Chapter 1 reviews the evolution and the most recent advances in semiconductor packaging, leading to the requirement for extremely fine pitch interconnection, and Chapter 2 reviews different technologies for direct copper interconnection, with advantages and disadvantages for various applications. Chapter 3 offers an in-depth review of the hybrid bonding technology, outlining the critical processes and solutions. The area of materials for hybrid bonding is covered in Chapter 4, followed by several chapters that are focused on critical process steps and equipment for copper electrodeposition (Chapter 5), planarization (Chapter 6), wafer bonding (Chapter 7), and die bonding (Chapter 8). Aspects related to product applications are covered in Chapter 9 for design and Chapter 10 for thermal simulation. Finally, Chapter 11 covers reliability considerations and computer modeling for process and performance characterization, followed by the final chapter (Chapter 12) outlining the current and future applications of the hybrid bonding technology. Metrology and testing are also addressed throughout the chapters. Business, economic, and supply chain considerations are discussed as related to the product applications and manufacturing deployment of the technology, and the current status and future outlook as related to the various aspects of the ecosystem are outlined in the relevant chapters of the book.

The book is aimed at academic and industry researchers as well as industry practitioners, and is intended to serve as a comprehensive source of the most up-to-date knowledge, and a review of the state-of-the art of the technology and applications, for direct copper interconnection and advanced semiconductor packaging in general.

Direct Copper Interconnection for Advanced Semiconductor Technology

Edited by
Dongkai Shangguan

CRC Press
Taylor & Francis Group
Boca Raton London New York

CRC Press is an imprint of the
Taylor & Francis Group, an **informa** business

First edition published 2024
by CRC Press
6000 Broken Sound Parkway NW, Suite 300, Boca Raton, FL 33487-2742

and by CRC Press
4 Park Square, Milton Park, Abingdon, Oxon, OX14 4RN

CRC Press is an imprint of Taylor & Francis Group, LLC

ISBN: 9781032528236 (hbk)
ISBN: 9781032528403 (pbk)
ISBN: 9781003408673 (ebk)

DOI: 10.1201/9781003408673

Typeset in Times New Roman
by KnowledgeWorks Global Ltd.

Contents

Preface....................vii
Acknowledgments....................ix
Editor....................xi
Contributors....................xiii

Chapter 1 Advanced Packaging Landscape for Heterogeneous Integration....................1

Ravi Mahajan, Sairam Agraharam, Debendra Mallik, and Vijaya Boddu

Chapter 2 Direct Copper Interconnection for Die/Wafer Bonding: Overview....................19

Tzu-Heng Hung and Kuan-Neng Chen

Chapter 3 Hybrid Bonding Process Technology....................54

Guilian Gao and Laura Mirkarimi

Chapter 4 Materials for Hybrid Bonding....................116

Andrea Chacko, Dongshun Bai, and Rama Puligadda

Chapter 5 Copper Electrodeposition for Advanced Packaging and Hybrid Bonding....................131

Bryan Buckalew

Chapter 6 Planarization for Advanced Packaging and Hybrid Bonding....................168

Brian J. Brown, Ekaterina Mikhaylichenko, Jianshe Tang, Jun Qian, and Yen-Chu Yang

Chapter 7 Permanent and Temporary Wafer Bonding....................201

Viorel Dragoi, Thomas Uhrmann, and Paul Lindner

Chapter 8 Die-to-Wafer Hybrid Bonding for Direct Copper Interconnections....................222

Jonathan Abdilla and Stefan Schmid

Chapter 9 Design for Hybrid Bonding and Chiplets....................268

Brandon Wang and Rong Bao

Chapter 10 Thermal Modeling and Simulation for Advanced 3DIC Systems...... 299

Norman Chang, David Geb, Akhilesh Kumar, C.T. Wu, and Bo Ren

Chapter 11 Characterization, Modeling, and Reliability for Direct Copper Interconnection 361

Stéphane Moreau and Dongkai Shangguan

Chapter 12 Applications of Hybrid Bonding and Chiplets for Heterogeneous Integration 428

Chandrasekhar Mandalapu, John Wuu, Raja Swaminathan, and Liwei Wang

Index 447

Preface

I still remember the day when I got to see electronics "under the hood"—literally and figuratively—for the first time, as I walked into the Electronics Division at Ford Motor Co. in Dearborn, Michigan. Since then, I have had the opportunity to witness the evolution of the semiconductor and electronics technology over the past three decades—toward miniaturization and functional densification, with increasing performance and reliability, and reduced cost.

Over the years, I have been amazed by the advancement in the transistor technology following Moore's law, and more and more, I appreciate the growing importance of the "interconnect"—from the transistor to the system. In the "More than Moore" era, performance for leading edge semiconductor devices demands extremely fine pitch interconnection in semiconductor packaging.

Today, semiconductor packaging is about to enter a new era—with interconnect pitches down to 1 μm (or even below). Direct copper interconnection has emerged as the technology of choice for fine pitch interconnection, with significant benefits for interconnect density and device performance.

While the direct copper interconnection technology shows great promise, its broad adoption—in both scale and scope—depends on a number of critical enabling technologies.

This book offers a comprehensive review and in-depth discussion of the key topics in this critical new technology. Chapter 1 reviews the evolution and the most recent advances in semiconductor packaging, leading to the requirement for extremely fine pitch interconnection, and Chapter 2 reviews different technologies for direct copper interconnection, with advantages and disadvantages for various applications. Chapter 3 offers an in-depth review of the hybrid bonding technology, outlining the critical processes and solutions. The area of materials for hybrid bonding is covered in Chapter 4, followed by several chapters that are focused on critical process steps and equipment for copper electrodeposition (Chapter 5), planarization (Chapter 6), wafer bonding (Chapter 7), and die bonding (Chapter 8). Aspects related to product applications are covered in Chapter 9 for design and Chapter 10 for thermal simulation. Finally, Chapter 11 covers reliability considerations and computer modeling for process and performance characterization, followed by the final chapter (Chapter 12) outlining the current and future applications of the hybrid bonding technology. Metrology and testing are also addressed throughout the chapters. Business, economic, and supply chain considerations are discussed as related to the product applications and manufacturing deployment of the technology, and the current status and future outlook as related to the various aspects of the ecosystem are outlined in the relevant chapters of the book.

The book is aimed at academic and industry researchers as well as industry practitioners and is intended to serve as a comprehensive source of the most up-to-date knowledge, and a review of the state of the art of the technology and applications, for direct copper interconnection and advanced semiconductor packaging in general.

It is my sincere hope that this book will play a meaningful role in stimulating further exploration in the field of advanced semiconductor technology.

Dongkai Shangguan
San Jose, CA
October 2023

Acknowledgments

First, I would like to thank the chapter authors—from many companies and institutions over three continents—for their contributions and for their confidence in me.

My industry colleagues worldwide deserve special thanks, as they have been the source of inspiration for my work in this area.

I would also like to thank Thermal Engineering Associates, Inc., for supporting me in this endeavor.

Finally, it is the love from my family and friends that constantly motivates me to keep marching forward on this truth-seeking journey that may in some small way benefit our society.

Dongkai Shangguan
San Jose, CA
October 2023

Editor biography

Dongkai Shangguan, IEEE Fellow and IMAPS Fellow, is president of Thermal Engineering Associates, Inc., a provider of analog ASIC chips for thermal testing. He is also a strategic advisor to innovative companies in the global semiconductor and electronics industry. Previously, he served as corporate vice president for Technology & Engineering at Flex (formerly Flextronics), a global player for electronics design and manufacturing, and as chief marketing officer at STATS ChipPAC (currently part of JCET), a leading provider of semiconductor assembly and test services. Early in his career, he held various technical and management responsibilities at Ford and Visteon on automotive electronics.

Dr. Shangguan has published several books and authored/coauthored over 200 technical papers, and has been issued 33 U.S. patents and a number of international patents. He is an IEEE EPS Distinguished Lecturer, and has given numerous keynotes, seminars, and lectures globally.

Dr. Shangguan has served on the iNEMI board of directors, the IEEE EPS board of governors, and the IPC board of directors. He has received a number of recognitions for his contributions to the industry, including the Electronics Manufacturing Technology Award and the Outstanding Sustained Technical Contribution Award from IEEE EPS, the William D. Ashman Achievement Award from IMAPS, the President's Award from IPC, and the Total Excellence in Electronics Manufacturing Award from the Society of Manufacturing Engineers.

Dr. Shangguan received his B.Sc. in Mechanical Engineering from Tsinghua University, China, MBA degree from San Jose State University, and Ph.D. in Materials from the University of Oxford, UK. He conducted postdoctoral research at the University of Cambridge and The University of Alabama, and has served as a guest professor at several universities.

Dr. Shangguan is based in San Jose, California, and travels globally in service of the industry.

Contributors

Jonathan Abdilla
BESI
Radfeld, Austria

Sairam Agraharam
Intel
Chandler, Arizona, USA

Dongshun Bai
Brewer Science, Inc.
Rolla, Missouri, USA

Rong Bao
Cadence Design Systems
Milpitas, California, USA

Vijaya Boddu
Intel
Santa Clara, California, USA

Brian J. Brown
Applied Materials
Santa Clara, California, USA

Bryan Buckalew
Lam Research
Tualatin, Oregon, USA

Andrea Chacko
Brewer Science, Inc.
Rolla, Missouri, USA

Norman Chang
Ansys
San Jose, California, USA

Kuan-Neng Chen
National Yang Ming Chiao Tung University
Hsinchu, Taiwan

Viorel Dragoi
EV Group
Sankt Florian am Inn, Austria

Guilian Gao
Adeia
San Jose, California, USA

David Geb
Ansys
San Jose, California, USA

Tzu-Heng Hung
National Yang Ming Chiao Tung University
Hsinchu, Taiwan

Akhilesh Kumar
Ansys
San Jose, California, USA

Paul Lindner
EV Group
Sankt Florian am Inn, Austria

Ravi Mahajan
Intel
Chandler, Arizona, USA

Debendra Mallik
Intel
Chandler, Arizona, USA

Chandrasekhar Mandalapu
AMD
Austin, Texas, USA

Ekaterina Mikhaylichenko
Applied Materials
Santa Clara, California, USA

Laura Mirkarimi
Adeia
San Jose, California, USA

Stéphane Moreau
University Grenoble Alpes
CEA-Leti
Grenoble, France

Rama Puligadda
Brewer Science, Inc.
Rolla, Missouri, USA

Jun Qian
Applied Materials
Santa Clara, California, USA

Bo Ren
Ansys
San Jose, California, USA

Stefan Schmid
BESI
Radfeld, Austria

Dongkai Shangguan
Thermal Engineering Associates, Inc.
Santa Clara, California, USA

Raja Swaminathan
AMD
Austin, Texas, USA

Jianshe Tang
Applied Materials
Santa Clara, California, USA

Thomas Uhrmann
EV Group
Sankt Florian am Inn, Austria

Brandon Wang
Synopsys
Sunnyvale, California, USA

Liwei Wang
AMD
Austin, Texas, USA

C.T. Wu
Ansys
San Jose, California, USA

John Wuu
AMD
Fort Collins, Colorado, USA

Yen-Chu Yang
Applied Materials
Santa Clara, California, USA

1 Advanced Packaging Landscape for Heterogeneous Integration

Ravi Mahajan, Sairam Agraharam, Debendra Mallik, and Vijaya Boddu

1.1 INTRODUCTION

The trajectory of integrated circuits (ICs), first described in the now classic Moore's law paper [1], has served as a reliable self-fulfilling prophecy [2, 3], fueling the digital revolution and driving fundamental changes in many aspects of modern society [4]. The transistor is at the heart of this revolution and the 75th anniversary of the transistor in 2022, was celebrated by a series of articles and discussions on the future of Moore's Law innovation [5]. While the transistor is the kernel of the digital revolution, there is increasing realization that a systems approach, implemented using System-Technology-Co-Optimization (STCO) techniques, is needed for continued power, performance, and cost optimization [3]. The elements of STCO from a hardware perspective include collaborative development of transistors, interconnects, process technology, hard IP, memory technologies, advanced packaging architectures, and physical system architectures. System hardware optimization must be complemented by software optimization for all targeted workloads.

In this chapter, we focus on one of the key elements of STCO, i.e., the attributes and scaling vectors of advanced packaging architectures with a primary focus on die-to-die and die-to-package interconnects. For purposes of this chapter, advanced packaging architectures are defined only from an interconnect perspective, as architectures with (die-to-die and die-to-package) interconnect pitches of $\leq$100 μm and interconnect lines and spaces $\leq$9–12 μm. Heterogeneous integration (HI) refers to the integration of separately manufactured components into a higher level assembly that, in the aggregate, provides enhanced functionality and improved operating characteristics [6]. In addition to on-package interconnects, advanced packaging architectures also need to provide improved off-package connectivity, signaling performance, and power delivery [7–10].

The structure of the chapter is as follows. The evolution of packaging over the past five decades is described showing the major technological advances that have ensured continued progress of Moore's Law. The motivations for increased interest in HI, where advanced packaging architectures are used to create a compact

DOI: 10.1201/9781003408673-1

platform for high-performance HI, are described next. This is followed by a review of the current state of the art and projected scaling of on-package interconnects in advanced packaging architectures, establishing the need for hybrid bond interconnects (HBI). High-level overviews of process, materials, and equipment characteristics needed for HBI are then described. The chapter concludes with a discussion of how product architectures can benefit from HBI.

1.2 THE EVOLUTION OF MICROELECTRONICS PACKAGING

In the very early stages of IC evolution, the primary functions of packaging were to provide a reliable enclosure for intrinsically fragile silicon devices and provide electrical connectivity between the die and the system motherboard. Ceramic packages, typically employing wire-bond interconnects between pads on the die periphery and package pads, were used to provide hermetic protection. In lower cost segments, organic packages, typically with lead-frame-based peripheral interconnects, were used to provide non-hermetic protection (Figures 1.1 and 1.2). Since the interconnect count was low, wire-bond and lead-frame interconnects, which were essentially on the die periphery, provided adequate connectivity. In the early 1990s, product architectures, taking greater advantage of Moore's Law scaling, increased in functionality and performance. In parallel, semiconductor product volumes grew exponentially as increased customer interest in both the consumer and business markets for computers led them to become mainstream [11]. As a result, packaging technologies underwent significant changes to meet performance, volume, cost, and system form-fit constraints [12]. Some of these changes were as follows:

- As die-to-package interconnect counts increased, area array interconnect technologies (primarily solder-based C4/flip-chip interconnects) replaced peripheral array interconnects (e.g., wire-bonding) in high-performance applications.

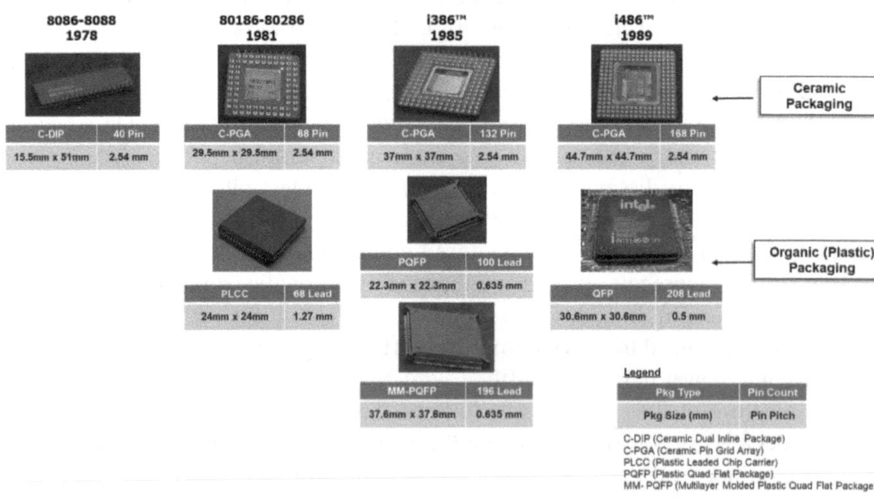

FIGURE 1.1 Representative packaging technologies used in the 1970s and 1980s.

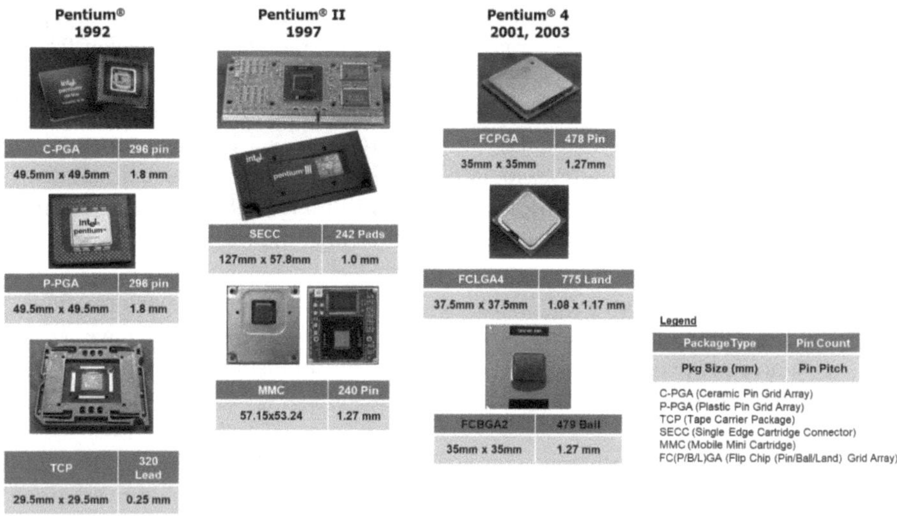

FIGURE 1.2 Representative packaging technologies used in the 1990s and early 2000s. Notice the steady increase in pin counts, the transition from ceramic packaging technologies to organic (plastic) packaging technologies, and the increasing diversity in form factors.

- Organic substrates replaced ceramic substrates enabling improved performance because lower resistance copper (Cu) interconnects replaced tungsten (W) based interconnects (used in high-temperature co-fired ceramic packages), and lower processing temperatures in substrate manufacturing likely reduced manufacturing costs.
- Increasing performance led to increased power requiring greater focus on thermal management and high-efficiency power delivery networks [8]. Thermal management technologies such as improved thermal interface materials (TIMs), heat-spreaders, active and passive heat sinks (including fan-sinks, heat pipes), and power delivery technologies (such as improved passives, voltage regulators) became mainstream and were integrated into or with the package [13].
- As signaling speeds increased, there was greater emphasis on understanding and improving the package interconnect performance by optimizing design layouts and use of lower loss materials [9, 14].
- Interconnect technologies for area array package to the motherboard, including sockets (PGA, LGA) and surface mount technologies (BGA) increased pin counts.
- With increasing diversity especially in consumer applications (as evidenced by the popularity of laptops, tablets, cell phones, and other handheld devices), physical system form factors imposed significant size, power, and weight constraints on package form factors leading to the development of thin and light packages [15].

- Packages became environment-friendly as lead-free metals were used in flip-chip and BGA joints and the organic package substrates became halogen-free.
- With increasing product volumes and a focus on cost reduction, a greater emphasis was placed on manufacturing processes, equipment, materials, and yield optimization, and this continues to date.
- Test methods and their associated equipment, which are critical for performance characterization and improved manufacturing yields, continue to receive tremendous focus and have become increasingly sophisticated [16].
- Key enabling technologies (such as materials, modeling, performance/process/reliability characterization methods and metrologies) have also concurrently evolved to support the increasing sophistication of packaging architectures.

The concurrent and complementary evolution of these changes has led to significant growth in the field of packaging, assembly, and test to support continued growth in computing and communication performance.

1.2.1 The Package as a Platform for HI

There is a long history of using the package as a platform for multi-chip integration, one of the best-known early examples being the IBM Thermal Conduction Module [17]. As shown in Figure 1.3, the multi-chip package has been used for memory and controller integration.

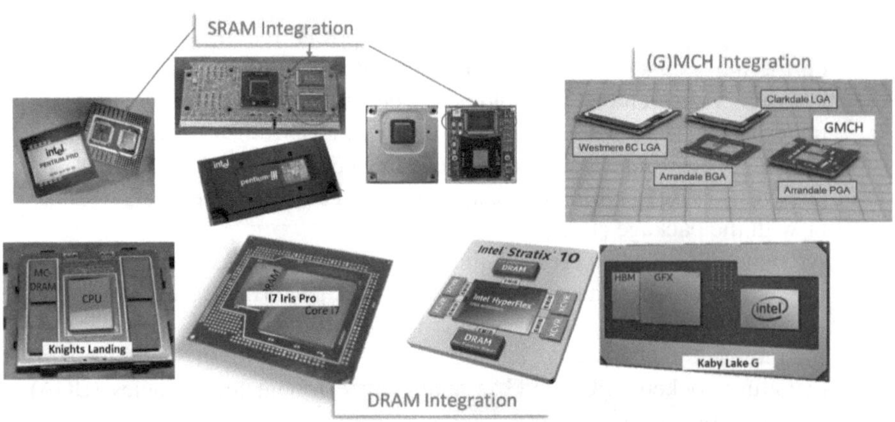

FIGURE 1.3 Three historical examples where the package is used as a compact platform for multi-chip integration. In the SRAM example, both packaged and unpackaged SRAMs were integrated on package mainly for faster time to market. Since SRAMs can be integrated with logic, they were integrated in the logic die in later instantiations of the similar products and they became single-chip configurations. (G)MCH and DRAM integration with the logic die is not as feasible or likely and these configurations persisted longer.

FIGURE 1.4 Motivations for the package as a compact HI platform.

In recent years, the package is increasingly used as a platform for compact HI for the reasons illustrated in Figure 1.4, and such use has sparked significant broad-based interest [18].

A number of concurrent advances in key enabling technologies in the last decade have helped realize the benefits of the package as a HI platform and have set the trajectory for HI technology evolution. Some of the important enabling technologies are as follows:

- The maturing of advanced packaging technologies that provide significantly higher die-to-package and on-package interconnect densities [19–21]
- The development of advanced assembly technologies for attaching die with high and variable bump densities to the package [22]
- The availability of high-density memories [23–25] and a focus on standardizing low-power, high-bandwidth (BW) interfaces [26]

These technologies have enabled some truly technologically impressive products, e.g., the Intel® Data Center GPU Max Series (Ponte Vecchio) (Figure 1.5). The idea that dies (also referred to as chiplets) could be integrated on package using standardized interfaces has been tangibly demonstrated in a number of very interesting use cases and is anticipated to continue evolving [27–30].

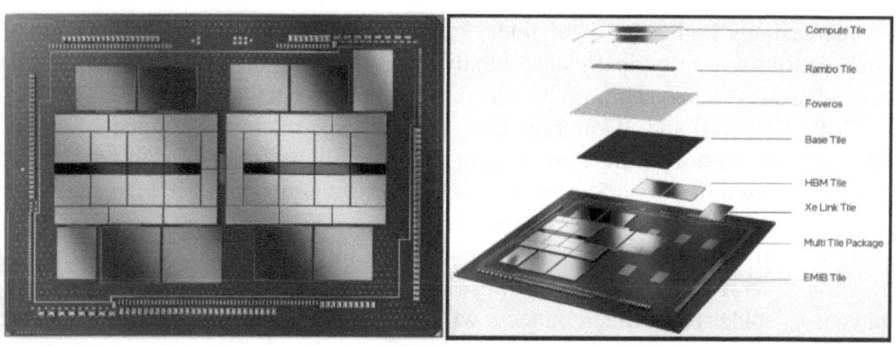

FIGURE 1.5 The Intel® Data Center GPU Max Series (Ponte Vecchio) package combines 47 tiles, integrates die from five different process nodes, and employs high-density 2.5D and 3D interconnects.

1.2.2 The Role of Advanced Packaging in Driving the Evolution of Technologies

Advanced packaging, implemented using various methods, including the use of high-density organic interconnects (e.g., ultrahigh density (UHD) fanout redistribution layers (RDL), etc.) and silicon-based interconnects (e.g., silicon bridges and silicon interposers), today offers a wide range of (die/die-stack) to package and on-package interconnect densities [31, 32]. Each of the technologies described in Refs [31, 32] demonstrates the unleashing of considerable innovation once the need for increased HI has become mainstream and each is optimized for cost, performance, and system fit.

In the next section, we will examine the status and future trajectory for the key attribute enabled by these innovations, i.e., package interconnects, with a focus on their physical attributes.

1.2.3 On-package Physical Interconnects

A package supports six broad types of interconnects [33]:

a. Die-to-die interconnects between stacked dies
b. On-package die-to-die lateral interconnects
c. Die-to-package interconnects
d. Within package interconnects
e. Package-to-motherboard interconnects
f. Package-on-package interconnects

Of these, we will focus on the first three types of interconnects. Two physical metrics (Figure 1.6) describe interconnect capability:

a. Linear interconnect density, i.e., the number of wires per unit die edge length and per layer of the package (if a silicon interposer or bridge is used, then "package layer" will be replaced by "silicon interconnect layer)
b. Areal interconnect density, i.e., the number of bumps per unit die area

Increasingly higher levels of linear interconnect density can be achieved using high-density wiring typically used in silicon far backend metallization, and higher areal densities can be achieved by shrinking the bump pitch [34].

Both 2.5D [34] and 3D architectures [35] use fine-pitch interconnects for links that require high IO/mm^2 (see Figure 1.7 that shows solder-based EMIB and Foveros interconnects at 55 μm and 36 μm minimum bump pitch, respectively, for the Intel® Data Center GPU Max Series (Ponte Vecchio) Ponte Vecchio architecture [36]). While solder-based interconnects can be scaled down to 10 μm [37], there is general consensus that a transition to HBI will be needed with shrinking pitches as solder-based interconnects will be increasingly limited in their current-carrying capability and increasingly challenging to build due to increased planarity requirements, reduced solder volumes, and reduced bump gaps (which will limit underfilling), even though various technical (e.g., die size, thickness, and

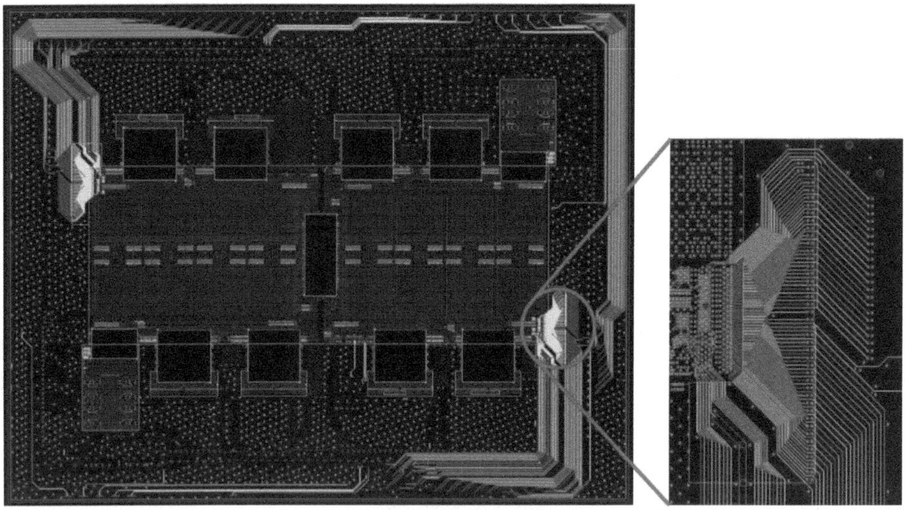

(a)

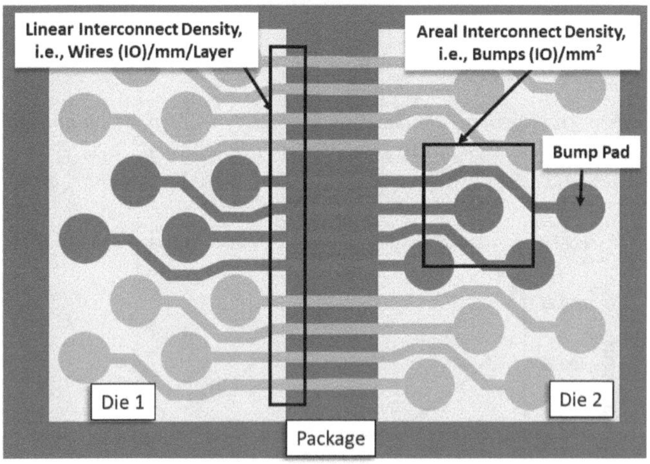

(b)

FIGURE 1.6 (a) Top view of a package routing layer. A bump on the die lands on a pad on the package and wires connected to the pad fanout to provide interconnect space transformation. (b) Two key interconnect density metrics, i.e., linear interconnect density (number of wires per unit die edge length per layer of the package) and areal interconnect density (number of bumps per unit area). (c) Example calculation of linear and areal interconnect densities assuming a single wire can be placed at minimum pitch. See Ref. [34] for technology capabilities. (d) Bumps/mm^2 as a function of minimum bump pitch (calculated as reciprocal of the square of the minimum bump pitch). (IO/mm^2 is sometimes used interchangeably with bumps/mm^2. In the strictest sense, IO/mm^2 is signals/mm^2, i.e., excluding power/ground, whereas bumps/mm^2 includes all bumps, i.e., electrical terminals.) *(Continued)*

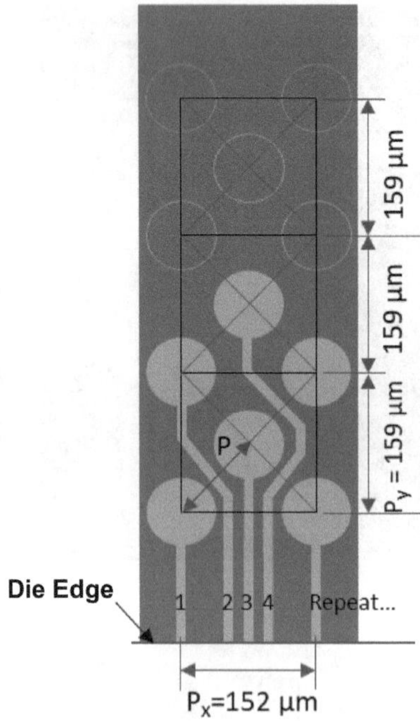

For (Line Width = 9µm, Space = 12µm, Pad Diameter = 77µm, Minimum Pitch = 110 µm)
→ Wires/mm/Layer = 26.3 and Bumps/mm² = 82.7

(c)

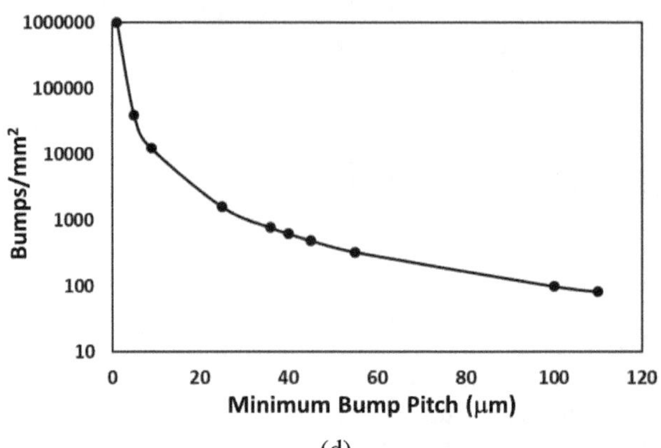

(d)

FIGURE 1.6 *(Continued)*

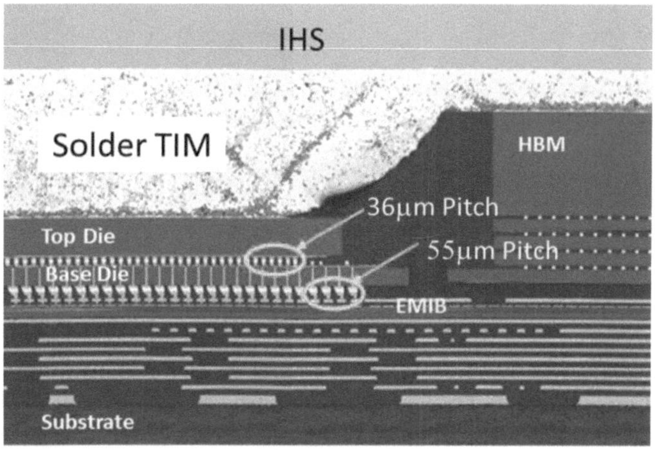

FIGURE 1.7 Cross section of Ponte Vecchio.

warpage impacts) and economic factors make it difficult and unproductive to try and pinpoint an exact pitch at which a transition from solder to HBI becomes inevitable.

HBI offers significantly higher IO/mm^2 with reducing pitches, which make them an attractive technology to investigate. In the next section, we will describe how HBI are created and in Section 1.4, we will examine why interconnect density scaling and hence HBI is needed for HI.

1.3 HYBRID BONDING

Hybrid bonding is the foundation for enabling future 3D IC packaging delivering significantly higher interconnect scaling, since it can enable sub-10 μm die-to-die direct Cu–Cu interconnect pitches with acceptable process thermal budgets of 200–400°C and low CTE mismatch. These fine interconnect pitches open up many avenues for die disaggregation, improved functionality, higher product performance/W and higher performance/$, where different IPs can be separated from or added as companion die to the SOC and can be built on more optimal fab nodes/design rules with better performance, functionality, and lower cost [38]. Fine-pitch HBI significantly reduce the latency and power compared to the looser pitch solder interconnects. The ability to reduce the die size on leading edge nodes through HBI enabled disaggregation results in improvement in fab yields which leads to significant cost savings.

While wafer-to-wafer hybrid bonding has been used for 3D packaging applications especially in CMOS image sensors, the chip-to-wafer hybrid bonding with the ability to select known good die (KGD) prior to bonding enables cost-effective HI and opens up a wide variety of applications and 3D construction options in advanced 3D IC packaging [39].

Hybrid bonding involves the formation of a strong interconnection between two surfaces where each surface is comprised of two or more materials [40–45]. Typically, each surface contains an array of conducting Cu pads surrounded by a sea

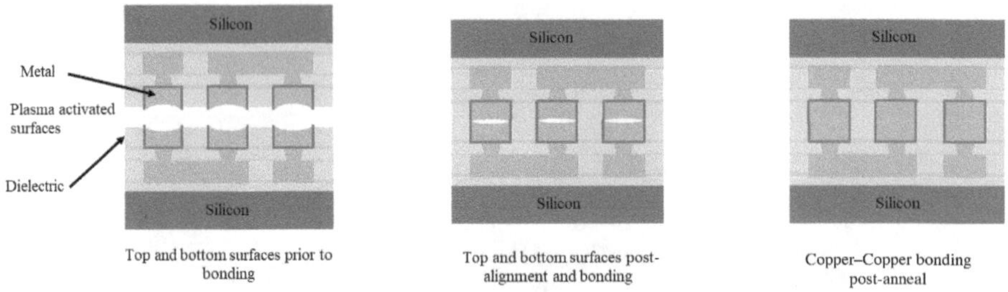

FIGURE 1.8 Hybrid bond formation.

of insulating dielectric such as SiO_2, SiN, and SiCN. The simplified process flow for hybrid bonding is illustrated in Figure 1.8.

There are multiple process flows to enable hybrid bonding such as wafer-to-wafer (W–W) bonding, chip-to-wafer bonding (C–W), chip to reconstituted wafer bonding (C–rW), reconstituted wafer-to-wafer bonding (rW–W), reconstituted wafer-to-reconstituted wafer bonding (rW–rW), and chip-to-chip (C-C) bonding. In addition to the process flows, there are variations in the bonding stack where we can have face-to-face hybrid bonding and face-to-back hybrid bonding. The choice of process flow and bonding stack is dependent on the application, wafer yields, product die disaggregation needs, product functionality/redundancy, and costs. Furthermore, the W–W process flow can be repeated multiple times to create 3D multi-die stacks.

W–W bonding has been used widely for more than a decade especially in CMOS image sensors and photonics applications [46–48]. W–W bonding has been demonstrated in production down to 1 μm minimum pad pitch [49] and there is active research to reduce the pitch further down to 400 nm [50]. The process flow for a W–W bonding is shown in Figure 1.9.

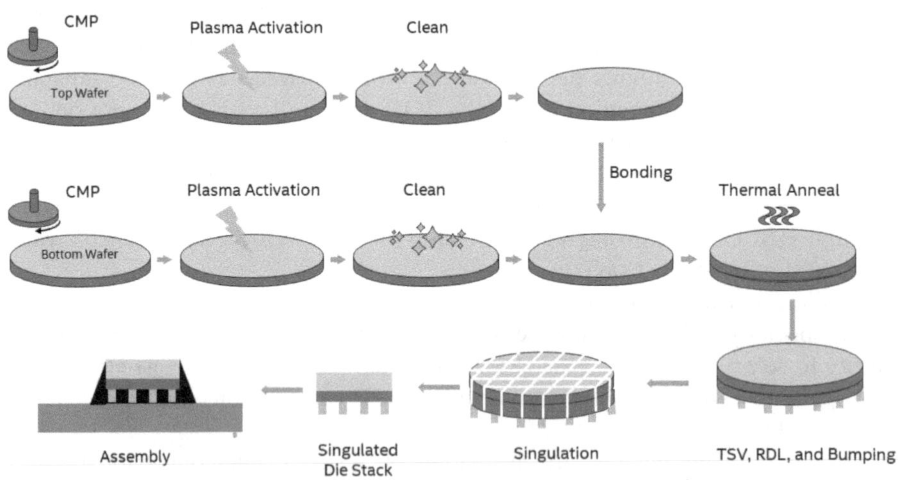

FIGURE 1.9 Wafer-to-wafer hybrid bonding process flow.

Both top and bottom wafers need to be either sorted or electrically tested prior to hybrid bonding to ensure that the wafers had good yields (i.e., known good wafer (KGW)) with no fab process excursions. In this flow, it is not possible to segregate the bad die prior to bonding and, therefore, there is stacked die yield loss associated with bad die locations on both top and bottom wafers. It is imperative that both the top and bottom wafers have very high yields, and this flow is ideally suited for designs with high levels of redundancy/repair. Second, the top die and bottom die have to be of the same size in a W–W flow which is restrictive for certain products. In a W–W process flow, the top and bottom KGWs are bonded to one another using the same sequence of process steps described in the previous section, and one of the wafers needs through silicon via (TSV) connections to connect the final hybrid stacked die to the package. Typically, there are two common approaches to make the TSVs:

1. *TSV-middle flow:* TSVs are made as part of fab backend prior to hybrid bonding and these TSVs are revealed after hybrid bond stacking.
2. *TSV-last flow:* TSVs are created after hybrid bond stacking.

After the TSVs are revealed or created, the wafer stack goes through redistribution layers and bumping. Subsequently, the stacked wafer is singulated into individual stacked die, which are then assembled on packages and tested.

The C–W bonding flow is shown in Figure 1.10. The process flow is very similar to W–W flow except that the top wafer has to be sorted, singulated, and only the KGD is attached to the bottom wafer.

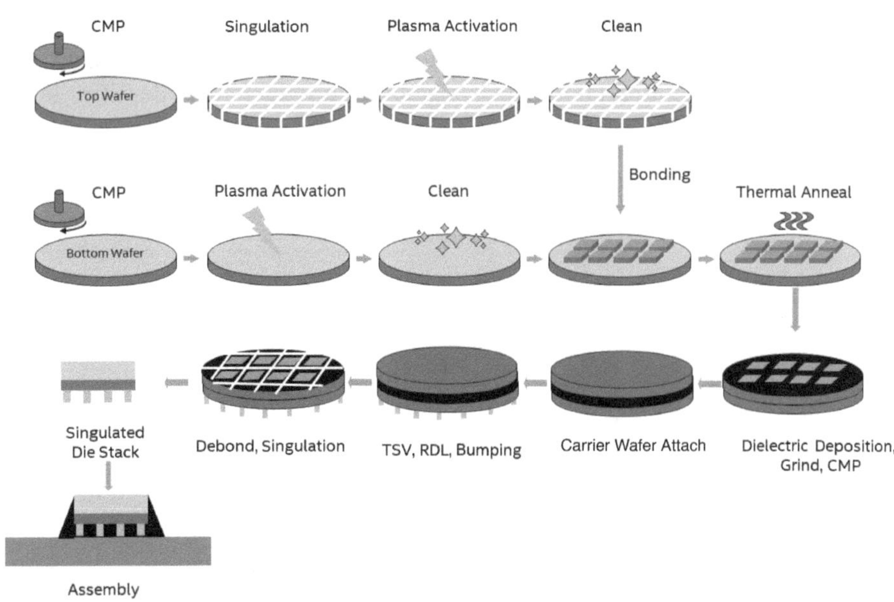

FIGURE 1.10 Chip-to-wafer hybrid bonding process flow.

In this section, we have covered key process attributes and limitations of the HBI flow. A considerable amount of work has been done in maturing the HBI technology and expanding the scope of its applicability. In the next section, we will examine how HBI enables HI.

1.4 PRODUCT ARCHITECTURES ENABLED BY HYBRID BONDING INTERCONNECTS

In Section 1.2, we made the case for the package as a compact platform for HI and showed how HBI enable significantly higher interconnect density for heterogeneously interconnected die. While it is beyond the scope of this chapter, it is understood that scaling interconnect pitch must be accompanied by scaling of silicon metal layers and TSVs to take advantage of the increased interconnect density.

In this section, we will show how HBI interconnects enable improved die-to-die link performance.

There are two important product trends currently driving interest in HI:

a. AI, ML, and HPC applications are BW constrained (Figure 1.11).
b. There is increasing recognition of the value of HI with open architectures for enabling increased and cost-optimized performance. The idea of "chiplets," i.e., die with IP optimized on different processes, and possessing common, standardized interfaces that can be integrated on the package, has gained considerable momentum in recent years since it offers design flexibility, modularity, and faster time to market [51, 52]. The industry focus on standardized interfaces is evidence of both interest in HI and collaboration to accelerate and streamline development of HI architectures [53–56] (Figures 1.12 and 1.13). HBI have been used for memory integration taking advantage of high interconnect densities for high BW [50, 57].

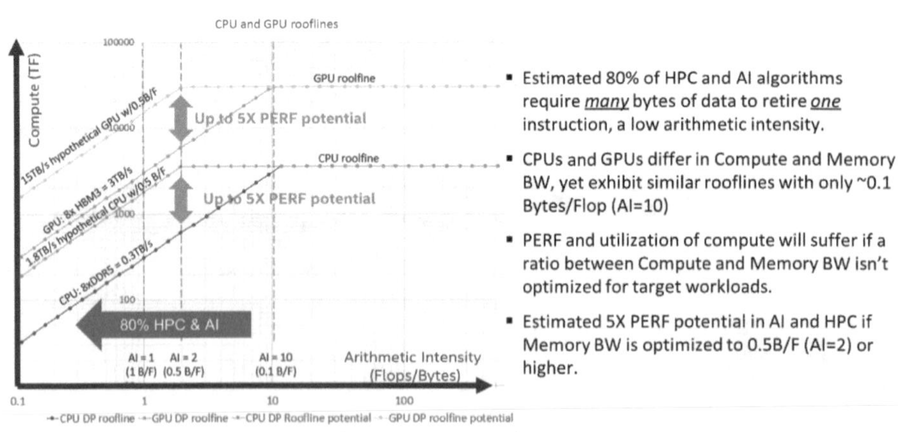

FIGURE 1.11 CPU and GPU rooflines. (Courtesy of Eduard Roytman, Intel.)

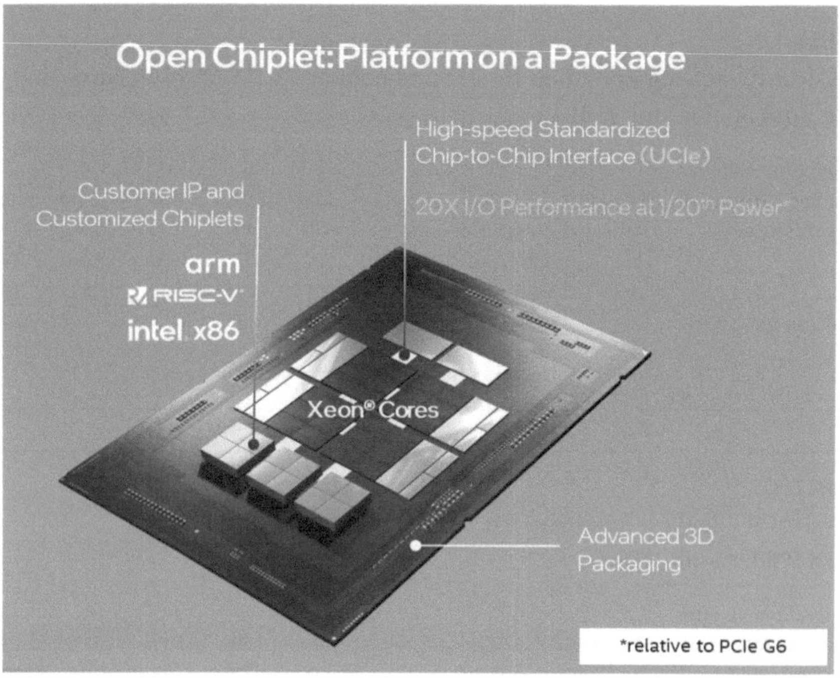

FIGURE 1.12 Modularity with standardized interfaces. (Courtesy of Bob Brennan and Debendra Das Sharma, Intel.)

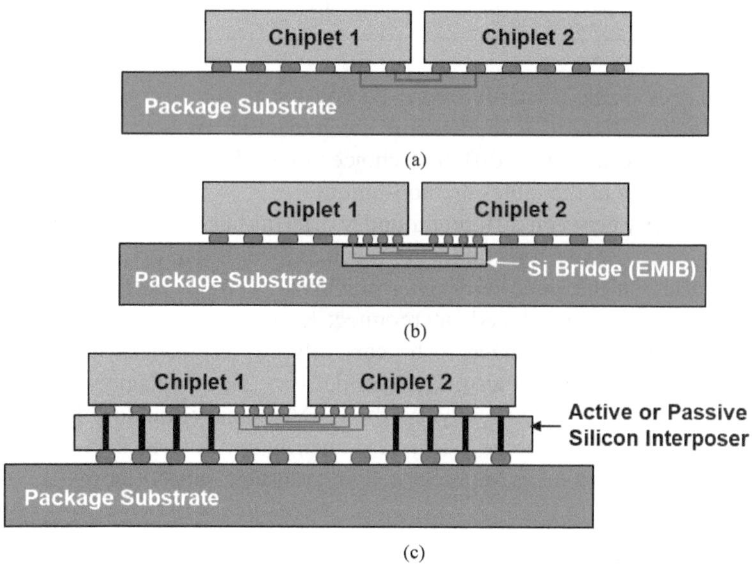

FIGURE 1.13 (a) 2D packages with lateral interconnects. (b) Enhanced 2D packages with lateral interconnects. (c) 3D and lateral 2D interconnects (see Ref. [33] for a detailed description of package nomenclature).

TABLE 1.1
Physical IO Scaling Roadmap for 2D and Enhanced 2D Architectures that Use Both Solder and HBI

Generation Number		1	2	3	4	5
Raw linear BW density (GBps/mm)		125	250	500	1,000	2,000
Package technology	Minimum bump pitch (μm)	55	40	30	20	10
	Linear escape density (IO/mm)	500	667	1,000	2,000	4,000
	Areal escape density (IO/mm^2)	331	625	1,111	2,500	10,000
Signaling speed (Gbps)		2	3	4	4	4

Note: See Ref. [33] for details. Generations 4 and 5 are likely to be implemented in HBI.

TABLE 1.2
Physical IO Scaling Roadmap for 3D Architectures that Use Both Solder and HBI

Generation Number		1	2	3	4	5
Raw areal BW density ($GBps/mm^2$)		125	250	500	1,000	2,000
Package technology	Minimum bump pitch (μm)	40	30	20	15	10
	Areal escape density (IO/mm^2)	625	1,111	2,500	4,444	10,000
Signaling speed (Gbps)		1.6	1.8	1.6	1.8	1.6

Note: See Ref. [33] for details. Generations 4 and 5 are likely to be implemented in HBI.

HBI can provide significantly improved BW for both lateral and vertical die-to-die links (Tables 1.1 and 1.2). It should be noted that the BW targets shown in Tables 1.1 and 1.2 can be met with different choices of packaging technologies, i.e., the bump pitch choices in the tables are non-unique.

Interoperability between different foundry offerings need to be ensured with fine-pitch HBI to realize the open chiplet approach and build interoperable die from different foundries on the same packaging platform. 3D unlocks the full potential of HBI-based links due to reduced interconnect lengths, and chiplets built to be 3D stacked with other chiplets need to be carefully co-designed taking into account various considerations (e.g., thermal, power delivery, signal integrity, etc.) to ensure the ability to reuse them seamlessly. As the HBM memory standard has illustrated, designers can usefully leverage a standard/open lateral interface to a 3D stack of chiplets, which is supplied as an integrated and reusable subassembly.

1.5 SUMMARY

In this chapter, we have described the evolution of packaging from being a simple protective enclosure providing space transformation with limited interconnect capabilities to a high-performance platform for the HI of compute and communication elements. A key element of advanced packaging platforms today and in the anticipated future is their high density of interconnects that enable high-performance HI using high BW and lower power interfaces. HBI are a key element in enabling continued scaling of interconnects in advanced packaging technologies, which need to scale along other vectors as well [58].

ACKNOWLEDGMENTS

The authors would like to gratefully acknowledge help from Nitesh Nimkar, Daniel Bills, Zhiguo Qian, Kemal Aygun, Zhichao Zhang, Chris Baldwin, Je-Young Chang, Bob Brennan, Tanay Karnik, Vivek Rajan, Anand Iyer, Dipankar Das, Edi Roytman, Josh Fryman, Debendra Das Sharma, Robert Munoz, Adel Elsherbini, and Rajiv Mongia from Intel Corporation.

REFERENCES

1. G. E. Moore, “Cramming More Components onto Integrated Circuits,” in Proceedings of the IEEE, vol. 86, no. 1, pp. 82–85, 1998.
2. G. E. Moore, “No Exponential Is Forever: But “Forever” Can Be Delayed! [Semiconductor Industry],” IEEE International Solid-State Circuits Conference, Digest of Technical Papers, San Francisco, CA, 2003, vol. 1, pp. 20–23.
3. A. B. Kelleher, “Celebrating 75 Years of the Transistor! A Look at the Evolution of Moore’s Law Innovation,” International Electron Devices Meeting (IEDM), San Francisco, CA, 2022, pp. 1.1.1–1.1.5.
4. State of the Semiconductor Industry, 2022 Semiconductor Industry Association Report.
5. https://spectrum.ieee.org/celebrate-transistor-anniversary-with-ieee
6. https://eps.ieee.org/technology/heterogeneous-integration-roadmap.html
7. R. Mahajan et al., “Co-packaged Photonics for High Performance Computing: Status, Challenges and Opportunities,” in Journal of Lightwave Technology, vol. 40, no. 2, pp. 379–392, 2022.
8. K. Radhakrishnan, M. Swaminathan, B. K. Bhattacharyya, “Power Delivery for High-performance Microprocessors: Challenges, Solutions, and Future Trends,” in IEEE Transactions on Components, Packaging and Manufacturing Technology, vol. 11, no. 4, pp. 655–671, 2021.
9. C. S. Geyik, Y. S. Mekonnen, Z. Zhang, K. Aygün, “Impact of Use Conditions on Dielectric and Conductor Material Models for High-speed Package Interconnects,” in IEEE Transactions on Components, Packaging and Manufacturing Technology, vol. 9, no. 10, pp. 1942–1951, 2019.
10. Z. Zhang, X. Wang, Y. Yan, “A Review of the State-of-the-art in Electronic Cooling,” e-Prime: Advances in Electrical Engineering, Electronics and Energy, vol. 1, 2021, 100009.
11. https://www.computerhistory.org/timeline/computers/
12. R. Mahajan, “Quiet Revolutions: How Advanced Microelectronics Packaging Continues to Drive Heterogeneous Integration,” 19th IEEE Intersociety Conference on Thermal and Thermomechanical Phenomena in Electronic Systems (ITherm), Orlando, FL, 2020, pp. 1408–1412. doi: 10.1109/ITherm45881.2020.9190247.
13. Debendra Mallik, Ravi Mahajan, Nachiket Raravikar, Kaladhar Radhakrishnan, Kemal Aygun, Bob Sankman, “Flip-chip Packaging for Nanoscale Silicon Logic Devices: Challenges and Opportunities,” in Nanopackaging: Nanotechnologies and Electronics Packaging, Jim Morris (ed.), 2nd edition, Springer, 2018, pp. 921–957.

14. C. S. Geyik, Z. Zhang, K. Aygün, "Improved Package Modeling and Correlation Methodology for High-speed IO Design," IEEE 66th Electronic Components and Technology Conference (ECTC), Las Vegas, NV, 2016, pp. 985–991.
15. M. Manusharow, S. Muthukumar, E. Zheng, A. Sadiq, C. Lee, "Coreless Substrate Technology Investigation for Ultra-thin CPU BGA Packaging," IEEE 62nd Electronic Components and Technology Conference, San Diego, CA, 2012, pp. 892–896.
16. https://eps.ieee.org/images/files/HIR_2021/ch17_test_final.pdf
17. https://www.glennsmuseum.com/items/ibm_mcm/
18. https://www.whitehouse.gov/wp-content/uploads/2021/06/100-day-supply-chain-review-report.pdf
19. R. Mahajan et al., "Embedded Multi-die Interconnect Bridge (EMIB): A High Density, High Bandwidth Packaging Interconnect," IEEE 66th Electronic Components and Technology Conference (ECTC), Las Vegas, NV, 2016, pp. 557–565.
20. M. Sunohara, T. Tokunaga, T. Kurihara, M. Higashi, "Silicon Interposer with TSVs (through Silicon Vias) and Fine Multilayer Wiring," in Proceedings of the IEEE Electronics Components Technology Conference (ECTC), Lake Buena Vista, FL, 2008, pp. 847–852.
21. K. Oi et al., "Development of New 2.5D Package with Novel Integrated Organic Interposer Substrate with Ultra-fine Wiring and High Density Bumps," in Proceedings of the IEEE 64th Electronics Components Technology Conference, Orlando, FL, 2014, pp. 348–353.
22. A. Eitan, K.-Y. Hung, "Thermo-compression Bonding for Fine-pitch Copper-pillar Flip-chip Interconnect: Tool Features as Enablers of Unique Technology," IEEE 65th Electronic Components and Technology Conference (ECTC), San Diego, CA, 2015, pp. 460–464.
23. J. T. Pawlowski, "Hybrid Memory Cube (HMC)," 2011 IEEE Hot Chips 23 Symposium (HCS), Stanford, CA, 2011, pp. 1–24. doi: 10.1109/HOTCHIPS.2011.7477494.
24. https://www.jedec.org/document_search?search_api_views_fulltext=jesd235
25. https://www.jedec.org/news/pressreleases/jedec-updates-groundbreaking-high-bandwidth-memory-hbm-standard
26. https://www.intel.com/content/dam/www/public/us/en/documents/white-papers/accelerating-innovation-through-aib-whitepaper.pdf
27. https://www.darpa.mil/program/common-heterogeneous-integration-and-ip-reuse-strategies
28. D. Greenhill et al., "3.3 A 14 nm 1 GHz FPGA with 2.5D Transceiver Integration," IEEE International Solid-State Circuits Conference (ISSCC), San Francisco, CA, 2017, pp. 54–55. doi: 10.1109/ISSCC.2017.7870257.
29. https://old.hotchips.org/wp-content/uploads/hc_archives/hc29/HC29.22-Tuesday-Pub/HC29.22.50-FPGA-Pub/HC29.22.523-Hetro-Mod-Platform-Shumanrayev-Intel-Final.pdf
30. https://eri-summit.darpa.mil/docs/Shumarayev_Sergey_CHIPS_Final.pdf
31. J. H. Lau, "Recent Advances and Trends in Multiple System and Heterogeneous Integration with TSV Interposers," in IEEE Transactions on Components, Packaging and Manufacturing Technology, vol. 13, no. 1, pp. 3–25, 2023. doi: 10.1109/TCPMT.2023.3234007.
32. J. H. Lau, "Recent Advances and Trends in Advanced Packaging," in IEEE Transactions on Components, Packaging and Manufacturing Technology, vol. 12, no. 2, pp. 228–252, 2022. doi: 10.1109/TCPMT.2022.3144461.
33. https://eps.ieee.org/images/files/HIR_2021/ch22_2D-3D.pdf
34. R. Mahajan et al., "Embedded Multidie Interconnect Bridge: A Localized, High-Density Multichip Packaging Interconnect," in IEEE Transactions on Components, Packaging and Manufacturing Technology, vol. 9, no. 10, pp. 1952–1962, 2019. doi: 10.1109/TCPMT.2019.2942708.

35. W. Gomes et al., "8.1 Lakefield and Mobility Compute: A 3D Stacked 10 nm and 22FFL Hybrid Processor System in 12×12 mm^2, 1 mm Package-on-package," IEEE International Solid-State Circuits Conference (ISSCC), San Francisco, CA, 2020, pp. 144–146. doi: 10.1109/ISSCC19947.2020.9062957.
36. W. Gomes et al., "Ponte Vecchio: A Multi-tile 3D Stacked Processor for Exascale Computing," IEEE International Solid-State Circuits Conference (ISSCC), San Francisco, CA, 2022, pp. 42–44. doi: 10.1109/ISSCC42614.2022.9731673.
37. Z. Li et al., "Scaling Solder Micro-bump Interconnect Down to 10 µm Pitch for Advanced 3D IC Packages," IEEE 71st Electronic Components and Technology Conference (ECTC), San Diego, CA, 2021, pp. 451–456.
38. A. Elsherbini et al., "Enabling Next Generation 3D Heterogeneous Integration Architectures on Intel Process," International Electron Devices Meeting (IEDM), San Francisco, CA, 2022, pp. 27.3.1–27.3.4. doi: 10.1109/IEDM45625.2022.10019499.
39. A. Elsherbini et al., "Enabling Hybrid Bonding on Intel Process," IEEE International Electron Devices Meeting (IEDM), San Francisco, CA, 2021, pp. 34.3.1–34.3.4. doi: 10.1109/IEDM19574.2021.9720586.
40. Daniel W. Fischer, Sarah Knickerbocker, Daniel Smith, Robert Katz, John Garant, Jorge Lubguban, Vilmarie Soler, Norman Robson, "Face to Face Hybrid Wafer Bonding for Fine Pitch Applications," IEEE 70th Electronic Components and Technology Conference, Orlando, FL, 2020, pp. 595–600.
41. Hoaxing Ren, Yu-Tao Yang, Guangqi Ouyang, Subramaniam S. Iyer, "Mechanism and Process Window Study of Die-to-Wafer (D2W) Hybrid Bonding," ECS Journal of Solid State Science and Technology, vol. 10, no. 6, 2021. doi:10.1149/2162-8777/ac0a52
42. J. A. Theil, L. Mirkarimi, G. Fountain, G. Gao, R. Katkar, "Recent Developments in Fine Pitch Wafer-to-Wafer Hybrid Bonding with Copper Interconnect," International Wafer Level Packaging Conference (IWLPC), San Jose, CA, 2019, pp. 1–6. doi: 10.23919/IWLPC.2019.8913862.
43. B. Brandstätter et al., "High-speed Ultra-accurate Direct C2W Bonding," IEEE 70th Electronic Components and Technology Conference (ECTC), Orlando, FL, 2020, pp. 1943–1949. doi: 10.1109/ECTC32862.2020.00303.
44. S. Moreau, J. Jourdon, S. Lhostis, D. Bouchu, S. Ayoub, L. Arnaud, H. Fremont, "Review—Hybrid Bonding-based Interconnects: A Status on Last Robustness and Reliability Achievements," in ECS Journal of Solid State Science and Technology, vol. 11, p. 024001, 2022.
45. S. C. Chong, I. C. Daniel, V. N. Sekhar, S. Lim, V. Srinivas, "Chip-to-Wafer Hybrid Bonding for High Performance 2.5D Applications," IEEE 24th Electronics Packaging Technology Conference (EPTC), Singapore, 2022, pp. 19–23. doi: 10.1109/EPTC56328.2022.10013161.
46. S. Lhostis et al., "Reliable 300 mm Wafer Level Hybrid Bonding for 3D Stacked CMOS Image Sensors," IEEE 66th Electronic Components and Technology Conference (ECTC), Las Vegas, NV, 2016, pp. 869–876. doi: 10.1109/ECTC.2016.202.
47. Y. Kagawa et al., "Novel Stacked CMOS Image Sensor with Advanced Cu2Cu Hybrid Bonding," IEEE International Electron Devices Meeting (IEDM), San Francisco, CA, 2016, pp. 8.4.1–8.4.4. doi: 10.1109/IEDM.2016.7838375.
48. K.-W. Lee, A. Noriki, K. Kiyoyama, T. Fukushima, T. Tanaka, M. Koyanagi, "Three-dimensional Hybrid Integration Technology of CMOS, MEMS, and Photonics Circuits for Optoelectronic Heterogeneous Integrated Systems," in IEEE Transactions on Electron Devices, vol. 58, no. 3, pp. 748–757, 2011.
49. D. B. L. Yolanda, "Wafer to Wafer Bonding to Increase Memory Density," China Semiconductor Technology International Conference (CSTIC), Shanghai, China, 2022, pp. 1–4.

50. Han-Jong Chia, Shih-Peng Tai, Ji James Cui, Chuei Tang Wang, Chih-hang Tung, Kuo-chung Yee, Douglas C.H. Yu, "Ultrahigh Density Low Temperature SOIC with Sub-0.5 μm Bond Pitch," IEEE 73rd Electronics Components and Technology Conference, pp. 1–4, 2023.
51. F. Sheikh et al., "2.5D and 3D Heterogeneous Integration for Emerging Applications," in IEEE Solid-State Circuits Magazine, vol. 13, no. 4, pp. 77–87.
52. https://youtu.be/wnjiUAs8l08
53. https://github.com/chipsalliance
54. https://www.chipsalliance.org/news/aib-2-0-draft-specification/
55. D. D. Sharma, "System on a Package Innovations with Universal Chiplet Interconnect Express (UCIe) Interconnect," in IEEE Micro, vol. 43, no. 2, pp. 76–85, 2023.
56. D. Das Sharma, G. Pasdast, Z. Qian, K. Aygun, "Universal Chiplet Interconnect Express (UCIe): An Open Industry Standard for Innovations with Chiplets at Package Level," in IEEE Transactions on Components, Packaging and Manufacturing Technology, vol. 12, no. 9, pp. 1423–1431, 2022.
57. https://community.amd.com/t5/business/advanced-packaging-enabling-moore-s-law-s-next-frontier-through/ba-p/503428
58. https://www.spiedigitallibrary.org/conference-proceedings-of-spie/12751/0000/Directions-challenges-and-opportunities-in-heterogeneous-integration/10.1117/12.2688377.full

2 Direct Copper Interconnection for Die/Wafer Bonding: Overview

Tzu-Heng Hung and Kuan-Neng Chen

2.1 INTRODUCTION

With the advancement of packaging technology, Cu–Cu bonding has gained increasing significance as a replacement for conventional solder joints in packaging applications. This shift is attributed to the favorable properties of Cu–Cu interconnections, such as lower contact resistance, reliable electrical performance, and the potential for submicrometer scaling. However, achieving high-quality Cu–Cu bonding often requires thermal treatment at temperatures up to 350°C and more. Unfortunately, such high temperatures can lead to stress and reliability issues, which can further reduce fabrication yields for temperature-sensitive components, thereby limiting the flexibility of heterogeneous integration platforms. Consequently, there has been significant focus on reducing the temperature requirements for Cu–Cu bonding, making it a crucial area of development in Cu interconnection.

This chapter offers a comprehensive overview of Cu interconnection technologies. It outlines various approaches that have emerged in recent years to achieve low-temperature Cu–Cu bonding and discusses key factors that determine the quality of Cu–Cu bonding based on recent research and development. The focus lies on introducing the principles behind these approaches and how they enable successful low-temperature Cu–Cu bonding, rather than merely presenting their bonding conditions.

2.2 PRINCIPLE OF Cu–Cu BONDING

Figure 2.1 illustrates the model of Cu bonding mechanism when two clean Cu layers come into contact and bond [1]. As depicted, Cu atoms in the contact region diffuse and convert the two separate Cu surfaces into a grain boundary, driven by stress gradients and the capillary effect. Eventually, a significant portion of the Cu–Cu bonding interface becomes a grain boundary, and the stress-induced bonding process stops. During the annealing process, voids at the bonding interface and the grain boundaries gradually evolve, reaching a state of minimum total system energy. Figure 2.2 presents images of the Cu–Cu bonding interface at annealing

DOI: 10.1201/9781003408673-2

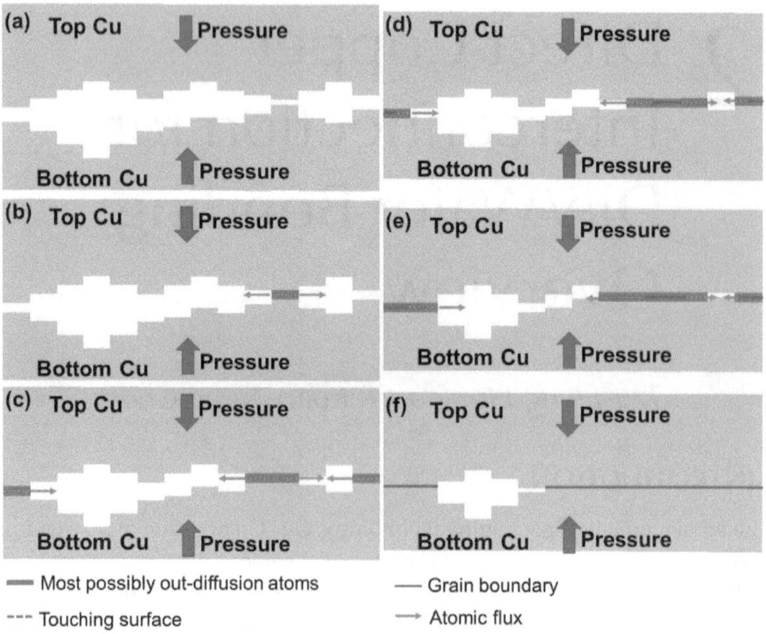

FIGURE 2.1 Schematic of Cu–Cu bonding mechanism [1].

temperatures of 100, 200, 300, and 400°C for 2 hours, revealing the evolution of the bonding interface [2]. At 200°C annealing, small wedges begin to appear at the bonding interface, extending into the grain boundary on the opposite side of the interface. With further annealing, the initially flat bonding interface adopts a zigzag shape with a triple junction angle of approximately 120°, as depicted in Figure 2.3.

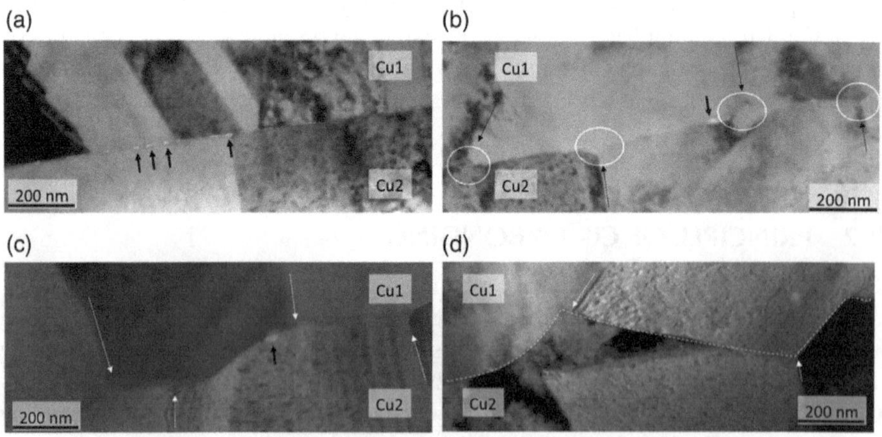

FIGURE 2.2 TEM images of Cu–Cu bonding interface with (a) 100°C, (b) 200°C, (c) 300°C, and (d) 400°C annealing for 2 hours [2].

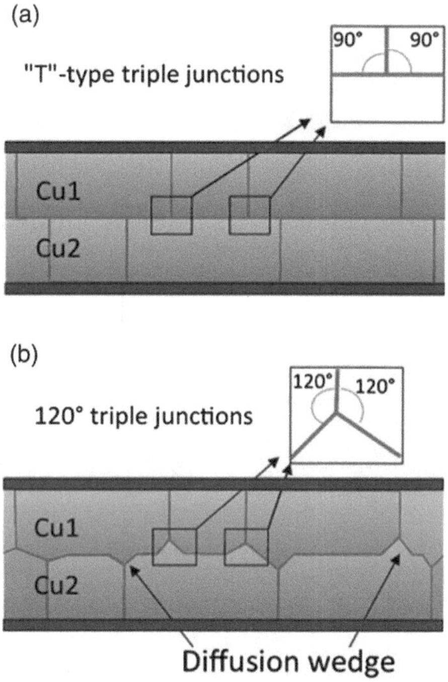

FIGURE 2.3 Illustration of Cu–Cu bonding interface (a) before and (b) after annealing process [2].

2.2.1 Cu Oxide Barrier

To achieve a successful Cu–Cu bonding, it is crucial to ensure a clean Cu surface when two Cu films/pads come into contact. However, Cu is highly susceptible to oxidation. During the oxidation process, the Cu surface gradually transforms into Cu_2O. Furthermore, even when the surface has completely oxidized into Cu_2O, the outward diffusion of Cu atoms from the subsurface and the inward diffusion of O atoms continuously increase the thickness of the oxidized Cu layer, resulting in the oxidation of the entire Cu structure [3–5]. If there are no pure Cu atoms remaining in the Cu_2O layer, in the environment with higher temperatures, further oxidation still occurs, leading to the formation of CuO [5, 6], as depicted in Figures 2.4 and 2.5. Consequently, the resistivity of the oxidized Cu film significantly increases, as shown in Figure 2.6.

In most cases, Cu surfaces are cleaned to remove Cu oxides before bonding. Various methods are employed to minimize the amount of Cu oxides, such as in situ cleaning processes or controlling the waiting time between cleaning and bonding. It has been reported that residual Cu oxides on the surface can be dissolved into Cu during annealing under a vacuum environment [7]. In the case of partially oxidized Cu films, the concentration of Cu_2O begins to decrease with annealing temperatures ranging from 200°C to 300°C. On the other hand, fully oxidized Cu films require higher annealing temperatures for Cu_2O dissolution. Therefore, numerous research

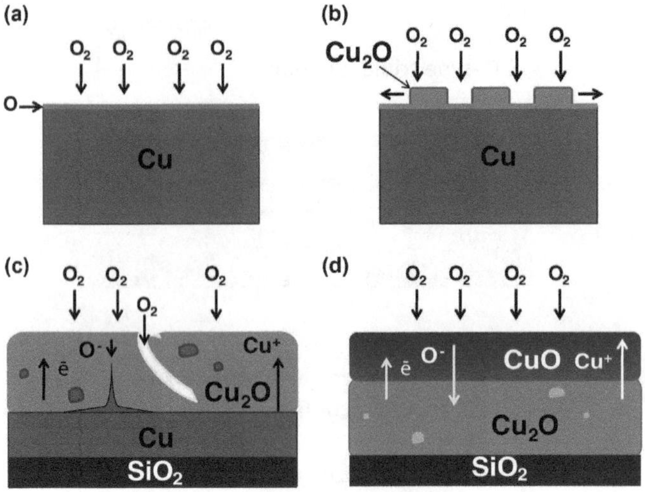

FIGURE 2.4 Schematic of the Cu oxidation mechanism [5].

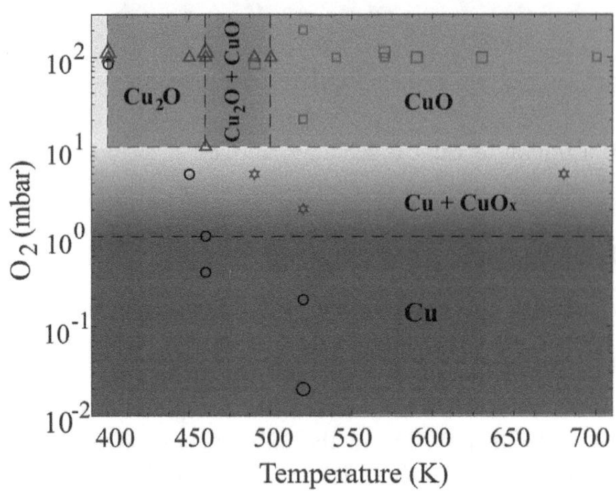

FIGURE 2.5 Phase diagram of Cu oxidation [6].

efforts have been devoted to reducing the amount of surface Cu oxides to lower the required process temperature for Cu–Cu bonding.

2.2.2 Grain Size

Since the Cu–Cu bonding process relies on Cu atom diffusion at the bonding interface, the distribution of Cu grains, which affects the diffusion behavior, can influence the Cu–Cu bonding process. It has been suggested that a smaller Cu grain size is advantageous for Cu–Cu bonding because it leads to a higher density of grain boundaries at the bonding interface, providing more channels for diffusion.

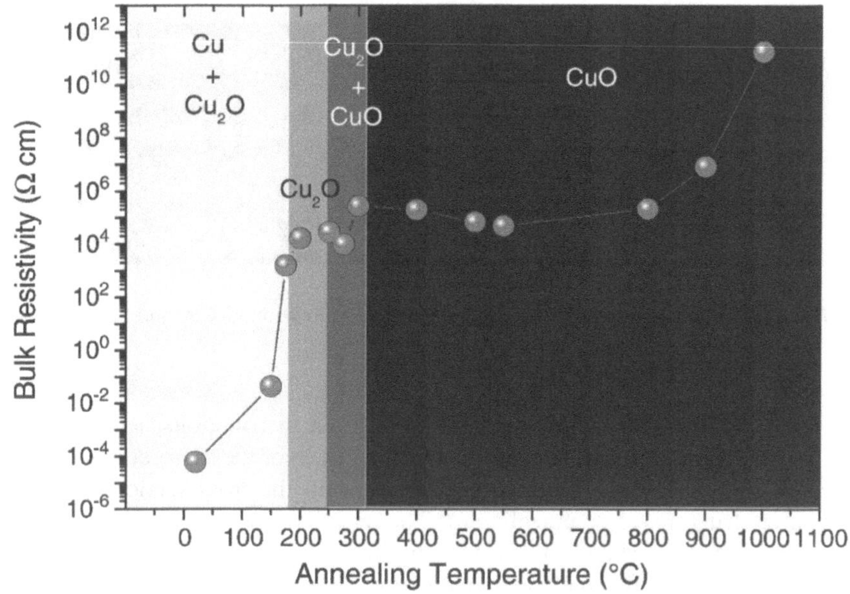

FIGURE 2.6 The resistivity phase diagram of oxidized Cu film [5].

Furthermore, the phenomenon of grain growth also contributes to the formation of a strong Cu–Cu contact [8, 9]. Figure 2.7 presents the results of Cu–Cu bonding with and without a prebonding annealing process. The annealing process promotes Cu grain growth, resulting in a larger Cu grain size for the annealed Cu film. In the TEM images, it is evident that for Cu–Cu bonding with a prebonding annealing process, a distinct bonding interface is present. Conversely, for Cu–Cu bonding without a prebonding annealing process, the bonding interface is unclear, and some Cu grains have grown through the bonding interface.

Based on this concept, a bonding approach utilizing small Cu grains has been demonstrated [10]. Figure 2.8 illustrates the electron backscatter diffraction (EBSD)

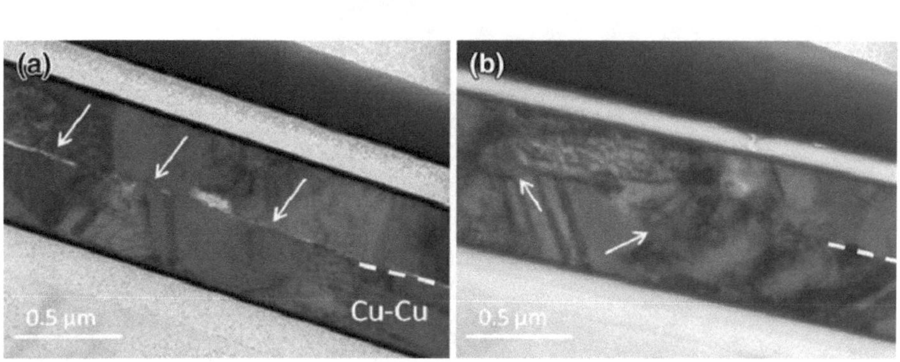

FIGURE 2.7 TEM images of bonded Cu layer (a) with and (b) without prebonding annealing process [8].

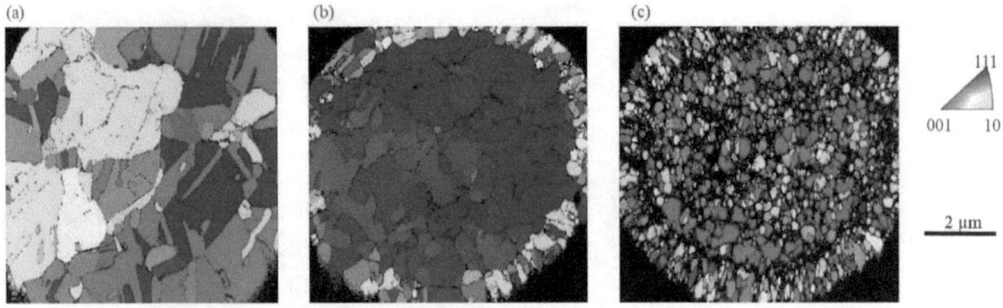

FIGURE 2.8 EBSD images of (a) regular Cu, (b) nanotwinned Cu, and (c) fine-grained Cu [10].

analysis of examined Cu films with different grain distributions, namely, regular Cu, Cu with nanotwins, and Cu with fine-grains. Each of these structures is bonded separately and annealed at 200°C. Figure 2.9 presents the cross-sectional inspection of the three structures after annealing for 2 hours and 6 hours. In the case of Cu with fine-grains, a distinct change at the bonding interface is observed compared to the outcomes of regular Cu and Cu with nanotwins. This change is attributed to the fast diffusion occurring in Cu with fine-grains.

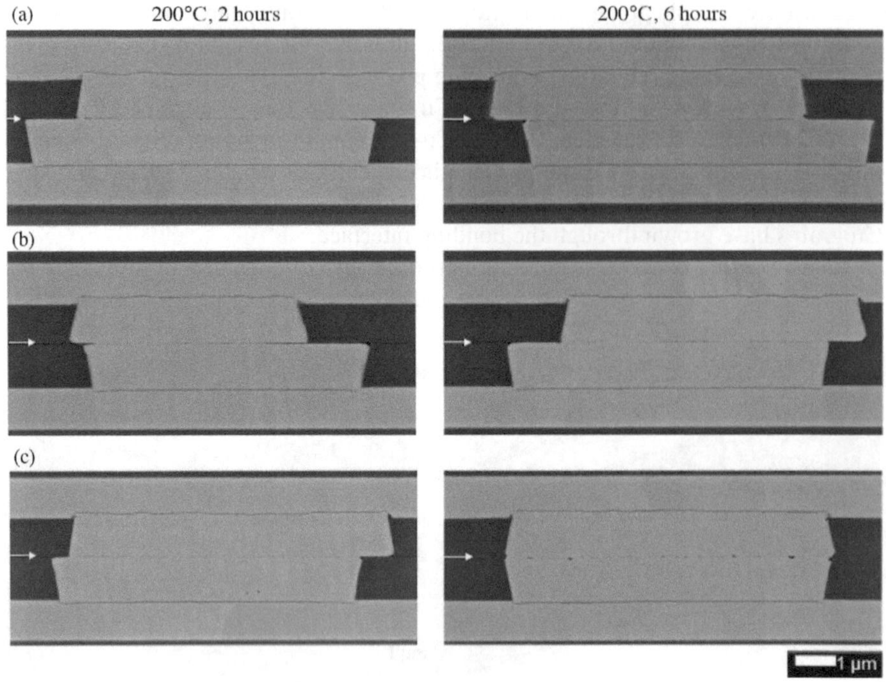

FIGURE 2.9 SEM images of bonded (a) regular Cu, (b) nanotwinned Cu, and (c) fine-grained Cu after 200°C annealing [10].

TABLE 2.1
Calculated Cu Surface Diffusivity on (111), (100), and (110) Planes [16]

$D_{Surf.}$\Temperature	(111)	(100)	(110)
300°C	1.51×10^{-5}	1.48×10^{-8}	1.55×10^{-9}
250°C	1.22×10^{-5}	4.74×10^{-9}	3.56×10^{-10}
200°C	9.42×10^{-6}	1.19×10^{-9}	5.98×10^{-11}
150°C	6.85×10^{-6}	2.15×10^{-10}	6.61×10^{-12}

2.2.3 Orientation

The influence of Cu grain orientation in the interconnect has been extensively discussed due to the anisotropic nature of Cu grains. Various aspects such as thermal expansion behavior [11, 12], oxidation behavior [13–15], and more have been explored. In terms of Cu–Cu bonding, the high surface diffusivity and low oxidation rate of Cu grains with a (111) orientation have been the topic of extensive research [16]. Table 2.1 provides a comparison of surface diffusivity between Cu grains with (111), (100), and (110) orientations at different temperatures. The high surface diffusivity exhibited by (111) Cu is advantageous for Cu–Cu bonding when two Cu grains come into contact, as depicted in Figure 2.1.

2.2.4 Surface Flatness

During the Cu bonding process, the peak regions of the two Cu films are more likely to come into contact initially. In Cu films with nanometer-scale flatness, a larger contact area can be achieved, and more Cu free surfaces can transform into grain boundaries through surface diffusion. Simultaneously, the void ratio at the bonding interface can be reduced. As depicted in Figure 2.10, these voids will move and converge during the annealing process [17]. In the structure of Cu interconnection, these voids typically migrate to the interface between Cu and the barrier, rather than remaining at the Cu–Cu interface [17]. However, achieving a coplanar condition is crucial for the successful fabrication of Cu interconnections. In such cases, the overall surface topography, including dishing, erosion, or warpage, must be delicately optimized. The requirements for these specifications may vary from one structure to another.

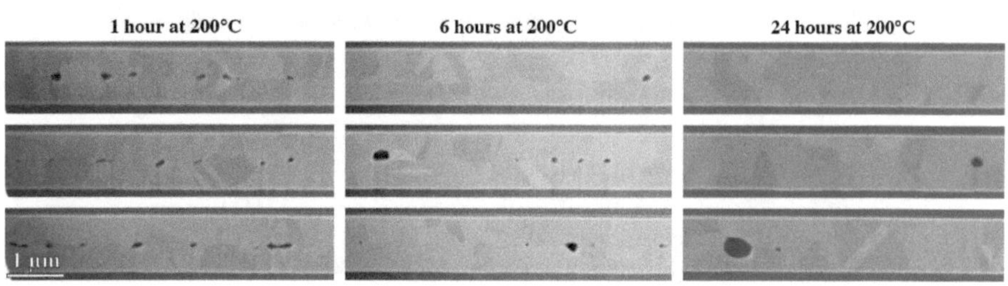

FIGURE 2.10 Cross-sectional images of Cu–Cu bonding interface after 1, 6, and 24 hours annealing at 200°C [17].

For the region of Cu–Cu bonding, it is important to ensure that the two Cu pads/vias can make contact with each other during both the bonding and annealing processes.

2.2.5 Thermal Compression Bonding (TCB)

TCB is a commonly employed method for bonding. In the TCB process, external force and heat are applied simultaneously when two metal films come into contact. For metals with a low melting point, solid–liquid interdiffusion (SLID) can occur at the bonding interface, resulting in the formation of a eutectic bonding joint with intermetallic compounds (IMCs) at the interface. Eutectic bonding through TCB offers advantages such as fast processing, low cost, and low temperature requirements. Consequently, it has found widespread use in packaging applications. In the case of Cu–Cu bonding, the bonding mechanism during TCB relies on solid-state interdiffusion, as mentioned earlier. Conventional Cu–Cu bonding through TCB typically requires bonding temperatures ranging from 350°C to 400°C when no other pretreatment is applied for Cu–Cu bonding.

2.2.6 Room Temperature Bonding with Post-annealing

In addition to the TCB process, a method of achieving Cu–Cu bonding through room temperature bonding with post-annealing has been proposed [18–21]. To meet the requirements of this method, the Cu surface prior to bonding should be well-prepared. The root mean square roughness of the Cu surface should be less than 0.4 nm, and the surface should exhibit high hydrophilicity with a water contact angle of less than 7°. Under this condition, Cu–Cu bonding can be accomplished with an annealing temperature of 200°C. Figure 2.11 depicts the evolution of the Cu–Cu bonding

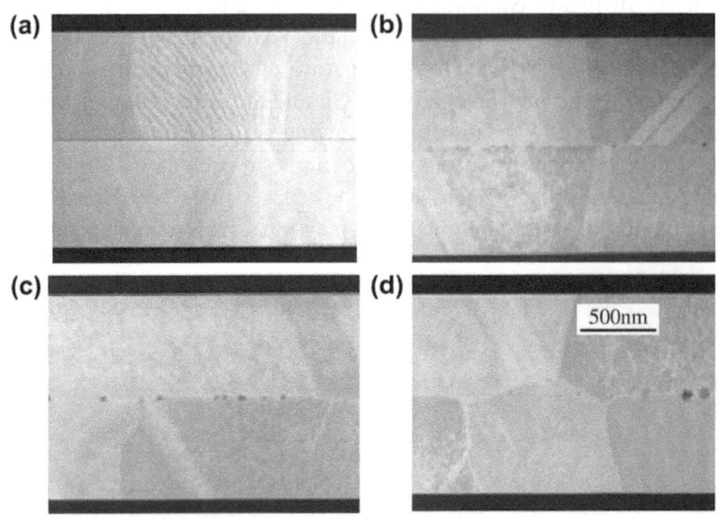

FIGURE 2.11 TEM images of Cu–Cu bonding interface (a) as bonded and with post-annealing at (b) 200°C, (c) 300°C, and (d) 400°C [20].

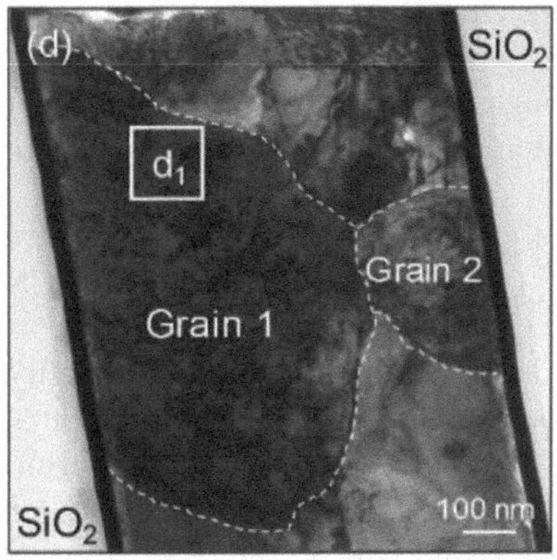

FIGURE 2.12 TEM image of Cu–Cu bonding by the co-hydrophilization method after 200°C annealing [23].

interface achieved through room temperature bonding and subsequent annealing. With annealing at 200°C, the thin Cu oxide becomes unstable, and grain boundaries can successfully develop during the annealing process.

For surface preparation, a method involving the optimization of pretreatment processes to create a co-hydrophilic surface has been proposed for Cu/SiO_2 hybrid bonding [22, 23]. Although the demonstration was conducted through the TCB process due to dishing issues from the CMP process, the principle of the co-hydrophilic surface aligns with the method of achieving Cu–Cu bonding through room temperature bonding with post-annealing. The pretreatment process involves the use of Ar/O_2 plasma and NH_4OH immersion. On the Cu surface, CuO is transformed into –OH and –NH_2 through the following reactions:

$$CuO + H_2O \rightarrow Cu(OH)_2 \quad (2.1)$$

$$Cu^{2+} + NH_4OH \rightarrow Cu(NH_3)_4^{\ 2+} + OH^- \quad (2.2)$$

During the bonding and annealing process, the –NH_2 and –OH groups facilitate dehydration–condensation reactions for Cu–Cu bonding at 200°C. Figure 2.12 demonstrates clear grain growth phenomena across the Cu–Cu bonding interface. This demonstration provides an insight into how Cu–Cu bonding can be achieved through room temperature bonding with post-annealing.

2.3 DEVELOPMENT OF LOW-TEMPERATURE Cu–Cu BONDING TECHNOLOGY

In the subsequent sections, various technologies developed in recent years for low-temperature Cu–Cu bonding will be discussed. Most of these approaches focus on reducing the presence of Cu oxides on the bonding surfaces to enable Cu–Cu bonding at lower temperatures. These technologies can be briefly categorized into different types, including cleaning the oxidized surface, providing additional surface coverage for protection, altering surface properties to mitigate oxidation, and breaking Cu oxides through the application of higher external stress. By employing these technologies, the required temperature for Cu–Cu bonding can be significantly reduced from 350°C to below 200°C, and even close to room temperature.

2.3.1 Wet Pretreatment

To address the issue of Cu oxides and achieve high-quality Cu–Cu bonding, several methods have been proposed. One of the most common approaches is the wet pretreatment process using acid solutions [24–32]. The principle behind acid solution cleaning is to react with the Cu oxides using the hydrogen in the solution, forming salts or soluble compounds that can be removed from the Cu surface. Numerous types of acid solutions have been proposed for Cu surface cleaning, including hydrochloric acid [24], sulfuric acid [25–27], acetic acid [28, 29], and others [30–32]. For instance, the reactions involving acetic acid are as follows [28]:

$$CuO + 2CH_3COOH \rightarrow Cu(CH_3COO)_2 + H_2O \tag{2.3}$$

$$Cu_2O + 4CH_3COOH \rightarrow 2Cu(CH_3COO)_2 + H_2O + H_2 \tag{2.4}$$

The effectiveness of Cu oxide removal in the acid solution cleaning process depends on various parameters, such as acid concentration, solution temperature, and the type of acid used. To properly evaluate the effect of Cu oxide removal, surface analyses such as XPS (X-ray photoelectron spectroscopy) and AES (Auger electron spectroscopy) are suitable methods for comparing the Cu spectra before and after acid solution cleaning. Figure 2.13 illustrates the XPS spectra of the Cu 2p peak [32]. The peaks corresponding to Cu oxides should be reduced after the wet pretreatment process. Additionally, in the XPS spectra of Cu, the Cu_2O peak overlaps with the Cu metal peak. The presence of Cu_2O can be observed from the Cu metal using

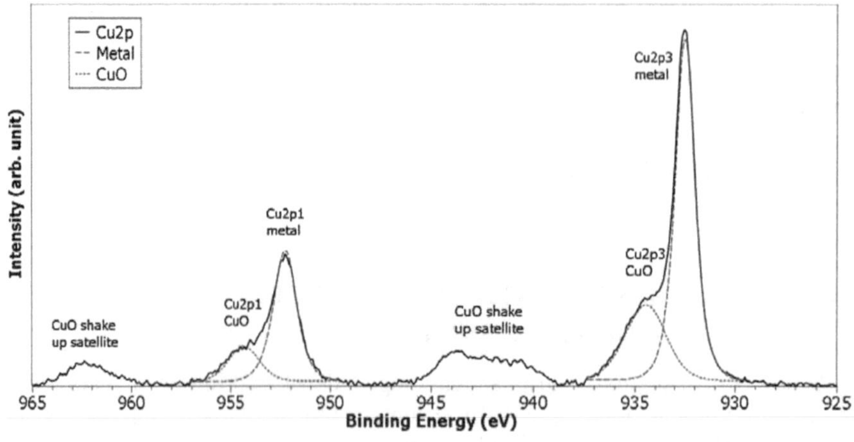

FIGURE 2.13 XPS spectra of Cu 2p peak after storage in cleanroom [32].

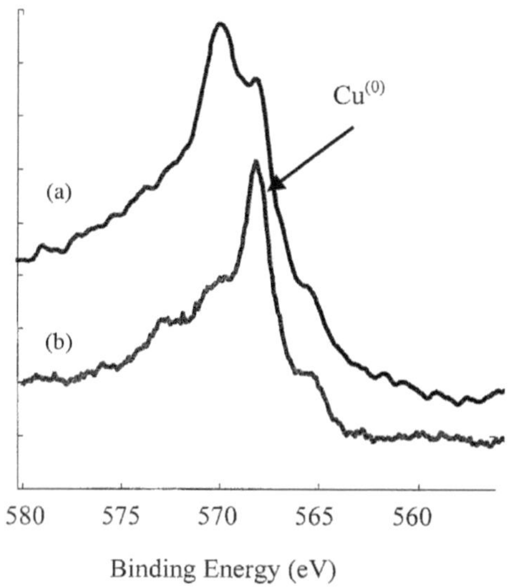

FIGURE 2.14 Auger electron spectra of Cu (LMM) for (a) original native copper oxide and (b) copper surface after 5-second acetic acid treatment at 35°C [28].

AES analysis. Figure 2.14 presents the Cu LMM spectra before and after acetic acid pretreatment [28].

Although acid solutions can effectively remove Cu oxides, native Cu oxides can reform within minutes when exposed to the atmosphere, as shown in Figure 2.15 [28]. The re-formation of Cu oxides can negatively impact Cu bonding quality.

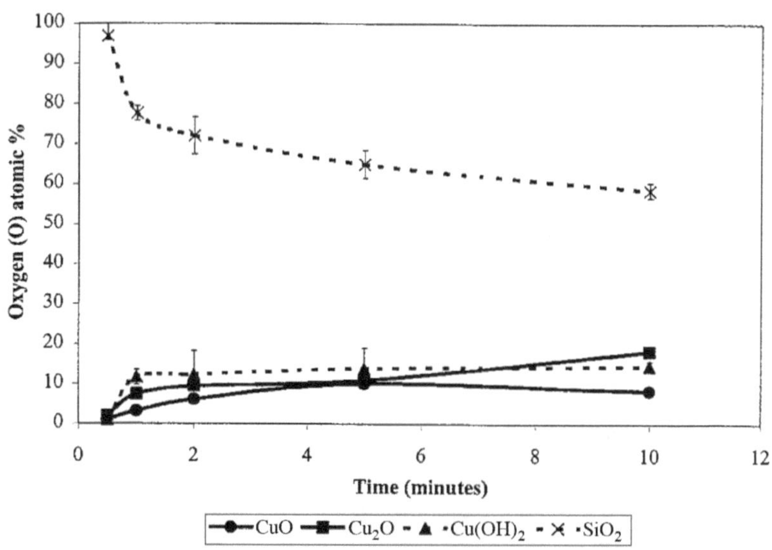

FIGURE 2.15 Oxygen concentration of various Cu oxides as a function of exposure to ambient air after 1-minute acetic acid treatment at 35°C [28].

Therefore, the wet pretreatment process is not suitable for achieving high-quality Cu bonding.

2.3.2 Dry Pretreatment

To address the issue of rapid Cu oxide re-formation in an atmospheric environment, an in situ pretreatment process has been proposed to remove Cu oxides in a vacuum following the bonding process, as depicted in Figure 2.16 [33]. The concept of dry pretreatment is similar to the wet pretreatment process, relying on the reaction between hydrogen and Cu oxides. In the dry pretreatment process, the by-products such as H_2O, CO_2, and H_2 desorb from the Cu surface in a vacuum environment. Various gas compositions can be used for the dry pretreatment process, including forming gas (mixture of H_2 and N_2) [32], formic acid (HCOOH)/N_2 [34–38], H_2 plasma [39, 40], H_2/Ar plasma [41], and others.

For achieving low-temperature Cu–Cu bonding by removing Cu oxides, both wet and dry pretreatment processes can be included in the process flow to ensure bonding quality. The wet process can remove the majority of Cu oxides in an atmospheric environment, followed by the dry process to address any remaining small amounts of regrown Cu oxides. Figure 2.17 demonstrates that the combination of citric acid

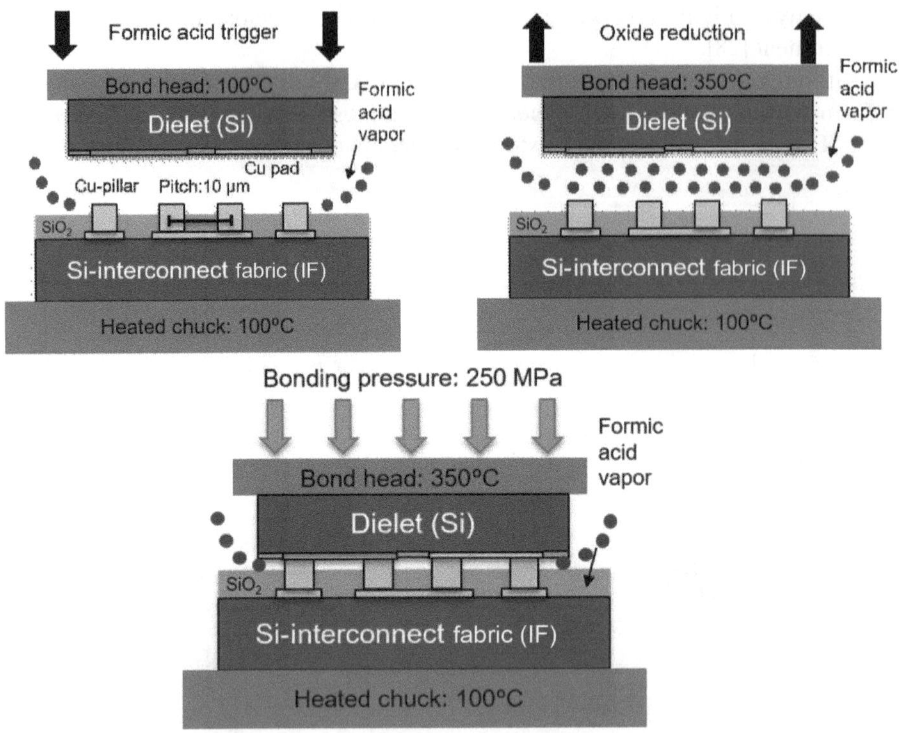

FIGURE 2.16 Schematic of the assembly process illustrating the steps of in situ cleaning and the TCB process [33].

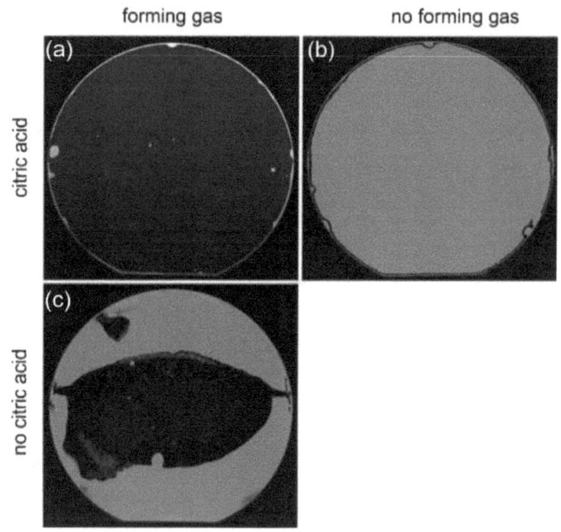

FIGURE 2.17 C-SAM images for different pretreatments of 175°C for 30 minutes bonded Cu wafer pairs after annealing at 200°C for 1 hour. (a) Combining forming gas treatment and citric acid treatment. (b) Only citric acid treatment. (c) Only forming gas treatment [32].

and forming gas pretreatment can achieve better bonding results under the conditions of 175°C bonding for 30 minutes and 200°C annealing for 1 hour [32].

In addition to forming gas, formic acid vapor is another commonly used source for the dry pretreatment process. Figure 2.18 presents a comprehensive scheme of the formic acid reduction mechanism [37]. This process involves the following [35]:

$$HCOOH \rightarrow HCOO_{(ads)} + H_{(ads)} \tag{2.5}$$

$$Cu_2O + HCOO_{(ads)} + H_{(ads)} \rightarrow Cu(HCOO)_{(ads)} + CuOH_{(ads)} \tag{2.6}$$

$$Cu(HCOO)_{(ads)} + CuOH_{(ads)} \rightarrow 2Cu + CO_2 + H_2O \tag{2.7}$$

It has been reported that the use of Pt as a catalyst can facilitate the decomposition of formic acid into CO_2 and H_2 [42, 43]. The presence of H_2 promotes the decomposition of copper formate and enhances the efficiency of copper oxide removal. Figure 2.19 indicates that formic acid pretreatment with Pt improves both the contact resistance and mechanical strength [42].

2.3.3 Surface-Activated Bonding (SAB)

As the dry pretreatment process can alleviate the issue of Cu reoxidation after surface cleaning, conducting the cleaning process under an ultrahigh vacuum (UHV) environment can greatly reduce the possibility of reoxidation. The main concept of SAB technology is to create an activated surface condition through ion/atom bombardment in UHV and directly bond the two surfaces through interatomic bonds,

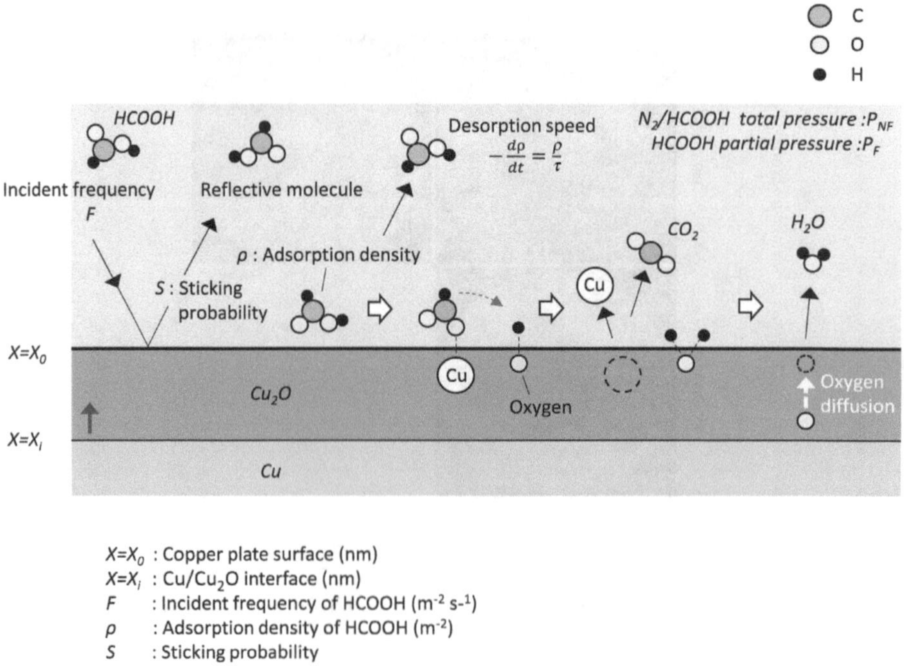

FIGURE 2.18 Formic acid reduction mechanism [37].

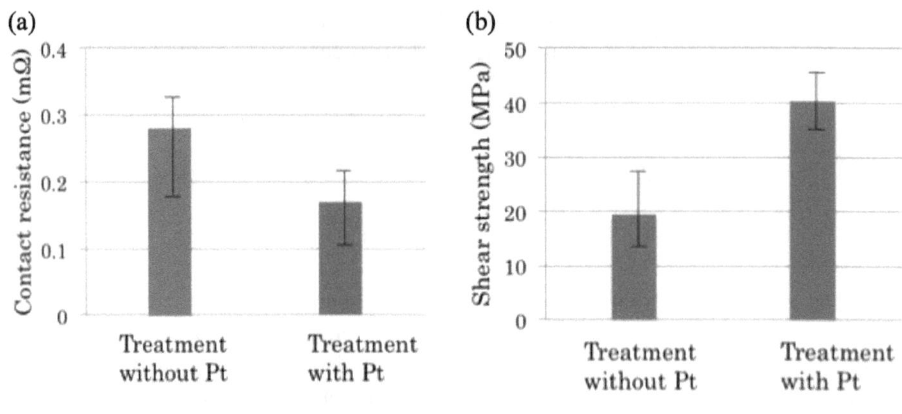

FIGURE 2.19 (a) Contact resistance and (b) shear strength of the bonded samples using formic acid treatment with and without Pt catalyst [42].

as depicted in Figure 2.20a [44]. Figure 2.21 illustrates the bonded area ratio of Cu bonding versus vacuum pressure, demonstrating the feasibility of SAB technology and the significance of the UHV environment [45]. Figure 2.20b presents an alternative approach to complete the bonding process at a relatively lower vacuum environment. By forming a nano-adhesion layer after UHV ion bombardment, the scope of feasible bonding materials with SAB technology can be extended. Although

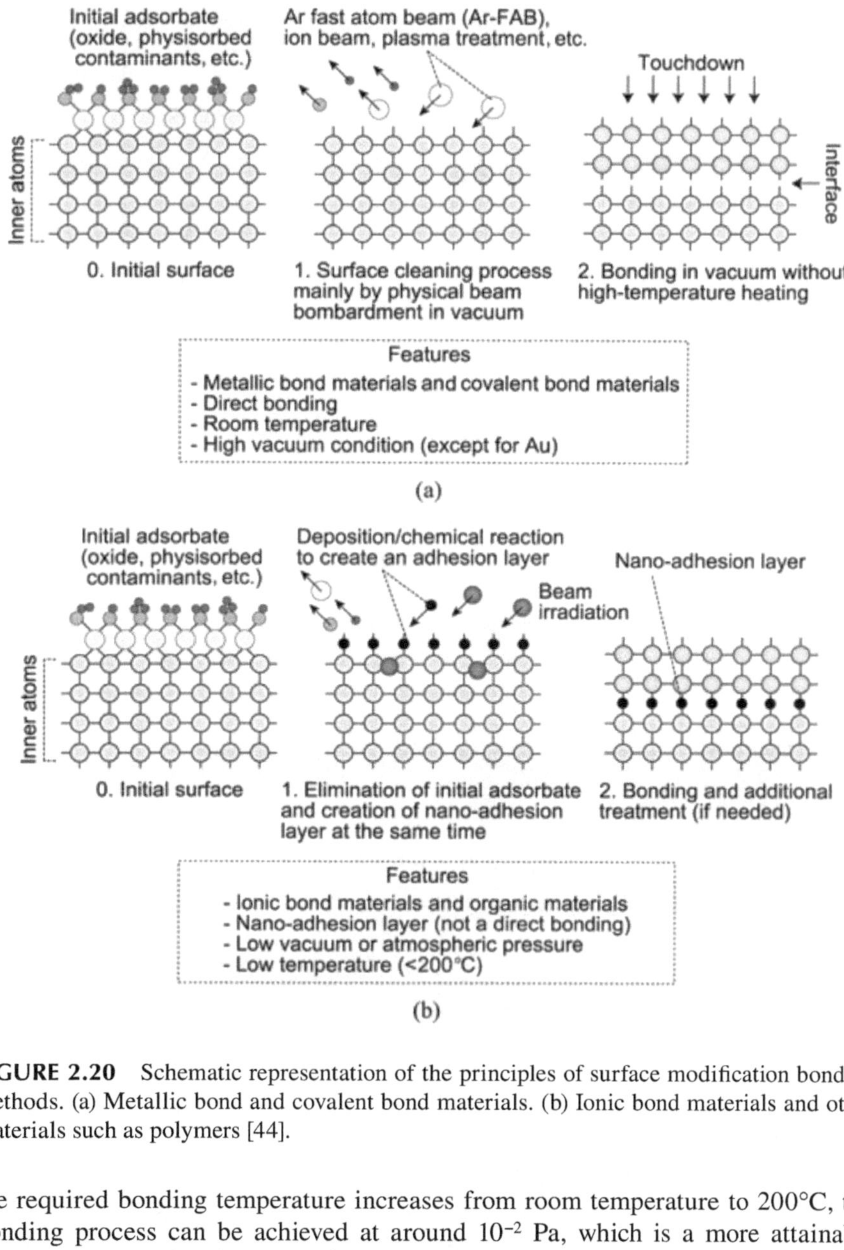

FIGURE 2.20 Schematic representation of the principles of surface modification bonding methods. (a) Metallic bond and covalent bond materials. (b) Ionic bond materials and other materials such as polymers [44].

the required bonding temperature increases from room temperature to 200°C, the bonding process can be achieved at around 10^{-2} Pa, which is a more attainable demand compared to the 5×10^{-6} Pa of UHV bonding [46].

With the SAB technology, seamless Cu–Cu bonding can be achieved. Figure 2.22 demonstrates the result of UHV bonding [47]. The Cu–Cu bonding interface is oxide-free and exhibits a clear interface because the bonding process is conducted at room temperature without Cu recrystallization. For Cu–Cu bonding with a nano-adhesion layer, the proposed material for the nano-adhesion layer is water, and the Cu–Cu bonding can be accomplished through the condensation reaction of Cu-OH.

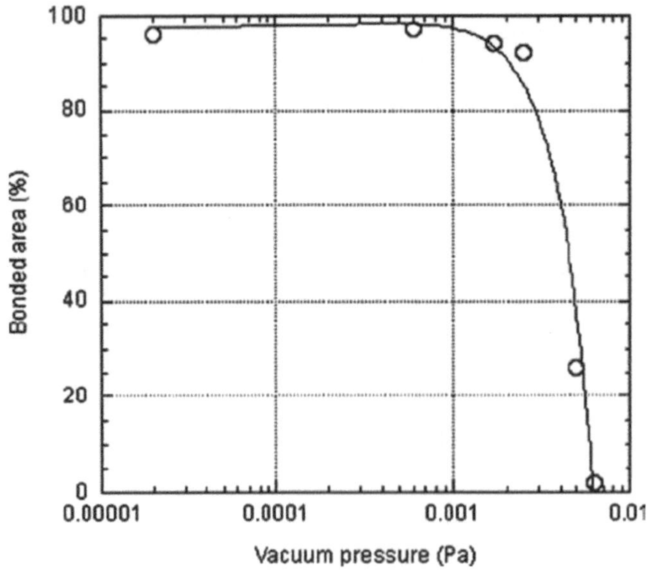

FIGURE 2.21 The change in the bonded area at various vacuum pressures by SAB technology [45].

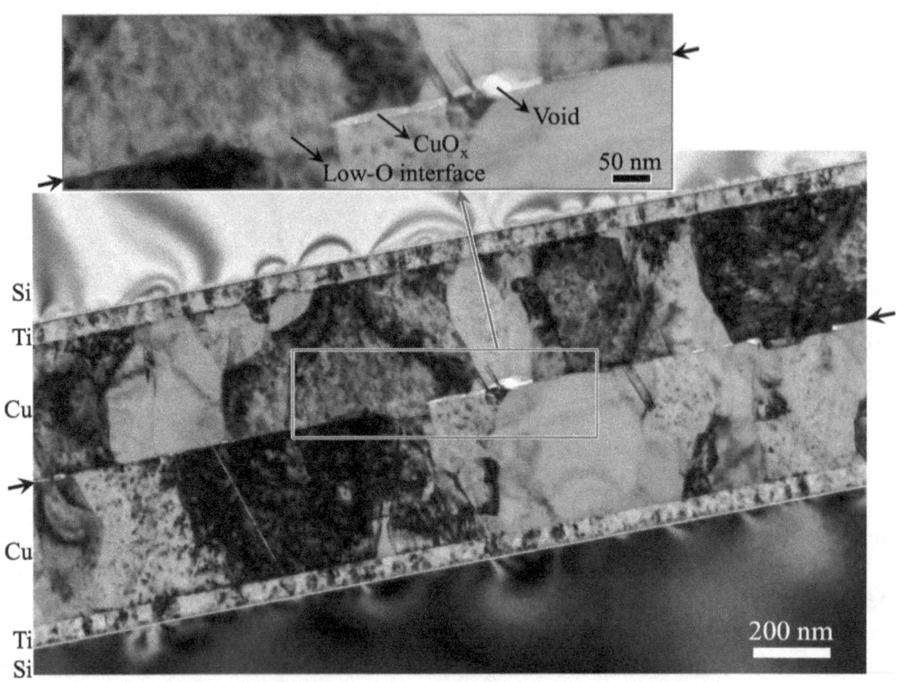

FIGURE 2.22 TEM image of the Cu–Cu hydrophilic bonded interface [46].

Consequently, a thin CuO_x layer may remain at the bonding interface, as shown in Figure 2.22 [46].

2.3.4 Self-assembled Monolayer (SAM)

In addition to removing the surface oxidation layer before Cu–Cu bonding, preventing the formation of Cu oxides can be another approach to improve low-temperature Cu–Cu bonding. Several protective approaches have been proposed, and one such technology is SAM, which utilizes a surface passivation layer to prevent Cu oxidation. In the SAM technology, organic compounds such as alkane-thiols or benzotriazole (BTA) are used as materials for surface passivation [48–52]. These organic compounds attach to the Cu surface, forming a hydrophobic layer. The SAM material and the Cu surface should have a high affinity to achieve a uniform protection layer. Additionally, the protection layer should be easily and completely desorbed from the Cu surface under controlled conditions to ensure a clean Cu surface during low-temperature bonding.

Among the SAM materials, alkane-thiols with different chain lengths have been widely studied, such as propanethiol (C3) [48], hexanethiol (C6) [49–51], decanethiol (C10) [49], dodecanethiol (C12) [50], and hexadecanethiol (C16) [52]. During the attachment process of alkane-thiol to Cu, the thiol group (–SH) binds strongly to Cu due to its high affinity, as shown in Figure 2.23 [52]. The methyl ($-CH_3$) terminal group provides the hydrophobic property. Figure 2.24 demonstrates the contact angle of the Cu surface before and after SAM passivation [50]. The formation of SAM passivation increases the measured contact angle, indicating enhanced hydrophobicity. For BTA, the bond strength with Cu might be lower than that of alkane-thiols

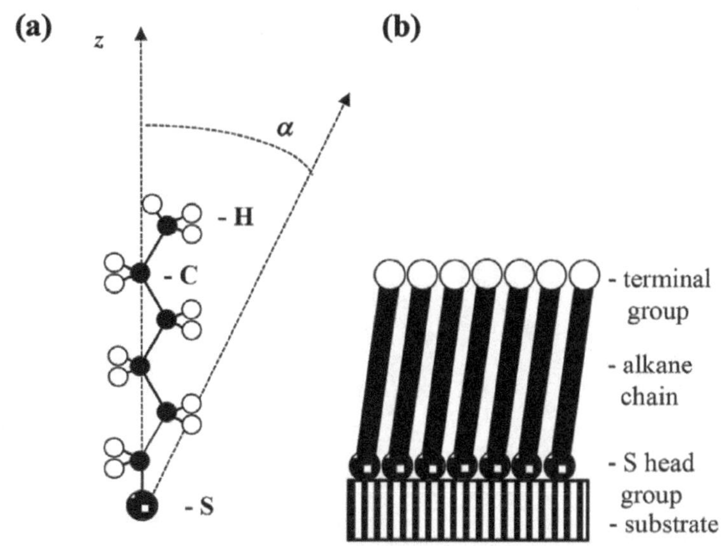

FIGURE 2.23 (a) 1-Hexanethiol with a tilt angle α. (b) Thiolates chemisorbed and arranged on the substrate [52].

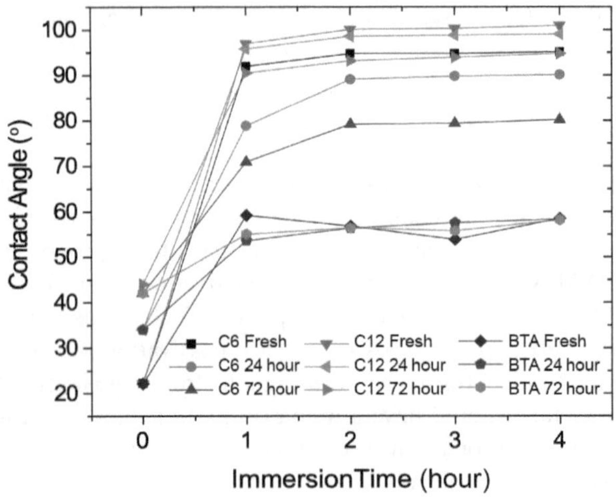

FIGURE 2.24 Comparison of the thermal stability of C6, C12, and BTA in ambient air versus immersion time in ethanolic solution [50].

and Cu due to the weaker electronegativity of the N atom compared to the S atom. Additionally, the aromatic ring of BTA results in less surface coverage compared to linear alkane-thiols. Among C6, C12, and BTA, BTA SAM exhibits the lowest contact angle. On the other hand, the higher contact angle observed for C12 compared to C6 suggests that alkane-thiols with longer chain lengths can provide better protection for the Cu surface. Figure 2.25 presents the antioxidation effect of C6, C12, and BTA through XPS analysis [50]. The lowest O/Cu ratio is observed for C12, indicating better protection compared to C6 and BTA. Similar results have been observed between C6 and C16 as well.

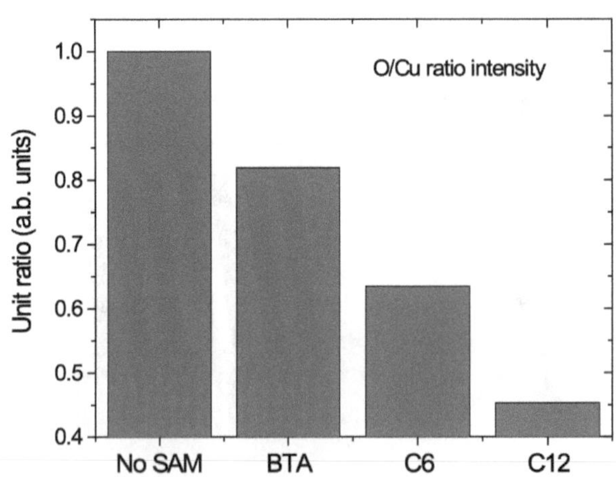

FIGURE 2.25 XPS data showing O content on wafer surface after five days [50].

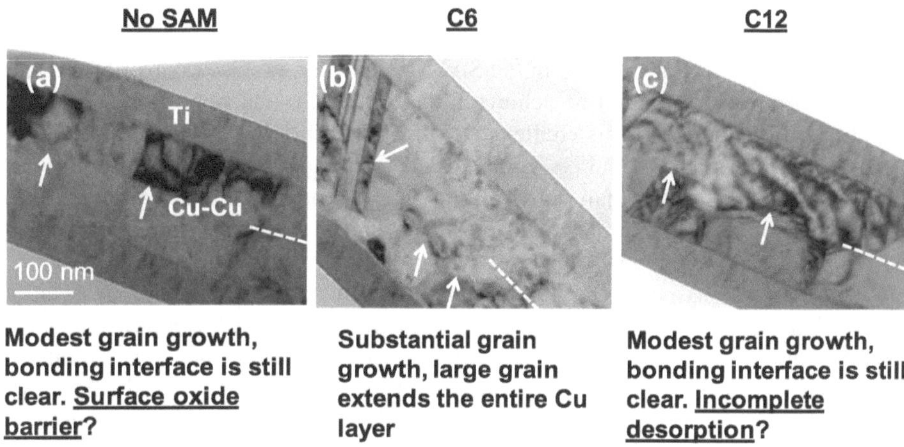

FIGURE 2.26 TEM micrographs of bonded Cu layers. (a) Without SAM passivation. (b) With C6-SAM passivation. (c) With C12-SAM passivation [50].

However, it has been reported that desorption of alkane-thiols with longer chain lengths might be more difficult, and the remaining organic compounds can hinder the quality of Cu–Cu bonding [50]. Figure 2.26 shows TEM images of Cu bonding with C6 and C12, respectively, as compared with no SAM protection. The specimens with SAM were annealed in an N_2 environment at 250°C for 30 minutes to desorb the SAM material. All the specimens were bonded at 250°C for 1 hour in a vacuum environment of 10^{-4} mbar. In this comparison, Cu bonding with C6 SAM protection resulted in the best bonding quality, with obvious Cu grain growth through the original bonding interface. Figure 2.27 demonstrates the antioxidation ability of the SAM technology [50]. With SAM protection, the XPS signal of oxygen is detected near the bonding interface but not throughout the Cu layer. Interface oxidation might be attributed to the fresh Cu surface after SAM desorption, resulting in oxidation prior to bonding.

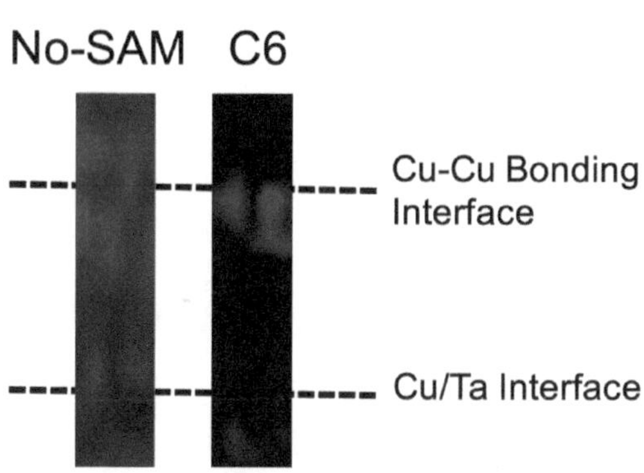

FIGURE 2.27 Oxygen concentration distribution (from EDX) in the bonded Cu layer: without and with C6 prebonding SAM passivation [50].

2.3.5 Metal Passivation

To address the reoxidation issue in the SAM technology, precise control of the vacuum environment is crucial to achieve high-quality Cu–Cu bonding at low temperatures. In response to this challenge, metal passivation bonding technology has emerged as a promising solution. Unlike SAM technology, metal passivation bonding utilizes metals rather than organic compounds to prevent Cu oxidation [53–62]. Due to the conductive and interdiffusion properties of metals, the impact of the passivation material on Cu bonding is relatively minor compared to SAM.

In the metal passivation bonding technology, there is no need for passivation desorption/removal steps. As shown in Figure 2.28, during the bonding process, metal bonding is achieved either through the bonding of the passivation material or through Cu–Cu bonding with Cu atoms diffusing through the passivation layer [53]. The latter is more likely to occur when the passivation layer is thin enough. Compared to conventional Cu–Cu bonding, solid-state bonding relies on the behavior of Cu lattice diffusion to bond the two Cu layers/pads. In passivation technology involving Cu diffusion through the antioxidation layer, the bonding process relies on grain boundary diffusion between Cu and the passivation material, which requires lower energy. Consequently, bonding temperature requirements can be significantly reduced with the passivation technology involving Cu diffusion.

To effectively prevent Cu oxidation, the ideal candidate for the metal passivation bonding technology is a noble metal such as Au [54–58], Ag [53, 59, 60], Pt [54, 61], or Pd [57, 62]. For metals that are prone to oxidation, the reduction in bonding temperature is not effectively achieved, and the contact resistance of the bonding structure tends to be high. For example, titanium (Ti) has been used as the passivation material for Cu–Cu bonding. As shown in Figure 2.29, the specific contact resistance using Ti passivation is approximately 10^{-3} Ω-cm^2 [62], which is relatively high compared to the typical range of 10^{-8}–10^{-9} Ω-cm^2 for Cu–Cu bonding.

Among the proposed metal passivation materials, Au and Ag exhibit the lowest bonding temperature requirements. With a 10-nm Au passivation layer, Cu–Cu bonding can be achieved at temperatures as low as 120°C, as demonstrated on a 4-inch wafer, as shown in Figure 2.30a [55]. Figure 2.30b demonstrates that with Ag passivation, bonding can be achieved at 70°C [53]. Additionally, Ag passivation

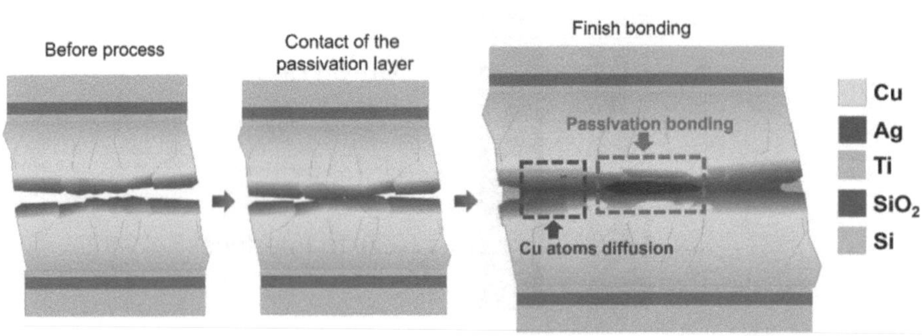

FIGURE 2.28 Bonding schematic of Cu–Cu bonding with Ag passivation structure [53].

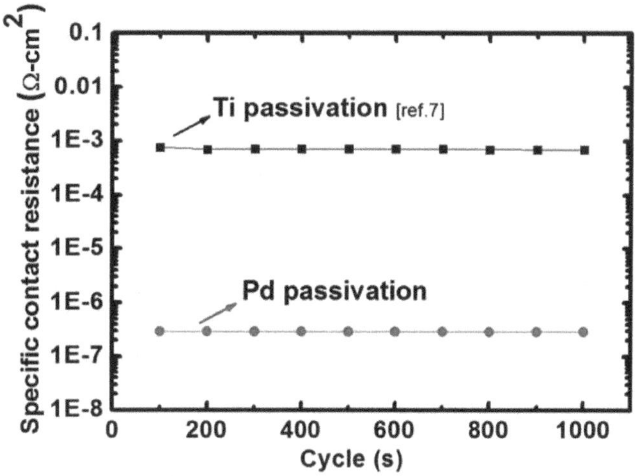

FIGURE 2.29 Electrical characteristic comparison of bonded Cu interconnects with the Pd and the Ti passivation under the current stressing test [62].

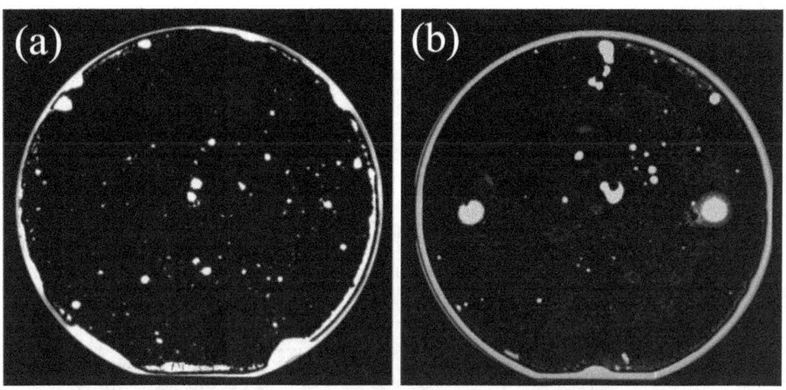

FIGURE 2.30 SAT images of Cu–Cu bonding with (a) Au passivation structure bonded at 120°C [55] and (b) ultrathin cluster-Ag passivation structure bonded at 70°C [53].

with hybrid bonding has been achieved at temperatures 120–150°C, as shown in Figure 2.31 [53]. To further decrease the required bonding temperature, a wetting layer is deposited between Cu and the passivation metal. The wetting layer reduces the surface roughness of the metal passivation and decreases the surface grain size. As a result, more grain boundaries allowing Cu diffusion are created, thus reducing the required bonding temperature. Figure 2.32 illustrates bonding with Au as the passivation layer and Ti as the wetting layer [58]. By incorporating a wetting layer, Cu–Cu bonding temperatures can be reduced to below 100°C with a 60-minute bonding duration or achieved at room temperature with 100°C post-bonding annealing.

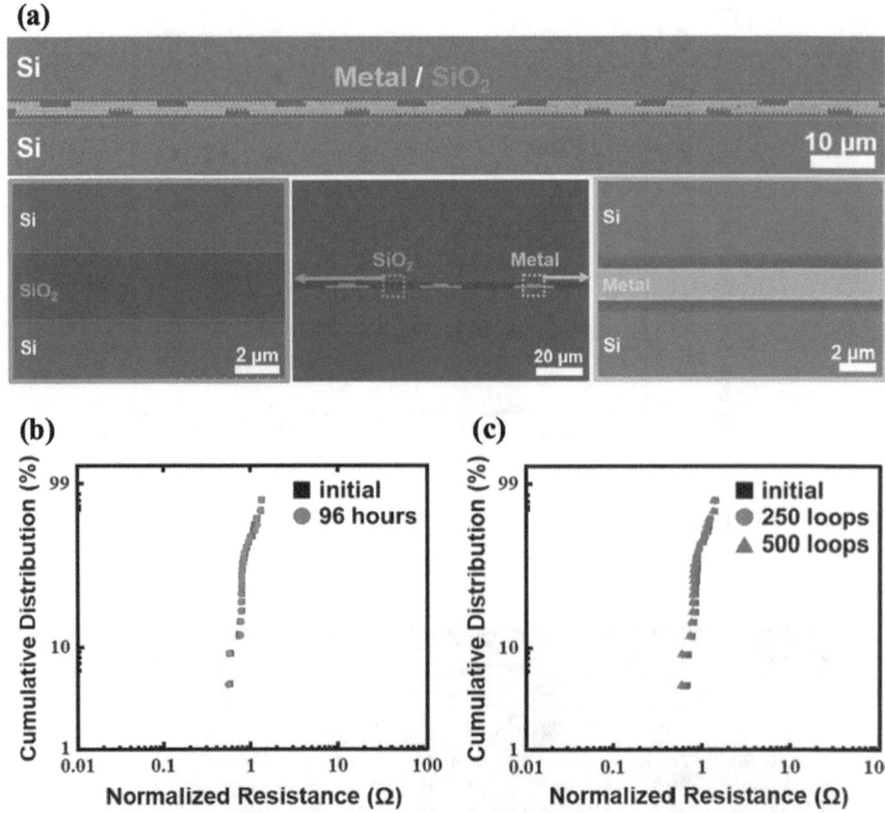

FIGURE 2.31 (a) SEM image of Cu/SiO_2 hybrid bonding with cluster-Ag passivation layer, (b) electrical performance after 96 hours of unbiased HAST, and (c) electrical performance after 500 loops of TCT [53].

FIGURE 2.32 The SAT and SEM cross-sectional images of Cu–Cu bonding with Au passivation and Ti wetting layer bonded at room temperature before and after annealing process [58].

2.3.6 Copper Nitride by Plasma Treatment

In comparison to SAM and metal passivation technologies, the plasma treatment method offers a simpler approach for protecting Cu without the need for surface cleaning processes or extra material deposition before bonding. By subjecting the Cu

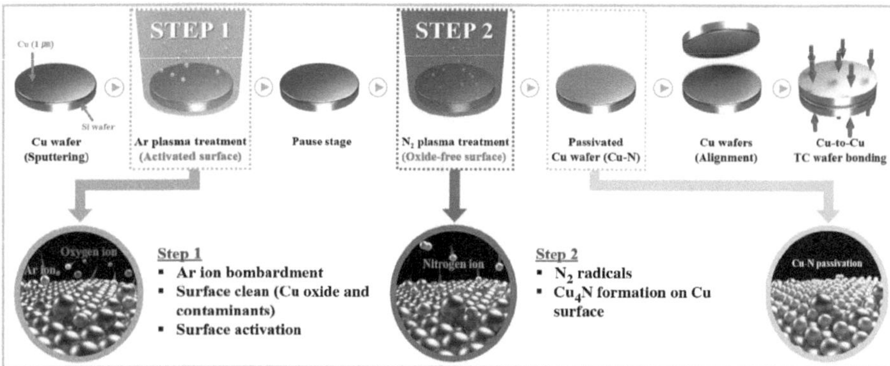

FIGURE 2.33 Schematics of sample preparation by two-step plasma treatment process forming copper nitride [63].

surface to appropriate plasma treatment, the properties of the Cu surface itself can be altered to form a protective layer of copper nitride. A two-step plasma treatment process has been proposed as a reliable method for transforming Cu into copper nitride [63–69].

The two-step plasma treatment process involves Ar plasma bombardment for surface cleaning and N_2 plasma treatment for the formation of Cu_4N, as depicted in Figure 2.33 [63]. Ar bombardment effectively removes Cu oxides from the surface and breaks the Cu crystalline structure, creating dangling bonds that aid in the formation of copper nitride during the subsequent N_2 plasma treatment. The N_2 plasma treatment results in the formation of a protective layer of copper nitride. A comparison of TEM images of the Cu film before and after the two-step plasma treatment is shown in Figure 2.34 [64]. Copper nitride exhibits structural stability

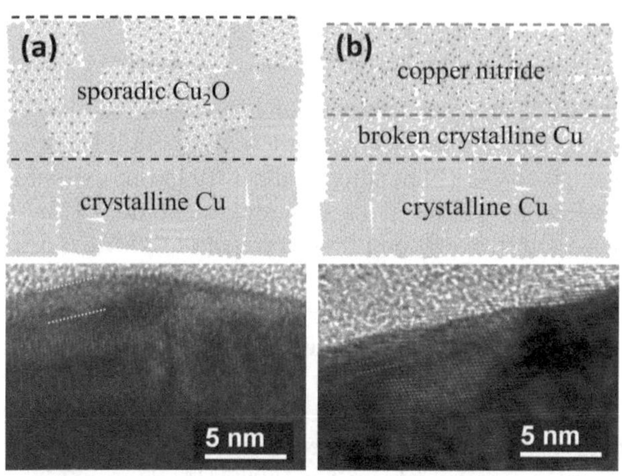

FIGURE 2.34 Schematic and cross-sectional TEM images of the Cu surface. (a) Before and (b) after two-step plasma treatment [64].

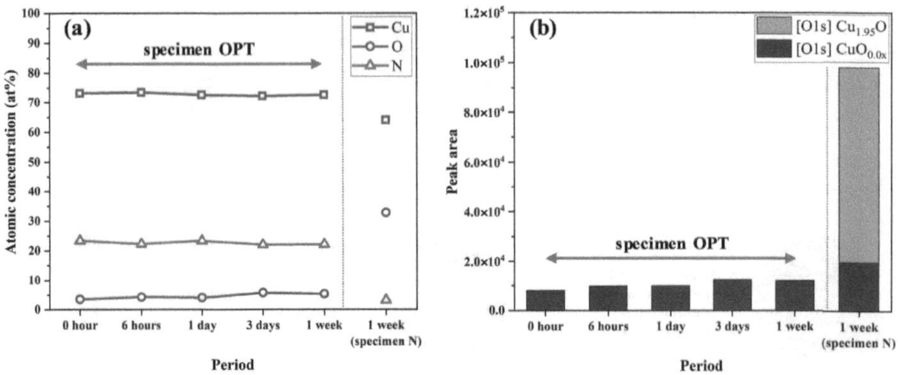

FIGURE 2.35 (a) Stoichiometric changes in the specimens (specimen OPT) treated under the two-step Ar/N_2 plasma with optimized N_2 plasma conditions during lifetime test, compared with a non-plasma-treated Cu specimen (specimen N). (b) Detailed analysis results of the Cu oxide peak area in the O1s for all specimens [65].

at room temperature and decomposes into Cu and N_2 at higher temperatures. Due to its stability at room temperature, copper nitride effectively protects Cu from oxidation. Figure 2.35 presents XPS results of Cu specimens with a one-week storage period at room temperature and atmospheric conditions [65]. The Cu specimens subjected to the two-step plasma treatment exhibit low oxygen concentrations, while the specimens with no pretreatment show obvious signs of oxidation. During the bonding process, at temperatures between 200°C and 300°C, the Cu_4N protective layer decomposes, enabling the formation of Cu–Cu bonding interface. Figure 2.36 demonstrates the results of C-SAM and SEM analysis for Cu–Cu bonding achieved at 260°C bonding temperature [65], showcasing the low-temperature bonding capability of the two-step plasma treatment process.

2.3.7 (111)-Oriented Nanotwinned Cu

In addition to surface cleaning and passivation methods, bonding using (111)-oriented nanotwinned Cu (nt-Cu) is a technique that leverages the inherent properties of the Cu structure to enhance Cu–Cu bonding quality. Cu grains exhibit anisotropic electrical and mechanical properties along different orientations [13, 70, 71]. (111)-Cu, in particular, has a lower oxidation rate compared to (200)-Cu or Cu with random orientation. Additionally, the surface diffusion coefficient on the (111) plane is higher than on the (100) and (110) planes. These properties make (111)-oriented Cu favorable for Cu–Cu bonding at low temperatures. Moreover, high-density nt-Cu exhibits increased tensile strength and higher resistance to electromigration compared to regular Cu [72]. Figure 2.37 illustrates the comparison between (111)-oriented nt-Cu and regular Cu in terms of interconnect performance. With (111)-oriented nt-Cu, Cu–Cu bonding can be achieved at 150°C [16]. Figure 2.38 shows the bonding result of two highly (111)-oriented nt-Cu films. With a bonding temperature of 150°C and a 1-hour thermal treatment, Cu bonding can be successfully achieved. As

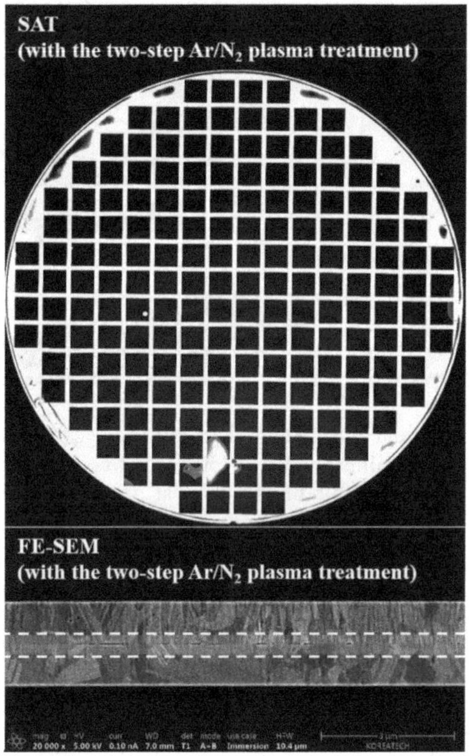

FIGURE 2.36 SAT and FE-SEM images of the 260°C Cu–Cu bonding interface using optimized two-step Ar/N_2 plasma-treated Cu wafers [65].

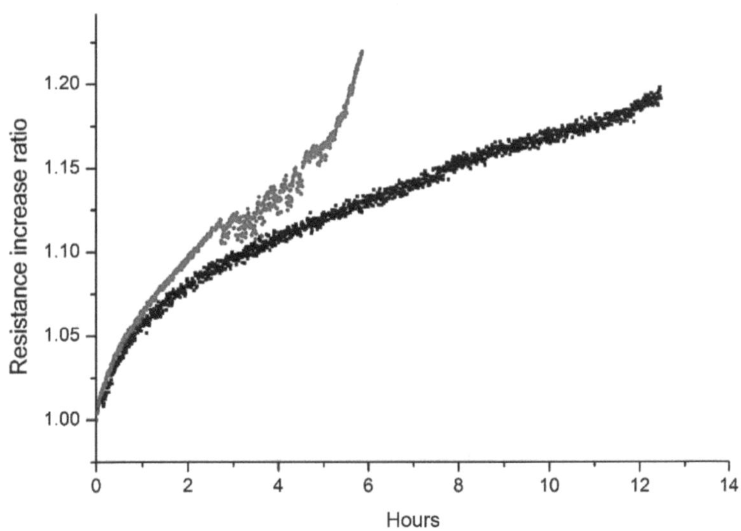

FIGURE 2.37 EM lifetime of nt-Cu (black) versus regular Cu (red) line [72].

FIGURE 2.38 (a) Bright-field and (b) dark-field TEM images obtained using the (111) diffraction for two (111) Cu thin films bonded at 150°C for 1 hour. (c) Bright-field and (d) dark-field TEM images of two (111) cu thin films bonded at 200°C for 30 minutes [16].

the bonding temperature increases, the required bonding duration can be reduced. TEM images in Figure 2.38 demonstrate a seamless bonding interface achieved at 200°C for 30 minutes.

However, achieving a high area ratio of (111)-oriented nt-Cu requires careful control of the deposition process. It has been proposed that both sputtering and electroplating processes can successfully deposit highly (111)-oriented nt-Cu films, but films prepared by the electroplating process exhibit lower bonding temperatures and shorter bonding durations [73]. The film quality of electroplated (111)-oriented nt-Cu is highly sensitive to process parameters such as additive composition [74], additive concentration [75], bath temperature [76, 77], deposition current density [77, 78], and the condition of the seed layer [79]. For instance, Figure 2.39 presents a comparison of (111)-oriented nt-Cu deposited on a stronger ⟨111⟩-oriented Cu seed layer versus a regular ⟨111⟩-oriented Cu seed layer. The analysis indicates that a stronger ⟨111⟩-oriented seed layer results in denser deposited nt-Cu, thereby improving the quality of low-temperature Cu bonding.

2.3.8 Pillar–Concave Bonding

Pillar–concave bonding is a bonding approach that aims to provide more energy for Cu diffusion over the barrier for Cu–Cu bonding. In this bonding structure, compared to conventional Cu pillar-to-pillar bonding, the pillar–concave structure exhibits a higher shear stress during the TCB process due to the reduced contact area. As a result, large deformation occurs during the bonding process. Figure 2.40 illustrates the phenomenon of plastic deformation and dislocation movement during the bonding process [80]. The deformation in the pillar–concave structure generates heat through internal friction, resulting in an exothermic reaction that provides energy for Cu atom diffusion. Additionally, the deformation process increases the

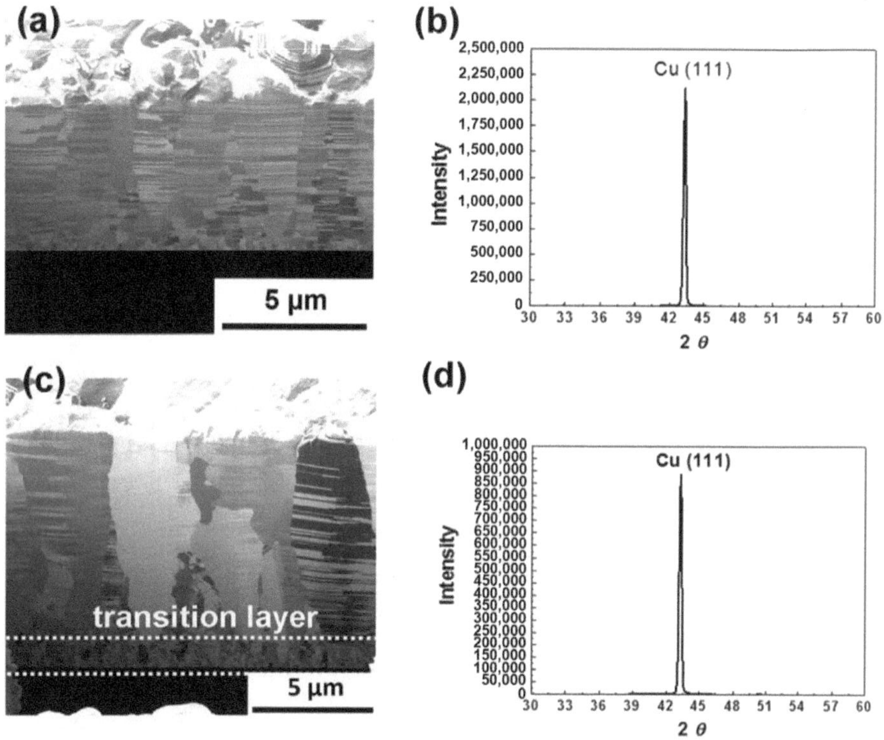

FIGURE 2.39 (a) Cross-sectional FIB image for the Cu film on the strong ⟨111⟩-oriented seed layer after 3,000 cycles and (b) its corresponding X-ray diffraction. (c) Cross-sectional FIB image for the Cu film on the regular ⟨111⟩-oriented seed layer after 4,500 cycles and (d) its corresponding X-ray diffraction [79].

presence of vacancy defects at the grain boundary, facilitating the diffusion of Cu atoms at the grain boundary under the TCB process. Consequently, the required heating temperature from the environment can be reduced.

It has been reported that using the pillar–concave structure, Cu–Cu bonding can be achieved at 150°C with stable electrical performance after reliability tests [81]. Figure 2.41 shows FIB images of the pillar–concave bonding interface, demonstrating a clear interface even at a bonding temperature of 150°C. The bonding structure exhibits high strength during shear testing, indicating a high resistance to lateral forces at the Cu–Cu interface in the pillar–concave structure.

To achieve the desired bonding result, sufficient stress needs to be generated during the pillar–concave TCB process. Figure 2.42 illustrates the relationship between several structural parameters, generated stress, and resulting deformation [82]. This simulation demonstrates that the pillar–concave Cu bonding process is influenced by various factors, such as the shape of the opening and the angle of the concave sidewall. Therefore, proper structure design and related simulations are crucial to ensuring high-quality Cu–Cu bonding.

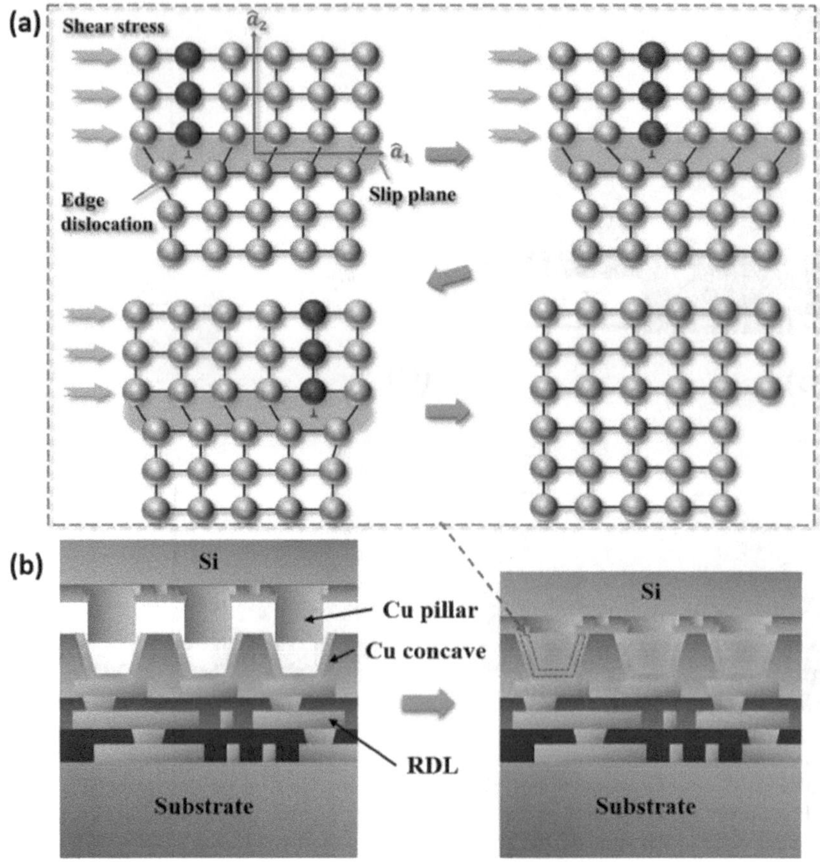

FIGURE 2.40 (a) Illustration of plastic deformation in crystalline scale. (b) Schematic of pillar–concave bonding scheme [80].

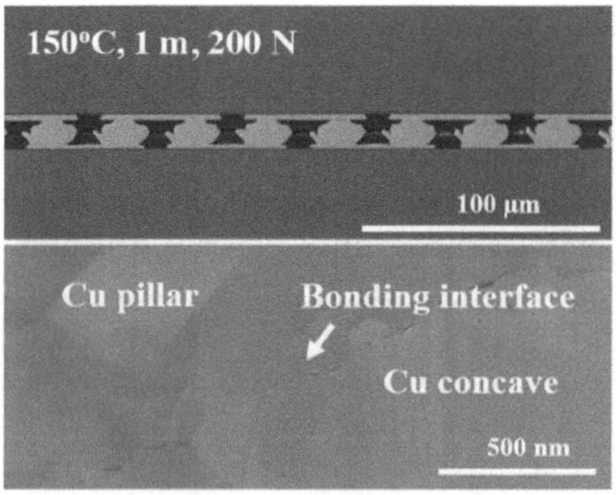

FIGURE 2.41 FIB images of Cu–Cu bonding with pillar–concave structure after 150°C bonding [81].

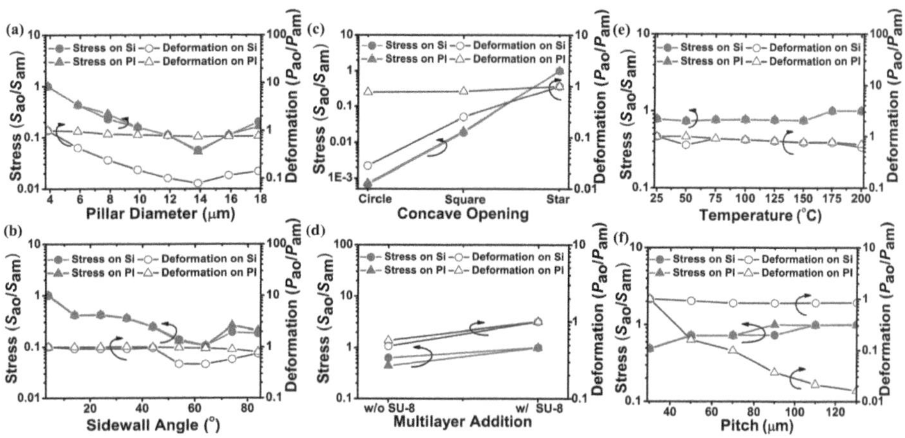

FIGURE 2.42 Simulation results with different parameters in pillar–concave structure [82].

2.4 Cu-BASED HYBRID BONDING STRUCTURE

In the case of Cu-based hybrid bonding structures, which involve bonding Cu with surrounding materials, it is crucial to consider the change in surface topography during the bonding and annealing process. The goal is to ensure that all regions of the structure come into contact and bond effectively.

Hybrid bonding can be achieved through either TCB process or room temperature bonding with post-annealing. In the TCB process, the bonding interface is heated and pressed under external force simultaneously, resulting in successful bonding across the hybrid structure. On the other hand, room temperature bonding with post-annealing allows extremely flat surfaces to be contacted and attached under room temperature, enhancing bonding quality during annealing.

For the method of room temperature bonding with post-annealing, taking the example of a Cu/SiO_2 structure, an ideal surface profile is one where the Cu surface is either at the same level or slightly lower than the SiO_2 surface, as depicted in Figure 2.43 [83]. This ensures that the Cu–Cu bonding can generate sufficient

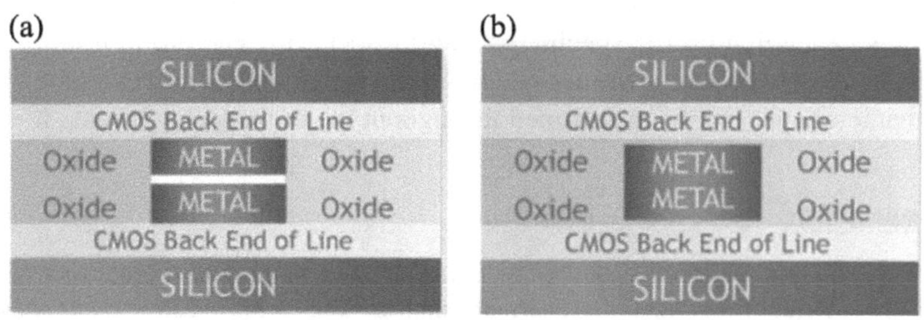

FIGURE 2.43 Process flow schematic for hybrid bonding with metal dishing [83]. (a) Oxide–oxide bonding at room temperature. (b) Formation of metal–metal interconnect during annealing.

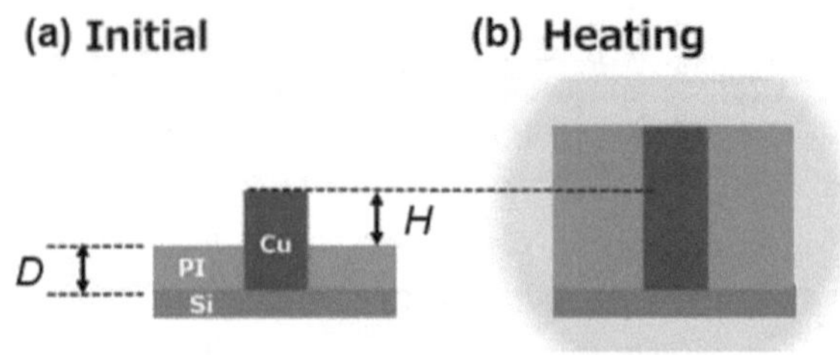

FIGURE 2.44 Schematic of volume change of Cu and PI when heated [84].

internal stress when heated. The internal stress arises from the mismatch in the coefficient of thermal expansion (CTE) between Cu and its surrounding materials. Excessive Cu dishing, where the Cu layer becomes significantly recessed compared to the surrounding material, can reduce the yield of Cu/SiO_2 hybrid bonding. It is important to control the surface profile to optimize the bonding process.

However, TCB process might be more suitable in other situations. For instance, there has been growing interest in Cu-based hybrid bonding with polymers, driven by the demand for flexible substrates and lower cost packaging solutions. Polymer films have different properties compared to SiO_2 or Si, and their composition can be adjusted to suit specific requirements. In some cases, the desired surface profile for Cu–polymer hybrid bonding may involve having the Cu surface at a higher level than the polymer, considering the polymer's greater expansion characteristics, as shown in Figure 2.44 [84]. Due to the topography of Cu protrusion, TCB process is more suitable to achieve Cu–polymer hybrid bonding.

Since the process flow for hybrid bonding can vary depending on the specific application, it is important to carefully evaluate the feasibility of the chosen bonding technology and consider the surface profile requirements for successful Cu-based hybrid bonding.

2.5 SUMMARY

In this chapter, we have discussed the recent developments in Cu–Cu bonding technology. We explored the mechanisms and principles of Cu–Cu bonding, as well as various low-temperature bonding technologies. The feasibility of these bonding methods varies depending on the specific packaging scenario. Therefore, it is important to understand how low-temperature bonding technologies work and evaluate whether their characteristics can be maintained throughout the prebonding process flow.

REFERENCES

1. K.-C. Shie, A. M. Gusak, K. N. Tu, and C. Chen, "A Kinetic Model of Copper-to-Copper Direct Bonding under Thermal Compression," *J. Mater. Res. Technol.*, vol. 15, pp. 2332–2344, 2021.
2. M. Martinez *et al.*, "Mechanisms of Copper Direct Bonding Observed by In-situ and Quantitative Transmission Electron Microscopy," *Thin Solid Films*, vol. 530, pp. 96–99, 2013.

3. B. Maack and N. Nilius, "Impact of Granularity on the Oxidation Kinetics of Copper," *Phys. Status Solidi B*, vol. 257, no. 6, p. 1900778, 2020.
4. Y. J. Kim, D. Kim, Y. Kim, Y. Jeong, B. Jeong, and J. Y. Park, "Effect of Water Vapor on Oxidation Processes of the Cu(111) Surface and Sublayer," *Int. J. Mol. Sci.*, vol. 24, no. 1, p. 810, 2023. doi: 10.3390/ijms24010810.
5. L. De Los Santos Valladares *et al.*, "Crystallization and Electrical Resistivity of Cu_2O and CuO Obtained by Thermal Oxidation of Cu Thin Films on SiO_2/Si Substrates," *Thin Solid Films*, vol. 520, no. 20, pp. 6368–6374, 2012, doi: 10.1016/j.tsf.2012.06.043.
6. B. Maack and N. Nilius, "Oxidation of Polycrystalline Copper Films: Pressure and Temperature Dependence," *Thin Solid Films*, vol. 651, pp. 24–30, 2018. doi: 10.1016/j.tsf.2018.02.007.
7. S. Y. Lee, N. Mettlach, N. Nguyen, Y. M. Sun, and J. M. White, "Copper Oxide Reduction through Vacuum Annealing," *Appl. Surf. Sci.*, vol. 206, no. 1–4, pp. 102–109, 2003.
8. L. Peng, D. F. Lim, L. Zhang, H. Y. Li, and C. S. Tan, "Effect of Prebonding Anneal on the Microstructure Evolution and Cu–Cu Diffusion Bonding Quality for Three-dimensional Integration," *J. Electron. Mater.*, vol. 41, no. 9, pp. 2567–2572, 2012.
9. B. Rebhan and K. Hingerl, "Physical Mechanisms of Copper–Copper Wafer Bonding," *J. Appl. Phys.*, vol. 118, no. 13, p. 135301, 2015.
10. L. Mirkarimi *et al.*, "The Influence of Cu Microstructure on Thermal Budget in Hybrid Bonding," in *2022 IEEE 72nd Electronic Components and Technology Conference (ECTC)*, San Diego, CA, IEEE, May 2022, pp. 162–167.
11. Z. Fan, X. Chen, Y. Liu, Y. Jiang, and Y. Zhang, "Effects of Anisotropy on the Reliability of TSV Microstructure," *Microelectron. Reliab.*, vol. 114, p. 113745, 2020.
12. S. B. Liang, C. Wei, C. B. Ke, S. Cao, M. B. Zhou, and X. P. Zhang, "Investigation of the Interaction Effect between the Microstructure Evolution and the Thermo-Mechanical Behavior of Cu-filled through Silicon Via," *IEEE Trans. Device Mater. Reliab.*, vol. 22, no. 2, pp. 267–275, 2022.
13. C.-H. Tseng, K. N. Tu, and C. Chen, "Comparison of Oxidation in Uni-directionally and Randomly Oriented Cu Films for Low Temperature Cu-to-Cu Direct Bonding," *Sci. Rep.*, vol. 8, no. 1, p. 10671, 2018.
14. Y. Li *et al.*, "Crystal-plane-dependent Redox Reaction on Cu Surfaces," *Nano Res.*, vol. 13, no. 6, pp. 1677–1685, 2020.
15. Y.-J. Lee *et al.*, "Completing the Picture of Initial Oxidation on Copper," *Appl. Surf. Sci.*, vol. 562, p. 150148, 2021.
16. C.-M. Liu *et al.*, "Low-temperature Direct Copper-to-Copper Bonding Enabled by Creep on (111) Surfaces of Nanotwinned Cu," *Sci. Rep.*, vol. 5, no. 1, p. 9734, 2015.
17. B. Rebhan, J. Svoboda, and M. Panholzer, "A Thermodynamic Study of Voiding Phenomena in Cu–Cu Thermo-compression Wafer Bonding," *Microsyst. Technol*, vol. 24, no. 1, pp. 815–822, 2018.
18. P. Enquist, G. Fountain, C. Petteway, A. Hollingsworth, and H. Grady, "Low Cost of Ownership Scalable Copper Direct Bond Interconnect 3D IC Technology for Three Dimensional Integrated Circuit Applications," in *2009 IEEE International Conference on 3D System Integration*, San Francisco, CA, IEEE, September 2009, pp. 1–6.
19. P. Gueguen *et al.*, "Copper Direct-bonding Characterization and Its Interests for 3D Integration," *J. Electrochem. Soc.*, vol. 156, no. 10, p. H772, 2009.
20. H. Moriceau *et al.*, "Low Temperature Direct Bonding: An Attractive Technique for Heterostructures Build-up," *Microelectron. Reliab.*, vol. 52, no. 2, pp. 331–341, 2012.
21. P. Gondcharton, F. Baudin, L. Benaissa, and B. Imbert, "Mechanisms Overview of Thermocompression Process for Copper Metal Bonding," *MRS Proc*, vol. 1559, p. mrss13-1559-aa08-07, 2013.

22. Q. Kang *et al.*, “Low-Temperature Co-hydroxylated Cu/SiO_2 Hybrid Bonding Strategy for a Memory-centric Chip Architecture,” *ACS Appl. Mater. Interfaces*, vol. 13, no. 32, pp. 38866–38876, 2021.
23. Q. Kang, G. Li, Z. Li, Y. Tian, and C. Wang, “Surface Co-hydrophilization via Ammonia Inorganic Strategy for Low-Temperature Cu/SiO_2 Hybrid Bonding,” *J. Mater. Sci. Technol,*, vol. 149, pp. 161–166, 2023.
24. K. N. Chen, C. S. Tan, A. Fan, and R. Reif, “Copper Bonded Layers Analysis and Effects of Copper Surface Conditions on Bonding Quality for Three-dimensional Integration,” *J. Electron. Mater*, vol. 34, no. 12, pp. 1464–1467, 2005.
25. N. Habbache, N. Alane, S. Djerad, and L. Tifouti, “Leaching of Copper Oxide with Different Acid Solutions,” *Chem. Eng. J*, vol. 152, no. 2–3, pp. 503–508, 2009.
26. J.-W. Kim, M.-H. Jeong, and Y.-B. Park, “Effect of HF & H_2SO_4 Pretreatment on Interfacial Adhesion Energy of Cu–Cu Direct Bonds,” *Microelectron. Eng.*, vol. 89, pp. 42–45, 2012.
27. Y.-S. Lee, J.-S. Yoon, Y.-R. Jo, H. Lee, and S.-K. Rha, “Dilute H_2SO_4 Solution for Copper Seed Cleaning in Electroplating,” *Trans. Nonferrous Met. Soc. China*, vol. 23, no. 2, pp. 562–566, 2013.
28. K. L. Chavez, and D. W. Hess, “A Novel Method of Etching Copper Oxide Using Acetic Acid,” *J. Electrochem. Soc.*, vol. 148, no. 11, p. G640, 2001.
29. E.-J. Jang, S. Hyun, H.-J. Lee, and Y.-B. Park, “Effect of Wet Pretreatment on Interfacial Adhesion Energy of Cu–Cu Thermocompression Bond for 3D IC Packages,” *J. Electron. Mater.*, vol. 38, no. 12, pp. 2449–2454, 2009.
30. C. Pernel, J. Farkas, and D. Louis, “Copper in Organic Acid Based Cleaning Solutions,” *J. Vac. Sci. Technol. B Microelectron. Nanometer. Struct.*, vol. 24, no. 5, p. 2467, 2006. doi: 10.1116/1.2335866.
31. O. J. Kwon, J. H. Bae, B. K. Cho, Y. J. Kim, and J. J. Kim, “Investigation of Cleaning Solution Composed of Citric Acid and 5-Aminotetrazole,” *Korean J. Chem. Eng.*, vol. 28, no. 7, pp. 1619–1624, 2011. doi: 10.1007/s11814-011-0023-7.
32. B. Rebhan, G. Hesser, J. Duchoslav, V. Dragoi, M. Wimplinger, and K. Hingerl, “Low-temperature Cu–Cu Wafer Bonding,” *ECS Trans.*, vol. 50, no. 7, pp. 139–149, 2013. doi: 10.1149/05007.0139ecst.
33. S. Jangam *et al.*, “Fine-pitch (≤10 μm) Direct Cu–Cu Interconnects Using In-situ Formic Acid Vapor Treatment,” in *2019 IEEE 69th Electronic Components and Technology Conference (ECTC)*, Las Vegas, NV, IEEE, May 2019, pp. 620–627. doi: 10.1109/ECTC.2019.00099.
34. W. Yang, M. Akaike, and T. Suga, “Effect of Formic Acid Vapor In Situ Treatment Process on Cu Low-temperature Bonding,” *IEEE Trans. Compon. Packaging Manuf. Technol.*, vol. 4, no. 6, pp. 951–956, 2014.
35. R. He, M. Fujino, M. Akaike, T. Sakai, S. Sakuyama, and T. Suga, “Combined Surface Activated Bonding Using H-containing HCOOH Vapor Treatment for Cu/Adhesive Hybrid Bonding at below 200°C,” *Appl. Surf. Sci.*, vol. 414, pp. 163–170, 2017.
36. W. Yang, Y. Lu, C. Zhou, J. Zhang, and T. Suga, “Study of Cu Film Surface Treatment Using Formic Acid Vapor/Solution for Low Temperature Bonding,” *J. Electrochem. Soc*, vol. 165, no. 4, pp. H3080–H3084, 2018.
37. N. Ozawa and M. Shibata, “Reduction Process of Copper Oxide Film by Formic Acid,” *Jpn. J. Appl. Phys*, vol. 60, no. 11, p. 115502, 2021.
38. J.-J. Jhan, K. Wataya, H. Nishikawa, and C.-M. Chen, “Electrodeposition of Nanocrystalline Cu for Cu–Cu Direct Bonding,” *J. Taiwan Inst. Chem. Eng*, vol. 132, p. 104127, 2022.
39. M. R. Baklanov *et al.*, “Characterization of Cu Surface Cleaning by Hydrogen Plasma,” *J. Vac. Sci. Technol. B Microelectron. Nanometer Struct.*, vol. 19, no. 4, p. 1201, 2001.

40. F. Brecelj and M. Mozetic, "Reduction of Metal Oxide Thin Layers by Hydrogen Plasma," *Vacuum*, vol. 40, no. 1–2, pp. 177–181, 1990.
41. J. Lee, T. S. Williams, and R. F. Hicks, "Atmospheric Pressure Plasma Reduction of Copper Oxide to Copper Metal," *J. Vac. Sci. Technol. A*, vol. 39, no. 2, p. 023001, 2021.
42. T. Suga, M. Akaike, and W. Yang, "Formic Acid Treatment with Pt Catalyst for Cu Direct and Hybrid Bonding at Low Temperature," in *2014 IEEE 64th Electronic Components and Technology Conference (ECTC)*, Orlando, FL, IEEE, May 2014, pp. 1143–1147.
43. W. Yang, M. Akaike, M. Fujino, and T. Suga, "A Combined Process of Formic Acid Pretreatment for Low-temperature Bonding of Copper Electrodes," *ECS J. Solid State Sci. Technol.*, vol. 2, no. 6, pp. P271–P274, 2013.
44. A. Shigetou and T. Suga, "Modified Diffusion Bonding for Both Cu and SiO_2 at 150°C in Ambient Air," in *2010 Proceedings 60th Electronic Components and Technology Conference (ECTC)*, Las Vegas, NV, IEEE, 2010, pp. 872–877.
45. A. Shigetou, T. Itoh, and T. Suga, "Direct Bonding of CMP-Cu Films by Surface Activated Bonding (SAB) Method," *J. Mater. Sci*, vol. 40, no. 12, pp. 3149–3154, 2005.
46. R. He, M. Fujino, A. Yamauchi, and T. Suga, "Combined Surface Activated Bonding Technique for Hydrophilic SiO_2–SiO_2 and Cu–Cu Bonding," *ECS Trans*, vol. 75, no. 9, pp. 117–128, 2016. doi: 10.1149/07509.0117ecst.
47. R. He, M. Fujino, A. Yamauchi, Y. Wang, and T. Suga, "Combined Surface Activated Bonding Technique for Low-Temperature Cu/Dielectric Hybrid Bonding," *ECS J. Solid State Sci. Technol.*, vol. 5, no. 7, pp. P419–P424, 2016. doi: 10.1149/2.0201607jss.
48. T. Ghosh, E. Krushnamurthy, C. Subrahmanyam, V. SivaRamaKrishna, A. Dutta, and S. G. Singh, "Room Temperature Desorption of Self Assembled Monolayer from Copper Surface for Low Temperature & Low Pressure Thermocompression Bonding," in *2015 IEEE 65th Electronic Components and Technology Conference (ECTC)*, San Diego, CA, IEEE, May 2015, pp. 2200–2204.
49. K. Tanaka *et al.*, "Investigation of Surface Pre-treatment Methods for Wafer-level Cu-Cu Thermo-compression Bonding," *Micromachines*, vol. 7, no. 12, p. 234, 2016.
50. D. F. Lim, J. Wei, K. C. Leong, and C. S. Tan, "Cu Passivation for Enhanced Low Temperature (≤300°C) Bonding in 3D Integration," *Microelectron. Eng.*, vol. 106, pp. 144–148, 2013.
51. L. Qiu, A. Ikeda, and T. Asano, "Effect of Coating Self-assembled Monolayer on Room-temperature Bonding of Cu Micro-interconnects," *Jpn. J. Appl. Phys*, vol. 52, no. 6R, p. 068004, 2013.
52. M. Lykova *et al.*, "Characterization of Self-assembled Monolayers for Cu–Cu Bonding Technology," *Microelectron. Eng.*, vol. 202, pp. 19–24, 2018.
53. Z.-J. Hong *et al.*, "Low-temperature Hybrid Bonding with High Electromigration Resistance Scheme for Application on Heterogeneous Integration," *Appl. Surf. Sci.*, vol. 610, p. 155470, 2023.
54. H. Kuwae, K. Yamada, W. Momose, S. Shoji, and J. Mizuno, "Cu–Cu Quasi-direct Bonding with Atomically Thin-Au and Pt Intermediate Layer Using Atomic Layer Deposition," in *2019 International Conference on Electronics Packaging (ICEP)*, Niigata, Japan, IEEE, 2019, pp. 207–211.
55. D. Liu, P.-C. Chen, T.-C. Chou, H.-W. Hu, and K.-N. Chen, "Demonstration of Low-temperature Fine-pitch Cu/SiO_2 Hybrid Bonding by Au Passivation," *IEEE J. Electron Devices Soc.*, vol. 9, pp. 868–875, 2021.
56. D. Liu, P.-C. Chen, Y.-W. Liu, H.-W. Hu, and K.-N. Chen, "Low-temperature (70°C) Cu-to-Cu Direct Bonding by Capping Metal Layers," *IEEE Electron Device Lett.*, vol. 42, no. 10, pp. 1524–1527, 2021.
57. Z.-J. Hong *et al.*, "Investigation of Bonding Mechanism for Low-temperature Cu–Cu Bonding with Passivation Layer," *Appl. Surf. Sci.*, vol. 592, p. 153243, 2022.

58. Z.-J. Hong *et al.*, "Room Temperature Cu–Cu Direct Bonding Using Wetting/Passivation Scheme for 3D Integration and Packaging," in *2022 IEEE Symposium on VLSI Technology and Circuits (VLSI Technology and Circuits)*, Honolulu, HI, IEEE, 2022, pp. 387–388.
59. Y. Kim, S. Park, and S. E. Kim, "The Effect of an Ag Nanofilm on Low-temperature Cu/Ag–Ag/Cu Chip Bonding in Air," *Appl. Sci.*, vol. 11, no. 20, p. 9444, 2021.
60. T.-C. Chou *et al.*, "Electrical and Reliability Investigation of Cu-to-Cu Bonding with Silver Passivation Layer in 3-D Integration," *IEEE Trans. Compon. Packaging Manuf. Technol.*, vol. 11, no. 1, pp. 36–42, 2021.
61. D. Liu *et al.*, "Investigation of Low-temperature Cu–Cu Direct Bonding with Pt Passivation Layer in 3-D Integration," *IEEE Trans. Compon. Packaging Manuf. Technol.*, vol. 11, no. 4, pp. 573–578, 2021.
62. Y.-P. Huang, Y.-S. Chien, R.-N. Tzeng, and K.-N. Chen, "Demonstration and Electrical Performance of Cu–Cu Bonding at 150°C with Pd Passivation," *IEEE Trans. Electron Devices*, vol. 62, no. 8, pp. 2587–2592, 2015.
63. H. Park, H. Seo, and S. E. Kim, "Characteristics of Copper Nitride Nanolayer Used in 3D Cu Bonding Interconnects," *Electron. Mater. Lett*, vol. 17, no. 5, pp. 392–398, 2021.
64. H. Seo, H. Park, and S. E. Kim, "Comprehensive Analysis of a Cu Nitride Passivated Surface that Enhances Cu-to-Cu Bonding," *IEEE Trans. Compon. Packaging Manuf. Technol.*, vol. 10, no. 11, pp. 1814–1820, 2020.
65. H. Park, H. Seo, Y. Kim, S. Park, and S. E. Kim, "Low-temperature (260°C) Solderless Cu–Cu Bonding for Fine-Pitch 3-D Packaging and Heterogeneous Integration," *IEEE Trans. Compon. Packaging Manuf. Technol*, vol. 11, no. 4, pp. 565–572, 2021.
66. H. K. Seo, H.-S. Park, and S. E. Kim, "Effects of Argon and Nitrogen Ion Bombardments on Sputtered and Electroplated Cu Surfaces for Cu Bonding Application," in *2019 International 3D Systems Integration Conference (3DIC)* Sendai, Japan, IEEE, October 2019, pp. 1–4.
67. H. Park, H. Seo, and S. E. Kim, "Anti-oxidant Copper Layer by Remote Mode N2 Plasma for Low Temperature Copper–Copper Bonding," *Sci. Rep.*, vol. 10, no. 1, p. 21720, 2020.
68. L. Hu *et al.*, "Time-dependent Evolution Study of Ar/N_2 Plasma-activated Cu Surface for Enabling Two-step Cu–Cu Direct Bonding in a Non-vacuum Environment," *ECS J. Solid State Sci. Technol.*, vol. 10, no. 12, p. 124001, 2021.
69. L. Hu *et al.*, "In-depth Parametric Study of Ar or N_2 Plasma Activated Cu Surfaces for Cu–Cu Direct Bonding," in *2021 IEEE 71st Electronic Components and Technology Conference (ECTC)*, San Diego, CA, IEEE, June 2021, pp. 420–425.
70. P. M. Agrawal, B. M. Rice, and D. L. Thompson, "Predicting Trends in Rate Parameters for Self-diffusion on FCC Metal Surfaces," *Surf. Sci.*, vol. 515, no. 1, pp. 21–35, 2002.
71. J.-Y. Juang *et al.*, "Copper-to-copper Direct Bonding on Highly (111)-Oriented Nanotwinned Copper in No-vacuum Ambient," *Sci. Rep.*, vol. 8, no. 1, p. 13910, 2018.
72. I.-H. Tseng, K. Shie, B. Tzu-Hung Lin, C.-C. Chang, and C. Chen, "Electromigration in 2 μm Redistribution Lines and Cu–Cu Bonds with Highly-oriented Nanotwinned Cu," in *2020 IEEE 70th Electronic Components and Technology Conference (ECTC)*, Orlando, FL, IEEE, 2020, pp. 479–484.
73. T.-F. Lu, T.-Y. Lai, Y. Y. Chu, and Y. S. Wu, "Effect of Nanotwin Boundary on the Cu–Cu Bonding," *ECS J. Solid State Sci. Technol*, vol. 10, no. 7, p. 074001, 2021.
74. Y. Zhang *et al.*, "All Copper Is Not Created Equal: Examples of Grain Engineering in Plating," in *2022 IEEE 72nd Electronic Components and Technology Conference (ECTC)*, San Diego, CA, IEEE, May 2022, pp. 851–855.
75. J. W. Huang, K. C. Shie, H. C. Liu, Y. J. Li, H. Y. Cheng, and C. Chen, "Copper-to-copper Direct Bonding Using Different (111) Surface Ratios of Nanotwinned Copper Films," in *2019 International Conference on Electronics Packaging (ICEP)*, Niigata, Japan, IEEE, 2019, pp. 52–55.

76. I.-J. Wang, C.-S. Ku, T.-N. Lam, E.-W. Huang, K. N. Tu, and C. Chen, "Tuning Stress in Cu Thin Films by Developing Highly (111)-Oriented Nanotwinned Structure," *J. Electron. Mater.*, vol. 49, no. 1, pp. 109–115, 2020.
77. H.-Y. Cheng, D.-P. Tran, K. N. Tu, and C. Chen, "Effect of Deposition Temperature on Mechanical Properties of Nanotwinned Cu Fabricated by Rotary Electroplating," *Mater. Sci. Eng. A*, vol. 811, p. 141065, 2021.
78. K.-J. Chen, J. A. Wu, and C. Chen, "Effect of Reverse Currents during Electroplating on the ⟨111⟩-Oriented and Nanotwinned Columnar Grain Growth of Copper Films," *Cryst. Growth Des.*, vol. 20, no. 6, pp. 3834–3841, 2020.
79. C.-M. Liu, H.-W. Lin, C.-L. Lu, and C. Chen, "Effect of Grain Orientations of Cu Seed Layers on the Growth of ⟨111⟩-Oriented Nanotwinned Cu," *Sci. Rep.*, vol. 4, no. 1, p. 6123, 2014.
80. T.-C. Chou *et al.*, "Investigation of Pillar–Concave Structure for Low-temperature Cu–Cu Direct Bonding in 3-D/2.5-D Heterogeneous Integration," *IEEE Trans. Compon. Packaging Manuf. Technol.*, vol. 10, no. 8, pp. 1296–1303, 2020.
81. T.-C. Chou *et al.*, "Non-planarization Cu–Cu Direct Bonding and Gang Bonding with Low Temperature and Short Duration in Ambient Atmosphere," in *2019 IEEE International Electron Devices Meeting (IEDM)*, San Francisco, CA, ⟨111⟩, 2019, pp. 5.5.1–5.5.4.
82. Y.-T. Yang *et al.*, "Low-temperature Cu–Cu Direct Bonding Using Pillar–Concave Structure in Advanced 3-D Heterogeneous Integration," *IEEE Trans. Compon. Packaging Manuf. Technol.*, vol. 7, no. 9, pp. 1560–1566, 2017.
83. G. Gao *et al.*, "Scaling Package Interconnects below 20 μm Pitch with Hybrid Bonding," in *2018 IEEE 68th Electronic Components and Technology Conference (ECTC)*, San Diego, CA, IEEE, May 2018, pp. 314–322.
84. T. Shirasaka *et al.*, "Comprehensive Study on Advanced Chip on Wafer Hybrid Bonding with Copper/Polyimide Systems," in *2022 IEEE 72nd Electronic Components and Technology Conference (ECTC)*, San Diego, CA, IEEE, May 2022, pp. 317–323.

3 Hybrid Bonding Process Technology

Guilian Gao and Laura Mirkarimi

3.1 INTRODUCTION

Demand for high-performance computing (HPC) in the server, gaming, artificial intelligence, and machine learning applications are growing rapidly [1], and the HPC packaging market was a $2.2 billion business in 2022 and is predicted to be $16 billion in 2028 with a compound annual growth rate (CAGR) of 40% [2]. At the same time, the semiconductor industry has experienced more than a decade of slowing of Moore's Law, as the cost and technical challenges to produce the next transistor node have risen sharply.

In response, the industry embraced advanced packaging with vertical stacking in 2.5D and 3D platforms to achieve higher compute performance, overcome the advanced node slowdown, and maintain product release timelines. Silicon interposer-based 2.5D and 3D packaging—leveraging through-silicon via (TSV) technology—has been in high-volume manufacturing (HVM) for more than ten years, as the ecosystem evolved with the boost from companies that brought stacks of memory on logic into products with support from foundries and outsourced semiconductor assembly and test providers (OSATs). The chiplet approach in 2.5D was shown to cost one-half of the comparable monolithic structure [3]. However, the adoption rate of these products has been relatively slow and limited to a few companies, in part due to technical challenges with the interconnect density and overall cost [4].

The standard method to construct a fine-pitch 2.5D or 3D IC (integrated circuits) structure is to prepare solder-capped microbumps with Cu pillar underneath on the die wafer before dicing, and then employ thermal compression bonding (TCB) to attach the die to the host wafer or another die. High-volume adoption has been slow due to some inherent challenges of the TCB technology. As shown in Figure 3.1, as the bonding pitch of a flip chip assembly decreases, the amount of height variations that can be tolerated across the different bumps on the device shrinks. Compared to more traditional flip chip technology, tolerance for bump height variation for high-yield manufacturing with microbumps is greatly reduced due to the reduced solder volume available during solder reflow, and the yield loss from solder open or bridging is difficult to overcome [5]. Furthermore, certain elements of this package structure limit the product performance. Formation of brittle intermetallics can reduce mechanical reliability. Underfill materials typically have low thermal conductivity and hence reduce thermal performance of 3D stacks with each additional layer. For multi-die stacking, the standard underfill process can only be performed after the entire stack is completed. If pre-applied underfill such as non-conductive paste

 DOI: 10.1201/9781003408673-3

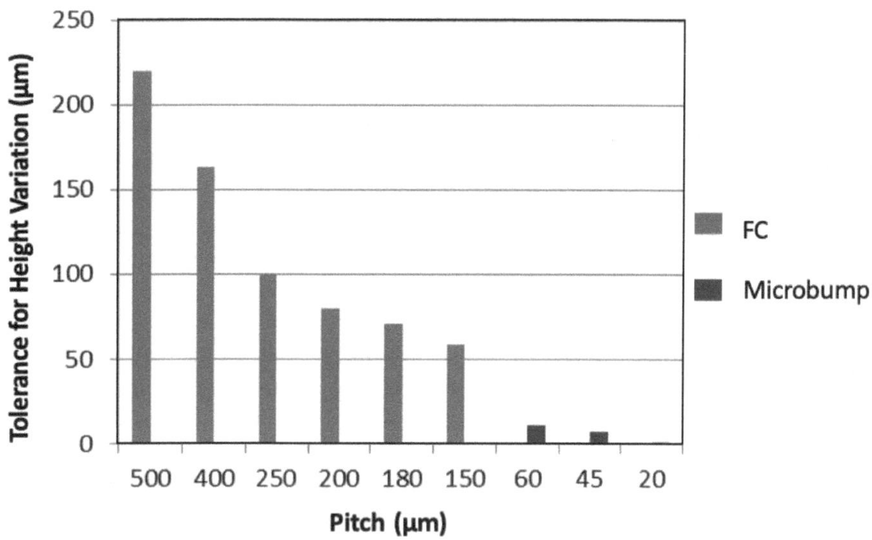

FIGURE 3.1 Tolerance for height variations for solder flip chip interconnects as a function of interconnect pitch. The number rapidly shrinks down as the microbump pitch becomes finer [5].

(NCP) or non-conductive film (NCF) is used, then additional process challenges must be addressed, and these assembly processes are relatively slow. Additionally, implementation of this technology below 35 µm pitch is challenging, particularly as the number of dies in the stack increases. Finally, shrinking the pitch to below 10 µm faces major yield problems for HVM. Alternatively, in the "mass reflow molded underfill" process that enabled higher yield for dynamic random access memory (DRAM) high bandwidth memory (HVM) modules [6], multiple layers of dies are placed on the substrate for bonding all at once through reflow, and then the gaps between the layers of dies or between the chip and the substrate are simultaneously filled with a molding material. However, this process does not enable pitch scaling to below 10 µm either.

The Direct Bond Interconnect (DBI®) hybrid bonding platform technology, an all-solid-state joining technology, enables pitch scaling of electronic packages to below 1 µm, by forming a direct dielectric-to-dielectric bond at room temperature and then establishing metal-to-metal interconnection (usually Cu-to-Cu) by a low-temperature batch annealing process (150–300°C). The initial dielectric bonding process happens spontaneously upon contact at one point between the two surfaces. Bonding takes place at ambient conditions, and no adhesive or other filler materials are required in the process. Room temperature bonding reduces the misalignment induced by differential temperatures of the bond surfaces for each component. The bond strength at room temperature is sufficient to hold the parts together for the subsequent annealing step with no need for external pressure. A batch anneal process with high throughput is carried out in a conventional oven to drive the Cu–Cu interdiffusion for bond formation.

3.2 BRIEF HISTORY OF HYBRID BONDING

The term "hybrid bonding" in advanced semiconductor packaging is used to describe the bonding of a dielectric material and a metal at the same bond interface. Depending on the dielectric material used, the technology is further divided into organic dielectric–metal hybrid bonding and inorganic dielectric–metal hybrid bonding. Since the inorganic dielectric–metal hybrid bonding technology has so far been embraced by the industry for HVM of hybrid bonded packages and is the most widely researched and published, this chapter is exclusively dedicated to the DBI technology which is based on inorganic dielectric–metal hybrid bonding [7].

Spontaneous bonding of flat, polished dielectric surfaces upon contact has been known for a long time [8, 9]. However, an anneal temperature of >800°C was required to achieve bonding strength comparable to the fracture strength of silicon. Such a high-temperature anneal is not practical for semiconductor device packaging and assembly because it will damage the device. In 2000, Tong et al. [10] discovered that activating the bonding surface with a slightly etching plasma greatly enhances the bonding strength. Bonding strength comparable to silicon fracture strength can be achieved at much lower anneal temperatures, such as 150°–300°C, or even after extended storage at room temperature. The discovery is an important breakthrough that makes direct bonding a real option for semiconductor packaging applications, and is the basis of the low-temperature direct dielectric bonding technology.

In 2005, DBI was developed by incorporating metal pads into the dielectric bonding interface [11]. It was first demonstrated with Ni as the interconnecting material (Ni DBI) which was a buildup process. In 2009, hybrid bonding with an all-Cu interconnect was demonstrated. Cu hybrid bonding is much easier for the semiconductor industry to adopt because of the extensive use of the damascene process that exists in the wafer foundries today. Although the DBI technology was first demonstrated in die-to-wafer (D2W) format, wafer-to-wafer (W2W) bonding was the first process adopted for high-volume production.

The technology was first adopted in 2016 for HVM of image sensors in W2W format [12] for backside illuminated image sensors, which revolutionized the digital imaging industry and cell phone cameras. In 2022, it was adopted for HPC in D2W format [13]. The technology has been demonstrated to sub-micrometer bonding pitches, and a 1 μm bond pitch has been implemented in HVM [14–16].

3.3 BONDING MECHANISMS

3.3.1 Dielectric Direct Bonding Mechanism

A silicon wafer surface is covered by a layer of oxide, and the oxide thickness can vary from a few nanometers in the case of native silicon oxide to several micrometers or tens of micrometers for a thermally grown or deposited film. The deposition techniques include physical vapor deposition (PVD), chemical vapor deposition (CVD), or plasma enhanced chemical vapor deposition (PECVD). Water molecules adsorb to the oxide surface, as shown in Figure 3.2 [17]. When the two ultrasmooth silicon oxide surfaces make contact, spontaneous bonding of the dielectric surfaces occurs first through van der Waals or hydrogen bonds.

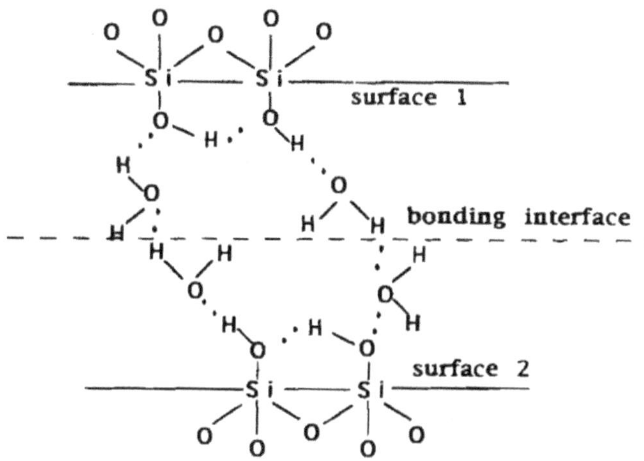

FIGURE 3.2 Schematic illustration of water bridge formation between two silicon oxide surfaces. Solid lines represent covalent bonds and dashed lines the hydrogen bonds. (© The Electrochemical Society. Reproduced with permission of IOP Publishing Ltd. from Ref. [17].)

The bonding mechanism is described by Masteika et al. [18] as follows:

1. The initial bond of surface OH groups to a water interlayer:

$$Si - OH \cdots nH_2O \cdots HO - Si \tag{3.1}$$

During annealing, which can be heated, or can happen naturally at ambient conditions, a condensation polymerization reaction occurs to produce covalent high-strength bonding:

$$Si - OH + HO - Si \leftrightarrow Si - O - Si + H_2O \tag{3.2}$$

As this reaction is reversible up to approximately 400°C, the released water is then able to diffuse through the relatively thin siloxane layer into the bulk Si to oxidize Si and release hydrogen:

$$-Si - Si - +H_2O \rightarrow -Si - O - Si - +H_2 \tag{3.3}$$

The released hydrogen causes void formation in the bonded structure, as documented by Tong et al. [19]. Two approaches were attempted to avoid the formation of annealing voids. The first was to use one or more thermally oxidized wafers as part of the wafer sandwich. It was well understood that excess water in the interface after annealing at >110°C tended to oxidize available bulk silicon or hydrolyze oxide, both actions resulting in a release of excess hydrogen. By using wafers coated with 1 μm thick thermal oxide, the intention was for the porous thermal oxide layer to adsorb this excess of hydrogen. The second approach was to anneal wafer pairs exhibiting voids at 1,000°C for 1 hour. This allowed the trapped hydrogen to diffuse away from the interface causing the bulk of the voids to close. Neither of these approaches is suitable for bonding semiconductor devices due to the excessive thermal budget.

A gas plasma treatment prior to bonding in ambient conditions was shown to enhance the bond energy (BE) of the bonded Si pairs at room and low temperatures. The early work in the field was done on planar silicon surfaces or oxides grown on silicon. Tong et al. [10] advanced the art of direct bonding by depositing a polishable dielectric layer at a relatively low temperature, planarizing the bonding surface, then using chemical mechanical polishing (CMP) to polish the surface to subnanometer roughness with high coplanarity, activating the surface with a slightly etching plasma, and initiating contact between the wafers to bond, and achieved void-free high BE surfaces. The discovery formed the foundation of low-temperature direct bonding for semiconductor applications. It will be shown later in this chapter that BE >2,000 mJ/m^2 can be routinely achieved at the bonded silicon oxide interface after 150°C anneal for 15 minutes (Figure 3.14).

3.3.2 Hybrid Bonding Mechanism

The technology further evolved into two-step hybrid bonding by embedding metal contacts inside the bonding dielectric layer. As shown in Figure 3.3, the metal pads at the bonding interface are either flush with or slightly recessed from the dielectric surface. When the two surfaces make contact, a spontaneous bond initiates between the dielectric surfaces. Once the bonding is initiated, a bond wave emanates from the contact point and travels along the entire surface of a die or a wafer. Since the metal surface is either flush with or slightly lower than the dielectric surface, it does not interrupt the traveling wavefront of the bond. Once the dielectric bonding is complete, the metal surfaces are isolated from the ambient environment.

Gao et al. [20] evaluated the effectiveness of the bonded dielectric as a hermetical seal using bonded silicon wafers with etched cavities. After 300°C/1 hour anneal, the wafer pair was diced into parts with a single cavity. The diced parts have an oxide bond line width at the outer edge of the cavity varying from 8 μm to 280 μm. The results are shown in Figure 3.4. A bond line as thin as 8 μm can provide a helium leak rate below 10^{-11} atm-cc/s, far below the leakage rate of 5×10^{-8} atm-cc/s which is required for hermetic packaging [21]. The bonding strength immediately after bonding is strong enough to hold the bonded pair together with no need for an external fixture or pressure for the subsequent step of heated anneal.

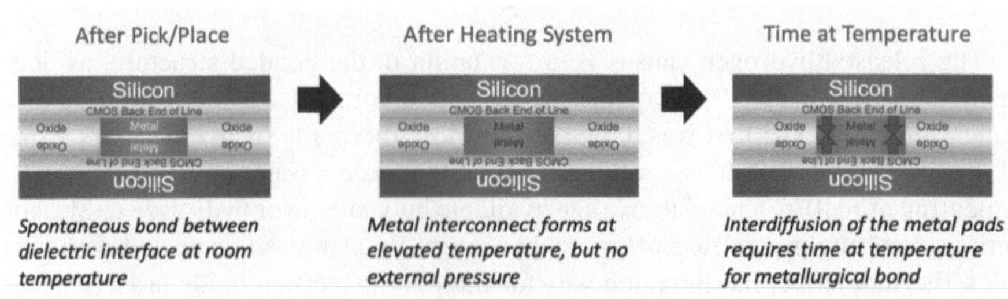

FIGURE 3.3 Illustration of the two-step bonding process for DBI.

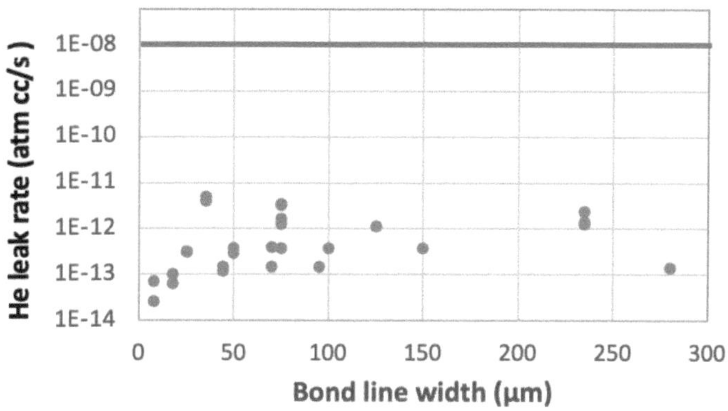

FIGURE 3.4 Helium leak rate of cavity seal versus the bond line width. The solid line marks the threshold for hermetic packages. (© 2021, IEEE. Reprinted with permission from Ref. [20].)

In the second step of the process, the bonded substrates are annealed in an oven at an elevated temperature (normally 150-300°C). In this process, the bonded dielectric interface is strengthened through the formation of covalent bonds, as described in Equations 3.2 and 3.3. The metal contacts on both sides of the bonding interface expand more than the surrounding organic dielectric layer due to the differential coefficient of thermal expansion (CTE). For example, the CTE of Cu is 16.7 ppm/°C, whereas the CTE of SiO_2 is 0.5 ppm/°C. As a result, the metal on both sides of the bonding surface expands to contact each other, and then metal atoms diffuse across the interface to form a permanent metallic bond. The hermetic seal provided by the bonded dielectric material surrounding the metal features prevents oxygen from diffusing to the interface and negatively affecting the Cu metal interdiffusion.

3.4 HYBRID BONDING LAYER FABRICATION

3.4.1 Device Side Bonding Layer Fabrication Process

Most wafer fabrication processes today finish the wafer with an aluminum metal layer, and a passivation dielectric layer covers the majority of the die surface. The metal contact and test pads are exposed through lithographic patterning and dielectric etch. As illustrated in Figure 3.5, the surface has topography that does not meet the requirement for direct bonding or hybrid bonding. The first step to prepare a bonding layer is to deposit an inorganic dielectric film onto the wafer surface and planarize the surface to meet the long-range flatness requirement, as shown in step 1 of Figure 3.5. The dielectric layer is commonly deposited through PVD, CVD, or PECVD, although other deposition methods can also be used. After the dielectric deposition, a CMP process is performed to planarize the surface, and more information on the CMP process is covered in a separate chapter of this book. The wafer topographical requirements for bonding will be provided later in this chapter. If the wafer is prepared for direct bonding of the dielectric layer only, the surface needs to be polished to meet the surface roughness (subnanometer) and long-range flatness requirements for bonding.

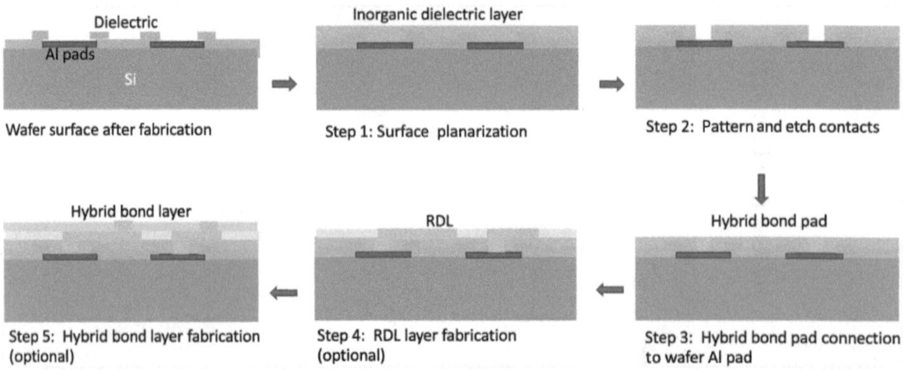

FIGURE 3.5 Illustration of module steps for fabrication of a hybrid bonding layer. Step 1: Planarize the surface with an inorganic dielectric layer (blue). Step 2: Pattern and etch dielectric to expose metal contacts. Step 3: Electroplating to fill the contact openings and CMP to remove metal on the dielectric surface. Step 4: RDL layer fabrication (light blue). Step 5: Bonding layer fabrication (light green).

Next, the planarized dielectric layer needs to be lithographically patterned and etched to expose electrical interconnect features on the functional wafer. A damascene process [22] is normally used to complete pad formation shown in steps 2 and 3 of Figure 3.5. The most common metal used today for the hybrid bond interconnect is Cu because it is readily available in wafer fabrication foundries for the back end of line (BEOL) application. In the damascene process, after patterning the dielectric, a barrier layer, an adhesion layer, and a Cu seed layer are deposited over the patterned surface. Next, Cu is plated to fill the openings and cover the entire surface. The excess metal above the unetched dielectric area is referred to as "overburden." The Cu overburden, the barrier layer material, the adhesion layer, and the Cu seed layer in the field area, all are removed by the CMP process steps. Different CMP chemistries are required for the Cu removal and barrier removal to achieve the best results for hybrid bonding.

This layer as described above can optionally be used as the bonding layer. If this layer is used as the bonding layer, the surface finish needs to meet the dielectric roughness, metal recess, and long-range flatness requirement for bonding.

Alternatively, if a redistribution layer (RDL) is required between the finished wafer and the hybrid bond pads, additional dielectric deposition, patterning, plating, and CMP process cycles can be performed to form the RDL layer, as illustrated in step 4 of Figure 3.5. The process can be repeated to fabricate the via layer and the bonding layer, as shown in step 5 of Figure 3.5, either through two single damascene processes or through a dual damascene process.

In summary, wafers currently fabricated with Al as the last metal layer need to be planarized, and a bonding layer needs to be added for hybrid bonding. In the future, additional planarization and bonding layer fabrication steps can be eliminated when the hybrid bonding requirement is incorporated into the wafer design to have the last metal of the BEOL meet hybrid bonding requirements. This is one example of how cost reduction is possible by designing for hybrid bond interconnect from the beginning.

Alternatively, Cu interconnect can be fabricated with a buildup process. In this case, the plating seed layer is deposited onto the wafer surface before the bonding dielectric layer is deposited. The surface is then patterned with a photoresist and plated. After the plating, the photoresist is stripped, the exposed seed layer is removed, the bonding dielectric material is deposited, and the surface is planarized by CMP.

Other metals may also be used for hybrid bonding with dielectrics. For example, the hybrid bonding technology was first demonstrated with Ni interconnects [7]. Ni interconnects will be more advantageous in applications that demand extremely high temperatures and radiation hardening. It is also easier to CMP Ni than Cu, since it requires only one slurry formulation for the CMP process, whereas Cu CMP requires at least two slurry formulations.

3.4.2 Dielectric Materials for the Bonding Layer

The most widely used dielectric material in hybrid bonding is PECVD silicon oxide, and the most widely used metal is Cu. Silicon oxynitride layers have also been used when there is a concern about Cu migration to cause device failure. Depending on the maximum temperature for the device wafer, dielectric deposition temperatures vary from less than 200°C to over 400°C. Silicon carbide nitride (SiCN) films have gained a lot of traction as the capping layer for bonding [23–26]. According to Wu et al. [27], SiCN can potentially offer high bonding energy at a low-temperature anneal condition. Moreover, no outgassing-induced voids were observed when the post-bond annealing temperature exceeded the deposition temperature for the case of SiCN–SiCN bonding. SiCN was initially evaluated as the dielectric in hybrid bonding because it was known as a better diffusion barrier against Cu electromigration than SiO_2.

After plasma activation, bonding, and anneal, the bond interface between two SiCN–SiCN layers is found to consist primarily of silicon oxide. As an example, Inoue et al. [28] showed the transmission electron microscopy (TEM) and electron energy loss spectroscopy (EELS) analysis results on the interface of SiCN–SiCN pairs (Figure 3.6), demonstrating that the interfacial layer is SiO_2.

Similarly, Ma [24] showed a TEM–EELS characterization of the SiCN–SiCN bonding interface. The SiCN layers converted to anapproximately 6 nm thick SiO_x layer at the interface, likely formed during the pre-bond plasma activation and hydration process.

In this work, a TEM and EDX (energy dispersive X-ray spectroscopy) analysis of a bonded SiCN–SiCN interface also shows the presence of approximately 10 nm thick SiO_x at the bonding interface, as shown in Figure 3.7.

Based on this evidence, it is more accurate to note that the high BE measured at the SiCN–SiCN interface is attributed to SiO_2 layers, not to the bonding between the SiCN layers. Another consideration with regard to the efficacy of the SiCN layer as a diffusion barrier layer is that the interface is primarily SiO_2 with residual N

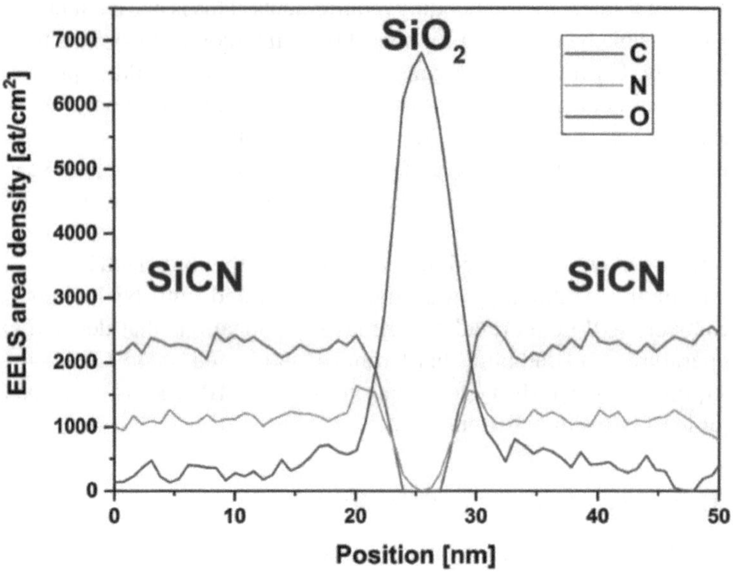

FIGURE 3.6 TEM–EELS elemental analysis for SiCN–SiCN bonding interface. (© The Electrochemical Society. Reproduced with permission of IOP Publishing Ltd. from Ref. [28].)

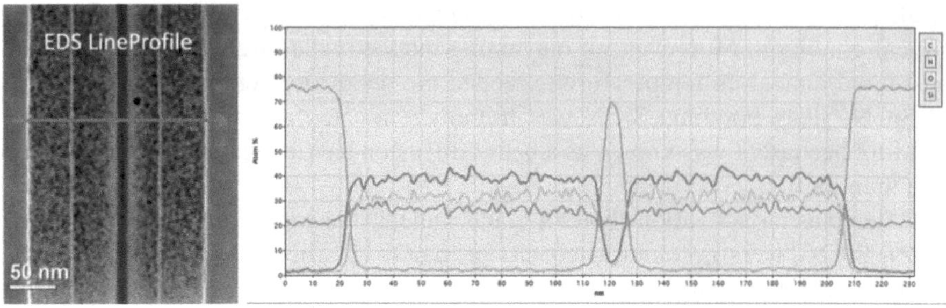

FIGURE 3.7 TEM image and EDX scan of a SiCN–SiCN bonded interface. Approximately 10 nm thick SiO_x layer was formed between SiCN.

beyond the interface, and the Cu would need to diffuse through the SiONC layer which will assist in blocking diffusion; however, if nature effectively gives SiO_2 at the interface, a multilayer dielectric of SiCN with SiO_2 on top can also be built up to achieve what is effectively happening after activation. It is important, however, to optimize the plasma activation processes for the various dielectrics. As for the perceived advantage of SiCN as a better diffusion barrier against Cu electromigration along the bonding interface, Cu diffusion along the bonding interface of SiCN–SiCN and SiO_2 should be very similar, due to the presence of a SiO_2 layer at the interface. To date, the perceived benefit of SiCN over SiO_2 for higher Cu electromigration resistance has not been validated experimentally in the literature.

3.4.3 Backside Bonding Layer Fabrication for W2W Bonding

Fabrication of the bonding layer on the backside of a device wafer for W2W (W2W) bonding is typically carried out after the wafer has been bonded to another wafer. The bottom wafer serves as the supporting wafer during the backside surface fabrication of the top wafer, and the need for a temporary carrier wafer is eliminated.

A TSV is required for front-to-back communication in a tri-wafer stack (or higher). Figure 3.8 shows the simplified process flows for TSV-last, TSV-middle, and TSV-first fabrication. For the TSV-last process, after the wafer bond and anneal, the top wafer is thinned to the desired thickness. The thinning process normally involves

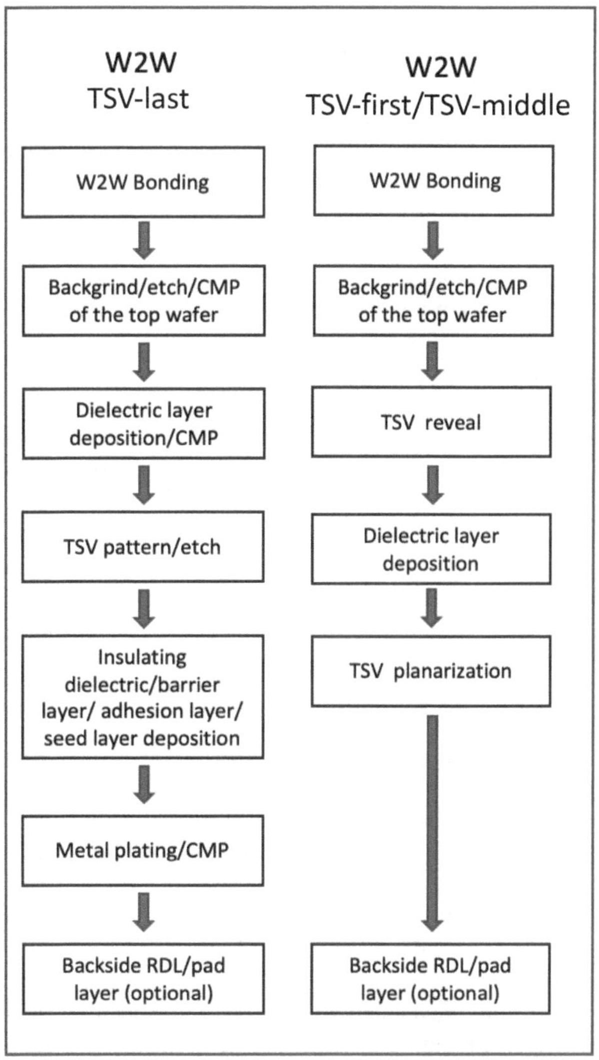

FIGURE 3.8 TSV-last, TSV-first, and TSV-middle process flows for the backside bonding layer fabrication.

mechanical grinding followed by CMP. The plasma etch process can optionally be added between mechanical grinding and CMP to remove damaged layers in the Si. An inorganic dielectric layer is then deposited onto the back surface with a PECVD deposition process. In the next step, the backside surface is patterned lithographically and etched with a plasma process to form TSV openings. An insulating dielectric layer is then deposited onto surfaces inside the TSV opening, followed by the deposition of an adhesion layer to promote adhesion between the metal and the dielectric, a barrier layer to prevent Cu migration into the insulating dielectric layer, and a Cu seed layer for electroplating of Cu. In the next step, the TSV is filled by Cu electroplating. After electroplating, the Cu overburden is removed by CMP to meet the requirement of dielectric roughness, Cu recess, and long-range flatness for hybrid bonding. Now the backside surface of the wafer is ready for bonding to another wafer. The TSV can be directly bonded to the bond pads on another wafer or die. If the wafer does not require any RDL layer and the TSV will be bonded directly to a mating bond pad, the surface will be cleaned, activated, and then bonded in a wafer bonder that has the appropriate alignment tolerance for the pad/pitch dimensions of the bonded interconnects. If a backside RDL and/or bonding pad layer is desired, the process for RDL and pad layer fabrication described for the device side bonding layer fabrication may be repeated on the wafer backside. After bonding a third wafer to the stack, the TSV and backside bonding layer fabrication process may be repeated to stack additional wafers.

In the TSV-first and TSV-middle processes, the TSVs are embedded in a wafer prior to W2W bonding. After bonding, the top wafer is thinned to reveal the TSV. The thinning process normally consists of mechanical grinding/plasma etch and CMP. The wafer is processed to a thickness where the tips of all TSVs are protruding above the surface, and the insulating dielectric layer around the TSV remains intact. This step is called the TSV reveal step. An inorganic dielectric layer is then deposited over the entire back surface, followed by a CMP process to remove the dielectric layer over the TSVs, in the TSV planarization step. Again, after the CMP process, the planarized oxide and TSVs may be directly bonded to the pad layers on another wafer or die. If backside RDL and bonding pad layers are desired, the process for RDL and pad layer fabrication described for the device side bonding layer fabrication may be repeated on the backside.

3.4.4 Backside Bonding Layer Fabrication for D2W Bonding

In contrast to W2W bonding, for D2W bonding, the bond layer on the backside of the die wafer needs to be fabricated prior to wafer singulation and die bonding. A TSV-middle process is commonly used in this case.

Figure 3.9 shows a simplified process flow for the fabrication of the backside bonding layer for die stacking with TSVs. After completion of the bond layer on the device side of the wafer, the device side is bonded to a temporary support wafer, or a carrier wafer as it is often called. Detailed information on temporary bonding and debonding is given in another chapter of this book. Currently, temporary bonding materials and systems are developed with the target application of solder interconnect die. Because hybrid bonding has much more stringent requirements for surface cleanliness and tight control of Cu recess, care must be taken to evaluate the temporary bonding materials and process chemistry and thermal excursions to avoid Cu corrosion and bond surface contamination.

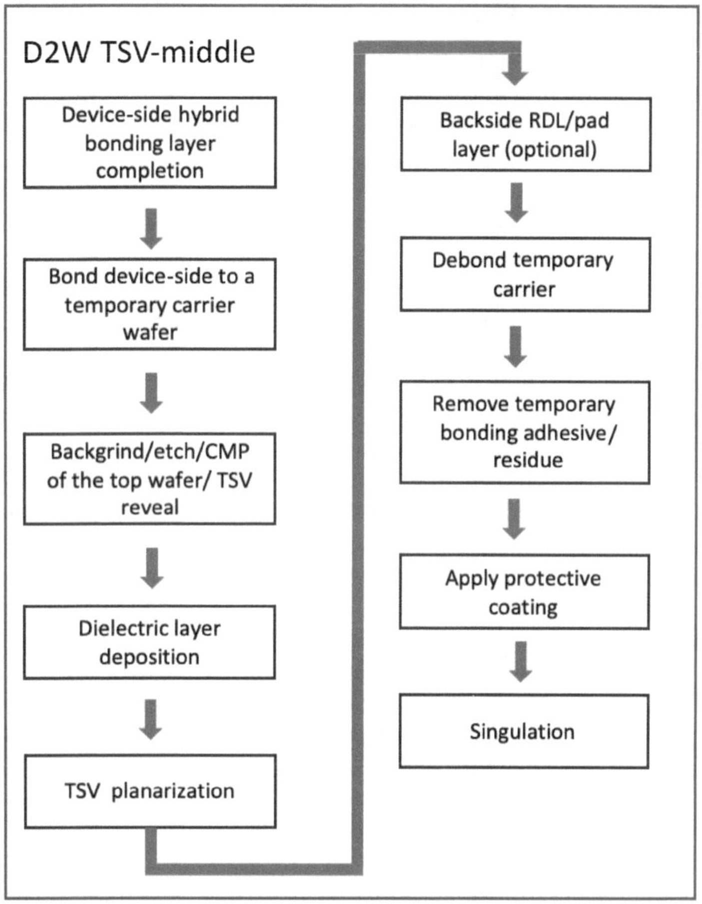

FIGURE 3.9 A simplified process illustration of backside bonding layer fabrication for a D2W bonding process with a TSV-middle process.

Once the wafer is bonded to the carrier wafer, the backside thinning and TSV reveal process is the same as for the W2W bonding. However, the process for dielectric deposition may differ since the temporary bonding materials currently available cannot withstand temperatures for standard PECVD dielectric film deposition. A low-temperature film deposition process must be evaluated and proven to be compatible with a carrier wafer process. After the completion of backside planarization, the carrier is removed, and the temporary bonding adhesive needs to be removed from the functional wafer surface. Depending upon the temporary bonding process, this may involve mechanical and chemical removal processes. Both the chemical and thermal aspects of the process must be considered to preserve a quality Cu interconnect surface for bonding. Custom cleaning processes may be needed to obtain high-yield hybrid bonding processes, as some materials may not be "residue-free" as claimed. Once the temporary bonding adhesive and residue are removed, the wafer is coated with a protective coating to protect the bonding surface from further contamination during the dicing process.

3.5 THE CMP PROCESS

The last step of the bonding layer fabrication is a CMP process. The CMP process polishes the dielectric film to sub-nanometer roughness while adding a slight recess to the metal features to enable direct bonding. Multiple CMP process steps with different slurries may be required. This process is critical for successful DBI bonding. Additionally, the metal recess from the dielectric surface must be tightly controlled because it influences the anneal thermal budget (temperature and duration) required to form the metal–metal bond. Since Cu is much softer than the surrounding dielectric material, the CMP process tends to cause deeper recess on large pads than small pads on the same wafer. Therefore, special recipes must be developed to achieve tight control on Cu recess for hybrid bonding.

In this work, through extensive design of experiments (DOE) to examine multiple factors that affect the metal recess in the CMP process (including slurry chemistry, flow rate, CMP pad material, down pressure, rotation rate, polishing time, etc.), multiple CMP recipes on commercial CMP tools have been developed to achieve very shallow and uniform metal recess across the entire wafer of 200 mm and 300 mm diameters, as shown in Figure 3.10.

The surface roughness and metal recess are measured using atomic force microscopy (AFM), which will be discussed in depth later in this chapter. AFM measurement methodologies developed have enabled accurate characterization of surface topology during CMP recipe development in this work.

Figure 3.10 shows the standard deviation of Cu recess over polished wafers of 200 mm and 300 mm diameters. The process is highly reproducible, as multiple recipes have been developed for wafers with very different metal feature sizes and pitches (ranging from 1.9/3.9 μm pad/pitch to 20/40 μm pad/pitch), and consistent metal recess over multiple lots of wafers polished have been obtained over the period of several months at multiple sites.

The excellent control of Cu recess has been verified with a much larger sample size, as shown in Figures 3.11 and 3.12 [30], which tested the stability of the process by polishing over 1,000 total wafers (structured and blank wafers) on a single set of

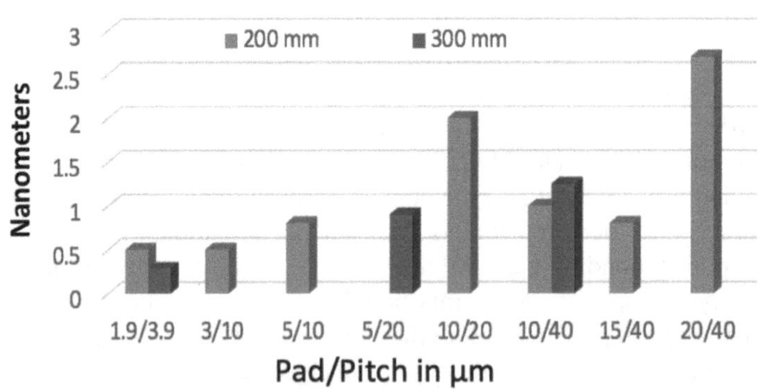

FIGURE 3.10 Uniformity of Cu recess on 200 mm and 300 mm wafers after CMP for different size Cu pads. Recess deviation across a 200 mm or 300 mm wafer is less than 3 nm [29].

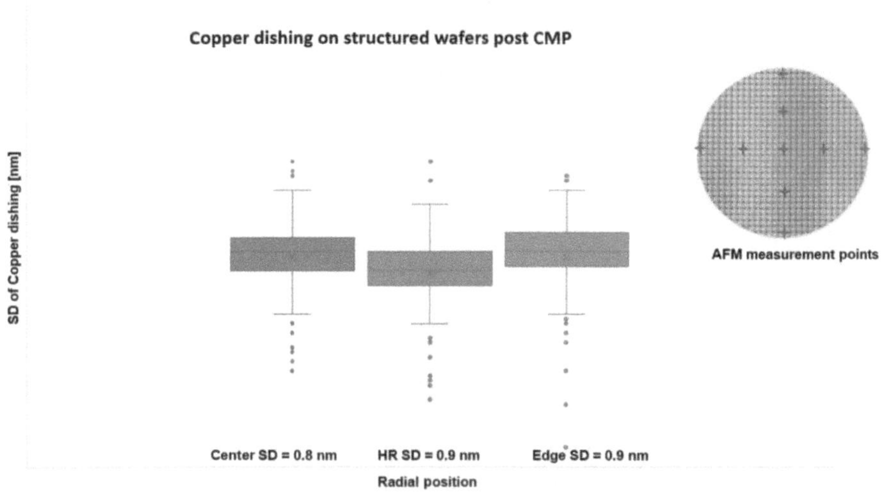

FIGURE 3.11 Distribution of copper dishing measured on >100 structured wafers on nine measurement points per polished wafer. (Reprinted with permission from Ref. [30].)

polishing pads, using the recipes noted above. Over 100 of the wafers polished in the experiment are structured wafers with Cu pad structures typical for hybrid bonding. Figure 3.11 shows AFM results of the standard deviation (SD) of copper dishing as a function of position on the wafer. The distribution of the copper dishing across the wafer is <1 nm with a standard deviation of 0.8 nm at the wafer center, 0.9 nm at approximately half radius, and 0.9 nm at the wafer edge.

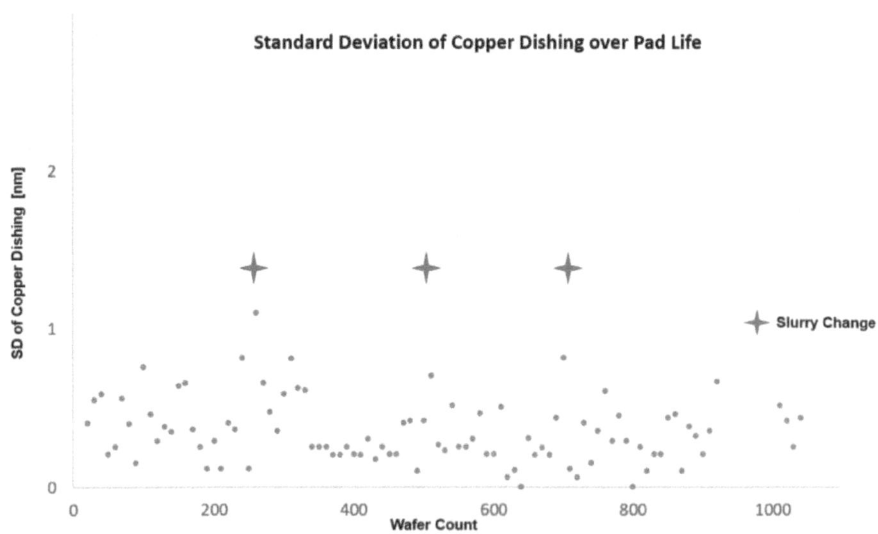

FIGURE 3.12 Standard deviation of Cu dishing measured on >1,000 total wafers (including >100 structured wafers) over the life of the pad. (Reprinted with permission from Ref. [30].)

Figure 3.12 shows the deviation of the Cu dishing from baseline values over the pad lifetime of more than 1,000 wafers. The break-in of a new pad and the change of slurry barrels during polish cause a small deviation of the Cu recess; however, the dishing deviation remains less than 1 nm which does not affect the hybrid bonding results.

3.6 W2W HYBRID BONDING

Figure 3.13 shows a simplified process flow for W2W dielectric direct bonding and hybrid bonding.

3.6.1 PLASMA TREATMENT

Both dielectric direct bonding and hybrid bonding use plasma treatments in the wafer preparation process. Oxygen plasma is used to remove potential organic contaminants from the bonding surface, and another plasma chemistry is used for surface activation by modifying the bonding surface to achieve higher bonding strength. The plasma chemistry is determined by the type of dielectric material to be bonded. For example, nitrogen plasma is commonly used for activation of SiO_2 surfaces [31] and SiCN surfaces [28]. The addition of fluorine species is also shown to enhance the bonding strength of SiO_2 [32].

The plasma activation process optimization is an important surface preparation step for successful bonding, as the plasma chemistry, chamber configuration, and process conditions all can affect the bonding strength [33]. Table 3.1 shows the results of a DOE on the activation of silicon oxide surfaces by a nitrogen plasma with three parameters: chamber pressure, plasma duration, and plasma power. After bonding, the wafer pairs were annealed at 150°C for 15 minutes. The BE measurement was accomplished with the double cantilever method, using a test fixture with an infrared (IR) camera positioned over the test sample to achieve high-accuracy measurements. A low annealing temperature and a short annealing time were chosen

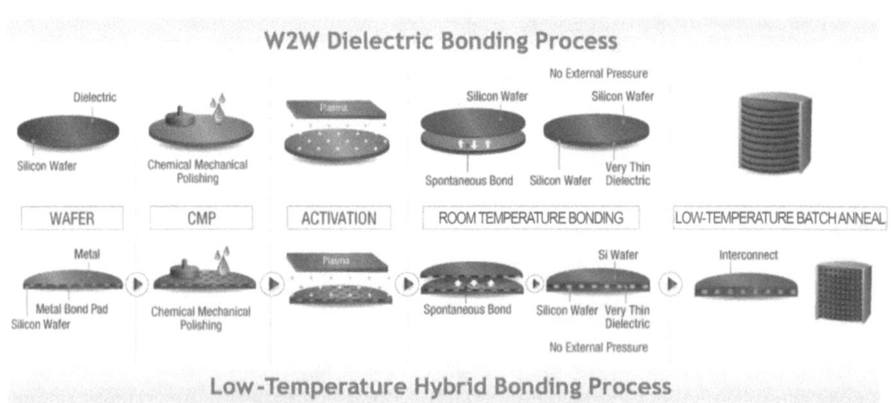

FIGURE 3.13 Illustration of W2W bonding process for dielectric direct bonding and hybrid bonding.

TABLE 3.1
BE Measured from a Plasma Activation DOE after 150°C Anneal for 15 Minutes

Chamber Pressure	Plasma Duration	Plasma Power	Bond Energy (mJ/m^2)
Low	Low	Low	1,747
Low	Low	High	2,032
Low	Middle	Low	1,883
Low	Middle	High	2,032
Low	High	Low	1,883
Low	High	High	1,747
High	Low	Low	1,883
High	Low	High	2,032
High	Middle	Low	2,032
High	Middle	High	1,747
High	High	Low	1,883
High	High	High	2,032

because the BE under this condition is more sensitive to process variations and is ideal for process optimization, and the BE at this condition is lower than the value obtained after a higher temperature anneal. Additionally, Cioccio et al. [34] showed that by increasing the annealing temperature from 150°C to 400°C, BE for oxide-to-oxide bonding increased by about 30% while the toughness for oxide–Cu bonding more than quadrupled. (As a side note, when comparing BE data from different sources in the literature, it is important to note the annealing temperature to avoid misleading conclusions.) While a BE of 2,000 mJ/m^2 (after full anneal) is desirable for many applications, 1,500 mJ/m^2 after a 150°C, 15 minutes anneal is considered adequate in this work.

All conditions in Table 3.1 met the requirement of 1,500 mJ/m^2, indicating that the activation process window is very wide. When the process is optimized, a BE of >2,000 mJ/m^2 was achieved with a low-temperature and short-duration anneal.

It is important to regularly monitor the plasma activation process to ensure good bonding results. As shown in Figure 3.14, BE measured over an 11-week time span [35] stayed consistently above 1,500 mJ/m^2 after a 15-minute anneal at 150°C.

The plasma activation process modifies the bonding surface by creating dangling bonds. The wafer is then exposed to an ambient environment after activation since bonding takes place in ambient conditions, and the dangling bonds on the surface are terminated by adsorption of water molecules due to exposure to moisture in the air, and/or due to deionized (DI) wafer rinse which is commonly used to remove particles generated in the plasma process.

The physical mechanism of plasma enhancement in BE is not well understood, even though BE above 2,000 mJ/m^2 is routinely achieved on SiO_2–SiO_2 bonding with low anneal temperatures, as shown in Figure 3.14. It appears that a lack of knowledge in the industry on plasma activation optimization has led to the misconception that SiO_2–SiO_2 bonding suffers from low BE.

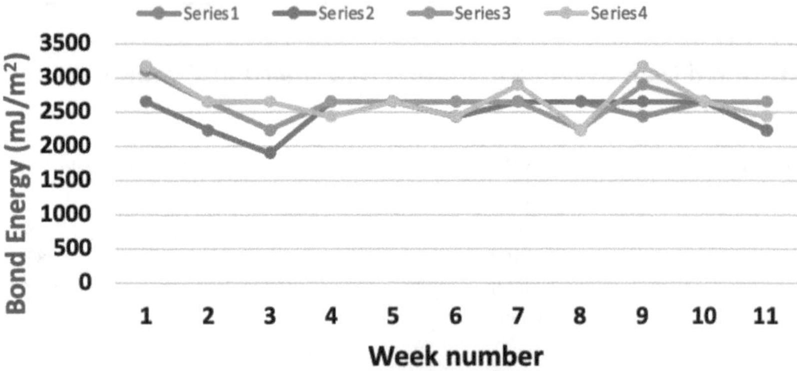

FIGURE 3.14 Example of bond energy (mJ/m^2) measured on weekly process monitoring wafer pairs (SiO_2–SiO_2 bonding). All wafer pairs were annealed at 150°C for 15 minutes [35].

3.6.2 Bonding

As illustrated in Figure 3.15, when the two wafers are placed over each other in an ambient condition, an air cushion is trapped between them, preventing direct contact of the wafers. Relying on gravity to expel the air cushion naturally takes a long time. The problem is overcome by applying a downward force on the top wafer to force contact between the two wafers at one point (the bond initiation point). Depending on the size and geometry of the wafers to be bonded, the bond initiation point is likely to be either the center of the wafer for 200 mm and 300 mm wafers, or the flat edge of 150 mm or smaller wafers. Once the bonding process is initiated, adhesion forces allow the spreading of the bonding wave over the entire wafer surface.

Since the bonding is at the molecular scale and there is no gap to accommodate particles, the bonding surface needs to be pristine to obtain void-free bonding. Processing the wafers in a Class 1,000 clean room environment (or better) can minimize the bonding void risk.

Best bonding results are obtained when the bonding wave propagates from one location only. The initiation of multiple bonding waves can lead to entrapped air and bonding voids. Therefore, the wafer bow needs to be controlled to avoid multipoint initiation of bonding waves.

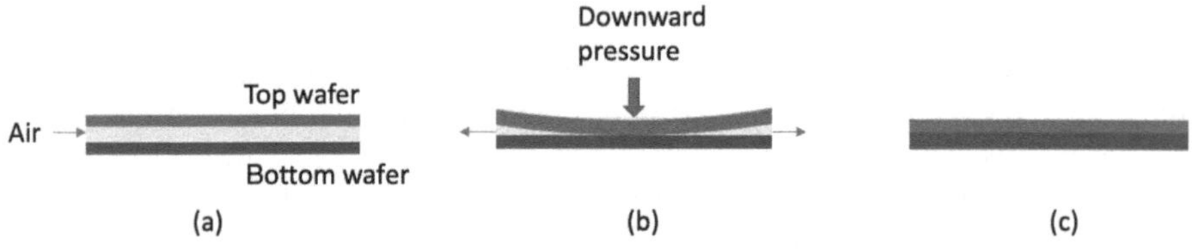

FIGURE 3.15 Schematic illustration of W2W bonding initiation and completion. (a) Top and bottom wafers are separated by a thin air cushion. (b) A downward force expels the air between the two wafers at one point to initiate the bonding. Spreading of the bonding by the adhesion force expels air from the cushion. (c) Bonded wafer pair with no air remaining between two wafers.

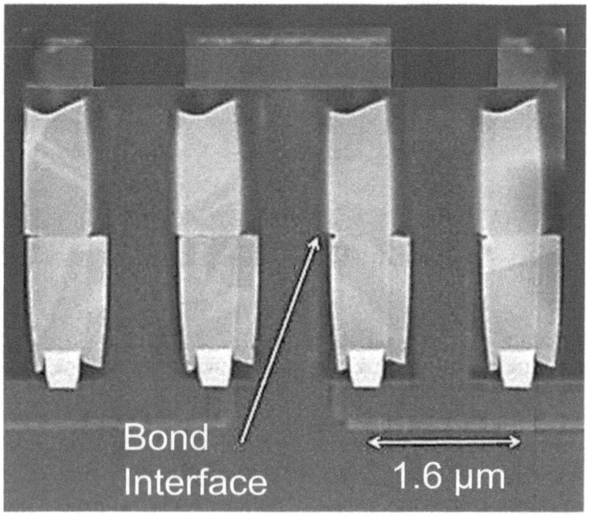

FIGURE 3.16 Cross section of DBI bonded joints at 1.6 μm pitch [33].

Pitch scaling in W2W hybrid bonding is limited by the bonder alignment accuracy. Bonding at 1.6 μm pitch was demonstrated in 2009 (Figure 3.16) [33]. The current state-of-the-art pitch capability is 1 μm, and the industry is actively pursuing sub-micrometer bonding pitch [14, 15, 24, 26].

Netzband et al. [15] have shown that by understanding the many facets of the bond process, high alignment accuracy can be achieved, as the bond pad pitch and pattern layout, CMP process, and the bond front travel speed all can affect displacement after bonding. The impact of some factors may be negligible at a larger bonding pitch; however, when the displacement requirement across the entire wafer is less than 100 nm, these factors become significant. Bonding process parameters such as bond initiation pressure, timing, vacuum zone configuration, pressure and release timing, gap heights, etc., all need to be optimized to minimize displacement. By optimizing the process parameters, <50 nm displacement on 0.5 μm pitch hybrid bonded wafers has been demonstrated [15].

One advantage of hybrid bonding is the elimination of heat from the initial dielectrics bonding process. If the wafers are heated, a slight difference in temperature between the two wafers can induce significant thermal runout and jeopardize yield in ultrafine pitch interconnect structures.

Bond interface voids are commonly characterized using acoustic imaging. The most commonly used mode, confocal scanning acoustic microscopy (CSAM), generates acoustic waves of certain frequencies and performs complex analyses of the echoes generated at the bonding interfaces. Depending on the algorithm chosen for the signal analysis, well-bonded areas normally show up as black and voids at the bonding interface normally show up as white (Figure 3.17).

CSAM images of real products are more complex due to interference of the device structure and features with the acoustic waves. For example, thick metal layers can

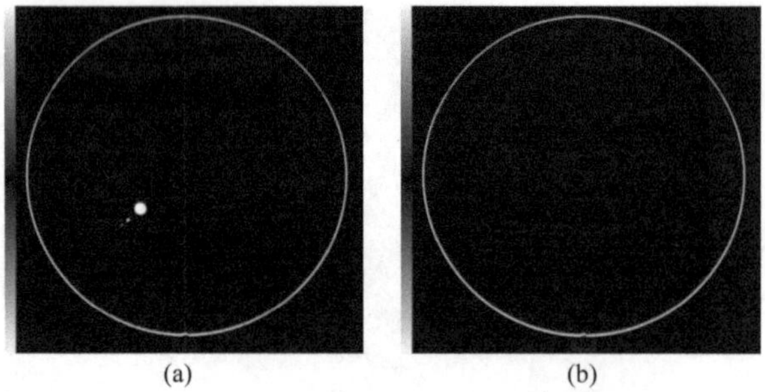

FIGURE 3.17 Examples of CSAM images of bonded blank Si wafer pairs. (a) Image showing bonding voids at the interface. (b) Image showing void-free bonding.

block the transmission of acoustic signals and shield the void defect from detection. Other features such as cavities in a MEMS device reflect sound waves the same way as voids at the bonding interface. These factors must be taken into consideration when interpreting CSAM images from bonded device wafers.

3.6.3 Post-bond Anneal

After the wafers are bonded in ambient conditions, they are placed into an oven for heated anneal. The heating process greatly increases BE. As shown in Figure 3.18, after the wafer pairs were bonded, BE increased very slowly during the 30 minutes aging at the room temperature. After 30 minutes, the wafer pair was annealed for 15 minutes at 150°C, and the anneal caused the BE to increase from 200 mJ/m^2 to 2,500 mJ/m^2. The strengthening of the bonding interface provides better protection for the metal features from oxidation and provides sufficient force to hold the parts together while metal features on both sides expand and push against each other,

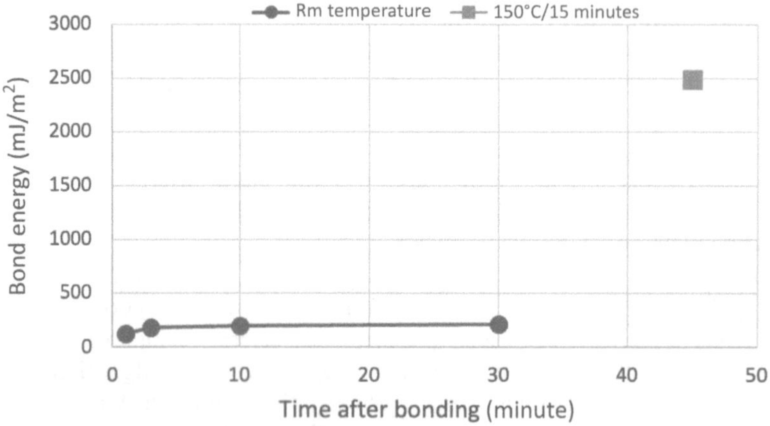

FIGURE 3.18 BE measured as a function of time after bonding, before and after heated anneal at 150°C/15 minutes.

allowing the metal atoms on the surface to interdiffuse to form a permanent metallic bond. As the dielectric bonding layer has a much lower CTE than Cu, when the wafer pair is heated, Cu expands more than SiO_2, and the differential volume increase bridges the small gap between the top and bottom bonding pads. Cu atoms on both sides of the bonding interface then interact with each other to form a permanent metallic bond.

The W2W hybrid bonding takes place in an ambient environment, and the initial dielectrics bonding does not require vacuum, controlled environment, heat, or pressure. As a result, the throughput of the initial dielectrics bonding is high since it normally takes less than a minute to bond a pair of wafers. Batch anneal also has a high throughput since many wafers can be annealed at the same time. Thus, the W2W hybrid bonding process has a high throughput which is well-suited for HVM.

W2W hybrid bonding is well-suited for products where the top and bottom wafers have the same wafer size and the same die size. Additionally, the yield on both wafers must be high to minimize compound yield loss. Two examples in HVM today are CMOS image sensors [36] and flash memory [37–39].

3.7 D2W HYBRID BONDING

3.7.1 D2W Bonding Introduction

Hybrid D2W bonding complements W2W bonding and addresses market segments where W2W bonding is not economical. For example, DRAM wafers normally have a lower yield, require speed binning, and therefore would suffer from high compound yield loss with a W2W approach. Another important market segment that is better suited economically for D2W is the chiplet market, where multiple small dies are bonded to one large base die or interposer.

There are several key differences between W2W and D2W hybrid bonding. (1) D2W bonding requires the processing of one wafer of the bonding pair into individual dies with thicknesses that may differ from the starting wafer. The dicing process generates a large number of particles and other contaminants. It can also generate edge defects that can interfere with direct bonding. Since hybrid bonding requires a pristine surface, special processes need to be developed to protect the die surface prior to dicing and return the surface to pristine condition after dicing. (2) W2W bonding normally takes place between two wafers of full thickness or with sufficient thickness to avoid wafer breakage during bonding. Wafer thinning normally takes place after W2W bonding and anneal. D2W bonding, however, requires the handling of thin die in certain applications, where die damage could easily happen during the die preparation process. (3) The bonding process is sequential with one die at a time; therefore, throughput consideration is vital to assembly cost for HVM. (4) In the HVM bonding process, the pick and flip tool touches every die and can potentially cause surface contamination. By comparison, in the case of W2W bonding, the bonding tool only touches a few supporting spots on the outer perimeter of the wafer during alignment.

The D2W bonding technology developed in this work, DBI® Ultra, addresses all these factors [40]. The simplified process is illustrated in Figure 3.19. A surface coating is applied prior to dicing to protect against contamination of the bonding

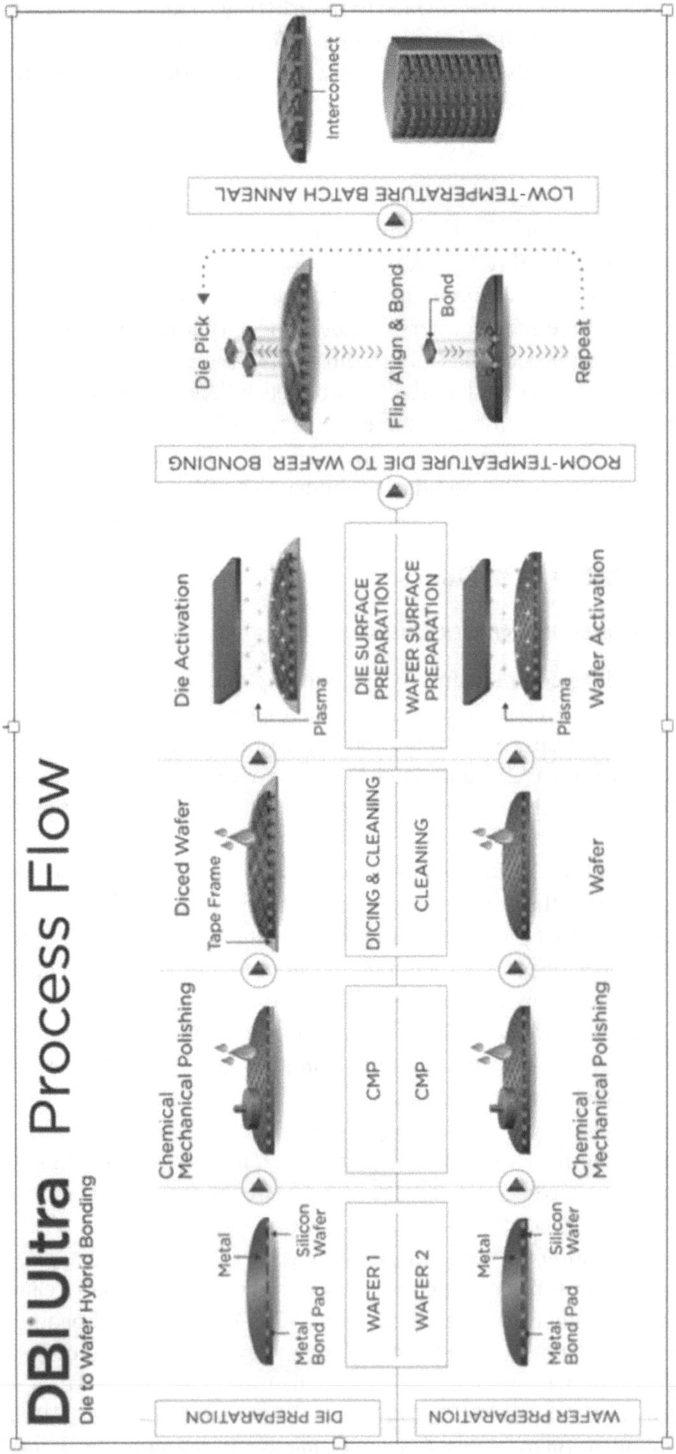

FIGURE 3.19 D2W hybrid bonding technology process flow.

surface. The entire die preparation process is carried out with the die staying on a tape frame, in order to minimize die handling induced damage. The fully processed dies are loaded into the bonder for direct pick and place (P&P) of known good die (KGD) based on a wafer map. The process is similar to flip chip assembly and has throughputs comparable to flip chip assembly.

3.7.2 Test Vehicle Designs

Most of the D2W development in this work was carried out using a daisy chain die size of 8 × 12 mm (~100 mm^2) because this is approximately the die size of a memory die and chiplet-based logic die which were anticipated to be the likely early adopters of D2W hybrid bonding. Additionally, this die size is large enough to check for differences in yield between center and corner daisy chain arrays due to die rotation and distortion relative to the substrate. Multiple designs with bond pad/bond pitch (ranging from 15/40 μm to 2/4 μm pad/pitch) were fabricated, bonded and electrically tested. Many other die sizes beyond the 100 mm^2 area were evaluated for compatibility with hybrid bonding, including die with dimensions as small as 0.4 × 0.4 mm and as large as 26 × 32 mm, full reticle size, as shown in Figure 3.20.

Figure 3.21 shows an example of one such test vehicle [41]. The interconnect design uses 3.5 μm pads at a 7 μm pitch. Daisy chain arrays cover the entire bonding surface and are divided into five regions: center, top, bottom, lift, and right. The central array covers an area of 50 mm^2 (~53% of the total die area) and contains 517,000 links. The arrays are further divided into 11 sub-arrays that can be independently tested for continuity. The top and bottom arrays have areas of 6 mm^2 with 63,000 links each for continuity test. Each array is further divided into three subarrays. The left array has an area of 8 mm^2 with 78,000 links with two interlaced chains that are individually testable for continuity. The right array is designed for testing leakage current between serpentine chains across 39,000 links.

The test wafers were fabricated using a Cu damascene process, and all wafers meet stringent requirements for Cu recess and long-range flatness for this work. The test vehicles were used for the work in the following sections.

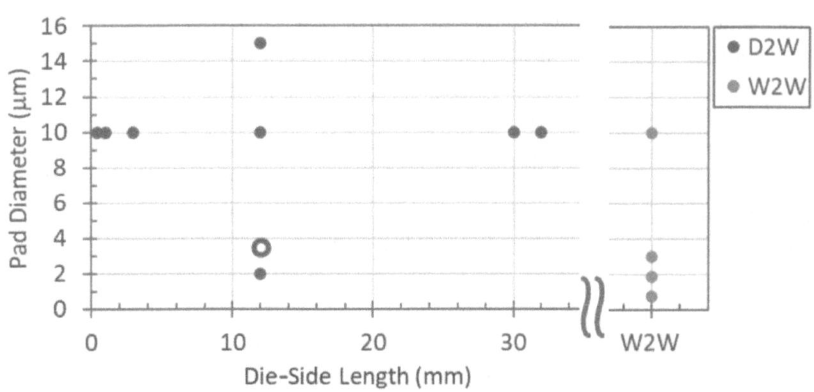

FIGURE 3.20 Die sizes in this work span a large range for hybrid bonding test vehicles.

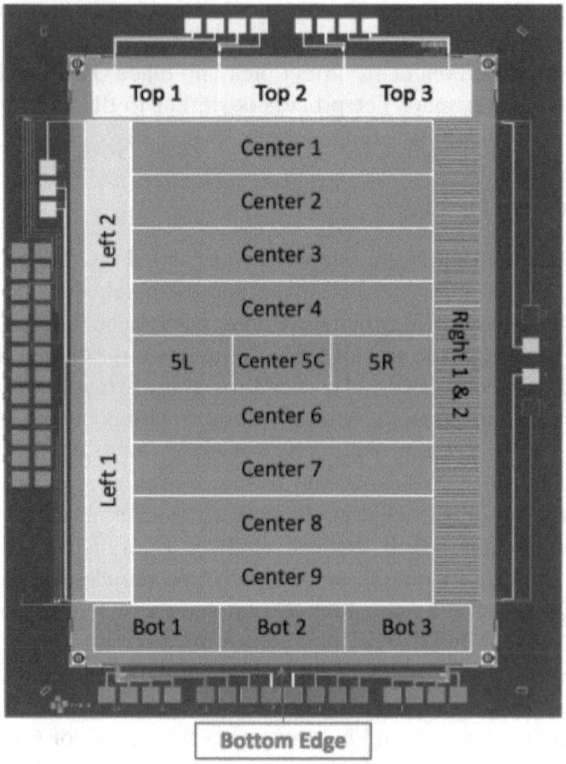

FIGURE 3.21 Example of hybrid bonding test chip layout used in this work, showing the six main daisy chain arrays and their subchain divisions. (© 2023, IEEE. Reprinted with permission from Ref. [41].)

3.7.3 Wafer Thinning and Dicing

3.7.3.1 Wafer Thinning

The die wafer is commonly thinned to a different thickness than the host wafer before bonding. If the design calls for a full thickness die, then the back grinding step can be ignored. Most of the bonding studies in this work used die of 200 μm and 50 μm in thickness.

As noted by Kawai [42], the backgrinding process includes two steps: coarse grind and fine grind. The coarse grind removes most of the materials, and the fine grind removes the last 40–100 μm to reach the final thickness. A fine grind finish may be accomplished with a 2000 to 4000 grit finish, and a 8000 grit finish further improves post-grinding surface roughness, die strength, and overall process quality.

When the wafer thickness is below 200 μm, a post-fine-grind stress relief process may be required to improve die strength and reduce wafer warpage. The process can be dry polish, CMP, or plasma etch.

For the backside processing of wafers bonded to a temporary carrier, edge trimming may be necessary to prevent edge chipping and improve yield.

3.7.3.2 Blade Dicing

After the wafer is thinned or debonded from a temporary carrier, it is mounted onto a tape frame to be cut into individual dies, commonly referred to as "dicing" or "singulation." There are four types of dicing methods: blade dicing, laser ablation, laser stealth, and plasma dicing.

Blade dicing is the most popular wafer singulation method with a widely installed infrastructure across the industry. Figure 3.22 is an illustration of blade dicing [43]. Dicing blades are metal blades that use synthetic diamonds as the abrasive grit. The fast-rotating dicing blade impacts the cutting front of the wafer and generates a large quantity of finely broken silicon particles in the process. It also causes damage at the die edge, commonly referred to as "chipping." The wafer is cooled by DI (deionized) water during the process, with CO_2 added to mitigate electrostatic charge generation. For hybrid bonding, a protective coating is applied to the wafer surface prior to dicing to protect against contamination and copper corrosion during the dicing process. There are normally test structures in the dicing street of the wafer, and care must be taken to avoid leaving protruded defects on the edge when cutting through the testing structures.

Blade dicing quality can be evaluated with optical inspection of the die before and after removal of the protective coating, and with D2W bonding. After bonding the die, the bond quality is analyzed with CSAM to detect interface voids. Process optimization to reduce edge chipping is necessary to qualify blade dicing for D2W hybrid bonding. In this work, with blade-diced daisy chain dies ranging from 0.4 × 0.4 mm to 26 × 32 mm in size, the voiding and electrical test yield after D2W bonding are comparable to dies singulated by laser stealth dicing and plasma dicing.

3.7.3.3 Laser Stealth Dicing

In recent years, laser Stealth Dicing™ (SD) has gained wide acceptance for thin and large dies such as DRAM dies due to higher throughputs and superior edge quality over mechanical saw dicing. It is a dry process requiring no water [44]. As illustrated in Figure 3.23, by focusing a short pulse laser inside the wafer, it generates a modified layer internally along the dicing street. The number of modified layers needed for good die separation depends on the die thickness. For example, only one layer is required for a 50 μm thick die, but multiple layers are needed for a 200 μm thick

FIGURE 3.22 Image of blade dicing of Si wafer. (© Disco Corp. Reprinted with permission from Ref. [42].)

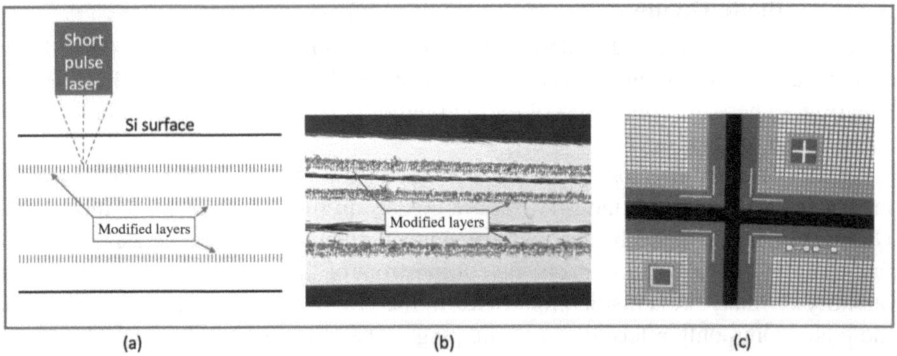

FIGURE 3.23 Illustration of laser stealth dicing. (a) Illustration of generation of modified layers inside the wafer by a short pulse laser beam. (b) Optical image of the die sidewall after separation, showing the modified layers. (c) Optical image of the corners of the four neighboring dies with smooth edges and sharp corners after tape expansion.

die. When the wafer with the internal modified layers is expanded on a tape frame, it separates along the dicing streets. Compared to blading dicing, SD is much cleaner and has better edge quality. In addition, the dicing street width can be reduced since the process does not physically remove silicon from the dicing street. This means that more dies can be produced from each wafer. Studies have shown that SD can achieve a higher die strength compared to blading dicing for thin wafers under 100 μm thick.

SD is most cost-effective when the die is relatively large. Furthermore, die separation is very challenging for very small dies. It is likely that the process can be further optimized for small die, but it may not be the best option.

3.7.3.4 Plasma Dicing

Plasma dicing has gained popularity for dicing very small dies. Figure 3.24 is an illustration of the plasma dicing using the Bosch process [45]. The wafer surface needs to be patterned with a mask to have only the dicing street exposed. A different plasma chemistry is required to etch away the BEOL dielectric layers to expose the Si underneath. For the Si etch using the Bosh process, an SF_6 plasma process is used for an isotropic etch of the Si in the exposed street (step 1), and a C_4F_8 plasma process

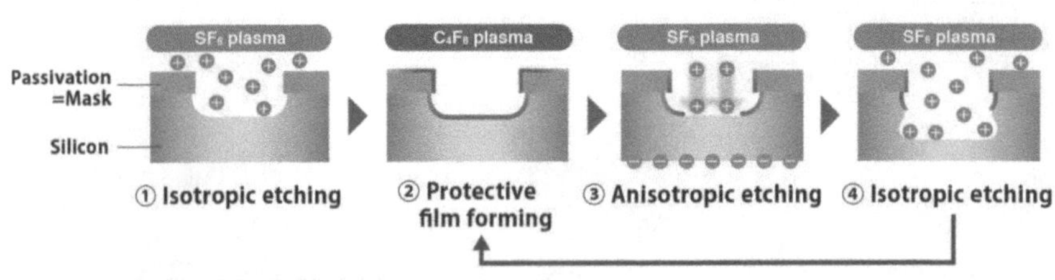

FIGURE 3.24 Illustration of the plasma dicing by the Bosh process. (© Disco Corp. Reprinted with permission from Ref. [44].)

is then used to form a polymeric protective film along the etched surface (step 2). A SF_6 plasma is again used to etch off the polymer film on the bottom but leaves the film on the side wall (step 3). Once the film on the bottom is etched off, SF_6 will continue etching the exposed Si (step 4). Alternating cycles of C_4F_8 and SF_6 plasmas are used until the dicing street area is etched through. After dicing, the mask on the surface needs to be removed. Plasma dicing is best suited for small dies. Because the etching is a parallel process, it has a very high throughput for dicing small die. It also increases the die count per wafer by minimizing the dicing street width. It does not generate debris during the dicing process, and the die edge quality from plasma is also superior to saw dicing. This emerging technology is quite promising; however, it currently has a very limited supply chain for outsourcing service.

In this work, equivalent bond results were obtained for 8 × 12 mm dies, and all three dicing methods resulted in comparable voiding results from CSAM inspection and electrical test yield from the bonded daisy chains dies, although the die edge quality of blade dicing is not optically as good as stealth dicing and plasma dicing.

3.7.4 Wafer and Die Preparation for D2W Bonding

The host wafer preparation for D2W bonding is the same as for W2W bonding; however, the diced wafer process is much more complex. Die handling during preparation can add contamination if the process is not well designed and controlled. There are three approaches to prepare dies for D2W bonding: die on tape, Collective D2W, and die holder.

The die-on-tape process, or the DBI Ultra process, focuses on the entire die preparation process for bonding that happens on a tape frame, including picking and placing the die directly from the tape frame using a KGD wafer map. It was developed to be compatible with the semiconductor chip assembly equipment sets in use for conventional packaging. The DBI Ultra process has many advantages: (1) minimizing die breakage, (2) eliminating the need for die size-specific trays or fixtures for die cleaning and handling in a flip chip bonder, and (3) utilizing supply chain equipment process flow with tape frame as the die handling system. Large thin dies (such as 30–50 μm thick DRAM dies) are prone to mechanical damage from die handling. Leaving the die on tape is the lowest risk option to protect the die, and die handling in a tape frame is the incumbent technology used for P&P in HVM. Bonding by direct P&P from a tape frame requires essentially the same equipment as flip chip bonding and constitutes the lowest technology adoption barrier, as the packaging industry has expertise in the process. A protective coating can be applied to the wafer surface prior to singulation, and after the singulation, the coating is removed with a wet chemical process while the dies are still on the tape frame. Dies of different sizes (ranging from 0.4 × 0.4 mm to 26 × 32 mm) on tape have been processed with good cleaning results [46]. Figure 3.25a shows a diced wafer at the completion of the chip preparation step and is ready for bonding, and Figure 3.25b shows a partially picked wafer with defective dies left behind in the picked area (top portion of the frame). Unused KGD left behind on the bottom portion of the wafer are picked and placed onto another host wafer in another bonding run, and the dies are stored on the tape in a dry cabinet between the two runs.

(a) (b)

FIGURE 3.25 (a) Picture of a dicing frame with dies fully prepared and ready for DBI bonding. (b) A frame after KGD dies on the wafer are partially picked.

3.7.5 D2W Bonding

3.7.5.1 Single-layer Bonding

With the DBI Ultra process flow, a flip chip bonder retrofitted with a cleanroom kit to achieve a Class 1 environment or a specially built direct bonder may be used to P&P KGD dies directly from a dicing frame onto the host wafer. In this work, many lots have been done on a retrofitted flip chip bonder with an alignment accuracy of ±7 μm @ 3σ, and another bonder with a specified alignment tolerance of ±3 μm @ 3σ, with die preparation and bonding in a Class 1,000 clean room environment. Both the dies on the tape frame and the substrate wafers are manually loaded into the bonder equipped with a Class 1 cleanroom kit. Both the host wafer and the die are exposed to the ambient temperature and environment, and the process is similar to a standard flip chip P&P process without flux dipping. Spontaneous bonding takes place as soon as the die surface touches the wafer surface, and the bonding process can run at the designated speed of the bonder. The entire wafer is populated before annealing, and if die stacking is required, all the layers are placed prior to the final anneal. All hybrid bonding joints through the stack are formed at the same time during the final anneal.

Figure 3.26 shows post-bond alignment accuracy obtained for four consecutive lots, using an infrared microscope to measure the misalignment for each bonded die. After bonding each lot, the programmable offset was finetuned; as a result, the misalignment was reduced slightly. Using a bonder with specified (x, y) alignment accuracy of ±3 μm @ 3σ, a slightly tight distribution of 2.8 μm (x, 3σ) and 2.0 μm (y, 3σ) was achieved in this work.

Since D2W hybrid bonding is new to the industry, design guidelines need to be developed. One parameter examined is the electrical yield of daisy chains versus its distance from the die edge. The electrical test yield of individual daisy chain rows is plotted as a function of the distance to the die edge in Figure 3.27. The yield stayed constant up to the distance of 71 μm from the die edge, and a slight drop was seen for the first row, 50 μm from the edge. This result is consistent with previous work on a test vehicle with 15 μm diameter pads that showed no degradation in D2W hybrid bonding yield as close as 85 μm from the die edge [47].

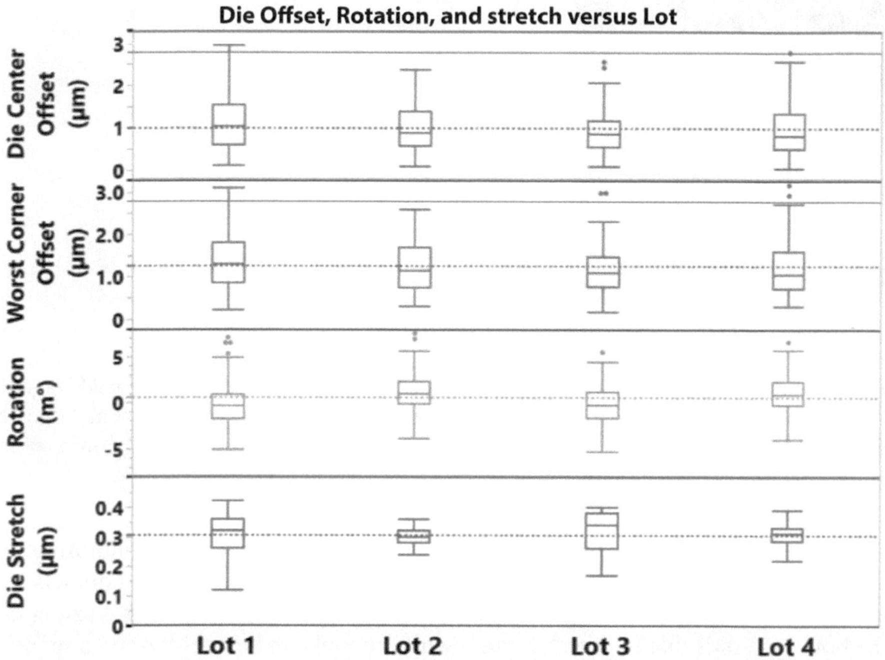

FIGURE 3.26 Post-bond alignment accuracy of four bonded lots, showing distribution of die center offset, worst corner offset, rotational angle, and the die stretch relative to the substrate, for all lots. Dotted lines are the global average, and red lines at the 2.8 µm offset indicate the maximum offset expected to yield electrically. (© 2023, IEEE. Reprinted with permission from Ref. [41].)

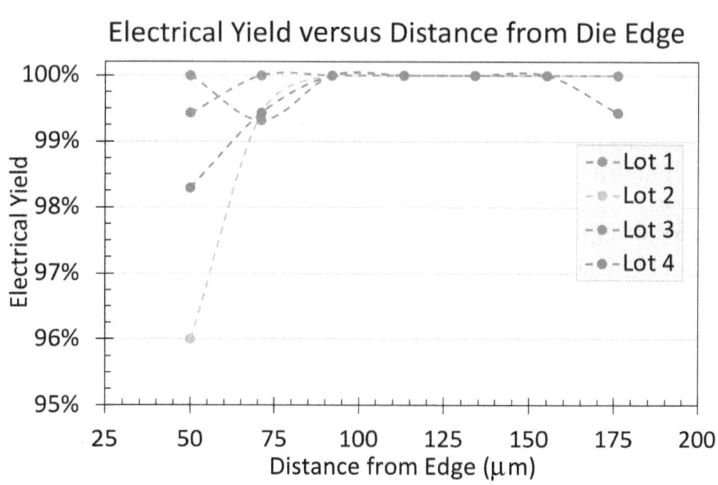

FIGURE 3.27 Electrical yield as a function of the distance of the chain from the die edge. (© 2023, IEEE. Reprinted with permission from Ref. [41].)

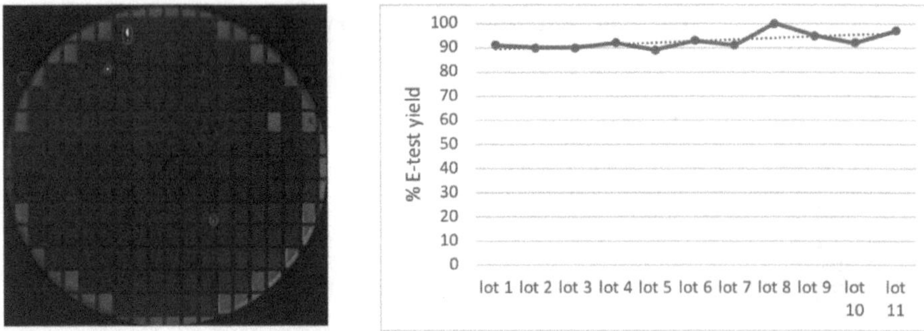

(a) (b)

FIGURE 3.28 (a) The CSAM image of a bonded wafer, with bond interface voids on only 1.7% of the bonded die. (© 2023, IEEE. Reprinted with permission from Ref. [40].) (b) Electrical test yield of a large daisy chain with 32,000 links and covering >50 mm^2 area for 11 consecutive runs in a prototyping laboratory with manual operations [29].

One concern for the DBI Ultra process flow is the potential contamination of the die bonding surface by the pick and flip tool, which directly contacts the bonding surface. With the optimization of the process and proper tool selection, the bonding voids due to direct contact with the tool can be avoided. As shown in Figure 3.28, >90% electrical yield has been consistently achieved in a prototype laboratory with manual operation on a large daisy chain containing 32,000 links and covering >50 mm^2 die area [29]. Although 90% yield is not acceptable for HVM and a yield of >99% is required, the fundamental yield hit in this study is attributed to particle contamination, due to manually operated HVM equipment modules, unlike the planned cluster tools with robotic handling systems and FOUPs (Front Opening Unified Pods) that can handle 300mm die on tape frame. The focus of the study was on understanding the root cause of the yield hit at each process step for process improvement.

Kamei et al. [48] also showed that the bonding voids can be eliminated with optimal pretreatment and bonding process. With process optimization, void-free bonding and 100% electrical test yield were achieved with a large die with >400 mm^2 surface area bonded to a 300 mm host wafer, as shown in Figure 3.29.

3.7.5.2 TSV Integration with D2W Hybrid Bonding

Stacking with D2W hybrid bonding involves integration of TSVs into the process flow. Fabrication of the backside bonding layer with a TSV-middle process is shown in Figure 3.9. Die stacking with hybrid bonding is more efficient than die stacking with solder-capped microbumps because the planarized TSV can be bonded directly to the bond pad of the next die. Gao et al. [20] reported their 5-Hi die stacking work in 2021, and Theil et al. [47] reported their 8-Hi stacking work in 2022. Figure 3.30 shows images of the test vehicle die: front side (Figure 3.30a), backside (Figure 3.30c), and cross section (Figure 3.30b), prior to bonding. The bond pad on the front is oversized to 15 μm to relax the alignment accuracy requirement during bonding, and the TSV diameter on the backside is 5 μm. The planarized TSV will be bonded directly to the hybrid bond pads

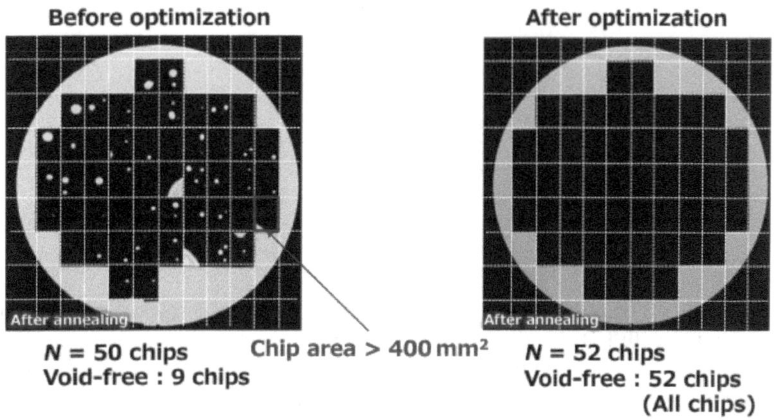

FIGURE 3.29 D2W bonding voids without and with optimal pretreatment process and bonding process. (© 2023, IEEE. Reprinted with permission from Ref. [48].)

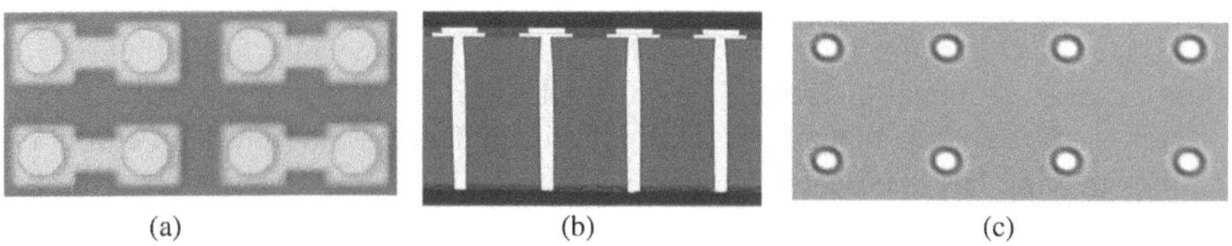

FIGURE 3.30 The daisy chain test die with TSV. (a) Optical image of the front-side DBI bonding pad and RDL layer for daisy chain connection. (b) SEM cross-sectional image showing the front and backside bonding surfaces. (c) Optical image of the planarized TSV on the backside. (© 2021, IEEE. Reprinted with permission from Ref. [20].)

on the front-side surface. The host wafer for bonding has a complementary RDL layer to complete the formation of the daisy chain circuit after bonding [20].

The assembly and bonding of dies with or without TSV is the same. The dies were directly picked from a tape frame and bonded to the host wafer or die. On the front-side surface of the die (shown in Figure 3.30a), 15 μm hybrid bond pads were designed to relax the alignment accuracy requirement of the P&P bonder available for this test vehicle. After completing the bonding of each layer, the backside of the bonded die was cleaned prior to the direct bond of the next die layer. A total of seven TSV die layers and one cap die with hybrid bond pad and RDL only were stacked to complete the 8-Hi stacked module shown in Figure 3.31. The solid Cu–Cu bonding was formed after the 200–250°C/1 hour anneal [47].

Only the dielectric layer is bonded at the room temperature for each die. After all the die layers are bonded, the bonded parts need to be annealed to form direct Cu–Cu bonding. For stacked parts, direct Cu–Cu bonding through the entire stack occurs in this single anneal process. The annealing temperature can be adjusted for the thermal budgets of the different products. For memory products such as DRAM, the annealing temperature must stay below 250°C to avoid degrading the device's

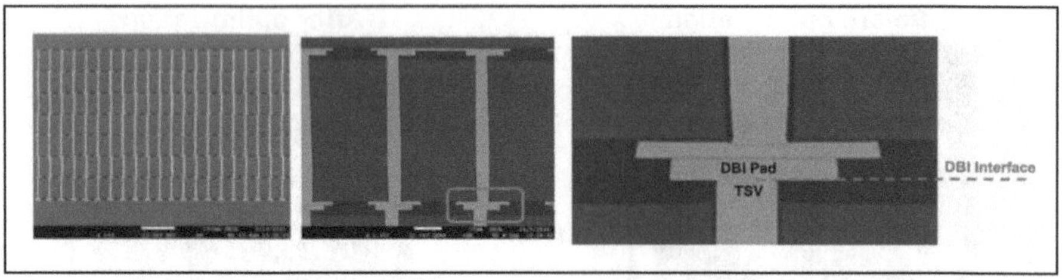

(a) (b) (c)

FIGURE 3.31 (a) SEM cross-sectional image of an eight-layer stacked part showing the TSVs integrated with DBI bonding and daisy connection. (b and c) A representative SEM image of TSV to DBI pad joints cross-section (at different magnifications). The blue lines mark the bonding interfaces. (© 2022 IEEE. Reprinted with permission from Ref. [46].)

performance. If the device can tolerate higher anneal temperatures, it is preferred to anneal at higher temperatures to promote faster Cu diffusion. After the anneal, CSAM was conducted to characterize bonding interface voiding, electrical resistance measurement was conducted for continuity, and cross-sectional microscopy analysis was conducted for quality of the Cu–Cu bonding interface. Figure 3.31 shows a cross-sectional image of the 8-Hi stack after 250°C/1 hour anneal.

Across the industry, eight-die stacking has been demonstrated by Kim et al. [49], 12-die stacking has been demonstrated by Tsai et al. [50], and 16-die stacking has been demonstrated by Kim et al. [51].

Theil et al. [47] examined hybrid bond interconnect performance as a function of the proximity to the die edge for single-layer, two-die stacking, and five-die stacking configurations. For the single-layer and two-die stacking samples, the electrical test yield of bond pads 85 μm to the die edge showed no difference from the pads further in from the die edge. For the five-die stacking configuration, consistent electrical yield was obtained with bond pads up to 220 μm from the die edge, and a drop in the yield was detected with bond pads 150 μm from the die edge. Theil et al. [47] also examined void-free die yield in the five-die stacking configuration, and found no significant difference between the layers. These findings are important for assessing the future of hybrid bonding for die-stacking applications in HVM.

3.7.5.3 Small Die Processing and Bonding

While the need for hybrid bonding in memory and HPC is driven by the demand for ultrafine pitch and electrical performance enhancement, that is not the only market segment that will benefit from hybrid bonding. Hybrid bonding can greatly enhance the thermal and electrical performance of many small form factor, low pin count devices, and can help reduce form factor and device height. Hybrid bonding also offers superior reliability for devices serving in high-temperature, harsh environments. Figure 3.32 is an illustration of different applications that hybrid bonding can be applied at different die sizes.

The suitability of DBI Ultra for D2W bonding for die sizes ranging from 0.4 × 0.4 mm to 32 × 26 mm has been demonstrated in this work [46]. Figure 3.33 shows pictures of a D2W bonded multi-chip demonstration module. Singulation of all three

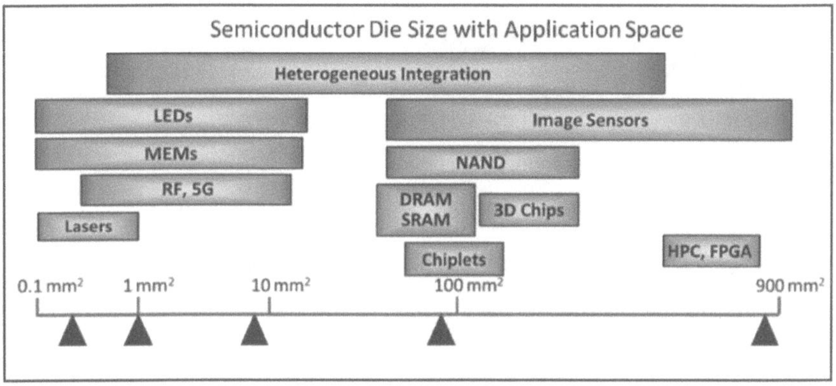

FIGURE 3.32 Illustration of different applications of hybrid bonding at different die sizes.

(a) (b)

FIGURE 3.33 Demonstration of a multi-chip module built with DBI Ultra D2W process. (a) Whole wafer image. (b) Magnified site images.

dies, 3 × 3 mm, 1 × 1 mm, and 0.4 × 0.4 mm, was done by mechanical saw. All three dies were prepared and bonded using the process shown in Figure 3.19, and the die transfer rate was 100%. The annealing temperature used for this build was 1 hour at 300°C. These results demonstrated the wide range of die sizes that the DBI Ultra die-on-tape process can handle.

The overlay of CSAM and electrical test yield of the bonded daisy chain dies for the above module is shown in Figure 3.34. The electrical test yield is 100% for the 0.4 × 0.4 mm die, 99% for the 1 × 1 mm die, and 95% for the 3 × 3 mm die, with a combined site yield of 94% for all three die. 50% of the failures were due to metal pattern defects; this is expected because the host wafer was fabricated in a lab environment. Bonding voids account for only 3% of the failures.

3.7.6 Other D2W Bonding Approaches

3.7.6.1 Collective D2W Bonding Approach

Suhard et al. demonstrated an alternative die preparation and bonding process flow: Collective D2W bonding [52, 53]. The principal idea of this flow is to P&P the die onto a temporary carrier and carry out the die surface cleaning after die P&P to avoid touching the die bonding surface after the final cleaning. After the final surface

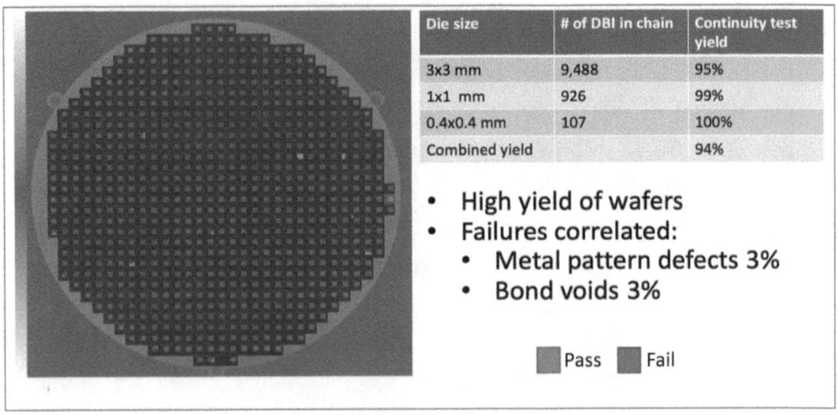

FIGURE 3.34 Overlay of CSAM and electrical test yield of the bonded daisy chain dies. (© 2022, IEEE. Reprinted with permission from Ref, [46].)

cleaning, all the dies on the wafer are bonded to the target wafer through a thermal compression W2W bonding process. The W2W bonding process involves heat and pressure, which is undesirable to achieve ultrafine pitch bonding.

Table 3.2 shows a comparison of die preparation and bonding module steps for the Collective D2W bonding flow and the DBI Ultra bonding for a 50 μm thick

TABLE 3.2
Side-by-side Comparison of Process Steps for the Collective D2W Bonding and Direct D2W Bonding (DBI Ultra)

Process Module Step	Collective D2W	Direct D2W (DBI Ultra)
Bond top wafer to carrier 1	YES	YES
Thin top wafer	YES	YES
Bond thin wafer to carrier 2	**YES**	**NO**
Debond carrier 1	YES	YES
Remove temp bond adhesive 1	YES	YES
Coat device side surface	YES	YES
Debond carrier 2	**YES**	**NO**
Remove temp bond adhesive 2	**YES**	**NO**
Dice	YES	YES
P&P to Carrier 3	**YES**	**NO**
Organic coating removal	YES	YES
W2W TCB to target wafer	**YES**	**NO**
P&P to target wafer	**NO**	**YES**
Debond carrier 3	**YES**	**NO**
Remove temp bond adhesive 3	**YES**	**NO**
Final anneal	YES	YES
Total process module steps	**15**	**9**

die. For the Collective D2W process, the top wafer (die wafer) was mounted on a Si carrier (first temporary bonding, W2W) and thinned down to 50 μm. Then, the thinned wafer was bonded to the second temporary carrier (second temporary bonding, W2W) and debonded from the first temporary carrier (first debonding) to expose the device surface and add protection for dicing. Although no details were given in the papers [52, 53], the residue of the first temporary bonding material needs to be removed (first cleaning process for temporary bonding material) before the protective coating is applied, as a protective layer is needed for protecting the dies from corrosion and contamination during dicing. The thinned wafer with the protective layer was then debonded on tape (second debonding) and diced. Similarly, the residue from the second temporary bonding material needs to be removed before the next process step (second cleaning for temporary bonding material 2). Subsequently, the dies were picked, flipped, and placed (first P, F&P) on a silicon carrier wafer (third temporary bonding, D2W) with a bonder. The temporary bonding material was stripped in between the dies (third cleaning for temporary bonding material 3), and the idea is to avoid "cross-talking" between the dies (i.e., to reduce adhesion between the die and the temporary bonding adhesive layer) and to improve the die transfer. After the protective layer removal, the carrier wafer with dies and the target wafer were activated. Then, the carrier wafer with dies was bonded to the target wafer with a TCB process (W2W bonding, thermal compression, at 13 kN pressure, 250°C, 30 minutes), and the carrier wafer was then removed mechanically (third debond). Finally, the transferred dies to the target wafer were annealed. The entire process included three temporary bonding steps with three carrier substrates; three temporary debonding and adhesive removal steps; pick, flip and placement of dies to the third carrier substrate; and a W2W TCB step. The Collective D2W bonding approach requires four W2W bonds to achieve die bonding of 50 μm thick die.

In comparison, the DBI Ultra D2W process involves only one temporary bonding step and one cleaning step, compared to three each for the Collective D2W. To accomplish the same D2W bonding task, the DBI Ultra D2W process involves only one temporary bonding and debonding step; one cleaning step; and one pick, flip and place step. As shown in Table 3.2, compared to the Collective D2W process, the DBI Ultra process requires six fewer module steps. Although this comparison was specified for die that must be thinned to 50 μm, it is important to note that no temporary bonding is required for die at a thickness of 200 μm or greater that are bonded in the DBI Ultra process. On the contrary, the Collective D2W process still requires a temporary bonding process and a W2W TCB process to bond dies at any thickness.

3.7.6.2 Die Holder Approach

The third approach of preparing dies for D2W is the die holder approach [54]. In this process flow, the KGD's are picked from the dicing frame and placed into custom die holders for cleaning and activation, as illustrated in Figure 3.35.

3.7.6.3 Comparison of the Three Processes

Table 3.3 compares the three D2W bonding process flows side by side for a bonding process (excluding all process steps prior to dicing). For die handling, the DBI Ultra D2W process only picks and places each die once with one flip, and bonding takes

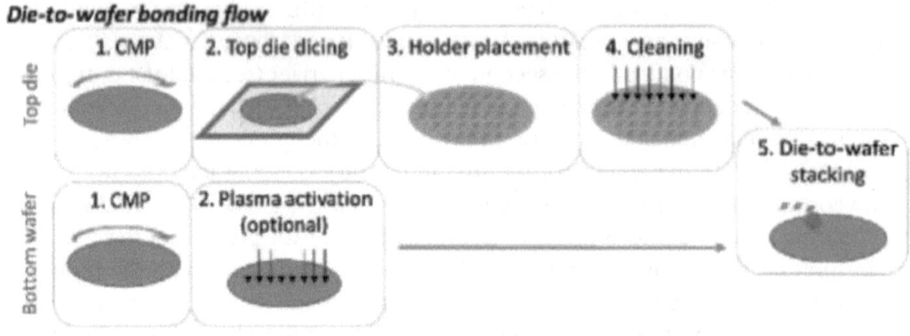

FIGURE 3.35 The die holder approach illustration for D2W bonding. (© 2021, IEEE. Reprinted with permission from Ref. [54].)

TABLE 3.3
Comparison of Three Different Process Flows for D2W Hybrid Bonding

Process	Process description	Complexity	Cost	Alignment Accuracy	Bonding Throughput	Maturity
DBI® Ultra	Die-on-tape prep P&P to target wafer from tape	Low	Low	P&P bonder limit	1,640 UPH demonstrated	HVM ready
Collective D2W	P&P to temp carrier, clean + W2W bonding + debond temp carrier	High	High	P&P bonder limit + W2W bonder limit	Data not available	Research
Die holder	P&P die to the die holder, clean, P&P to the target wafer from the die holder	High	High	P&P bonder limit	Data not available	Research

place at ambient conditions with no heating or thermal compression. The Collective D2W process picks and places the die with one flip, followed by thermal compression W2W bonding, which can potentially damage the bonding interface and cause displacement due to differential temperatures between the top wafer and the bottom wafer, and finally a debonding process after the TCB. The die holder process picks and places the die twice, first time into the die holder, and the second time to the target wafer for bonding. Additional P&P process and cleaning process in a die holder increases the risk of die breakage for thin and large die. Thus, the Die Holder approach is not practical for processing small dies such as the 0.4 × 0.4 mm die or the 1 × 1 mm die shown in Figure 3.33.

In terms of process complexity and investment, the DBI Ultra D2W process does not need additional bonders or fixtures for die preparation. The Collective D2W process requires investment in temporary bonding and debonding equipment, plus a high accuracy W2W bonder for the die transfer process. The die holder approach requires a dedicated fixture for each die size. The process complexity increases capital investment and processing cost and reduces process throughput.

One topic of concern in the industry for D2W bonding is the queue time of the die and host wafer after plasma activation. In this work, we routinely demonstrate that a host wafer may be bonded up to 30 days after activation and the die on tape for up to 2 weeks, with a high bonding yield. It is possible that the industry concern has originated from problematic wafer and die preparation and bonding processes that contribute to the poor bondability.

Pitch scaling in D2W hybrid bonding is currently limited by the bonder alignment capability. Brandstätter et al. [55] showed alignment capability of <200 nm @ 3σ for C2W alignment at a throughput >2,000 unit per hour (UPH), and IR microscopy measurements showed post-bond alignment accuracies of ~300 nm @ 3σ for C2W bonds. The pitch limit is estimated to be 3 µm, based on the rule-of-thumb of "bonding pitch = 10 × post-bonding displacement". Industry roadmap suggests that the 100 nm alignment accuracy machine may be available in 2023 and there is a pathway to 50 nm alignment accuracy [56], which can potentially enable 1 µm pitch D2W bonding.

3.8 PARAMETERS AFFECTING THERMAL ANNEAL AND Cu–Cu INTERCONNECT FORMATION

As described earlier in the chapter, the Cu–Cu interconnect in hybrid bonding is formed during heated anneal by bridging the nanometer-level gap between the top and bottom pads of a bonded device through differential thermal expansion and interface diffusion. Anneal temperatures of 300–400°C were commonly applied in the early hybrid bonding work. However, some product segments (such as DRAM) cannot tolerate such high temperatures, and reducing the overall anneal thermal budget is very important for further proliferation and performance enhancements of hybrid bonding.

Gao et al. [40] reported the successful formation of Cu–Cu joints with 200°C/1 hour anneal on D2W bonded samples with 10 µm circular bond pads in 2019. As shown in Figure 3.36, they annealed multiple sample pieces from multiple assembly lots spanning several months at 200°C for 1 hour. Four out of the five sample pieces showed 100% electrical test yield on void-free C2W pairs, and one sample set showed an electrical test yield of 92%. Figure 3.36 shows a cross-sectional image from Sample 1-1. The overlapping Cu surfaces showed no visible gap between the top and the bottom Cu bonding pads, indicating good electrical bonding.

As the bonding pads shrink with pitch scaling, the Cu pad expansion becomes more and more restricted, as shown by Beyne et al. [57]. They experimentally measured the expansion of sub-micrometer Cu pads (650 nm thickness) as a function of pad size and temperature, and extrapolated the data to pad sizes below the experimental resolution. The results, shown in Figure 3.37, clearly indicate a very limited expansion of the Cu pad at sub-micrometer pad dimensions.

Build	Sample ID	Anneal Temperature (°C)	# Die Pass/# Die No Defect	% Yield
1	#1-1	200	24/26	92
2	#2-1	200	20/20	100
3	#3-1	200	19/19	100
3	#3-2	200	13/13	100
3	#3-3	200	15/15	100

(a)

(b)

FIGURE 3.36 (a) Electrical test results for samples annealed at 200°C for 1 hour. (b) Cross-sectional image from sample 1-1 showing good Cu–Cu bonding. (© 2019, IEEE. Reprinted with permission from Ref. [40].)

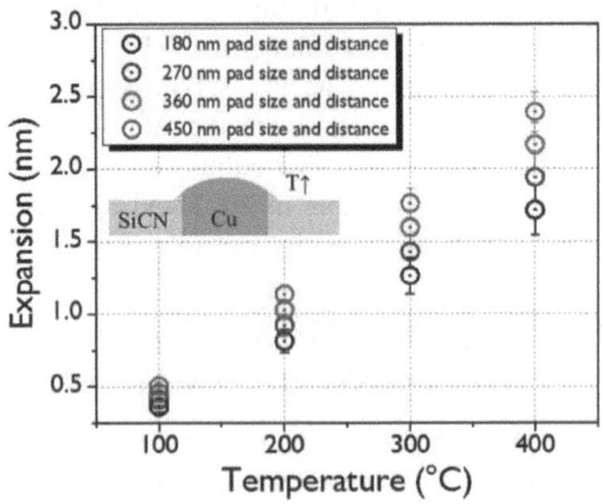

FIGURE 3.37 Extrapolated Cu expansion as a function of pad size and annealing temperature. (© 2017, IEEE. Reprinted with permission from Ref. [57].)

Parameters that affect the optimal anneal temperature for hybrid bonding include pad geometry design, dielectric selection, Cu composition and microstructure, CMP process optimization for topography control, die and wafer preparation procedures, and bonding process control. Since the CMP process optimization, die and wafer preparation, and bonding have been addressed earlier in the chapter, the Cu microstructure and pad geometry will be addressed below.

The impact of the Cu microstructure on the minimum hybrid bonding anneal temperature is a hot research topic. Much work has focused on nano-twin (NT) Cu due to the promise of the ⟨111⟩ crystallographic orientation with a theoretical diffusivity that is 1,000× higher than in other directions. Mirkarimi et al. [58] conducted a thorough study of Cu microstructure impact and compared three Cu microstructures in their study, including standard BEOL Cu (STD), NT copper, and fine grain (FG) Cu with targeted dopant content in the film. The samples were annealed at temperatures ranging from 150°C to 270°C and varying annealing times. As shown in Figure 3.38, the median resistance of the daisy chain varies with the anneal temperature. At lower annealing temperatures, the value is higher than the theoretical

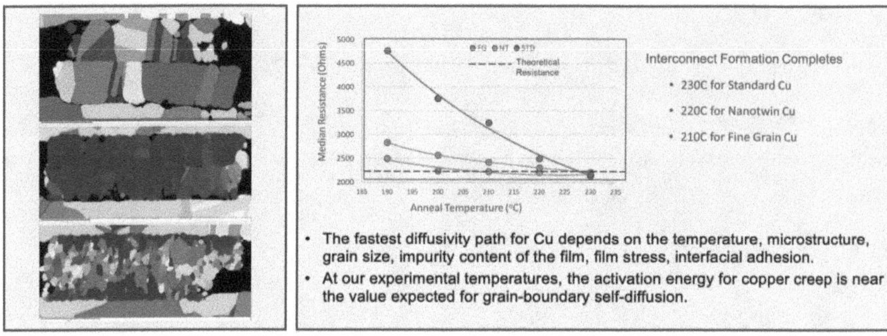

FIGURE 3.38 The median resistance is plotted as a function of the anneal temperature for three microstructures. (© 2022, IEEE. Reprinted with permission from Ref. [58].)

resistance value calculated based on the daisy chain design due to incomplete formation of the Cu–Cu bond. When the measured resistance value reaches the theoretical value, the Cu–Cu interconnect formation is considered complete and the annealing temperature is designated as T_c. The T_c value is 210°C, 220°C, and 230°C for the FG-Cu, NT-Cu, and STD-Cu, respectively. The results show that using NT-Cu can lower the anneal temperature compared to STD-Cu. The improved performance may be attributed to the higher diffusivity of the Cu along the ⟨111⟩ direction. However, FG-Cu can lower the annealing temperature even further. Mirkarimi [58] attributed the thermal budget reduction to the presence of a high concentration of fast diffusion paths along the grain boundaries and a higher creep rate in the FG-Cu films.

Changing the surface area ratio of the mating bonding pads between the top and the bottom devices impacts Cu expansion and leads to reduced anneal temperatures. The use of different size bonding pads across the bonding interface with one side slightly recessed and the other side slightly protruded was first illustrated by Enquist [59] and Tong et al. [60]. Recently, Beyne et al. [57] used unequal top and bottom pad sizes as shown in Figure 3.39, and optimized the CMP process to produce Cu protrusion on the

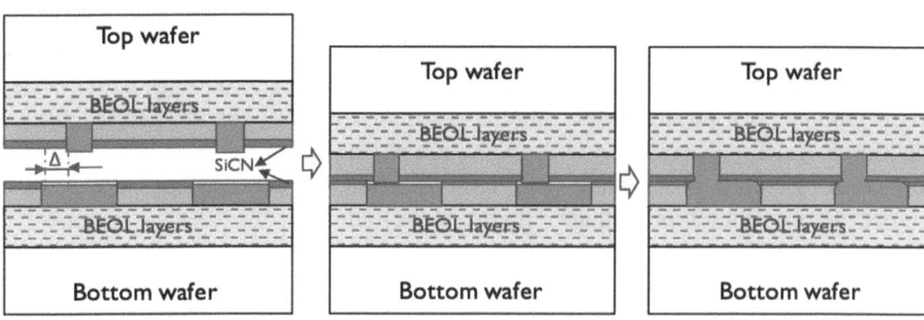

FIGURE 3.39 The direct hybrid bonding approach by Beyne et al. [57], with the top protruding Cu-pad landing on a larger bottom Cu-recessed pad. Size difference is defined by the W2W overlay tolerance. After bonding, both Cu pads are in near contact before anneal, thus reducing the anneal temperature needed for Cu–Cu bonding. (© 2017, IEEE. Reprinted with permission from Ref. [57].)

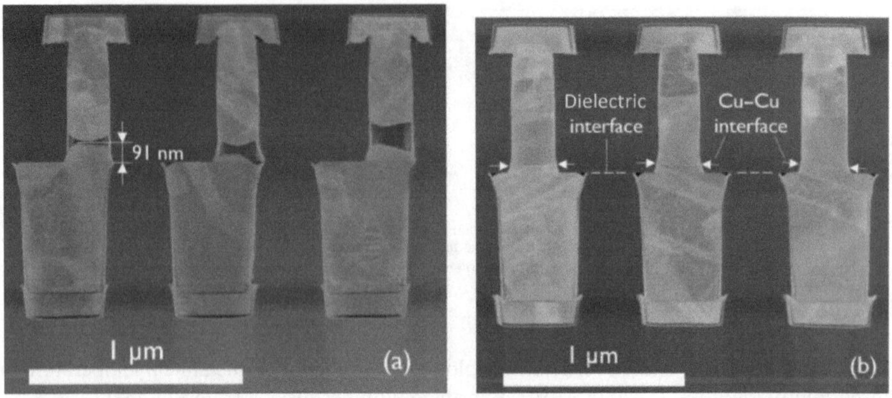

FIGURE 3.40 TEM cross section of 200/400 nm pad combination arrays. (a) Defective connections showing up to 91 nm of "bulge-out" of the bottom pad into the cavity of the corroded top pad. (b) Excellent connections achieved due to the same bugle-out mechanism, as evidenced by the raised Cu–Cu interface. (© 2023, IEEE. Reprinted with permission from Ref. [61].)

top pads and Cu recess on the bottom pads. They demonstrated feasibility at 1.44 μm pitch and even 0.9 μm and 0.72 μm pitches for W2W bonding after 350°C anneal.

A Cu "bulge-out" phenomenon reported with the differing top and bottom pad *x–y* dimensions has the potential to further reduce the thermal anneal budget. Messemaeker et al. [61] reported expansion of the bottom pad into the small cavity of corroded top pads, which is significantly more than predicted by an elastic expansion model (Figure 3.40a). Figure 3.40b shows how the bulge-out enabled excellent connections in a 700 nm pitch array, revealed by the Cu–Cu bonding interface which is raised by tens of nanometers above the dielectric bonding interface.

Further investigation of the Cu bulge-out phenomenon revealed [61] that the bulge-out occurs at temperatures as low as 230°C. The bulge-out is insensitive to the pad thickness over the range of 250–1,000 nm; however, they found a significant influence of the pad size and pad-to-cavity size ratio. The 400 nm pads bonded to 200 nm cavities (2:1 ratio) resulted in higher bulge-out than 600 nm pads bonded to 200 nm cavities (3:1 ratio). When the ratio was fixed at 2:1, 600 nm pads bonded to 300 nm cavities produced more bulge-out than 400 nm pads bonded to 200 nm cavities. Regarding correlation with Cu crystal plane orientation, they found that the low-index orientations ⟨111⟩ and ⟨100⟩ tend to produce low bulge-out values, while higher index orientations produce a higher bulge-out.

3.9 HYBRID BONDING RELIABILITY

As stated previously, several attributes of hybrid bonded parts favor superior thermal-–mechanical reliability, corrosion resistance in harsh environments, and superior electromigration resistance. CTE mismatch is the driving force behind failures in temperature cycling and high temperature storage, and removing solder and underfill from the package eliminates the CTE mismatch between the die and materials of much higher CTE. The bonded dielectric layer around the copper interconnect forms a hermetic seal around the Cu, providing protection against oxidation and corrosion of the copper from the environment. Additionally, monolithic Cu resists electromigration much better than a solder joint.

The reliability of hybrid bonded image sensors is proven in service since it has been in high-volume production since 2016. Further reliability data from D2W bonding are reported below.

Gao et al. reported data on single-layer D2W daisy chain parts with 5/10 μm pad/pitch in 2018 [62]. As shown in Table 3.4, the packages were tested according to the temperature cycling test specification JEDEC JESD22-A104D Condition M (from −40°C to 150°C). The JEDEC requirement is 1,000 cycles with an electrical resistance change of less than 10%. In this test, the samples showed no resistance increase after 1,000 cycles; instead, the value dropped slightly. To demonstrate the high reliability of the parts, the test was continued to 2,000 cycles, and there was still no increase in resistance; instead, the resistance dropped further slightly (1.5–1.7% drop in most dies, with one exception of 3.4% drop in one die). Cross-sectioning a post-2,000 cycle package showed no sign of crack initiation at the bonding interface. As shown in Figure 3.41b, the bonding interface after 2,000 cycles looks very similar to a new package with no temperature cycling exposure. This is understandable because eliminating the CTE mismatch between the top and bottom dies eliminated the driving force for crack initiation and growth. Furthermore, the bonded oxide shields the Cu joints from any oxygen and other corrosive species in the environment, effectively preventing Cu oxidation. As shown by the slight reduction in the

TABLE 3.4
Reliability Test Results for Single-layer D2W Samples and Five-die TSV-stacked DBI Samples.

	Single-layer DBI		Five-die TSV-stacked DBI	
Test	**Test Condition**	**Results**	**Test Condition**	**Results**
Temperature cycling (Board level)	−40°C to 150°C, 2,000 cycles, 45 samples	Pass	−40°C to 125°C, 1,250 cycles, 44/45 samples*	Pass, test continuing to 2,000 cycles
High-temperature storage	225°C, 2,000 hours, 275°C, 2,000 hours, 22 samples each	Pass	150°C, 2,000 hours, 22 samples each	Pass
Autoclave	121°C/100%RH, 15 psi, 168 hours, 22 samples	Pass	121°C/100%RH, 15 psi, 168 hours, 22 samples	Pass
Moisture sensitivity MLS 3	24 hours prebake + 30°C /60% RH for 192 hours + 3× reflow, 22 samples	Pass	24 hours prebake + 30°C /60% RH for 192 hours + 3× reflow, 22 samples	Pass

* 1/45 samples showed infant mortality intermittent failure.

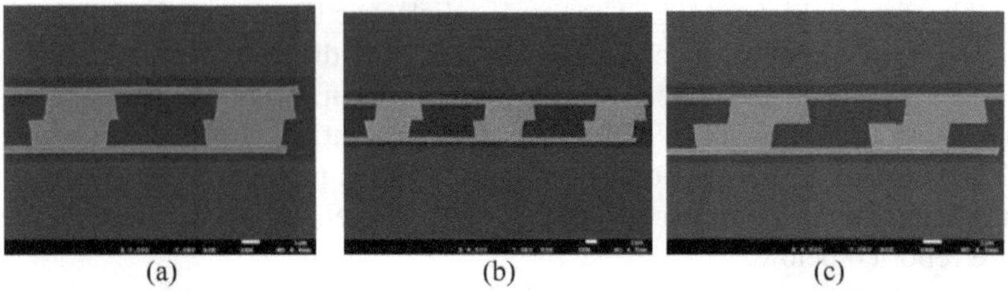

FIGURE 3.41 Cross-sectional image of Cu-to-Cu joints. (a) A bonded part at time zero, with no exposure to environmental stress testing. (b) A bonded part after 2,000 cycles of temperature cycling test. (c) A bonded part after 2,000 hours of HTS at 275°C.

daisy chain resistance, the temperature cycling serves as additional annealing for the Cu joints, potentially reducing micro-voiding at the bonding interface and causing a small decrease in the daisy chain resistance.

High-temperature storage tests were also conducted on two sample sets with 22 units each, according to JEDEC standard JESD22-A103. One test was conducted at 225°C and another at 275°C. Resistance measurements were conducted at 1,000 hours and 2,000 hours. All 44 parts showed a slight resistance drop of at least 1.2–1.4% after 2,000 hours of testing (twice the required duration). A cross section of one sample from the 275°C test showed a pristine Cu interconnect (Figure 3.41c).

Another set of 22 samples was tested through 168 hours of autoclave test at 121°C and 15 psi vapor pressure. All parts showed no resistance increase and no sign of Cu corrosion.

In 2021, Gao et al. [20] reported reliability test results on 5-die stacked daisy chain parts with TSV, and compared the results to the single-layer data. The data is also shown in Table 3.4. At the time of publication, the temperature cycling test passed 1,250 cycles, and the test was finally terminated at 2,000 cycles with no additional failure.

One part from the high-temperature storage test group was cross-sectioned. As shown in Figure 3.42, there is no sign of void growth or crack initiation at the bonding interfaces.

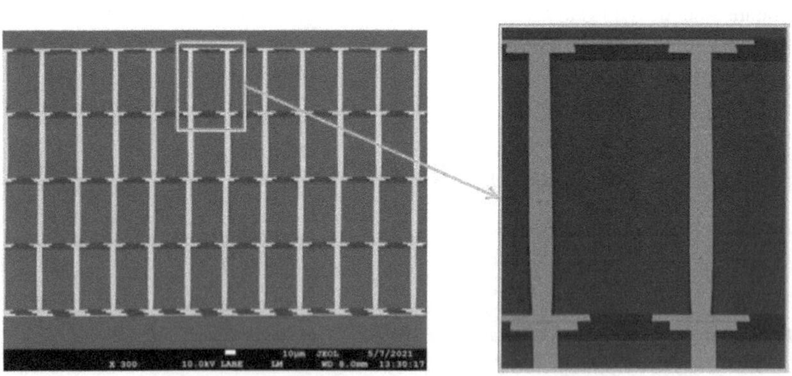

FIGURE 3.42 Cross-sectional image of a five-die stacked TSV sample after 2,000 hours of high-temperature storage test, showing no sign of crack initiation or growth.

3.10 METROLOGY

As discussed in the earlier sections of this chapter, there are two critical factors for successful hybrid bonding: bonding surface topography (including surface roughness and Cu recess), and bond strength of the dielectric interface. Metrology tools and procedures to quantify these two factors are critical to enable process optimization, for the CMP, dielectric materials qualification, dicing, and plasma activation processes. The tool commonly used to measure surface roughness and Cu recess is AFM, and the bonding strength of the dielectric interface is quantified by measuring the BE. This section summarizes tools and procedures of AFM and BE measurements.

3.10.1 AFM

AFM is a critical technique required to monitor the CMP process and ensure a robust manufacturing process. While AFM and related techniques have been known for decades, nanoscale- or sub-nanometer-scale characterization for hybrid bonding requires careful choice of the analysis configurations and parameters to avoid misinterpretation. Lee et al. [63] discussed key considerations for AFM analysis, including extraordinary AFM artifacts in the relative heights of the Cu and SiO_2 areas, and topographic characteristics of Cu/SiO_2 surfaces for successful hybrid bonding. Lee et al. [63] chose an AFM with a flexure scanner for the study and used first-order data flattening. Many traditional AFM systems have a tube scanner, which distorts the scanned profile, and data correction required with a tube scanner is more complicated. Given the need for nanometer accuracy over a relatively long scan range for hybrid bonding, AFM equipment with flexure scanners may be better suited.

A proper flattening protocol is still required for accurate measurement even with a flexure scanner, and the protocol should be chosen depending on the machine type. It is a good practice to check the machine's inherent background profile regularly by scanning a flat sample, such as a blank silicon wafer or an optical flat.

Lee et al. [63] addressed several important aspects in the data collection and analysis process to avoid or minimize artifacts due to improper data analysis, and went into details about data flattening procedures needed to build a 3D image of a hybrid bonding sample surface. In the examples shown in Figure 3.43, a commonly adopted data flattening procedure produced the artificial bright straps shown in Figure 3.43a, and an improved data flattening procedure produced the artifact-free image shown in Figure 3.43b.

Lee et al. [63] also explored different modes of data collection in the study, including tapping, non-contact, and contact modes, and found that the tapping mode AFM is a good compromise to obtain recess, roughness, and other surface characteristics from multiple areas with relatively good consistency. When an artifact is suspected, Lee et al. recommended replacing the tip or testing the sample in contact mode. Lee et al. also recommended replacement of the tip after a certain amount of use, before the tip is worn out because a still functioning worn tip will result in an artificially low surface roughness reading [63].

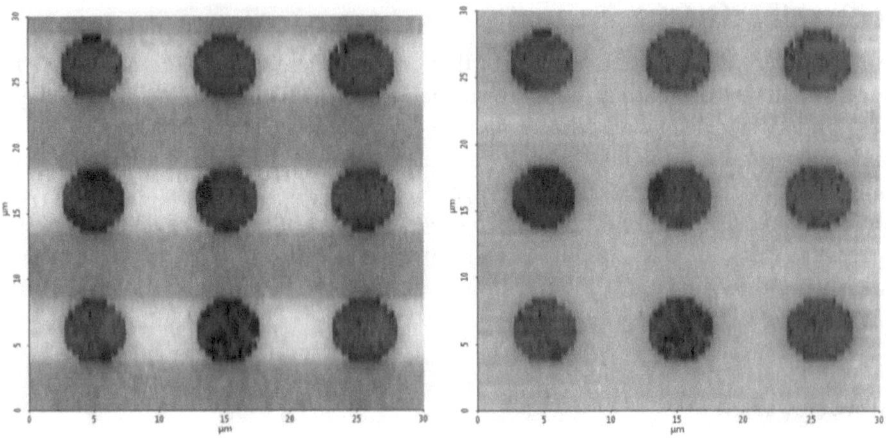

(a) (b)

FIGURE 3.43 A three-dimensional representation of a hybrid bonding sample surface from AFM scans. (a) Image with artificial bright stripes resulting from using a commonly adopted data flattening procedure. (b) Artifact-free image produced with improved data flattening procedure. (© 2019, IEEE. Reprinted with permission from Ref. [63].)

In addition, Lee et al. [63] discussed the issue of extraordinary artifacts that produced artificial Cu protrusion, as shown in Figure 3.44. This type of artifact is not widely known by most AFM users, and misinterpreting the data may result in the unnecessary rejection of good materials. Lee et al. did not pinpoint the origin of this artifact, believed to be related to the difference between the tip–Cu interaction and the tip–SiO_2 interaction in certain circumstances. As a precaution, Lee et al. recommend paying attention to the dielectric roughness over the course of measurements. If the data image is unusually blurry and the oxide roughness is much smaller than normal, it may be a sign of the artifact. Trouble-shooting approaches recommended

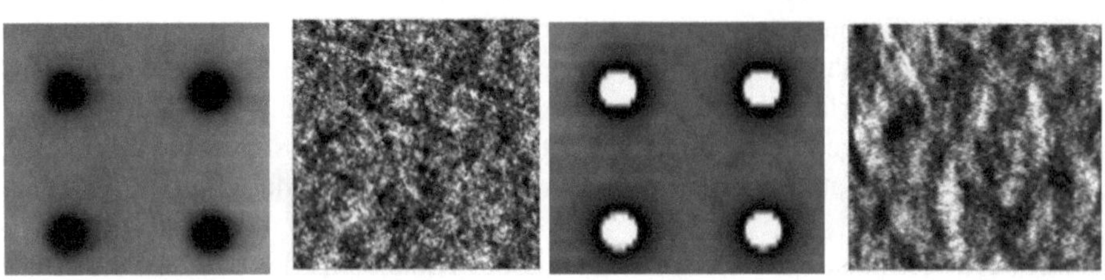

(a) (b) (c) (d)

FIGURE 3.44 AFM artifacts due to abnormal tip-to-sample interaction. (a) Normal AFM result from a DBI sample with recessed Cu areas. (b) Normal AFM result from a small SiO_2 area of the same sample. (c) Erroneous AFM result from the sample, showing protruding Cu areas. (d) AFM result from a small SiO_2 area taken after the artifact was observed. The image is blurry, and the resulting roughness is lower than normal. (© 2019, IEEE. Reprinted with permission from Ref. [63].)

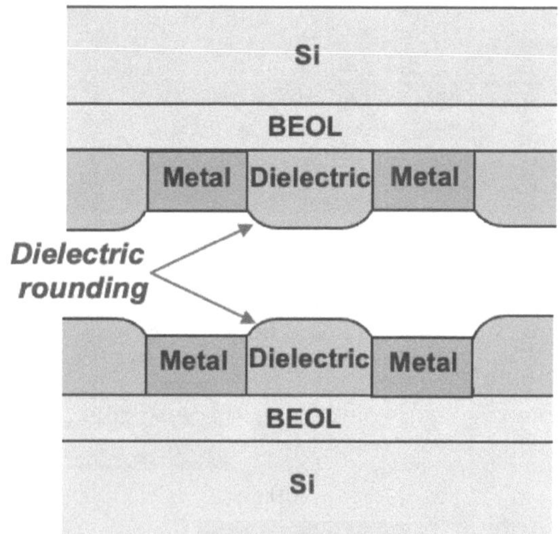

FIGURE 3.45 Illustration of DBI bonding surface with dielectric rounding at the DBI pad to dielectric edge. (© 2019, IEEE. Reprinted with permission from Ref. [63].)

by Lee et al. include replacing the tip or testing with contact mode to confirm if the protrusion is true or an artifact [63].

Another subject addressed in Lee et al.'s paper [63] is the erosion of oxide at the bonding pad edge. Lee et al. termed such erosion "dielectric rounding," as illustrated in Figure 3.45. Large dielectric rounding can reduce the area of bonded dielectrics and lead to bonding seams, as shown in Figure 3.46a and b. The dielectric rounding can be reduced and the bonding seams can be minimized or eliminated through the optimization of the CMP process.

AFM has a resolution of a fraction of a nanometer. However, this measurement is time-consuming and collecting large sample sizes is very expensive. As a compromise between speed and data accuracy, 3–10 points on a wafer can be measured to study the rough relationship between the Cu recess and the radial location on a wafer.

Mirkarimi et al. [41] reported a new technique using an optical profiler in the phase-shift interferometry mode to obtain the Cu pad recess statistics from approximately 10 million Cu pads, or ~3% of all Cu interconnects on the entire wafer used in the study. They first established correlation between the optical interferometer data and AFM data measured on the same sample, and discovered that the data measured by optical interferometry differ from that of AFM measurement, but those values showed a linear relationship to the true recess measured by AFM over the range of recesses measured. The cause of this offset is the different degree of phase change on reflection [41].

An example of optical interferometry data collected from a 200 mm wafer is shown Figure 3.47, with each data point representing the average recess within a small area (about 160 μm × 140 μm) that contains about 420 Cu pads. The maximum recess variation across the entire wafer was 4.5 nm, and the within-die recess variation was only 0.3 nm. The results show that optical interferometry has the potential

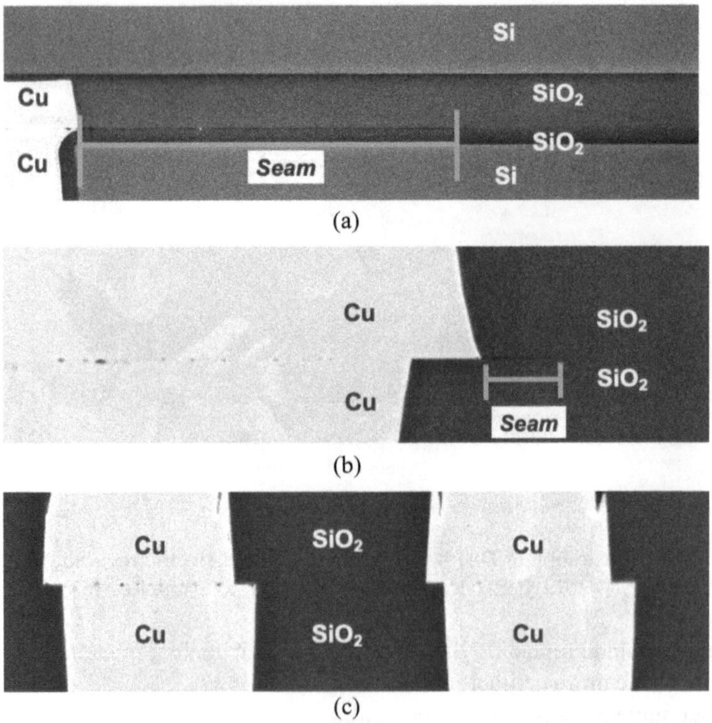

FIGURE 3.46 Cross-sectional SEM images of DBI pairs. (a) Significant non-bonded SiO_2 area (seam) near the Cu pads. (b) Minimal seam during bond. (c) DBI pair without visible seams due to optimized CMP conditions. (© 2019, IEEE. Reprinted with permission from Ref. [63].)

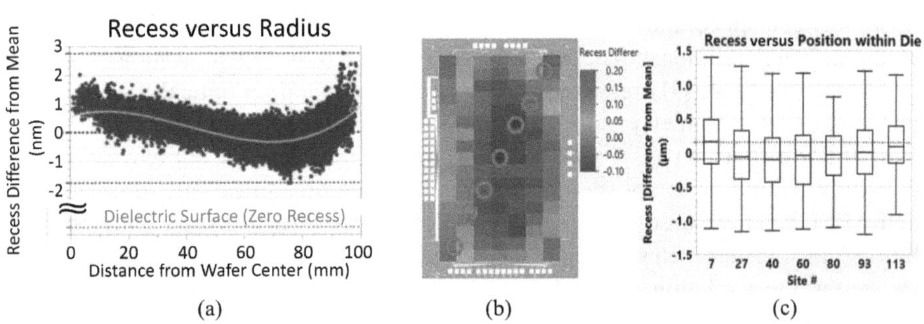

FIGURE 3.47 (a) Cu recess versus distance from the wafer center (within wafer variation) showing a 4.5 nm difference between the maximum (red line) and minimum (blue line) recess. (b) Map of sites measured within each die to determine within-die variation. (c) cu recess variation across a die diagonal, showing a 0.3 nm difference between the edge mean (red) and center mean recess (blue). Cu recess was measured using an optical profiler in phase-shift interferometry mode. (© 2023, IEEE. Reprinted with permission from Ref. [41].)

to be used as a quality control tool for HVM with a much larger sample size than what is possible with AFM.

3.10.2 BE Measurement

3.10.2.1 BE Measurement in W2W Bonding

From the very beginning of the development of wafer bonding, the bond strength, commonly referred to as BE, has been considered one of the most important characteristics to determine. BE measurement has contributed to the understanding of chemical reactions behind the bonding process.

Vallin et al. [64] did a comprehensive review of test methods for BE. The most used method is the double cantilever beam (DCB) method, first introduced to bonded wafers by Maszara et al. [65]. In this method, the bonded wafers are forced apart by insertion of a thin blade into the bonding interface (see Figure 3.48), and the resulting crack length is used to calculate BE. Si wafers are opaque in the visible range, but transparent in the infrared range; therefore, an infrared camera or infrared microscope is frequently used to detect the crack. A ruler is placed over the wafer to enable measurement of the crack length.

Based on Maszara et al.'s model [65], the fracture toughness γ, or the BE as it is often called, can be derived from the following equation:

$$\gamma = \frac{3E^* u^2 w^3}{8l^4} \tag{3.4}$$

Here:

E = effective Young's modulus
u = blade thickness
w = wafer thickness
l = crack length

Maszara et al.'s model [65] assumes that the insertion of the blade only gives rise to bending stresses in the beams, while neglecting the shear stress and the stress in the uncleaved part of the DCB specimen. Several augmented models that include the shear stress and the elastic restraint have been published, and Figure 3.49 shows a

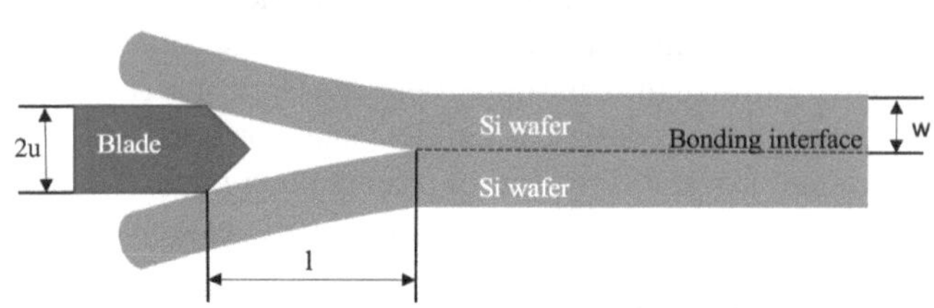

FIGURE 3.48 Illustration of a double cantilever beam test setup. The wafers are forced apart by insertion of a thin blade (the wedge) into the bonding interface. (2u is the blade thickness, l is the resulting crack length, and w is the wafer thickness.)

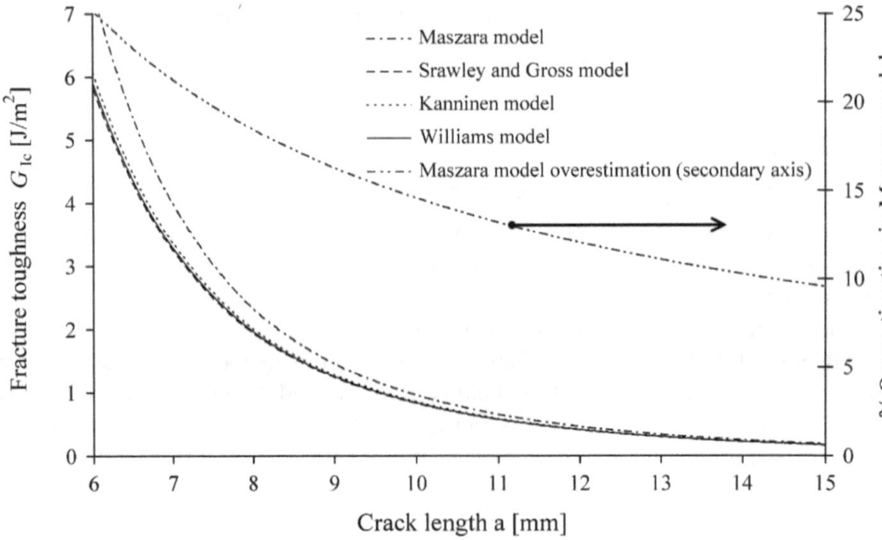

FIGURE 3.49 Comparison of fracture toughness (BE) calculated by five different models from the same measurement data. The models used are the Maszara model, the Srawley and Gross model, the Kanninen model, and the Williams model. (Used with permission of Elsevier from Ref. [64].)

comparison of BE values calculated using different models. The values were calculated for 525 μm thick {1 0 0} silicon with the beam and the crack in {100}-directions, and the blade thickness was 50 μm. The percentage of overestimation by Maszara et al.'s model is shown by the curve and axis to the right [64]. Compared to other augmented DCB models, Maszara et al.'s model overestimates BE by ~10% for BE values up to 3 J/m^2. To date, the most BE data in the literature are calculated using Maszara et al.'s model.

The DCB method is widely used due to the easiness of sample preparation and testing. A bonded pair of wafers can be directly tested without additional sample preparation. No expensive instrument is required for the blade insertion, measurements are fast, and large sample size can be measured in a short period of time with good repeatability. However, reproducibility among the published data is much harder to assess. Vallin et al. [64] analyzed the origin and mitigation of measurement errors. A good practice is to have a fixture for the blade insertion, and a setup with an infrared camera to minimize the time delay between blade insertion and crack length reading. When developing processes at multiple sites for bonding, such as cleaning and plasma activation processes, instead of relying on published data, samples with known BE should be used to calibrate the test method to establish baselines.

3.10.2.2 BE Measurement in D2W Bonding

While the bonding interface quality of W2W bonding BE can be analyzed relatively easily with BE measured with the DCB method, techniques to evaluate the quality of a D2W bonding interface are not well established. Lee et al. [66]

developed and evaluated multiple mechanical procedures to analyze the bond quality in D2W samples and MEMS cavity samples, including three-point bending coupled with finite element analysis (FEA); die shear test, die pull test, and die scratch test. They showed that a three-point bend test and subsequent FEA procedure could obtain the D2W BE with rectangular samples approximately the size of a DRAM die, while data from the other three tests either showed large scatter or are hard to quantify.

Lee et al. [66] verified that the new D2W method and the W2W DCB method resulted in equivalent BE values. They bonded two pairs of silicon wafers with thermal oxide on the bonding surface at the same time and annealed both pairs at 150°C for 15 minutes. They measured BE of both pairs by the DCB test and obtained an average BE of 3.2 J/m^2 with standard deviation = 0.6 J/m^2. The value is consistent with the values of about 2.5–3 J/m^2 after 150°C annealing, reported by Enquist [67], and is slightly higher than the fracture energy ~2.3 J/m^2 of bulk silicon [68]. After that, they took one pair of wafers and ground the top wafer to 200 μm thickness, then pattern and etched the top wafer into rectangular islands of 8 × 12 mm size, and diced the bottom wafer to generate samples shown in Figure 3.50. Using a three-point bending test combined with FEA, they obtained an average BE of 3.3 J/m^2, which matched the value measured on the same wafer pair by the DBC method. This study demonstrates that this 3-point bend test and FEA method can obtain valid and consistent results from the D2W geometry.

With the test method validated, Lee et al. [66] measured BE of D2W bonded samples with the same geometry and bonding dielectric. The measured BE turned out lower than the value from the W2W bonded samples. The lower BE can potentially be attributed to additional steps that the die went through, including protective coating, dicing, chemical cleaning, and P&P procedures.

The applicability of the three-point bending test is limited to medium die sizes, such as the 8 × 12 mm die used in Lee et al.'s test, and larger. It is not practical to do such a test on small die such as the 0.4 × 0.4 mm or 3 × 3 mm die bonded in this work. In addition, the test requires the bottom piece of the test sample to be significantly longer than the top piece. Therefore, it is not suitable for characterizing real products, which do not have proper geometry to perform the three-point bending.

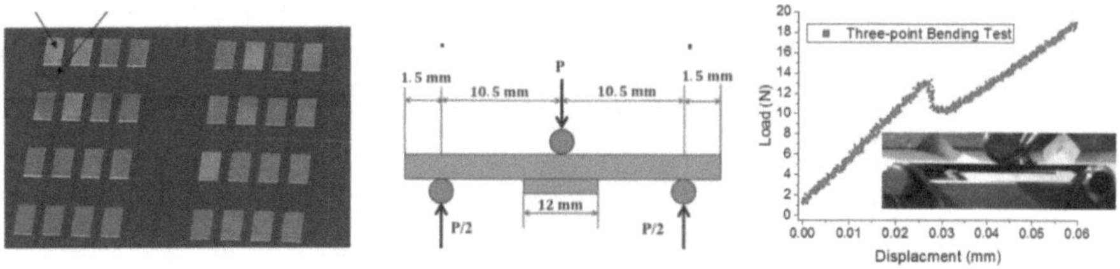

FIGURE 3.50 Bend tests for D2W BE analysis. (a) Photo of a D2W sample after bonding, annealing, and wafer cutting. (b) Three-point bend test configuration. (c) The typical load–displacement curve obtained from a three-point bend test. (© 2018, IEEE. Reprinted with permission from Ref. [66].)

3.11 HYBRID BONDING ADVANTAGES

3.11.1 Advantage of DBI Over Thermal Compression Solder Microbump Technology

As shown in Figure 3.1, the total variation of the solder bump height that can be tolerated by the microbump TCB technology is only 7 μm at 45 μm bonding pitch, and shrinks to 1 μm at a 20 μm bonding pitch. However, height variation of solder-capped microbumps across a wafer is on the order of several micrometers, inherent in its manufacturing process.

Compared to solder-capped microbumps, DBI bonding has three key advantages, as given by Gao et al. [69]: (1) superior planarity of bonding pads; (2) naturally formed hermetical seal around the interconnects, which prevents Cu oxidation during annealing and through the entire service life of the device; (3) comparable bonding throughput to mass reflow flip chip technology.

The total height variation between the hybrid bonding pads across a wafer is on the order of nanometers, which is 1,000× improvement over solder microbumps. As shown in Figure 3.51, for hybrid bonding, the reference surface is defined by the oxide-to-oxide bonded interface, and the height variation of Cu pads is the sum of metal recess for the top and bottom copper pads. As shown in Figures 3.10–3.12, the total variation adds up to a few nanometers, thanks to the optimized CMP process, and the gap is bridged by the expansion of the Cu pads during the final anneal process accompanied by simultaneous Cu diffusion across the interface to form permanent Cu-to-Cu bonding.

Compared to the TCB microbump technology, D2W hybrid bonding provides a pathway to generations of further pitch scaling, while offering additional technical advantages such as package height reduction; W2W hybrid bonding at 0.4 μm pitch has been demonstrated [14]. As we discussed earlier, the current pitch limit for D2W bonding is 3 μm, with a pathway to 1 μm in the near future [56].

Figure 3.52 shows a cross-sectional comparison of solder microbump and hybrid interconnects. The stand-off height of a solder microbump is 20–40 μm; however, with DBI Ultra D2W bonding, the height is reduced to zero.

The third advantage is improvement in reliability. The underfill material has much higher CTE than silicon and silicon oxide, and such a large CTE mismatch drives crack initiation and growth within the interconnect during temperature cycling. Additionally, intermetallic formation between the solder and the Cu leads to additional void growth under high current load. Hybrid bonding eliminates both problems by eliminating the solder and underfill from the bonded structure. The

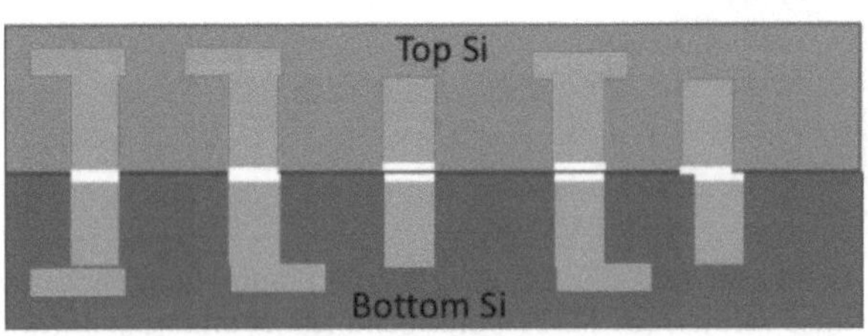

FIGURE 3.51 Illustration of the cu pad height variation at a hybrid bonding interface [70].

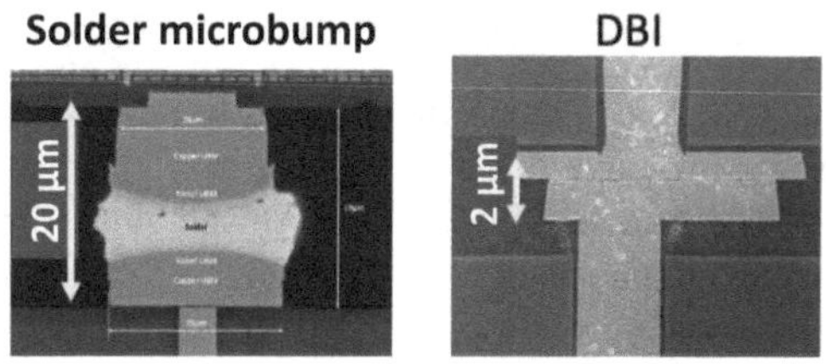

FIGURE 3.52 Comparison of a solder-capped microbump joint and a DBI joint.

hermetic seal formed naturally during hybrid bonding is also beneficial, as it keeps the Cu pads free from oxygen during the annealing process and reduces the annealing temperature for Cu–Cu bonding, and it also protects the Cu interconnect from further oxidation and corrosion in service.

Table 3.5 compares the fabrication steps to add the bonding layers for solder microbump and hybrid bonding surfaces for the wafer front side and the backside. Hybrid bonding offers significant process simplification. For the device side bonding surface fabrication, a CMP process replaces the following process steps in the microbump fabrication: Ni plating, solder plating, photoresist strip, wet etch, and solder reflow. For the backside processing, the following process steps are needed for the microbump die: adhesion layer deposition, seed layer deposition, the lithographic process for pad definition, Cu electroplating, Ni electroplating, photoresist strip, and wet etch for seed layer and adhesion layer removal. For hybrid bonding, none of the steps are necessary because the planarized TSV can be directly bonded; in total, four module steps are eliminated from the front-side fabrication and six module steps eliminated from the backside fabrication. The significant process savings in wafer-level fabrication coupled with the elimination of the underfill process is attractive from a cost perspective.

TABLE 3.5
Comparison of Process Steps for Fabrication of the Front and Back Bonding Layers for Microbump TCB versus DBI

Module	Frontside Process		TSV Side Processing	
Steps	**Microbump TCB**	**DBI**	**Mcrobump TCB**	**DBI**
1	Passivation opening	Yes	TSV reveal and planarization	Yes
2	Lithography	Yes	UBM (adhesion layer, seed layer)	NO
3	Adhesion layer, Cu seed layer	Yes	Lithography	NO
4	Cu electroplating	Yes	Cu electroplating	NO
5	Ni electroplating	NO	Ni electroplating	NO
6	Solder electroplating	NO	Resist strip	NO
7	Resist strip	NO	Wet etch (Cu + adhesion layer)	NO
8	Wet etch (Cu + adhesion layer)	NO		
9	Solder reflow	NO		
10	Cu CMP: NO	Yes		

3.11.2 Advantages of Hybrid Bonding over Cu–Cu TCB

A competing approach to low-temperature hybrid bonding is Cu–Cu TCB, which bonds Cu pillars protruding above the Si surface at high temperatures and with a high pressure. This approach is frequently confused with the solder microbump TCB; however, it has no solder at the bonding interface. It is sometimes confused with hybrid bonding because both approaches form direct Cu–Cu joints. However, the two processes are quite different.

Some fundamental challenge facing the Cu–Cu TCB is shown in Figure 3.53. Photoresist patterning is normally used for the fabrication of Cu pillars for TCB, and a height variation of >1 μm is common, inherent to the fabrication process. During the bonding process, the bond interface is defined by the highest pillar.

Unlike a soldering process, which relies on solder melting to compensate for the height differences, the Cu–Cu thermal compression process relies on deformation of Cu under higher temperatures (commonly above 350°C) and high pressure (commonly >250 MPa) for a duration of at least 15 minutes. Cu oxidizes rapidly at the temperatures required for TCB. At clean and dry ambient conditions, a clean Cu surface forms a layer of oxide which is approximately 2–3 nm thick [70]. However, at 300°C, 1-minute exposure results in a Cu oxide thickness of 300 nm, and 30 minutes of exposure resulted in 450nm thick oxide layers [71]. Therefore, an inert or reducing environment is required for bonding. This translates into high costs for HVM. The high pressure can also fracture thin dies due to the non-uniform distribution of stresses at high and low Cu posts.

The Cu pillar height variation can be reduced with additional processing such as diamond bit fly cutting [72]. However, even after the process, the Cu pillar height variation across a 200 mm wafer was still 600 nm, which is 100× of Cu height variations from a CMP process.

Another disadvantage of the TCB process is its long bonding cycle. Various attempts have been made to increase the bonder throughput of the Cu–Cu TCB process. Two-stage bonding process has been reported [73]. In this process, the die is placed at a relatively low temperature and a short duration sufficient for it to stick. After all the dies are placed on the wafer, a second-stage bonding process at a higher temperature and pressure is used to form the Cu–Cu contacts collectively across the entire wafer. This process change may enhance the bonding throughput somewhat, but has no effect in compensating for the Cu height variation.

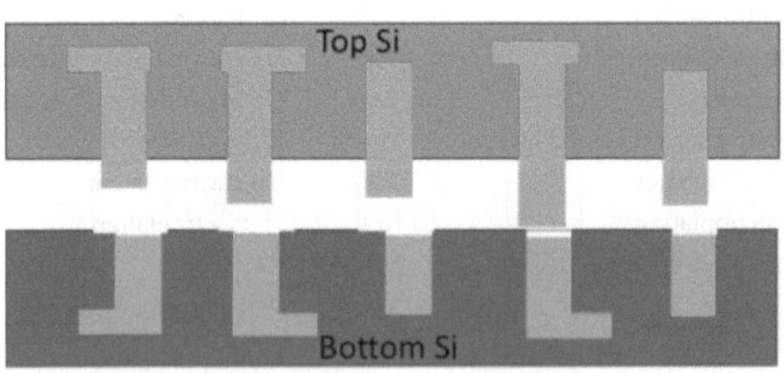

FIGURE 3.53 Illustration of Cu pillar height variation at the bonding interface of a TCB process [69].

3.11.3 Hybrid Bonding Performance Advantages

Performance advantages achievable through hybrid bonding are product dependent as well as bonding pitch dependent, as the degree of performance enhancement with hybrid bonding is largely tied to the bonding pitch. Early simulation and bonding results using hybrid bonding as a drop-in replacement for solder microbumps represent the minimum performance enhancement. As the bonding pitch shrinks, performance enhancement will greatly accelerate.

For 3D stacking applications with multiple layers of die with hybrid bonding, removing solder and underfill from the package results in enhanced electrical and thermal performance. In 2017, Agrawal et al.'s simulation [74] showed that a 5 μm hybrid bond pad has 96% less capacitance, 92% less inductance, and 64% less resistance than a typical 10 μm solder microbump and greatly reduces latency. In the simulation of 4- and 8-high dynamic random access memory (DRAM)-like configured stacks, the temperature differential (ΔT) between die 1 and die 8 is only 4°C in the hybrid-bonded stack; however, for the stack with microbumps, the ΔT is 28°C, as shown in Figure 3.54.

Kim et al. presented an 8 Hi HBM cube with hybrid bonding that showed a 22% reduction in thermal resistance compared to an 8 Hi HBM cube with solder microbump connections of the same bonding pitch [49]. Chen et al. [75] showed that when the bonding pitch is reduced from 40 μm with solder microbump to 9 μm with hybrid bonding, the interconnect density increases by 16×. At 2 GHz, the hybrid bonded package outperforms the microbump package by 12.5× in parasitic resistance, by 100× in self-inductance, and by 12.5× in parasitic capacitance. That translates into 156× improvement in RC delay, and 12.5× improvement in IR drop. As for thermal/power performance, the hybrid bonded package outperforms the solder microbump package by 11× in energy consumption/bit data, and by 1.25×–2× in thermal resistance, respectively, in normalized scale.

The above results are comparable to the 15× interconnect density increase and 3× energy efficiency performance (EEP) improvement reported by AMD on their V-cache with 9 μm pitch hybrid bonding [13].

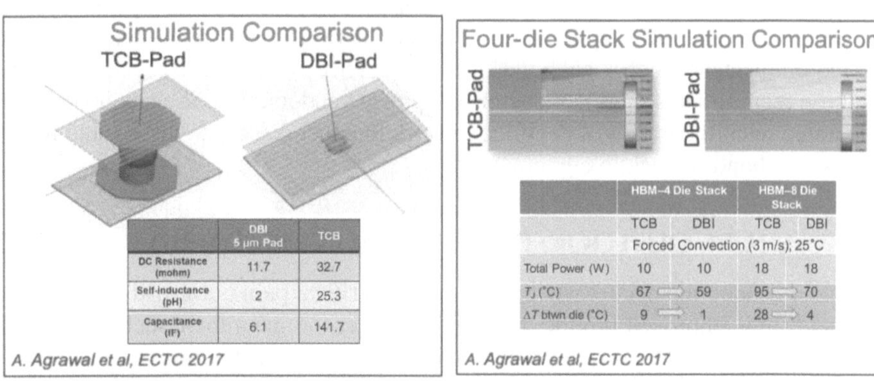

FIGURE 3.54 Simulation schematics for TCB solder microbumps and DBI interconnects with four-die and eight-die stacks.

As the pitch scaling progresses, more performance advantage is expected. As described by Chia et al. [14], the EEP will increase by another 2,824× when the bonding pitch is scaled down from 9 μm in the above analysis to 0.4 μm.

3.12 SUPPLY CHAIN READINESS FOR HVM ADOPTION

The ecosystem needed to ramp hybrid bonding technology into HVM is largely in place. The film deposition tools needed for the inorganic dielectric layer deposition are already in most foundries. The Cu Damascene process needed for the RDL and bonding layer fabrication is now the standard for wafer BEOL fabrication. Industry know-how to achieve tight control of surface roughness, dielectric erosion, and Cu recess is increasing rapidly. Plasma cleaning and activation tools suitable for hybrid bonding are readily available. Current dicing infrastructure including mechanical saw dicing, laser stealth dicing, and plasma dicing, can be used for D2W hybrid bonding. The bonder requirement for W2W bonding is very simple compared to other bonding processes that require heat, pressure, and a controlled atmosphere or vacuum. W2W bonding has been in HVM since 2015. For D2W bonding, cleaning systems for die preparation are available, and ultraclean, high-throughput bonders with submicrometer alignment accuracy have been introduced into the market.

Metrology tools and methodologies needed for hybrid bonding are also ready for deployment.

AFM for bonding surface qualification and monitoring is readily available, and measurement protocols for hybrid bonding surfaces have been developed [63]. BE measurement by DCB is commonly accepted by the industry and a large amount of published data exists. Optical inspection tools can be programmed to conduct defect inspection for D2W bonding. CSAM systems currently on the market work well for bonding void detection.

The areas for further development include increased metrology sampling throughout the assembly process for improved statistical analysis and data-driven process yield improvement. The integration of the hybrid bond layer on functional parts requires managing stress, warpage, as well as the topography requirements for high-yield bonding. This is a critical part of hybrid bond layer integration, essential for both W2W and D2W applications. The key is to improve the understanding of the requirements for high-yield bond and the influence of individual process modules on these requirements. Semiconductor equipment vendors, foundries, and vertically integrated manufactures will continue to work together closely to drive yield higher in this new bonding technology era.

3.13 FUTURE OUTLOOK: PROLIFERATION OF HYBRID BONDING

Technology adoption occurs when it addresses a critical pain point within the industry and brings improved product performance over the incumbent technology. The semiconductor industry is conservative and adopts new technologies that are costly only when the previous generation runs out of steam.

Since hybrid bonding is relatively new to the semiconductor industry, there were some initial potential engineering challenges to overcome and to ensure that the technology could be manufactured in step with the semiconductor industry supply chain and ecosystem. A few companies have led the charge to modify equipment sets for manufacturing D2W hybrid bonding today. These efforts will continue to expand as more vendors move to address the new requirements for hybrid bonding assembly lines.

Although the equipment sets are available today for manufacturing of hybrid bonding assembly of electronic devices, a deep understanding of the technology is not yet widespread within the industry. In fact, there are many misconceptions about what can be done today with hybrid bonding. As with any new technology, the industry's understanding of this technology will become stronger, more accurate, and unified with time.

The past few years have witnessed some general trends within the semiconductor industry on hybrid bonding adoption for manufacturing (Figure 3.55). The first wave of technology adoption was in W2W bonding where the ramp to manufacturing can be achieved within a relatively short period of time. Customers who practice W2W hybrid bonding often extend their suite of product applications into a D2W process technology. For example, adopting direct bond and hybrid bond interconnects has enabled Sony to offer enhanced performance for image sensors [76]; recently, Sony has implemented a D2W assembly technology for heterogeneous integration in their short-wave infrared sensor [77].

The adoption of hybrid bonding within the memory space is evident both in NAND and in DRAM. DRAM manufacturers began with D2W technology due to the main product focus. The intersection of the adoption of hybrid bonding for HBM is estimated to be in the 16-die stack module which adds complexity to the manufacturing ramp. Compound yield considerations in the qualification of this new technology present additional challenges for W2W hybrid bonding application. Nevertheless, Samsung reports that the only way to achieve the 16-die stack is with a hybrid bond approach because the thermal budget with the conventional microbump technology exceeds the allowed specification [78]. Kim et al. [78] explains that even with the enhanced thermal performance with hybrid bonding, power management will be required in the 16-die stack HBM planned for 2026.

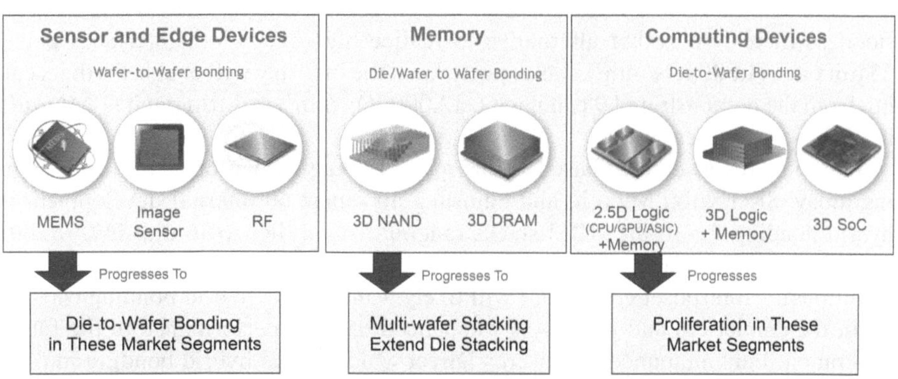

FIGURE 3.55 Hybrid bonding adoption trends.

The adoption of hybrid bonding in computing devices is gaining much attention within the industry. AMD is an early adopter of hybrid bonding in system-on-chip (SoC) modules, and the Ryzen and MI product modules tout 9 μm pitch D2W hybrid bond interconnects [79].

There are many features with hybrid bond interconnects that drive the implementation for various product lines due to enhanced performance. A few are discussed here as it lays the foundation for predicting the future application enabled with hybrid bonding. Image sensors, 3D NAND, and SoC (system on chip) all are driven by the advantages of separating circuit functions, optimizing separately and then recombining the functions through a hybrid bond interconnect. In the case of image sensors and 3D NAND, the CMOS logic is separated from the memory or image sensor wafer. Each wafer is optimized for the foundry process flow. Moving away from the monolithic approach frees the architect from the confines of the compromises of fabricating both CMOS and memory or image sensor on the same wafer. Kioxia [38] explained how they can bring 50% density increase to their eighth generation of NAND with hybrid bonding due to individual process improvements by separating the CMOS and NAND cell wafers (30%) and scaling improvement and architectural enhancements enabled with hybrid bonding (~15%) [38]. Along with the 50% higher density came an increased interface speed by 60% with 30% power efficiency improvement. Additionally, the interconnect between the logic and the NAND/sensor scales considerably over the prior technology, allowing for unique features such as pixel-level interconnect or high bandwidth in the case of memory–logic interfaces.

For D2W applications, the value proposition is slightly different. In addition to the benefit of disaggregation of the circuit functions which is important for SoC applications, there is a major thermal benefit to replace the microbump interconnect and underfill dielectric with the hybrid bond interconnect and the inorganic dielectric (with either SiO_2, Si_3N_4, or $SiO_xC_yN_z$). Both the scalability of the interconnect and the thermal performance enhancement result in a module with enhanced performance and the ability to reduce the mask steps in the preparation of the backside of the wafer.

We see adoption of hybrid bonding in four major packaging areas in advanced packaging in the next four years. Yole's report in High End Performance Packaging 2023 [2] shows that the hybrid bond interconnect primarily has a play in the finer interconnect in SoC and HBM memory stacking for packaging. This is well understood as there is no other alternative to realize high-bandwidth interfaces below 25 μm pitch (1,600 IOs/mm^2). It is expected that the industry will enable further scaling from the demonstrated 9 μm pitch (~12,000 IOs/mm^2) to 1 μm pitch (1 M/mm^2), in the next five to ten years.

Figure 3.56 was created based on the major packaging platforms in manufacturing today. SK Hynix, Micron, and Samsung all report on internal development in hybrid bonding for 16-die HBM stacks ranging from 4 dies to 16 dies [49, 78, 80]. Samsung has announced their intent to have hybrid bonded HBM products in 2026. Additionally, interposer companies will likely want to have hybrid bonding capability in the chiplet era and possibly for enhanced thermal performance in the future generation. Intel announced a Foveros Direct which brings hybrid bonding into the silicon bridge space as well [81].

Although the first example of a D2W hybrid bond product was a bumpless SoC assembled by TSMC at a 9 μm pitch [13], it appears that the proliferation profile is

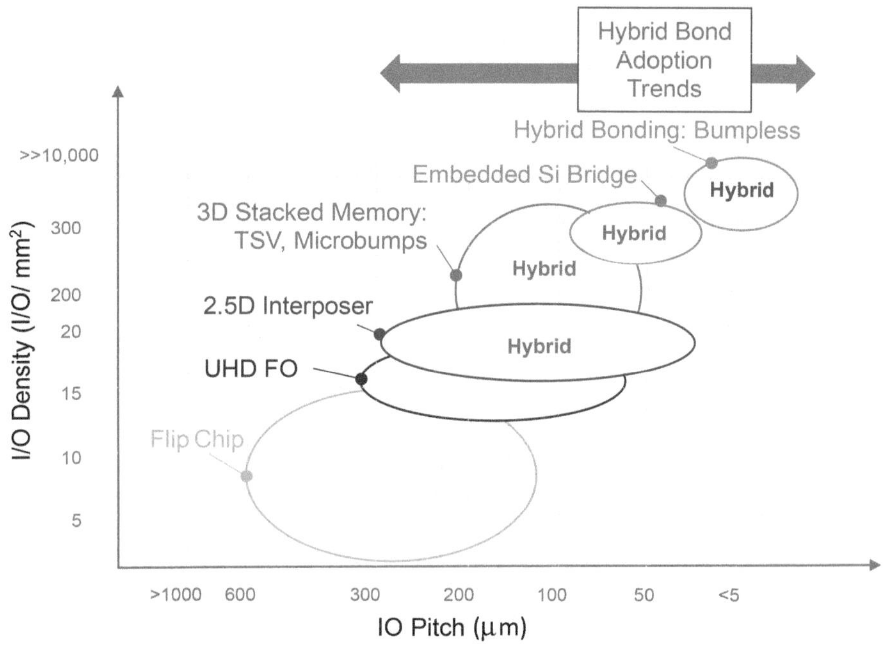

FIGURE 3.56 IO density versus IO pitch showing the major packaging options in manufacturing today and where hybrid bonding will participate. The blue text is where products exist today for very fine pitch applications, and the black text (including embedded Si bridge, 3D. and 2.5D interposer spaces, in the order of high to low IO pitch) are the planned adoption trends based on public disclosures.

toward both finer pitches in SoC and larger pitches to the 35 µm TSV interconnect in HBM devices and interposer. Once the hybrid bonding toolbox is adopted and well understood, further proliferation is expected due to the enhanced electrical, thermal, and reliability performance that can be achieved with this interconnect technology.

The chiplet era is expected to offer greater access of standard circuit functional chips and allow system architects to assemble modules for optimal system performance. Generally, access to these chipsets is cost-prohibitive, and thus only large semiconductor companies can access foundry services for their products. The important thing to note here is that there will be a mix of standard solder-capped microbump and hybrid bond interconnect; however, for performance reasons, the eventual migration to hybrid bonding will be substantial.

A large portion of the growth in the electronics industry is in HPC, AI chips for more efficient computation or decision-based computing, and memory–logic interfaces, due to the demand for more data, more bandwidth, and more compute power at the server and at the point of use (Figure 3.56). This also suggests the possibility of seeing hybrid bonding migration to some mobile products [2]. In the fourth phases of market adoption, we anticipate the hybrid bond interconnect extending beyond these applications to a pervasive heterogeneous integration phase that spans other segments within the semiconductor industry (Figure 3.57).

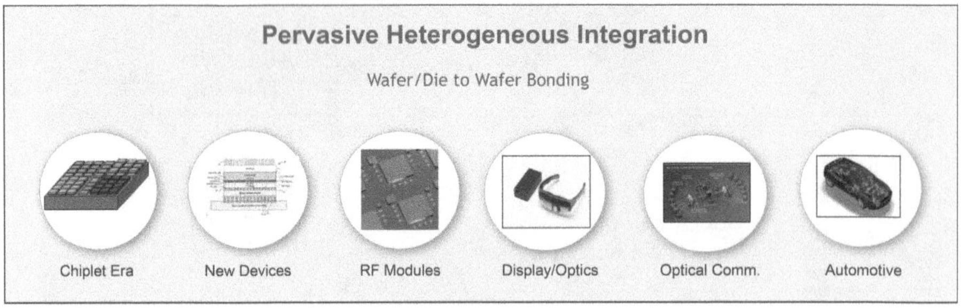

FIGURE 3.57 The next wave of hybrid bonding adoptions will be toward a pervasive heterogeneous integration which address many different market segments.

Another area where hybrid bonding could easily be adopted for the performance enhancements is in the lower IO density, high-speed modules in RF. The low inductance L, capacitance C, and resistance R of these chip-to-chip interconnects are attractive for the high-speed RF module space where the dies are small, IO are few, and the pad size is still relatively large. The desire to shrink the x–y footprint or squeeze in more functions in the same area for cost savings applies to all electronic products. Hybrid bonding allows significant x–y footprint saving with no underfill bleed-out and with the compact nature of hybrid bonding and zero standoff of the interconnect. RF will find many advantages to implement hybrid bond interconnects within that segment of the market.

The automotive industry will find value in a higher reliability performance of the hybrid bond interconnect as well as the high-temperature performance needed for electronics near hot surfaces such as batteries. Early work [7] focused on Ni metal interconnect that would sustain good operation at very high temperatures. Enhanced reliability, compute at the edge, and high-temperature performance are important in electronic components within the automotive industry.

Hybrid bond interconnection is well suited to enhance any heterogeneous integration applications. One area of interest in the past few years has been display applications, such as micro-LEDs, AR/VR, and more. In this application, the III–V material is bonded to the CMOS drivers to create a display. Additional heterogeneous integration application areas such as new devices with high density capacitors bonded to the functional CMOS circuits have been anticipated. If the compatibility with a CMOS process challenge is removed from the equation, architects may consider bringing new materials from completely different process flows together and bonding them at low temperatures to create new devices.

Overall, the implementation of a hybrid bond interconnect in the semiconductor industry will continue to grow. The market drivers are performance enhancements in product lines with adoption, reduced costs of process through supply chain management with disaggregation, and the enabling of novel devices. The segments which adopted this technology first are high-performance compute, artificial intelligence, and HBM where higher bandwidth, dense interconnect interfaces are required. 3D NAND is using hybrid bonding to overcome the limitations with monolithic stacking and secure future generations of performance improvement. Display technology challenges are solved by bringing CMOS circuits to drive III–V light emitters. Augmented reality (AR) and virtual reality (VR) have constraints that can be eliminated by disaggregated manufacturing of components and then assembled with hybrid bonding. Finally, novel devices will emerge in the hybrid bond era, in the future as system architects and process engineers separately optimize components, assemble at low temperatures to form an electrical interconnect at the bond interface.

The demand for enhanced performance and functionality is driving the semiconductor industry toward heterogeneous integration, and hybrid bonding is a key technology building block to usher in this new era.

REFERENCES

1. Laura Mirkarimi, "3D Chiplet Integration with Hybrid Bonding," Chip Scale Review, March–April, 2023. Available at: https://chipscalereview.com/wp-content/uploads/2023/04/ChipScale_Mar-Apr_2023-digital.pdf
2. Yole Intelligence, "High-end Performance Packaging 2023 Market and Technology Report," May 2023. Available at: https://www.yolegroup.com/product/report/high-end-performance-packaging-2023.
3. S. Naffziger, K. Lepak, M. Paraschou, M. Subramony, "2.2 AMD Chiplet Architecture for High-performance Server and Desktop Products," 2020 IEEE International Solid-State Circuits Conference (ISSCC), San Francisco, CA, 2020, pp. 44–45.
4. L. Mirkarimi, A. Nuruzzaman, "A New Era of Computing Performance with Hybrid Bonding," Chip Scale Review, July–August 2021. Available at: https://chipscalereview.com/wp-content/uploads/2021/02/ChipScale_July-Aug_2021-digital.pdf
5. G. Gao, G. Fountain, P. Enquist, C. Uzoh, L. Wang, S. McGrath, B. Lee, W. Subido, S. Arkalgud, L. Mirkarimi, "Direct Bond Interconnect (DBI®) Technology as an Alternative to Thermal Compression Bonding," Proceedings of the 2016 International Wafer Level Packaging Conference (IWLPC), October 2016, San Jose, CA.
6. https://www.eetimes.com/advanced-packaging-technologies-overcoming-the-memory-system-performance-and-capacity-limitation/
7. P. Enquist, "High Density Direct Bond Interconnect (DBI) Technology for Three Dimensional Integrated Circuit Applications in MRS Proceedings, vol. 970, p. 0970-Y01-04, 2006. doi: 10.1557/PROC-0970-Y01-04.
8. J. Obreimoff, "The Splitting Strength of Mica," in Proceedings of the Royal Society of London. Series A, Containing Papers of a Mathematical and Physical Character, vol. 127, no. 805, pp. 290–297, 1930.
9. L. Rayleigh, "A Study of Glass Surfaces in Optical Contact," in Proceedings of the Royal Society of London. Series A, vol. 156, pp. 326–349, 1936.
10. Q. Y. Tong, G. G. Fountain, P.M. Enquist, "Method for Low Temperature Bonding and Bonded Structure," US Patent 6,902,987, June 2005.
11. P. Enquist, "Low Cost of Ownership Scalable Copper Direct Bond Interconnect 3D IC Technology for Three Dimensional Integrated Circuit Applications", Proceedings of the 2009 IEEE International Conference on 3D System Integration, 2009.
12. https://www.3dincites.com/2015/03/ziptronix-licenses-dbi-hybrid-bonding-patents-sony-advanced-image-sensor-applications/
13. https://www.amd.com/en/technologies/3d-v-cache
14. H. J. Chia, S. P. Tai, J. J. Cui, "Ultra High Density Low Temperature SoIC with Sub-0.5 μm Bond Pitch," 2023 IEEE 73rd Electronic Component and Technology Conference (ECTC), p. 1, 2023.

15. C. Netzband, K. Ryan, Y. Mimura, "0.5 µm Pitch Next Generation Hybrid Bonding with High Alignment Accuracy for 3D Integration," 2023 IEEE 73rd Electronic Component and Technology Conference (ECTC), p. 1100, 2023.
16. T. Sherwood, R. Patlolla, J. Salfelder, "Aggressive Pitch Scaling (Sub-0.5 µm) of W2W Hybrid Bonding through Process Innovations," 2023 IEEE 73rd Electronic Component and Technology Conference (ECTC), p. 13, 2023.
17. Q. Y. Tong, U. Go¨sele, "A Model of Low-temperature Wafer Bonding and Its Applications," in Journal of the Electrochemical Society, vol. 143, no. 5, pp. 1773–1779, 1996.
18. V. Masteika, J. Kowal, N. St. J. Braithwaite, "A Review of Hydrophilic Silicon Wafer Bonding," *ECS Journal of Solid State Science and Technology*, vol. 3, no. 4, 2014.
19. Q.-Y. Tong, G. Cha, R. Gafiteanu, U. Go¨ sele, "Low Temperature Wafer Bonding," in Journal of Microelectromechanical Systems, vol. 3, no. 1, pp. 29–35, 1994.
20. G. Gao et al., "Low Temperature Hybrid Bonding for Die to Wafer Stacking Applications," 2021 IEEE 71st Electronic Components and Technology Conference (ECTC), p. 383, 2021.
21. M. Aimmerman, C. Lee, "Plastic Air Cavity GFQ Coupled with an Innovative Ultrasonic Lid Process Achieved True Hermetic Performance." Available at: https://iqlp.net/wp-content/uploads/2018/08/Plastic-Air-Cavity-QFN-Coupled-with-an-Innovative-Ultrasonic-lid-Process-Achieves-True-Hermetic-Performance-2007.pdf
22. P.C. Andricacos, C. Uzoh, J.O. Dukovic, J Horkans, H. Deligianni, "Damascene Copper Electroplating for Chip Interconnections," in IBM Research Report, vol. 42, no. 5, p. 587, 1988.
23. X. F. Brun, M. M. Hasan, Y. Yang, "Characterization of 300 mm Low Temperature SiCN PVD Films for Hybrid Bonding Application", 2023 IEEE 73rd Electronic Component and Technology Conference (ECTC), P548, 2023.
24. K. Ma, N. Bekiaris, S. Ramaswami, "0.5 µm Pitch Wafer-to-wafer Hybrid Bonding with 3D-IC Applications," 2023 IEEE 73rd Electronic Component and Technology Conference (ECTC), p1110, 2023.
25. S. H. Hahn et al., "Contamination-free Cu/SiCN Hybrid Bonding Process Development for Sub-µm Pitch Devices with Enhanced Bonding Characteristics," 2023 IEEE 73rd Electronic Components and Technology Conference (ECTC), p1390, 2023.
26. K. Ryan, C. Netzband, A. Gildea, "Electrical Analysis of Wafer-to-Wafer Copper Hybrid Bonding at Sub-Micron Pitches," 2023 IEEE 73rd Electronic Component and Technology Conference (ECTC), p114, 2023.
27. Z.C. Wu, Y. C. Lu, C. C. Chiang, "Advanced Metal Barrier Free Cu Damascene Interconnects with PECVD Silicon Carbide Barriers for 90/65-nm BEOL Technology," Digest. International Electron Devices Meeting, San Francisco, CA, pp. 595–598, 2002.
28. F. Inoue et al., "Influence of Composition of SiCN as Interfacial Layer on Plasma Activated Direct Bonding," ECS Journal of Solid State Science and Technology, vol. 8, no. 6, pp. 346–350, 2019.
29. G. Gao et al., "Die to Wafer Hybrid Bonding: Multi-Die Stacking with TSV Integration," 2020 International Wafer Level Packaging Conference (IWLPC), San Jose, CA, October 2020.
30. C. Rudolph, H. Wachsmuth, P. Gansauer, T. Werner, M. Junghaehnel, G. Fountain, J. A. Theil, L. Mirkarimi, "HVM CMP Process Development for Advanced Direct Bond Interconnect (DBI)," The International Conference on Planarization Technology (ICPT) 2022, Portland, September 27–29, 2022.
31. P Enquist, "Low Temperature Direct Bond Technology for 3D Microelectronics Integration and Wafer Scale Packaging," IMAPS 43rd International Symposium on Microelectronics, 2010.

32. Q. Y. Tong et al., "Fluorine-enhanced Low-temperature Wafer Bonding of Native-oxide Covered Si Wafers," Applied Physics Letters, vol. 85, no. 17, pp. 3731–3733, 2004.
33. L. Wang et al., "Direct Bond Interconnect (DBI®) for Fine-pitch Bonding in 3D and 2.5D Integrated Circuits," 2017 Pan Pacific Microelectronics Symposium (Pan Pacific), Kauai, HI, 2017.
34. L. D. Cioccioa, P. Gueguena, R. Taibib, "An Overview of Patterned Metal/Dielectric Surface Bonding: Mechanism, Alignment and Characterization," in *ECS Transactions*, vol. 33, no. 4, pp. 3–16, 2010.
35. G. Gao, et al., "Development of Hybrid Bond Interconnect Technology for Die-to-wafer and Die-to-die Applications," 2017 International Wafer Level Packaging Conference (IWLPC), San Jose, CA, October 2017.
36. S. Wuu et al., "A Review of 3-Dimensional Wafer Level Stacked Backside Illuminated CMOS Image Sensor Process Technologies," *IEEE Transactions on Electron Devices*, vol. 69, no. 6, pp. 2766–2778, 2022.
37. https://www.yolegroup.com/product/report/ymtc-232-layer-3d-nand-memory/
38. J. Ohshima, "The Future of Flash Memory," presented at Flash Memory Summit, Santa Clara, CA, 2023.
39. A. Ilkbahar, "Reinvigorating Virtuous Cycle of Growth for NAND Flash", presented at Flash Memory Summit, Santa Clara, CA, 2023.
40. G. Gao et al., "Low Temperature Cu Interconnect with Chip to Wafer Hybrid Bonding," 2019 IEEE 69th Electronic Components and Technology Conference (ECTC), p. 628, 2019.
41. L. Mirkarimi et al., "Fine Pitch Die-to-Wafer Hybrid Bonding," 2023 IEEE 73rd Electronic Component and Technology Conference (ECTC), p. 83, 2023.
42. A. Kawai, "Dicing, Grinding and Polishing," in Handbook of 3D Integration, 3D Process Technology, vol. 3, P. Garrou, M. Koyanagi, and P. Ramm, Eds., Wiley-VCH, pp. 241–260, 2014.
43. https://technology.discousa.com/method/dicing/
44. https://www.disco.co.jp/eg/solution/library/laser/stealth.html. (Stealth Dicing™ is a registered trademark of Hamamatsu Photonics.)
45. https://www.disco.co.jp/eg/solution/library/other/plasma.html
46. G. Gao et al., "Die to Wafer Hybrid Bonding for Chiplet and Heterogeneous Integration: Die Size Effects Evaluation-Small Die Applications," 2022 IEEE 72nd Electronic Components and Technology Conference (ECTC), p. 1975, 2022.
47. J. Theil, et al., "Analysis of Die Edge Bond Pads in Hybrid Bonded Multi-Die Stacks" 2022 IEEE 72nd Electronic Components and Technology Conference (ECTC), p. 130, 2022.
48. T. Kamei, et al., "Development of High Reliability 6 μm-pitch Cu–Cu Connections Using over-400 mm^2-large Chip on Wafer Bonding Process," 2023 IEEE 73rd Electronic Component and Technology Conference (ECTC), p. 556, 2023.
49. K. Kim et al., "C2W Hybrid Bonding Interconnect Technology for Higher Density and Better Thermal Dissipation of High Bandwidth Memory," 2023 IEEE 73rd Electronic Component and Technology Conference (ECTC), p. 1048, 2023.
50. C.H. Tsai et al., "Low Temperature SoIC™ Bonding and Stacking Technology for 12/16-Hi High Bandwidth Memory (HBM)" 2020 IEEE 70th Electronic Component and Technology Conference (ECTC), 2020.
51. M. Kim et al., "Multi-stack Hybrid Cu Bonding Technology Development Using Ultra-Thin Chips," 2023 IEEE 73rd Electronic Component and Technology Conference (ECTC), p. 1672, 2023.
52. S. Suhard et al., "Demonstration of a Collective Hybrid Die-to-wafer Integration," 2020 IEEE 70th Electronic Component and Technology Conference (ECTC), p. 1315, 2020.

53. S. Suhard et al., "Demonstration of a Collective Hybrid Die-to-wafer Integration Using Glass Carrier," 2021 IEEE 71st Electronic Component and Technology Conference (ECTC), p. 2064, 2021.
54. E. Bourjot et al., "Towards 5 μm Interconnection Pitch with Die-to-wafer Direct Hybrid Bonding," 2021 IEEE 71st Electronic Component and Technology Conference (ECTC), p. 470, 2021.
55. B. Brandstätter et al., "High-speed Ultra-accurate Direct C2W Bonding," 2020 IEEE 70th Electronic Components and Technology Conference (ECTC), p. 1943, 2020.
56. C. Scanlan, "Chip to Wafer Hybrid Bonding Opportunities and Challenges," presented at 2023 IEEE 73rd Electronic Component and Technology Conference (ECTC), 2023.
57. E. Beyne et al., "Scalable, Sub 2 μm Pitch, Cu/SiCN to Cu/SiCN Hybrid Wafer-to-wafer Bonding Technology," 2017 IEEE International Electron Devices Meeting (IEDM), San Francisco, CA, 2017, pp. 32.4.1–32.4.4.
58. L. Mirkarimi et al., "The Influence of Cu Microstructure on Thermal Budget in Hybrid Bonding," 2022 IEEE 72nd Electronic Component and Technology Conference (ECTC), p. 162, 2022.
59. P. Enquist, "Single Mask Via Device," US Patent 6867073, March 15, 2005.
60. Q.Y. Tong et al., "Method for Room Temperature Metal Direct Bonding," US Patent 6962835, November 8, 2005.
61. J. E. Messemaeker et al., "New Cu 'Bugle-Out' Mechanism Supporting Sub-micron Scaling of Hybrid Wafer-to-wafer Bonding," 2023 IEEE 73rd Electronic Component and Technology Conference (ECTC), p. 109, 2023.
62. G. Gao et al., "Development of Low Temperature Direct Bond Interconnect Technology for Die-to-wafer and Die-to-die Applications: Stacking, Yield Improvement, Reliability Assessment," 2018 International Wafer Level Packaging Conference (IWLPC), San Jose, CA, October 2018.
63. B. Lee et al., "Nanoscale Topography Characterization for Direct Bond Interconnect," 2019 IEEE 69th Electronic Component and Technology Conference (ECTC), p. 1041, 2019.
64. O. Vallin, K. Jonsson, U. Lindberg, "Adhesion Quantification Methods for Wafer Bonding," in *Materials Science and Engineering Reports*, vol. 50, no. 4-5, pp. 109–165, 2005.
65. W.P. Maszara, G. Goetz, A. Caviglia, J.B. McKitterick, "Bonding of Silicon Wafers for Silicon-on-insulator," in *Journal of Applied Physics*, vol. 64, pp. 4943–4950, 1988.
66. B. Lee et al., "Mechanical Strength Characterization of Direct Bond Interfaces for 3D-IC and MEMS Applications," 2018 IEEE 68th Electronic Component and Technology Conference (ECTC), p. 954, 2018.
67. P. Enquist, "Scalability and Low Cost of Ownership Advantages of Direct Bond Interconnect (DBI®) as Drivers for Volume Commercialization of 3-D Integration Architectures and Applications," Proceedings of the Materials Research Society Fall Meeting, 2008.
68. J.A. Hauch, D. Holland, M.P. Marder, H.L. Swinnery, "Dynamic Fracture in Single Crystal Silicon," in *Physical Review Letters*, vol. 82, p. 3823, 1999.
69. G. Gao et al., "Chip to Wafer Hybrid Bonding with Cu Interconnect: High Volume Manufacturing Process Compatibility Study," Proceedings of the International Wafer Level Packaging Conference (IWLPC), San Jose, CA, October 2019.
70. P. Keil, D. Lützenkirchen-Hecht, R. Frahm, "Investigation of Room Temperature Oxidation of Cu in Air by Yoneda-XAFS." Available at: http://www.slac.stanford.edu/econf/C060709/papers/144_WEPO79.PDF
71. H. K. Lee, H. C. Hsu, W. H. Tuan, "Oxidation Behavior of Copper at a Temperature below 300°C and the Methodology for Passivation," *Materials Research*, vol. 19, no 1, 2016.

72. F. Wei et al., "Ultra-precise Low-cost Surface Planarization Process for Advanced Packaging Fabrications and Die Assembly," 2016 IEEE 66th Electronic Components and Technology Conference (ECTC), p. 1740, 2016.
73. K. Shaoo, H Ren, S. S. Iyer, "A High Throughput Two-stage Die-to-wafer Thermal Compression Bonding Scheme for Heterogeneous Integration," 2023 IEEE 73rd Electronic Component and Technology Conference (ECTC), p. 362, 2023.
74. A. Agrawal, S. Huang, G. Gao, L. Wang, J. DeLaCruz and L. Mirkarimi, "Thermal and electrical performance of direct bond interconnect technology for 2.5D and 3D integrated circuits", 2017 IEEE 67th Electronic Components and Technology Conference (ECTC), p. 989, 2017.
75. F.C. Chen et al., "System on Integrated Chips (SoIC™) for 3D Heterogeneous Integration," 2019 IEEE 69th Electronic Component and Technology Conference (ECTC), p. 594, 2019.
76. https://www.sony-semicon.com/en/technology/is/cu-cu.html
77. https://www.sony-semicon.com/en/products/is/industry/swir.html
78. T. Kim et al., "Thermal Improvement of HBM with Joint Thermal Resistance Reduction for Scaling 12 Stacks and Beyond," 2023 IEEE 73rd Electronic Component and Technology Conference (ECTC), p. 767, 2023.
79. J. Wuu et al., "3D V-Cache: The Implementation of a Hybrid-bonded 64 MB Stacked Cache for a 7 nm × 86-64 CPU," IEEE International Conference on Solid-State Circuits (ISSCC), p. 428, 2022.
80. W. Zhou et al., "Critical Challenges with Copper Hybrid Bonding for Chip-to-wafer Memory," 2023 IEEE 73rd Electronic Component and Technology Conference (ECTC), p. 336, 2023.
81. D. B. Ingerly, S. Amin, L. Aryasomayajula, "Foveros: 3D Integration and the Use of Face-to-face Chip Stacking for Logic Devices," 2019 IEEE International Electron Devices Meeting (IEDM), 2019, pp. 19.6.1–19.6.4.

4 Materials for Hybrid Bonding

Andrea Chacko, Dongshun Bai, and Rama Puligadda

4.1 INTRODUCTION

Recent years have shown an explosion in the desire for and development of devices for 5G, Internet of Things, artificial intelligence (AI), and big data. The realization of such advanced technologies requires the continued drive toward smaller, faster, and more efficient devices. In addition to the continued shrinking of transistors through lithography advancements, advanced packaging technologies such as fan-out wafer-level packaging and 2.5D and 3D integrated circuits have been developed. Hybrid bonding is critical to achieving high-quality interconnects for fine-pitch integration, especially when the pitch is below 20 μm and the microbump connection will face alignment issues.

Materials for hybrid bonding can generally be divided into two categories: temporary materials and permanent materials. Temporary materials include all the materials used in the wafer thinning processes and die-attach adhesives to enable wafer-to-wafer and chip-to-wafer hybrid bonding. Permanent materials include those actively involved in the hybrid bond, including the dielectric material.

4.1.1 Temporary Materials for Wafer Thinning

Almost all advanced packages require the thinning of device wafers used in their manufacturing. Wafer thinning, or wafer backgrinding, is a critical step in the fabrication of semiconductor devices. It is the process of removing material from the backside of the wafer, reducing the thickness, and allowing for compact electronic devices. Wafer thinning is not a trivial process. The reduction in wafer thickness results in the tendency for a wafer to warp. Additionally, thinned substrates are often too fragile to withstand subsequent processes. In order to provide support to the device wafer, temporary adhesives may be used to bond the device wafer to a rigid carrier wafer throughout the thinning process and any additional process steps that require the device wafer to have good structural integrity.

The typical process for temporary bonding and debonding (TBDB) is shown in Figure 4.1. Briefly, an adhesive material is coated onto the device wafer. A release material, if necessary, is coated onto a carrier wafer. The two wafers are then bonded together in a bonder. The bonded wafer pair is then subjected to backgrinding in addition to any other backside processing. The thinned wafer is then mounted onto a film frame and the carrier wafer is removed. Several debonding technologies are

DOI: 10.1201/9781003408673-4

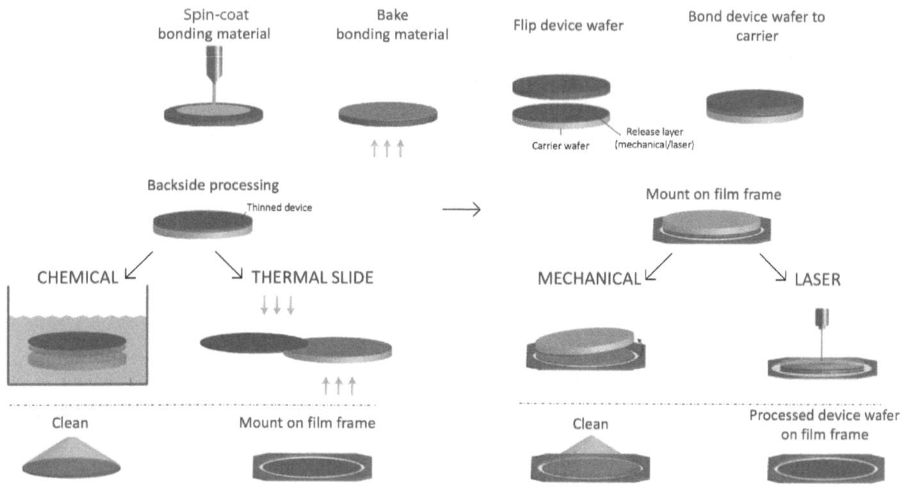

FIGURE 4.1 Generic process flow for temporary bonding and debonding for wafer thinning.

available. The selection of bonding material and debond technology is driven by several factors, including the target final thickness of the device wafer, the sensitivity of any components on the device wafer, the desired throughput of the process, the desired carrier wafer type, and the availability of the required tools.

Commercially available release technologies include chemical release, thermal slide release, mechanical release, and laser release. A comparison of these debonding technologies is shown in Table 4.1. For the purpose of hybrid bonding, we will focus on materials for mechanical and laser release in addition to emerging release technologies such as photonic debonding. Detailed information into the various hybrid bonding process flows as well as the requirements for the temporary materials used for each is described in Section 4.2. Detailed descriptions of the various debonding technologies are given below to provide an appropriate background for Section 4.2.

4.1.1.1 Mechanical Debonding

Mechanical release typically proceeds through the process flow described in Figure 4.1. Once the required processes are complete, and the wafer has been mounted onto a film frame, an initiator is inserted into the bonded wafer pair to initiate separation. The introduction of mechanical debonding allows for successful release of thinner wafers than chemical or thermal slide debonding without the need for chemical baths or heat during the debond process. The release layers for mechanical debonding are typically designed to provide a low-surface-energy interface which allows for separation. Still, the stress placed on the device by insertion of the initiator cannot be ignored. This stress limits this debond technology especially for ultrathin wafers.

4.1.1.2 Laser Debonding

Laser release also proceeds through the generic process flow shown in Figure 4.1, but a glass or other transparent carrier wafer is required. In order to initiate the separation,

TABLE 4.1
Comparison of Debonding Technologies

	Thermal	Chemical	Mechanical	Laser	Photonic
	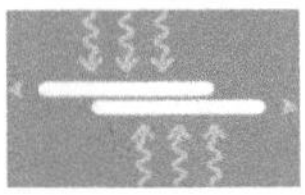	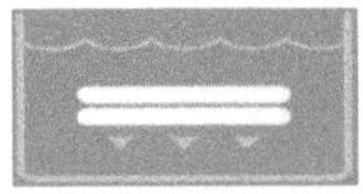	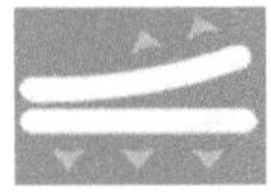		
Throughput (150-mm wafers)	20 wph	25 wph	20 wph	60 wph	>100 wph[1]
Stress	Low-middle	Low	Low	Low	Low
Processing temperature (°C)	100–200	Room temperature	Room temperature	Room temperature	Room temperature
Carrier	Glass or silicon	Perforated glass	Silicon	Glass	Glass
Equipment cost	Low	Low	Medium	High	Medium

The table compares throughput, stress, processing temperature, carrier types, and equipment cost for thermal slide, chemical, mechanical, laser, and photonic debonding methods. Thermal and mechanical debonding have the lowest cost of ownership but also have the lowest throughput. Mechanical release allows for room-temperature processing (unlike thermal slide) and the use of a silicon carrier. Laser debonding has higher throughput and higher equipment cost and requires a glass carrier. Photonic debonding has the highest throughput with moderate equipment cost and also requires a glass carrier.

[1] Estimate based on debonding equipment processing speed.

the wafer stack is exposed to a laser through the transparent carrier. Commonly used wavelengths include 248, 308, and 355 nm for UV laser and 1,064 nm for IR laser release. The laser release material undergoes either a photochemical or a photothermal process resulting in a loss of adhesion between the device and carrier wafers. The release layers for laser debonding must be designed to have good absorbance at the wavelength of the desired laser. Laser release offers many benefits, including higher throughput, the ability to release very thin wafers, and minimization of potential wafer damage. Some of the challenges facing laser debonding include the adoption of glass carrier use and the expense of laser debonding systems.

4.1.1.3 Photonic Debonding

Photonic debonding is an exploratory debonding technology which utilizes a high-energy flash lamp to initiate the debonding of the wafers [1]. Photonic debonding proceeds through a process flow similar to the one shown in Figure 4.1, with one major exception. A unique carrier system, which consists of a light-absorbing layer (LAL) coated on a glass wafer, is used for photonic debonding. In order to initiate the separation, the wafer stack is exposed to a high-energy pulse lamp through the transparent carrier. The LAL heats rapidly, resulting in a loss of adhesion between the LAL and the bonding layer. Photonic debonding can overcome many of the challenges seen by other release technologies. It is typically faster and cleaner than laser release and is capable of handling stressed wafers.

4.1.2 Permanent Materials for Hybrid Bonding

Fusion or direct bonding is a wafer bonding process resulting in the permanent connection of dielectric materials without the need for intermediate layers. Hybrid bonding is a permanent bond that combines fusion bonding of a dielectric with the bonding of embedded metal pads to form interconnections. There are several options of dielectric material for hybrid bonding. The selection of dielectric material will depend significantly on the process flow. Dielectrics used in hybrid bonding are typically plasma-enhanced chemical vapor deposition (PECVD) oxide, silicon nitride (SiN), silicon carbon nitride (SiCN), and silicon oxynitride (SiON). Alternatively, polymer dielectrics can also be used. Section 4.3 will give detailed information on both inorganic and polymer dielectrics.

4.2 ADHESIVE MATERIALS FOR HYBRID BONDING

In this section, we will describe the adhesive materials typically used in processes that prepare the die or wafers before hybrid bonding. Sections 4.2.1 and 4.2.2 describe the temporary adhesive materials in detail, while the dielectric materials are covered separately in Section 4.3.

Figure 4.2 describes the different adhesive materials and their use in hybrid bonding applications.

4.2.1 Temporary Bonding Materials for Thinning Wafers and Backside Processing

Die or wafers that are slated for heterogeneous integration utilizing hybrid bonding will have been prepared through a series of processing steps which include thinning and backside processing such as chemical mechanical polishing (CMP), the formation of redistribution layers (RDLs), through silicon via (TSV) formation, and various deposition steps. Thinning and reliable backside processing is enabled by a temporary bonding and debonding process during which the adhesive is in contact with the bonding side of the wafer/die, as described in Section 4.1.

A temporary bonding agent or adhesive and the related release layers that enable bonding and debonding processes must meet the following requirements:

- *Reliable and Simple Application Method:* The temporary bonding material must lend itself to a simple and reliable application process, such as spin-coating, slot-die coating, etc., to yield a highly uniform coating. The application process must be high-throughput compatible and preferably use known and well-accepted processes.
- *Good Adhesion to the Substrate:* The bonding layer must exhibit good wetting on a wide variety of surfaces that can exist on a wafer and good adhesion to the carrier wafer to avoid delamination or voiding in the bond line. Adhesion strength can be measured using standard methods such as cross hatch tape test (ASTMD3359), stud pull (ASTMD4541), four-point bending test (ASTMD4145), or Maszara razor blade test [2].

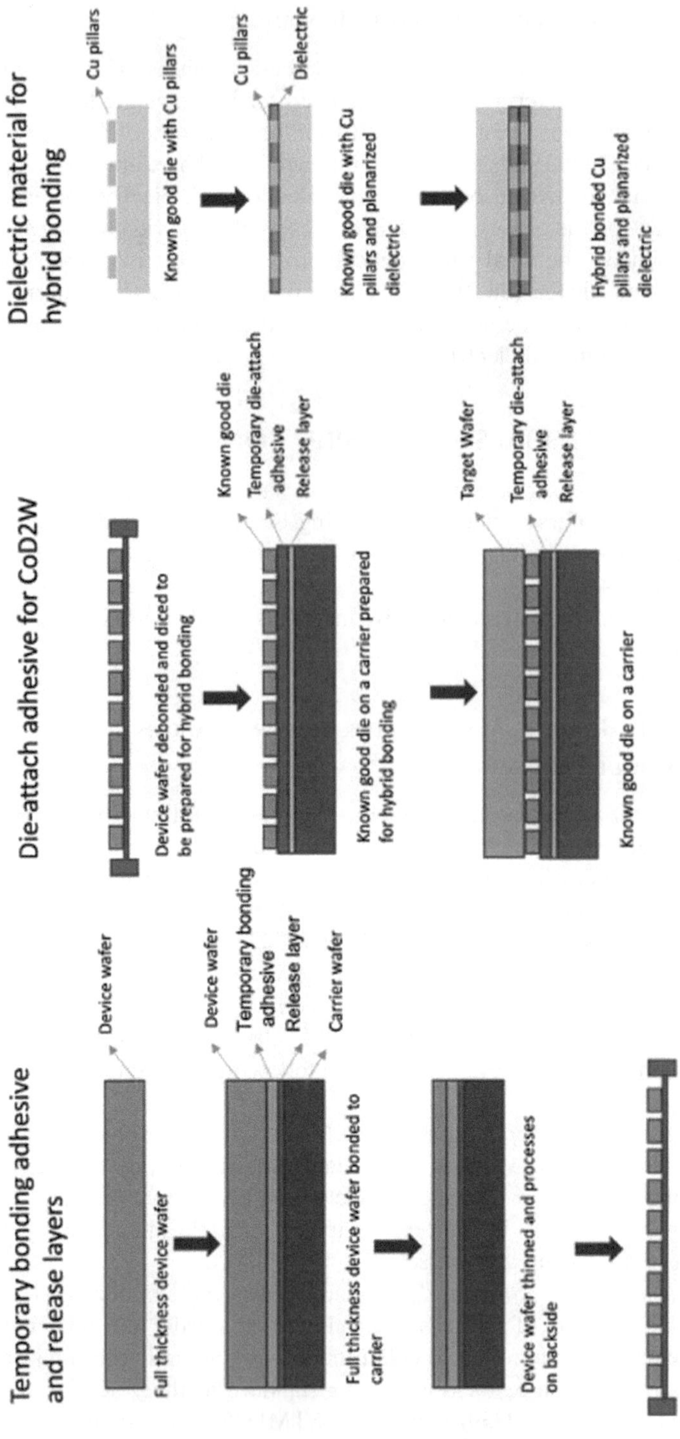

FIGURE 4.2 Overview of adhesive materials used in hybrid bonding.

- *Bond below Device-limited Temperatures:* A room-temperature or low-temperature bonding process is necessary for conserving thermal budgets of the device wafer.
- *High uniformity:* Uniformity of the bonding layer measured in total thickness variation (TTV) after coating and after bonding is a key factor that determines the nonuniformity of the thinned wafer. A bonding agent reflows during bonding process if the bonding temperature is well above its melt-flow temperature, allowing better planarity and uniformity. TTV can be measured using a surface measuring or infrared interference measuring tool [3].
- *Resistance to Process Chemicals:* The bonding material must survive all process chemicals such as cleaners, etchants, and corrosive electroplating solutions.
- *High Thermal and Mechanical Stability:* The bonding material and the adhesive layer(s) must not reflow, melt, decompose, or outgas while the device wafer is subjected to elevated temperatures during backside processes such as dielectric deposition, annealing, metal deposition, permanent bonding, and solder reflow. Curling or bowing of die or wafer during these processes will result in poor contact and defects in hybrid bonding steps. Scanning acoustic microscopy (SAM) can be used to check the bonding quality and calculate the bonding yield [25].
- *Complete Removal with No Residues:* Bonding agents and/or release layers used for preparing wafers and die for hybrid bonding in the downstream process must be removed completely from the wafer surface. This is a new requirement for materials used in packaging schemes using hybrid bonding to avoid difficulties in getting complete contact of the metal–metal and dielectric–dielectric interfaces between the two wafers or dies allowing void-free bonding. The hybrid bonding surfaces must be ultraclean, as even particles as small as 100 nm or the slightest residue could disrupt hundreds of connects and cause device failures or defective packages. This is discussed in more detail in Section 4.4.

4.2.2 Die Attach Materials for Collective Die-to-Wafer (D2W) Bonding

D2W bonding supports the integration of known good die (KGD), different die sizes, and different wafer types, all of which are not supported by wafer-to-wafer (W2W) bonding schemes. For each die, the wafers are first manufactured in a semiconductor fabrication facility before a hybrid bonding process is utilized to stack them vertically. D2W hybrid bonding results in high-quality packages as only KGD are used to achieve high yields. This approach uses a sequential process where KGD are bonded to the target wafer one at a time. An attractive alternative to the sequential D2W bonding is the collective D2W (CoD2W) bonding approach that enables processing on wafer-level tools, allowing cleaner bonding surfaces at the die level and faster throughput. This approach utilizes a carrier wafer coated with a release layer and a die-attach adhesive to populate with the die to be bonded. Once all the die are populated on the carrier, the target wafer is bonded to the die that are still supported

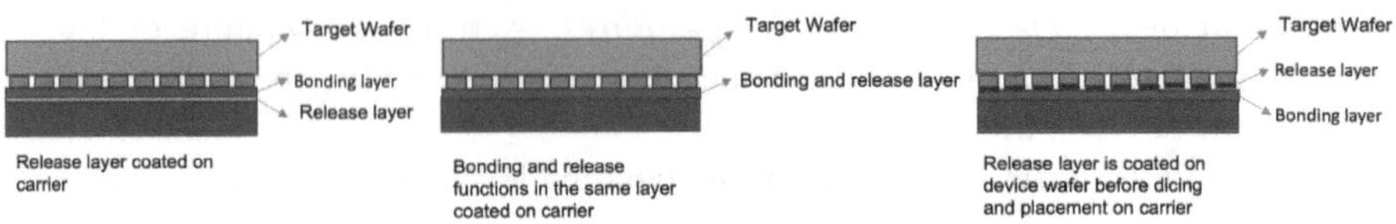

FIGURE 4.3 Die attach and release options for collective die-to-wafer hybrid bonding.

on the carrier wafer. The carrier wafer is then separated from the stack using the preferred debonding mechanism. The release layer and bonding layer are removed using appropriate solvents or cleaning agents. The die-transfer yield and the bond yield are determined by properties of the die-attach adhesive.

The die-attach adhesive material must meet the following requirements:

- Uniform coating on the carrier wafer that is already coated with a release layer: When coated on the release layer, the adhesive must form a continuous and uniform layer because high TTV will result in different heights of the die and cause poor or no contact of the die to the target wafer. In an alternative approach, die are already coated with a release layer on the face that is to be placed on the die-attach adhesive. In another viable alternative approach, the die-attach adhesive could itself be sensitive to laser ablation which would help eliminate the use of a separate release layer. This is a preferable option in situations where cost of ownership is a significant consideration. In all these approaches shown in Figure 4.3, the adhesive must form a uniform layer on the carrier. In a study comparing these three approaches, researchers at the Interuniversity Microelectronics Centre (imec) have shown that the first two approaches showed better process yield [4].
- Good adhesion to the die being populated on the carrier: In some cases, the die may be of diverse types and have different surfaces. The die must not shift during the entire process. Imec has demonstrated the use of IR microscopy to measure alignment accuracies during carrier population [22].
- The release layer and the die-attach adhesive must survive the plasma activation process followed by the hybrid bonding process. Any outgassing or deformation will result in bond failures and defective packages.
- The release layer and die-attach adhesive must be easily removed after the carrier wafer is debonded using the designated solvents and cleaning agents that do not adversely interact with device surfaces. The backside of the die that is now permanently bonded to the target wafer must be free of any residues.
- The die attach material should allow the die attach to be done at low temperature or even at room temperature to help with the alignment accuracy [22].

4.3 DIELECTRIC MATERIALS IN HYBRID BONDING

Over the past several years, many potential dielectric materials have been identified for the purpose of hybrid bonding, including thermally grown oxide, SiO_2, SiN, SiCN, SiON, and various polymer dielectrics. In addition to the dielectric properties of the material, bond strength, CMP performance, bonding temperature, and annealing temperature must also be considered when choosing an appropriate dielectric for a desired process flow.

4.3.1 Inorganic Dielectrics

To date, the most well-studied dielectrics for hybrid bonding are inorganic dielectrics. The typical process flow for inorganic dielectrics in hybrid bonding, shown in Figure 4.4, includes the following steps: deposition and planarization of the dielectric, patterning and etch of the dielectric, resist removal, deposition of barrier, liner, and seed layers within the via, and electroplating and planarization of the copper. Prior to joining the two substrates, the wafers or chips will be cleaned, usually through a combination of plasma and wet cleaning, and then bonded at room temperature. This is followed by post-bond annealing which strengthens the dielectric bonds between the two substrates and enables the copper–copper bonding. The anneal temperature can vary and is dependent on the type of dielectric and copper used in the process, the dishing of the copper, and the density of reactive groups on the surface of the dielectric. Annealing at too low a temperature may result in voids in the bond line, but too high an annealing temperature can impart stress into the wafer pair.

Chidambaram et al. presented a comparative study of different dielectric materials and their integration into hybrid bonding processes [5]. In their study, thermal oxide, PECVD TEOS SiO_2, SiN, and SiCN were characterized for surface energy, surface roughness after CMP, bond strength, and post-bond annealing.

Contact angle measurements were carried out prior to plasma treatment to determine the wetting angle of the dielectric films. Hydrophilic materials are more desirable for room-temperature bonding of the dielectrics. Among the studied dielectrics, the contact angles were found to increase in the following order: SiCN < high-T TEOS SiO_2 < low-T TEOS SiO_2 < SiN. This suggests that SiCN should have the highest bond energy. It should also be noted that contact angle may be lowered through different plasma treatments.

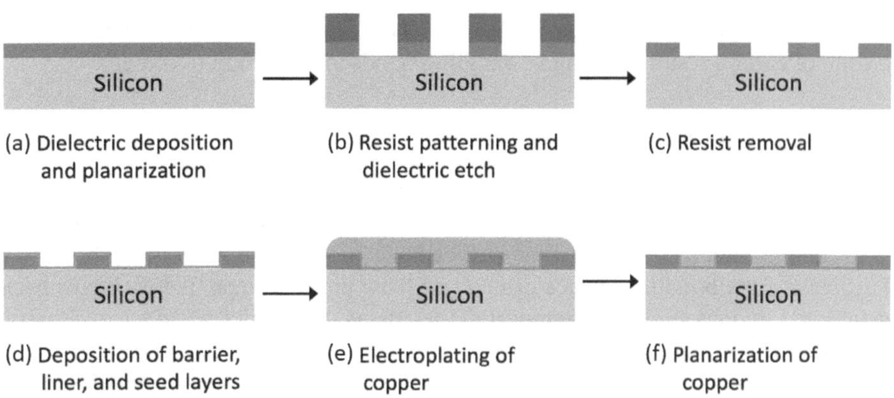

FIGURE 4.4 Typical process flow of inorganic dielectric and copper deposition for hybrid bonding.

The CMP process was also studied and optimized with the target of achieving exceptionally low surface roughness (R_a) for both Cu and dielectric (≤5 μm) and good roughness uniformity across the wafer (≤10%). CMP optimization often includes varying slurry composition, pad type, pad pressure, and time. It was found that the best roughness results were found for SiCN. Both high- and low-temperature TEOS SiO_2 showed the same roughness and SiN was unable to reach the roughness target post-CMP.

The bond strength was characterized by a razor blade insertion test and die shear testing (ASTMD4501). Both tests showed the same trend in bond strength. The bond strength trends inversely with the contact angle of the dielectric, confirming that hydrophilic films result in better bond strength.

The post-bonding anneal temperature was also studied. It was found that TEOS SiO_2 required a post-bond anneal temperature of at least 300°C and the strength of the bond was dependent on the temperature. SiCN has the lowest post-bond anneal temperature at 250°C, which may make it a more suitable dielectric for processes with temperature restrictions. It should be noted however, that low-temperature hybrid bonding processes are likely to require changes to the Cu composition as well.

Inoue et al. studied different compositions of PECVD-SiCN films to determine the role of different components on the dielectric bonding [6]. The composition of the SiCN films was controlled by tuning the precursors during the deposition process. Elastic recoil detection (ERD) analysis was used to determine the elemental content of the films. Hybrid bonding performance and bond energy were compared for standard SiCN, higher-carbon-containing SiCN, and higher-nitrogen-containing SiCN.

The effects of plasma activation were also studied. X-ray photoelectron spectroscopy (XPS) confirmed that when exposed to N_2 plasma, Si—C bonds are broken and form Si—O, C—O, or Si and C dangling bonds, which play a key role in wafer bonding [7]. SiCN films were found to have almost ten times higher density of dangling bonds than SiO_2. It is suggested that these dangling bonds, which are consumed during post-bond annealing, contribute to the chemical reaction and lead to a higher bond energy than SiO_2. The higher-carbon-containing SiCN was found to have the highest bond energy among the SiCN films. This is attributed to the higher concentration of carbon resulting in more dangling bonds after plasma treatment.

4.3.2 Polymer Dielectrics

One of the challenges facing the realization of hybrid bonding in high-volume manufacturing is the incredibly stringent requirement for wafer cleanliness. Any particle contamination on the bonding surface leads to void formation in the bond line. Particles on the bonding surface often arise from process steps like dicing or backgrinding. Previous studies have reported the effects that particles can have on hybrid bonding [8] and highlight the importance of maintaining clean surfaces throughout the entire process flow. Although inorganic dielectrics offer benefits like high-throughput and room-temperature prebonding, the yield loss due to contamination has encouraged the investigation of polymer dielectrics which are expected to suffer fewer consequences of surface contamination [9].

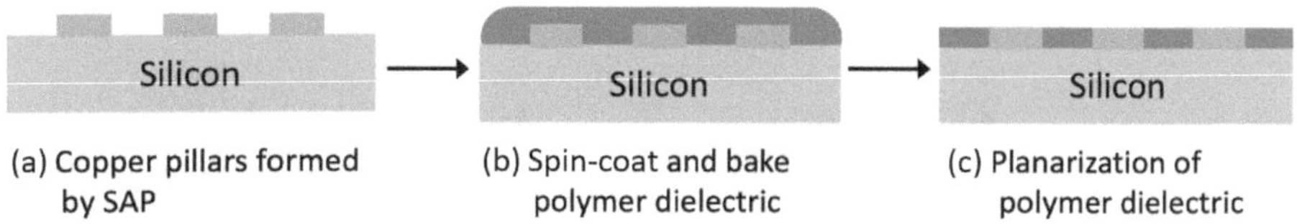

FIGURE 4.5 Typical process flow of copper and polymer dielectric deposition for hybrid bonding.

One possible process flow for the utilization of a polymer dielectric for hybrid bonding is shown in Figure 4.5. Copper pillars are first formed by the semi-additive process (SAP). A polymer dielectric is then spin-coated over the pillars and baked. After spin-coating, the polymer is subjected to a planarization process, such as CMP or flycut to reveal the copper pillars. Similar to the case with inorganic dielectrics, the wafers or chips are cleaned, typically with an acid rinse to remove CuO, prior to joining the two surfaces. First, the dielectric materials are bonded, typically at an elevated temperature. This is followed by post-bond annealing which strengthens the dielectric bonds between the two substrates and enables the copper–copper bonding.

Polymers used in this hybrid bonding process flow need to be able to fill the spaces between the copper pillars without voiding to provide proper insulation. They must also be able to be planarized to provide copper protrusion. Not all polymers are appropriate for planarization by techniques like CMP, which requires the polymer to be relatively hard to provide a smooth surface [10]. The height of the copper protrusion must be optimized for each different polymer dielectric due to the coefficient of thermal expansion (CTE) mismatch between the polymer and the copper. The polymer must also be stable to temperatures exceeding 300°C to survive the copper diffusion and be capable of bonding permanently with good reliability.

One demonstration of this process flow, reported by Shirasaka et al., utilized a polyimide as a dielectric [11]. A SAP was used to fabricate the copper pillars and the polyimide was coated and baked. CMP was utilized to planarize the material to a roughness of less than 1 nm and to reveal the copper pillars. A study into the copper protrusion height and its effect on yield was conducted. The substrates were acid-treated prior to the initial temporary bonding step. The copper protrusion height and acid treatment conditions have a significant role in the bond quality of the copper connections. The permanent bonding was optimized to provide good bond quality and strength between both the polyimide and the copper. After the permanent bonding step, the bond strength of the polyimide was found to be sufficient. One important consideration when choosing a polymer system is the requirements for curing as well as the conditions for the temporary bonding. Polyimides often require cure temperatures of over 300°C for several hours. They also typically exhibit high glass transition temperatures (T_g) which can limit their use in thermal-compression processes and their ability to embed particle defects.

Another potential process flow is shown in Figure 4.6. This process flow utilizes a photopatternable permanent bonding dielectric material to simplify the overall process flow. The polymer is spin-coated onto the substrate and baked to remove solvent. The polymer is then patterned using a mask aligner or stepper and developed

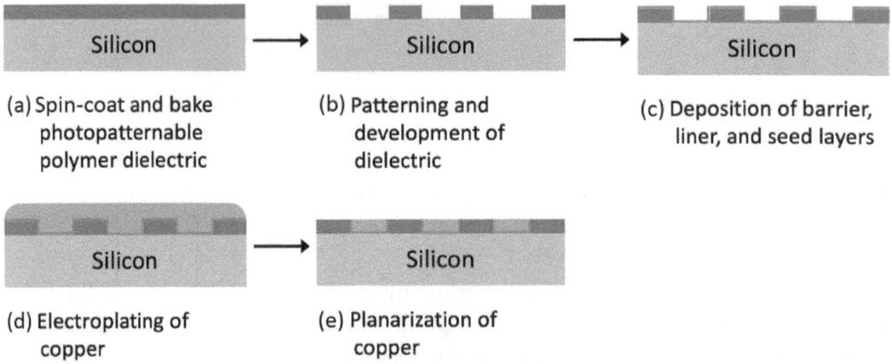

FIGURE 4.6 Alternative process flow of photopatternable dielectric and copper deposition for hybrid bonding.

to provide the desired pattern. Copper interconnections are then produced through a damascene process. The copper pillars are then planarized and smoothed by CMP. Surface cleaning and bonding proceeds the same as mentioned with the other polymer process flow above.

Polymers that can be used for this process flow have a similar list of requirements to the other process flows with a few key exceptions. First, the materials must be photopatternable to the resolution required for high-density interconnects. i-line (365 nm) patterning is commonly used in back-end-of-line (BEOL) manufacturing processes and will likely be able to provide the needed feature sizes. The polymer must be compatible with the damascene process and able to withstand thermal and chemical exposure. This process flow, however, does not require the material to be capable of gap-fill or planarization.

Huang et al. have published a photosensitive permanent bonding material (PS PBM) which may have applications in hybrid bonding [12]. The PS PBM is designed to be negative tone and is patternable with i-line exposure. After the material was coated and baked to remove solvent, the film was exposed using an i-line mask aligner and developed with solvent as the developer. A resolution of 5 μm line/space on a 5 μm film was achieved. A copper–tin metallization was completed on the patterned wafer through electroplating. The PS PBM itself could be bonded at room temperature. After final cure, wafer level bond strength was tested and found to be greater than that of inorganic dielectrics. Additionally, this material was thermally stable with a T_d of 373°C suggesting that it can survive the final cure needed to make the copper–copper connections.

This polymer offered many advantages over other polymers for hybrid bonding. The PS PBM could be cured at 180°C, which is a much lower temperature compared to other materials such as polyimides or polybenzoxazoles. Additionally, the material exhibited excellent dielectric properties when compared to other polymer dielectrics. The modulus and T_g of this material were low in comparison to the other materials, which may provide improved ability for particle embedding. The CTE, however, was high, which may cause issues through the hybrid bonding process.

4.4 SURFACE CLEANING FOR HYBRID BONDING

Surface cleanliness is crucial to hybrid bonding. Since the interface for hybrid bonding is solid state, even the tiniest of particles or thinnest of residues could disrupt the process flow and cause device failures [13, 14]. Intel's research showed a 50 nm particle causes a few failed connections at 20 μm pitch, but the same particle can cause >100 failed connections when pitch is <3 μm [15]. SET Corporation also claimed that a particle with a diameter of 1 μm between the chip and the substrate prevented a circular surface of a diameter of 5 mm (about 0.2 inch) to be bonded for a device thickness in the range of 525 μm [16]. Particles affect not only the bond accuracy but also the adhesion between chip and substrate [17]. So, it is critical to control every process step to reduce the particles and contamination from process, handling, tools, and environment. As EVG pointed out, hybrid bonding required substantially different manufacturing technologies from standard packaging processes, bringing it much closer to front-end manufacturing—especially in terms of cleanliness, particle control, alignment, and metrology precision [18].

While W2W bonding can usually count on a clean starting surface, D2W is a different story [19]. Gao et al. highlighted that singulation is a dirty process; therefore, maintaining a clean die surface for hybrid bonding is challenging [20]. A protection layer can be used to protect the wafer from particles during mechanical dicing and to reduce die edge chipping [21, 22]. After dicing, the dies are often repopulated on a cleaning carrier for cleaning and activation before final hybrid bonding. Laser stealth and plasma dicing can also be used to suppress particle contamination, chipping on die edge, and to enhance die break strength [23].

Besides particles, organic residues can also prevent good surface adhesion [16]. TBDB is widely used for wafer thinning and thin wafer handling. For surfaces that were in contact with TBDB materials during processing, the TBDB material needs to be removed completely to meet the hybrid bonding requirement [22, 24–26]. The remover used for cleaning must be compatible with the materials on the device and with the cleaning tools and parts. For devices that will see high-temperature processes, the cleaning of the TBDB material could be challenging. A heated remover with high-pressure spray can be helpful in these cases. Plasma treatment presents an additional option to remove the residues on the surface to meet the cleanliness requirement.

Polymeric dielectrics have also been tested for hybrid bonding [23, 24, 27, 28]. One of the major goals is to avoid particle sensitivity during hybrid bonding. When using polymeric materials in hybrid bonding, many factors must be considered and studied for good bonding with high yield, including the mobility of polymer materials at elevated temperatures, their capability for CMP, and the bonding strength and integrity.

4.5 CONCLUSIONS AND OUTLOOK

Hybrid bonding, enabled by innovative materials and processes, is a back-end process with front-end requirements that is bringing the two ends together and encouraging new collaboration and innovation. Recent technological advancements in upstream materials and processes that facilitate hybrid bonding have led to the increase in adoption of the technology for high-end applications such as high-performance computing (HPC) and AI. Further advancements in key areas such as metrology and test, alignment, plating, planarization and cleanliness of the bonding surfaces, and cost optimization of processes will lead to wider adoption in downstream applications such as micro-LED, automotive, and mobile applications.

REFERENCES

1. L. Prenger, X. Martinez, A. Chacko, V. Turkani, L. Reimnitz, V. Akhavan, and K. Schroder, "Photonic Debond: Scalability and Advancements," *2022 IEEE 72nd Electronic Components and Technology Conference (ECTC)*, San Diego, CA, 2022, pp. 1287–1293.
2. Y. Kwon and J. Seok, "An Evaluation Process of Polymeric Adhesive Wafer Bonding for Vertical System Integration," *Japanese Journal of Applied Physics, Part 1*, Vol. 44, No. 6A, 2005, 3893–3902.
3. Y. Liu, H. Tao, and X. Li, "An Investigation on the Total Thickness Variation Control and Optimization in the Wafer Backside Grinding Process," *Materials*, Vol. 15, No. 12, 2022, 4230.
4. S. Iacovo, F. Nagano, V. S. K. Channam, E. Walsby, K. Crook, K. Buchanan, A. Jourdain, K. Vanstreels, A. Phommahaxay, and E. Beyne, "Direct Bonding Using Low Temperature SiCN Dielectrics," *2022 IEEE 72nd Electronic Components and Technology Conference (ECTC)*, San Diego, CA, 2022, pp. 602–607.
5. V. Chidambaram, P. Lianto, X. Wang, G. See, N. Wiswell, and M. Kawano, "Dielectric Materials Characterization for Hybrid Bonding," *2021 IEEE 71st Electronic Components and Technology Conference (ECTC)*, San Diego, CA, 2021, pp. 426–431.
6. F. Inoue, L. Peng, S. Iacovo, A. Phommahaxay, P. Verdonck, P. Verdonck, J. Meersschaut, P. Dara, E. Sleeckx, A. Miller, G. Beyer, and E. Beyne, "Influence of Composition of SiCN as Interfacial Layer on Plasma Activated Direct Bonding," *ECS Journal of Solid State Science and Technology*, Vol. 8, No. 6, 2019, P346.
7. L. Peng, S.-W. Kim, S. Iacovo, F. Inoue, A. Phommahaxay, E. Sleeckx, J. De Vos, A. Miller, G. Beyer, E. Beyne, D. Zinner, T. Wagenleitner, T. Uhrmann, M. Wimplinger, B. Schoenaers, A. Stesmans, and V. V. Afanas'ev, "Advances in SiCN–SiCN Bonding with High Accuracy Wafer-to-Wafer (W2W) Stacking Technology," *2018 IEEE International Interconnect Technology Conference (IITC)*, Santa Clara, CA, 2018, pp. 179–181.
8. F. Nagano, S. Iacovo, A. Phommahaxay, F. Inoue, F. Chancerel, H. Naser, G. Beyer, E. Beyne, and S. D. Gerndt, "Void Formation Mechanism Related to Particles during Wafer-to-Wafer Direct Bonding," *ECS Journal of Solid State Science and Technology*, Vol. 11, No. 6, 2022, 063012.
9. H. Noda, A. Hiro, N. Nishiguchi, J. Derakhshandeh, J. Slabbekoorn, and E. Beyne, "An Investigation on Particle Embedding Capability of Wafer Level Spin-on Polymer Underfill Enabling Low Temperature Bonding of Hybrid Bonding System," *2023 IEEE 73rd Electronic Components and Technology Conference (ECTC)*, Orlando, FL, 2023, pp. 1677–1681, doi: 10.1109/ECTC51909.2023.00285.
10. Z. W. Zhong, Z. F. Wang, and Y. H. Tan, "Chemical Mechanical Polishing of Polymeric Materials for MEMS Applications," *Microelectronics Journal*, Vol. 37, No. 4, pp. 2006, 295–301.
11. T. Shirasaka, T. Okuda, T. Shibata, S. Yoneda, D. Matsukawa, M. Mariappan, M. Koyanagi, and T. Fukushima, "Comprehensive Study on Advanced Chip on Wafer Hybrid Bonding with Copper/Polyimide Systems," *2022 IEEE 72nd Electronic Components and Technology Conference (ECTC)*, San Diego, CA, 2022, pp. 317–323.
12. B. Huang, M. Dong, S. Fowler, A. Chacko, and R. Puligadda, "Wafer-level Polymer/Metal Hybrid Bonding Using a Photosensitive Permanent Bonding Material," *Chip Scale Review*, Vol. 26, No. 2, 2022, 49–53.

13. S. Hiebert, "Heterogeneous Integration Moved to Another Level Using Hybrid Bonding," *3D InCites Yearbook*, Vol. 5, 2023, 28–31.
14. J. Abdilla, "Die to Wafer Hybrid Bonding: The Next Generation of Advanced Packaging," presented at 3D & Systems Summit, Dresden, Germany, 2022.
15. A. Elsherbini, K. Jun, R. Vreeland, W. Brezinski, H. K. Niazi, Y. Shi, Q. Yu, Z. Qian, J. Xu, S. Liff, J. Swan, J. Yao, P. Liu, C. Pelto, S. Rami, A. Balankutty, P. Fischer, and B. Turkot, "Enabling Hybrid Bonding on Intel Process," *2021 IEEE International Electron Devices Meeting (IEDM)*, San Francisco, CA, 2021, pp. 34.3.1–34.3.4.
16. P. Metzger, N. Raynaud, A. Jouve, N. Bresson, L. Sanchez, F. Fournel, C. Castan, and S. Cheramy, "Die-to-Wafer Direct Bonding for Production Environment with a New Flip-Chip Bonder," *Micro/Nano-Electronics Packaging & Assembly, Design and Manufacturing Forum*, Grenoble, France, 2019.
17. P. Metzger, N. Raynaud, A. Jouve, N. Bresson, L. Sanchez, F. Fournel, and S. Cheramy, "Toward a Flip-chip Bonder Dedicated to Direct Bonding for Production Environment," *International Wafer-Level Packaging Conference*, San Jose, CA, 2017.
18. EV Group, "EV Group Achieves Die-to-Wafer Fusion and Hybrid Bonding Milestone with 100-Percent Die Transfer Yield on Multi-Die 3D System-on-a-Chip," July 12, 2022, https://www.evgroup.com/company/news/detail/ev-group-achieves-die-to-wafer-fusion-and-hybrid-bonding-milestone-with-100-percent-die-transfer-yield-on-multi-die-3d-system-on-a-chip.
19. K. Derbyshire, "The Darker Side of Hybrid Bonding," Semiconductor Engineering, December 17, 2020, https://semiengineering.com/the-darker-side-of-hybrid-bonding/.
20. G. Gao, L. Mirkarimi, T. Workman, G. Fountain, J. Theil, G. Guevara, P. Liu, B. Lee, P. Mrozek, M. Huynh, C. Rudolph, T. Werner, and A. Hanisch, "Low Temperature Cu Interconnect with Chip to Wafer Hybrid Bonding," *2019 IEEE 69th Electronic Components and Technology Conference (ECTC)*, Las Vegas, NV, 2019, pp. 628–635, doi: 10.1109/ECTC.2019.00100.
21. C. H. Fan, H. P. Ng, S. C. So, N. T. Man, C. Y. Lee, M. Li, T. Uhrmann, J. Burggraf, and M. Pires, "Direct Die to Wafer Cu Hybrid Bonding for Volume Production," *2023 IEEE 73rd Electronic Components and Technology Conference (ECTC)*, Orlando, FL, 2023, pp. 91–96, doi: 10.1109/ECTC51909.2023.00024.
22. K. Kennes, A. Phommahaxay, A. Guerrero, O. Bauder, S. Suhard, P. Bex, S. Iacovo, X. Liu, T. Schmidt, G. Beyer, and E. Beyne, "Introduction of a New Carrier System for Collective Die-to-Wafer Hybrid Bonding and Laser-assisted Die Transfer," *2020 IEEE 70th Electronic Components and Technology Conference (ECTC)*, Orlando, FL, 2020, pp. 296–302, doi: 10.1109/ECTC32862.2020.00056.
23. D. K. Mishra, V. N. Sekhar, C. S. Choong, and V. S. Rao, "Evaluation of Polymer Materials for Hybrid Bonding Application," *2022 IEEE 24th Electronics Packaging Technology Conference (EPTC)*, Singapore, Singapore, 2022, pp. 357–360.
24. V. N. Sekhar, D. K. Mishra, C. S. Choong, and V. S. Rao, "Thin Memory Chip Fabrication for Multi-stack Hybrid Bonding Applications," *2022 IEEE 24th Electronics Packaging Technology Conference (EPTC)*, Singapore, Singapore, 2022, pp. 589–593, doi: 10.1109/EPTC56328.2022.10013147.
25. K. Kennes, A. Phommahaxay, A. Guerrero, S. Suhard, P. Bex, S. Brems, X. Liu, S. Tussing, G. Beyer, and E. Beyne, "Carrier Systems for Collective Die-to-Wafer Bonding," *2022 IEEE 72nd Electronic Components and Technology Conference (ECTC)*, San Diego, CA, 2022, pp. 2058–2063, doi: 10.1109/ECTC51906.2022.00324.
26. V. N. Sekhar, M. D. Kumar, P. Lianto, S. C. Chong, and V. S. Rao, "Cu Damascene Process on Temporary Bonded Wafers for Thin Chip Stacking Using Cu–Cu Hybrid Bonding," *2023 IEEE 73rd Electronic Components and Technology Conference (ECTC)*, Orlando, FL, 2023, pp. 564–570, doi: 10.1109/ECTC51909.2023.00100.

27. Y.-M. Lin, P.-C. Chang, O.-H. Lee, W.-L. Chiu, T.-C. Chang, H.-H. Chang, C.-H. Lee, B. Huang, M. Dong, D. Tsai, C.-C. Lee, and K.-N. Chen, "A Hybrid Bonding Interconnection with a Novel Low-temperature Bonding Polymer System," *2022 IEEE 72nd Electronic Components and Technology Conference (ECTC)*, San Diego, CA, 2022, pp. 2128–2134, doi: 10.1109/ECTC51906.2022.00336.
28. S. K. Tippabhotla, L. Ji, and Y. Han, "Numerical Simulation of Cu/Polymer-Dielectric Hybrid Bonding Process using Finite Element Analysis," *2022 IEEE 72nd Electronic Components and Technology Conference (ECTC)*, San Diego, CA, 2022, pp. 1695–1703, doi: 10.1109/ECTC51906.2022.00267.

5 Copper Electrodeposition for Advanced Packaging and Hybrid Bonding

Bryan Buckalew

5.1 INTRODUCTION

Semiconductor device architecture consists of three main components:

1. The front-end of line (FEOL) consists of layers that form the transistor, capacitors, and resistor at the nanometer scale.
2. Backend of the line (BEOL) consists of the copper metallization, which consists of numerous layers of vias and trenches at the nanometer scale.
3. Advanced packaging consists of either pillars and micropillars or hybrid bonding (HB) pads at the micrometer scale.

The intrinsic speed limit of an integrated circuit (IC) is determined by the frequency at which its transistors can be turned on and off. Since smaller transistors have inherently higher clock frequencies, advances in IC speed historically have been achieved by downward scaling of feature sizes [1–3]. Several years ago, copper displaced aluminum for BEOL interconnects due to improved resistivity. This benefit of lower resistivity for copper helped decrease the interconnect RC delay, which, in turn, increases the IC speed.

The interconnects used for advanced packaging have also transitioned to the use of copper. The industry moved away from solder-based, C4 bump interconnects in favor of Cu pillar technology, as Cu pillar helped enable <100 μm interconnect pitches by reducing the solder bridging concerns at fine pitches (i.e., spacing between features). Copper also provided excellent electrical and thermal benefits compared to lead-free solder alloys [4–6].

Refer to Table 5.1 for a summary of interconnect metallization properties and bump pitch.

Advanced packaging has generally utilized semiadditive process (SAP) flow (which is generally less expensive than the damascene process flow) where copper pillars/micropillars are formed. These are capped with lead-free solder (typically Sn or SnAg), which allows the formation of an intermetallic compound during bonding. The SAP process flow has encountered scalability challenges, as the copper pillar technology is extendable only down to approximately 20 μm pitch [7].

The industry has recognized HB as an enabling technology to meet sub-10 μm interconnect pitch requirements [8, 9]. This approach utilizes the damascene process

DOI: 10.1201/9781003408673-5

TABLE 5.1
Interconnect Metallization Properties and Bump Pitch

Interconnect	Thermal Conductivity (W/m·K)	Electrical resistivity (μΩ·m)	Bump Pitch (μm)
Al	237	0.026	
Cu	400	0.017	≤20
Cu pillar + solder cap	300 (Effective)	0.025 (Effective)	≤100 ≥20
C4 bump (solder-only)	55–60	0.12–0.14	≥100

A list of interconnect metallization properties and bump pitch for Al, Cu, Cu pillar plus solder cap and C4 bump. For a Cu only interconnect, the thermal conductivity is the highest and the electrical resistivity is the lowest. A bump pitch of <20 μm can be achieved with Cu, while Cu pillar technology is restricted to >20 μm.

flow as opposed to the SAP process flow. See Figure 5.1. Note that although HB features are much smaller than pillar/micropillar features, they are still much larger in comparison to lower level BEOL vias and trenches. Typical dimensions for existing HB features range from 0.5 × 0.8 μm to 10 × 1.5 μm [9–11].

The following sections will review copper electrodeposition for semiconductor interconnects with a special emphasis on HB applications. A detailed review of copper electrodeposition fundamentals will be followed by discussions of the various components of the electrodeposition plating technology (including plating chemicals, reactor, fluidics handling, and dosing system). Finally, the critical film properties of a copper-plated film will be reviewed in the context of a HB interconnect.

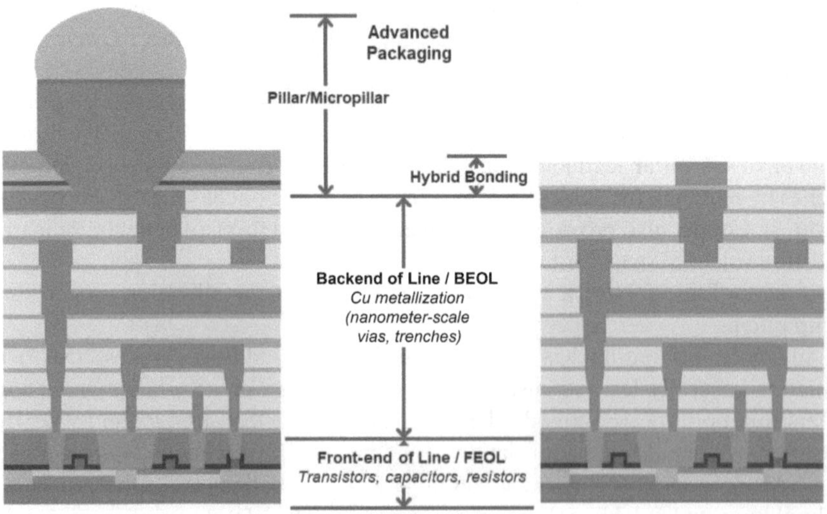

FIGURE 5.1 Semiconductor device architecture for two different advanced packing schemes: pillar/micropillar and HB pad.

5.2 FUNDAMENTALS

5.2.1 Basic Terms and Concepts

Electrodeposition refers to a film growth process which consists of the formation of a metallic coating onto a base material occurring through the electrochemical reduction of metal ions from an electrolyte [12]. The corresponding technology is often known as electroplating. The most common electroplating circuit consists of a copper anode and a copper wafer surface (cathode). Both electrodes are immersed in a vessel containing the electrolyte and are connected to an external power supply to make current flow possible. In the solution, the positively charged copper ions migrate to the negatively charged electrode (the cathode) under the influence of an electrical potential between the two electrodes. The copper ions are reduced to metallic copper at the cathode via the electrons provided by the power supply. This can be represented as the half-cell reaction:

$$Cu^{2+} + 2e^- \rightarrow Cu(\text{at the cathode}) \tag{5.1}$$

The circuit is completed by the oxidation of metallic Cu at the anode, releasing positive Cu ions into the electrolyte. The half-cell rection can be expressed as the following:

$$Cu \rightarrow Cu^{2+} + 2e^-(\text{at the anode}) \tag{5.2}$$

The reactions can be represented in a simplified copper electrodeposition circuit as shown in Figure 5.2.

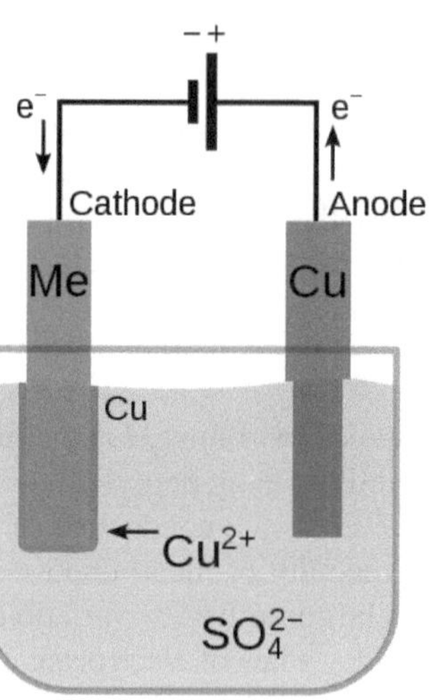

FIGURE 5.2 A simple copper electrodeposition circuit.

The reaction at the anode depends on the nature of the electrode material. Less common is the use of an inert anode where the electrode reaction consists of the decomposition of water into acid and oxygen gas according to

$$2H_2O \rightarrow 4H^+ + O_2 + 4e^- \text{(at the anode)} \tag{5.3}$$

In this case, the inert anode consists of a non-consumable anode such as inert carbon, platinum, etc.

5.2.2 Faraday's Law

The current source attached to an electrolytic cell is the source of electrons. In an electrochemical cell, electrons can be thought of as a reactant or a product of the chemical reaction, just like any other reactant or product. The current flowing through the cell is directly related to the chemical changes occurring in the overall cell reaction.

Faraday's law of electrolysis establishes the equivalence of electric charge and chemical change in a redox reaction [13]. Faraday's first law states that the quantity of elements separated by passing an electric current through a dissolved salt is proportional to the quantity of electric charge passing through the circuit, or simply stated as the following:

$$W \propto Q \tag{5.4}$$

where W is the mass of substance in grams and Q is the passed coulombs of electricity. Therefore, the mass of metal deposited during electrolysis is directly proportional to the charge on the cation, the current, and the length of time the cell runs. The unit of electric current is the ampere, which is equivalent to the charge carried in coulombs per second (1 Amp = 1 C/s).

The charge of a single electron is e, the fundamental unit of electric charge, negative because electrons have negative charge ($1e = 1.60218 \times 10^{-19}$ C/electron).

Faraday's second law states that the amount of current passing through different electrolytes connected in series, the mass of substance dissolved and deposited at the electrodes, is directly proportional to their equivalent weight [13].

$$W \propto M/n \tag{5.5}$$

where W is the mass of substance in grams, M is the molecular weight, and n is the number of electrons transferred per atom deposited.

Therefore, the current delivered to a conductive surface during electroplating is directly proportional to the quantity of metal deposited. Using the above relationships, the mass deposited can be controlled by variations of plating current and time. This expression is called Faraday's law of electrolysis.

$$W = ItA/(nF) \tag{5.6}$$

where W is themass in grams, t is the time in seconds, I is the current in amperes, F is the Faraday constant (96,500 C/mol), n is the number of electrons transferred per atom deposited, and A is the atomic weight. This equation is extremely important and widely used in practice to calculate the amount of metal deposited during electrolysis [14]. The mass of Cu deposited can be readily controlled through variations of plating current and time using this relationship.

5.2.3 Overpotential

Most electrolytic reactions require more voltage than the cell potential, E^0 cell, for the reaction to occur to an appreciable extent. Overpotential η is defined as the electrolytic cell's potential difference between the applied potential ΔV and the half-cell's thermodynamically determined reduction potential E^0 [15]:

$$\eta = \Delta V - E^0 \tag{5.7}$$

Overpotential is directly related to a cell's voltage efficiency and can be thought of as the penalty that must be paid to drive the reaction away from equilibrium. The farther from the equilibrium (i.e., the higher the current density), the larger the overpotential required. Overpotentials can be grouped into three main categories: activation, concentration, and ohmic (or resistance). These conditions influence the potential voltage drop.

Activation overpotential refers to the potential difference above the equilibrium value required to produce a current that depends on the activation energy of the redox event. The activation overpotential stems from the activity of the electrode, which is a function of the type of material used. It is the energy barrier that the reactants must overcome before they can be converted to products.

Concentration overpotential arises when the reaction is proceeding very fast, and it becomes difficult to supply the reactants to maintain a concentration of chemical species at the electrodes. This effect is very pronounced at high currents and acts as a cut-off for the current achievable for a given device.

Finally, the ohmic polarization is related to the cell design. These losses are dominated by the ohmic resistance within the plating cell due to the electrolyte conductivity, anode to cathode gap spacing, resistance of membrane, and other hardware components between the anode and the cathode [16]. The ohmic losses scale linearly with the current. Figure 5.3 provides a good illustration of overpotential.

5.2.4 Current–Potential Relationship

The Butler–Volmer (BV) equation relates the electrical current to changes in the metal potential caused by an external power source [17]. The BV equation is defined as follows:

$$I = i_0[e^{-\alpha nF\eta/RT} - e^{((1-\alpha)nF\eta/RT)}] \tag{5.8}$$

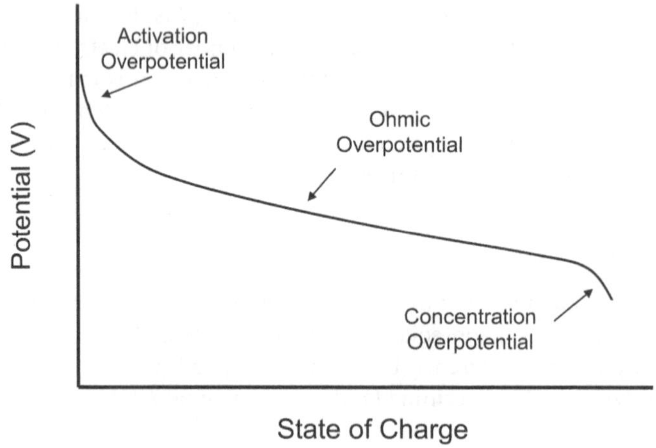

FIGURE 5.3 Potential loss due to overpotential contributors.

where I is the electrode current density (A/cm^2), i_0 is the exchange current density (A/cm^2), α is the charge transfer coefficient, n is the number of electrons in the electrode reaction (dimensionless), F is the Faraday constant, η is the activation overpotential (V), R is the gas constant (J/(mol K)), and T is the temperature (K). The BV equation reflects the reversibility of the electrochemical reactions, as the first part of the equation is the forward process, and the second is the reverse one. The sum of the forward and backward transfer coefficients is unity, so the reverse reaction is often written with a transfer coefficient of $(1 - \alpha)$. At high current densities and overpotentials, the I–V plot becomes non-linear, and this region is studied via Tafel analysis to relate the rate of an electrochemical reaction to its overpotential. The Tafel equation can be stated for a single electrode as follows [17]:

$$I = i_0 [e^{-\alpha n F \eta / RT}] \tag{5.9}$$

where I is the electrode current density (A/cm$_2$), i_0 is the exchange current density (A/cm^2), α is the charge transfer coefficient, n is the number of electrons in the electrode reaction (dimensionless), F is the Faraday constant, η is the activation overpotential (V), R is the gas constant (J/(mol K)), and T is the temperature (K). The Tafel equation describes how the electrical current flowing through an electrode depends on the voltage difference between the electrode and the electrolyte for a single-element redox potential. While Faraday's law (Section 5.2.2) is used to describe how the deposition rate is related to the current, the Tafel equation relates the deposition rate to the overpotential.

For an electrochemical reaction that occurs in two half-cell reactions on separate electrodes, the Tafel equation is applied to each electrode separately. The Tafel equation assumes that the concentrations at the electrode are equal to the concentrations in the bulk electrolyte, allowing the current to be expressed as a function of potential only. In other words, it assumes that the electrode mass transfer rate is much greater than the reaction rate, and that the reaction is dominated by the slower chemical reaction rate at

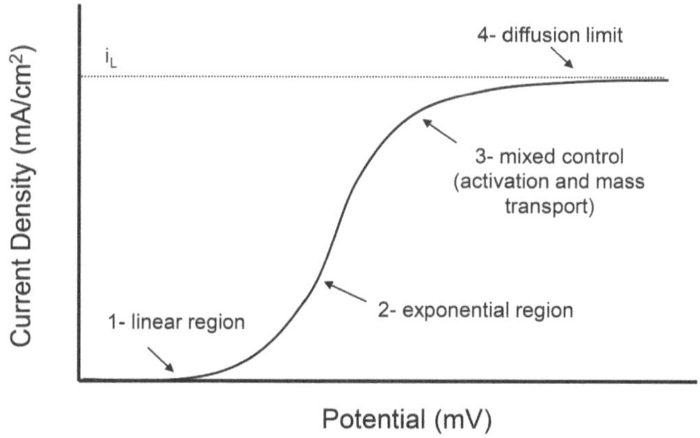

FIGURE 5.4 Current–potential relationship for a Cu electroplating cell.

the cathode. This strong dependence of current upon potential results in the need for plating cell designs which yield uniform potentials across the wafer surface.

Figure 5.4 shows an example of the current density and the voltage relationships for a Cu electroplating cell. The plating process is represented graphically by plotting the current density versus the voltage potential supplied to the electrolytic cell. This graph is used to indicate optimal working points for an electroplating process.

Region (1): This is the linear region from the BV equation. The initiation point is the rest potential of the circuit, where the potential equals zero. The rate is limited by the kinetics at the surface, with minimal current density being recorded. Most Cu plating processes do not operate in this region due to low overall productivity (i.e., wafer output).

Region (2): This is the Tafel region. In this region, the current is proportional to the potential per Ohms law. Mass transport effects are negligible under these conditions. Most copper plating processes are designed to operate in this region.

Region (3): This is the mixed control region where mass transport effects start to have an influence. Copper plating processes with higher productivity targets are generally designed to operate in the region between Regions (2) and (3).

Region (4): This is the region where the current density is mass transport limited. Some species may be depleted at the cathode in which case the limiting current i_L is reached. Approaches to raise the diffusion-limited current density include adopting reactors with high convection and chemistries with high ion concentrations. The fact that this region does not offer sufficient mass transport is a good reason to avoid this operating space for a Cu plating process.

5.2.5 Wagner Number

The dimensionless Wagner number (W_a) is used as a quantity that characterizes the equalizing influence of the overpotential on the current distribution [18]. The larger the Wagner number, the more even the current distribution, despite non-uniform geometry. The Wagner number is the ratio of the polarization resistance at the wafer surface to electrolyte resistance or conductance. The general equation for W_a is provided as follows:

$$W_a = d\eta_c / dj \cdot \kappa / l \tag{5.10}$$

where $d\eta_c/dj$ is the cathodic activation overpotential–current density dependence, κ is the conductivity of the electrolytic solution, and l is the distance between the electrodes.

The Wagner number can also be used as a guide to a uniform current distribution in a scale-up design [19]. Accordingly, the current distribution becomes more uniform for (1) a smaller characteristic length of the system (i.e., smaller die size, etc.), (2) higher solution conductivity, and (3) a steeper slope of the activation overpotential–current density curve. The Wagner number becomes more relevant for through-resist plating applications (wafers containing a photoresist) because the length scale l is a major driver for the current distribution in such applications [20-21].

An example which illustrates a current–potential relationship is shown in Figure 5.5. The curve labeled *No Additives* corresponds to an electrolyte containing sulfuric acid, chloride, and copper with no added organic additives. The plating current rises notably fast at low applied potentials. The curve labeled *Additives* corresponds to a chemistry containing an electrolyte, plus organic additives (accelerator, suppressor, and leveler). In this case, a higher applied voltage is necessary to achieve a given plating current. The shift in potential is evidence that the plating additives are adsorbing to and polarizing the electrode.

Given an operating current density of interest, the potential–current density slope can be qualitatively assessed for both cases. Since the W_a number for each case is

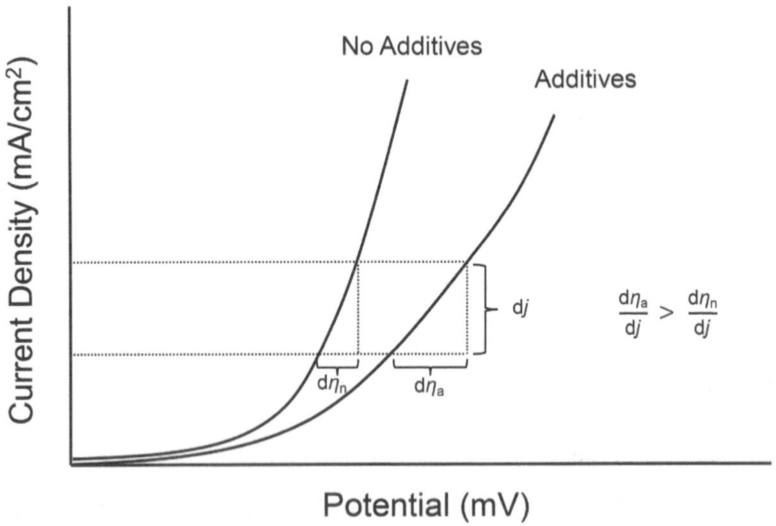

FIGURE 5.5 Current–potential relationship for a Cu electroplating cell showing Wagner number correlation.

directly proportional to $d\eta_c/dj$, we can conclude that *Additive* W_a > *No Additive* W_a. This means that the plating chemistry which contains the organic additives has a higher W_a which is generally desirable for most copper plating processes.

5.2.6 Kinetics and Mass Transfer

Concentration profiles of the ions in solution depend on the solution's hydrodynamics. Hydrodynamics is the forces acting on fluids, such as agitation. Without agitation, there are fewer reactants near the surface of the wafer; while with agitation, there are more reactants near the surface of the wafer. A typical cupric concentration profile as a function of cathode distance (i.e., wafer proximity) is shown Figure 5.6. The strong agitation profile maintains a higher copper ion concentration near the electrode, which is much closer to the bulk concentration. On the other hand, the weak agitation profile shows a much lower concentration near the electrode than a strong agitation profile. This results in a significantly longer distance between the electrode surface and the bulk copper concentration for the weak agitation profile. The distance from the electrode to where the concentration near the electrode meets the bulk concentration is called the Nernst diffusion layer thickness [22]. The slope of the line is used to determine diffusion layer thickness (δ). Note the difference in the diffusion layer thickness for strong agitation versus weak agitation; the diffusion layer is smaller for the strong agitation.

The limiting current is the maximum current that can be passed for a given electrode reaction. The limiting current value is approached as the rate of charge transfer

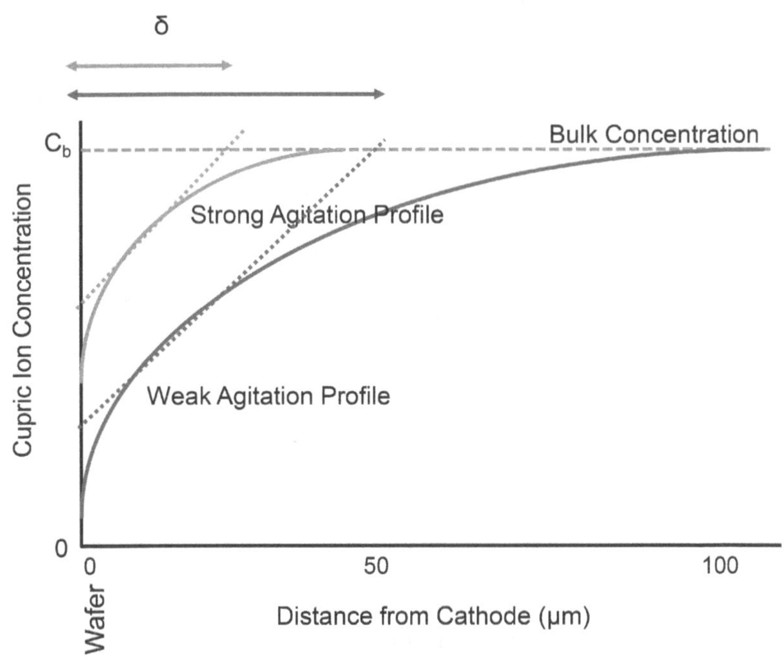

FIGURE 5.6 Cupric concentration profiles in solution at fixed current.

to an electrode is increased. However, the current flow is limited by the rate of mass transfer of ions to the surface. Typically, a high limiting current is desirable so that adequate copper film quality can be attained at the maximized productivity. The current flow is limited by the rate of mass transfer of ions to the surface and can be expressed as follows [23]:

$$i_l = nFDC_b / \delta \quad (5.11)$$

where i_l is the limiting current, n is the electrons per ion reduced, F is the Faraday constant, D is the diffusion coefficient of the reacting ion, C_b is the concentration of the ion in bulk solution, and δ is the diffusion layer thickness.

The limiting current value can be approached by increasing the electrode potential or by decreasing the rate of mass transfer to the electrode. A high concentration of copper ions in the electrolyte will allow the system to reach a higher limiting current.

5.3 CHEMISTRY

5.3.1 Overview

Modern-day copper plating formulations consist of a mixture containing a sulfate-based electrolyte and proprietary organic additives. The electrolyte contains sulfuric acid as a supporting electrolyte, copper sulfate pentahydrate ($CuSO_4{\cdot}5H_2O$) as a source of cupric ions, and chloride ion to modulate cupric ion oxidation and reduction reactions. The make-up solution for the plating bath is usually referenced as virgin makeup solution (VMS), and the VMS solution is supplied without any organic additives. Generally, the organic additives consist of a three-component system: accelerator, suppressor, and leveler additives, where each is added to the plating bath separately. The accelerator additive promotes bottom-up filling and grain refining; the suppressor is added to form a current-inhibiting film and enhance wetting characteristics; and the leveler is added to suppress the current and to control the mass transfer rate [24]. The industry practice is to mix the VMS and organic additives at the target concentrations in the bath reservoir of the plating tool. The plating chemical formulation is summarized in Figure 5.7 and is discussed in detail in this section.

5.3.2 Electrolytes

There are special considerations for the proper electrolyte selection. The electrolyte should provide adequate solution conductivity, support adequate metal ion concentration, and provide adequate diffusion and migration behavior. It should also provide minimal seed rate dissolution and support high current density and efficiency.

As it turns out, the electrolyte concentration varies significantly based on the target application and on-wafer specifications. Generally speaking, a low-acid electrolyte is used for damascene and HB applications. Low-acid electrolyte is a more resistive electrolyte and hence reduces the contribution of the thin copper seed layer

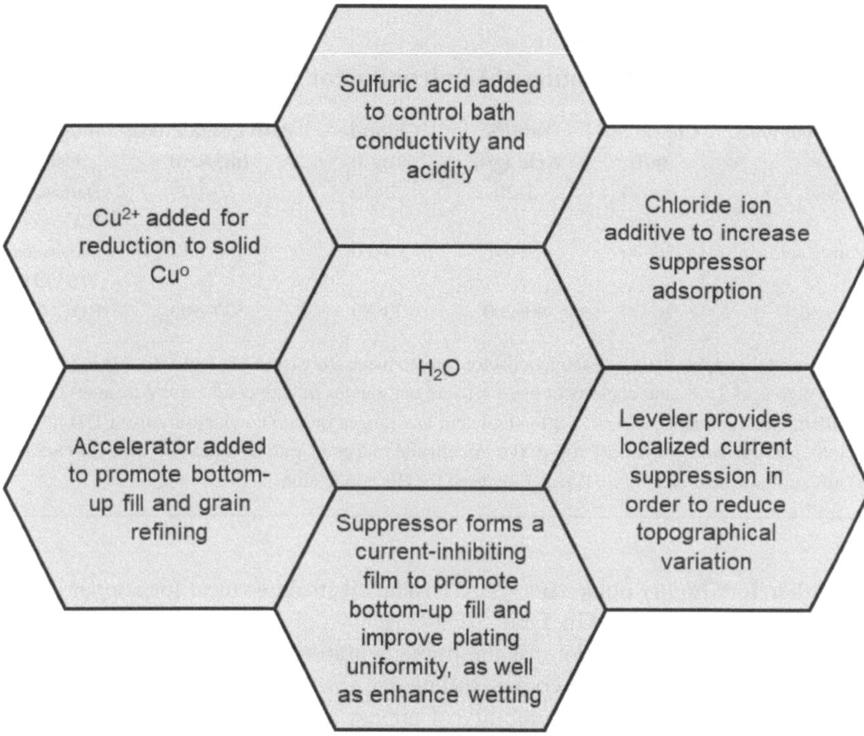

FIGURE 5.7 Plating chemical formulation component summary.

(i.e., terminal effect) and results in superior global plating uniformity [25]. The bottom-up filling characteristic of the plating chemistry is improved with the low-acid electrolyte. High-acid electrolytes are generally used for advanced packaging applications such as redistribution layer (RDL) and copper pillar (CuP). These applications involve copper electrodeposition through a photoresist mask where field and edge effects can dominate the thickness distribution. In this case, the adoption of high-conductivity electrolytes results in a higher overall Wagner number which is optimal for such plating applications. An intermediate (or medium) acid electrolyte is primarily used for advanced packaging applications such as through silicon via (TSV) and through glass via (TGV). These features are quite large in scale in comparison to damascene and HB features, and thus are dominated by copper diffusion to the bottom of the features. The copper concentration in the medium-acid electrolyte is higher than the low- and high-acid electrolytes, while the acid concentration falls within some intermediate concentration levels.

The chloride ion, added as HCl, is present at concentration of 20–80 ppm in copper plating baths. The primary requirement for their addition to the electrolyte is based on their interactions with the organic additives required to achieve desired metallurgy, defect rate, and fill performance. The chloride ion adsorbs strongly on the plated Cu surface and interacts with the suppressor additive, resulting in the formation of a well-adsorbed polymer film which blocks interfacial charge transfer

TABLE 5.2
Electrolytes for Copper Plating of Semiconductor Interconnects

Electrolyte Type	Cupric Ion (g/l)	Sulfuric Acid (g/l)	Chloride (mg/l)	Bath Conductivity (mS/cm)	Interconnect Used for
Low acid	20–50	5–20	20–80	30–60	Damascene, HB
Medium acid	30–70	20–80	20–80	150–250	Damascene, TSV, TGV
High acid	20–50	80–200	20–80	300–500	RDL, CuP

Electrolytes for copper plating of semiconductor interconnects consist of (1) low acid, (2) medium acid, and (3) high acid. Low acid consists of 5–20 g/l acid and ranges in bath conductivity from 30 mS/cm to 60 mS/cm. Medium acid consists of 20–80 g/l acid and ranges in bath conductivity from 150 mS/cm to 250 mS/cm. High acid consists of 300–500 d mS/cm and ranges in bath conductivity from 300 mS/cm to 500 mS/cm. Low-acid electrolyte is generally used for HB applications.

and is therefore highly polarizing [26, 27]. The electrolytes used for copper electrodeposition are summarized in Table 5.2.

One practical consideration for the implementation of a given electrolyte is the copper sulfate solubility. The copper sulfate can form a solid precipitate if the solubility limit is exceeded. The solubility of copper sulfate is reduced with the extra sulfate ions from the sulfuric acid, so this factor needs to be considered when selecting an electrolyte [28]. It is worth mentioning that the electrolyte is manufactured at a chemical production plant and then shipped to the semiconductor manufacturing facility, so the electrolyte must be stable for the shipment conditions. The copper and acid concentration are selected such that there is no $CuSO_4$ salting during shipment.

5.3.3 Accelerators

As mentioned above, organic additives are added to the electrolyte to achieve the desired metallurgy, defect rate, fill, thickness distribution, and reliability performance. The additive package varies between chemical suppliers and each exact molecule is proprietary in nature. The additive classes are commonly referred to as accelerator, suppressor, and leveler molecules based on the primary function of each of the three additive types. These are further described in the section below.

Accelerators are usually small molecules functionalized by disulfide or sulfonic head groups and are fast-diffusing species. Accelerators are commonly added to the electrolyte as the sodium or potassium salt of dimercaptopropone sulfonic acid (SPS) at very low concentrations corresponding to 1–100 ppm. In most cases, SPS or its reduced monomer, 3-mercaptopropylsulfonate (MPS), is utilized as accelerator for copper electrodeposition [29–31]. Figure 5.8 shows the chemical structure of the SPS and MPS molecules.

On a macroscale, accelerators act as brighteners and grain refiners for Cu plating. In the presence of adsorbed Cl^- and suppressor on the copper surface, accelerators

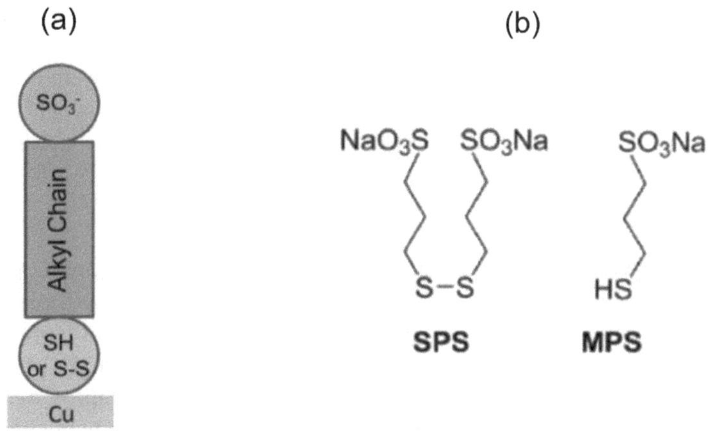

FIGURE 5.8 (a) Schematic of an adsorbed accelerator molecule to the copper surface, and (b) molecular structure of SPS and MPS molecules [32].

increase the local current density or the deposition rate by displacing suppressor molecules [33]. This property is critical for bottom-up fill process to occur, ensuring void-free plating of nominal to high-aspect-ratio HB features and other advanced damascene and TSV structures.

5.3.4 Suppressors

Suppressors are molecules added to plating solutions to improve the die/chip-level plating uniformity. Suppressors are water-soluble, long-chain, high-molecular-weight polymers. Molecular weight of suppressors generally ranges from about 1,000 to 100,000, and solution concentrations typically range from 100 to 10,000 ppm [34, 35]. Typically, polymers such as polyethylene glycol (PEG) or polypropylene glycol (PPG) or their copolymers are utilized as suppressors, as shown in Figure 5.9. Suppressors adsorb to the copper surface and polarize it, i.e., reduce the local current density at a given applied voltage. The degree of polarization is dependent on its molecular weight, with higher molecular weight polymers having higher current suppression. As stated earlier, the interaction of the suppressor with the adsorbed chloride ion greatly increases the suppressor adsorption strength. Furthermore, these properties are dependent on the type of molecule and its molecular structure (e.g., PEG versus PPG) [36, 37]. The behavior of the suppressor is also dependent on its

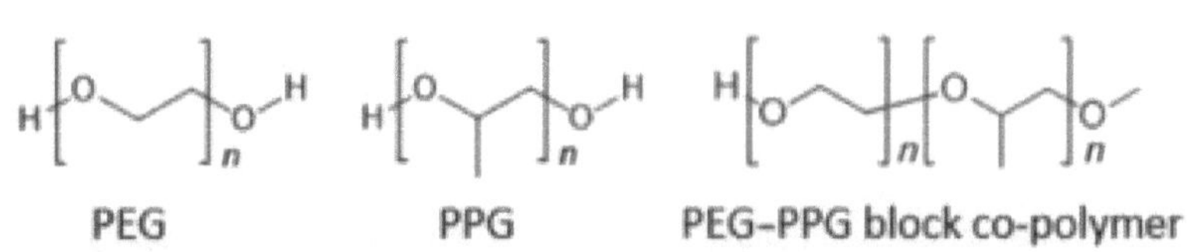

FIGURE 5.9 Example of molecules used as suppressors [39].

surface tension when added to an aqueous solution. The wettability of the solution decreases from about 70 dyne/cm for deionized water to 40–60 dyne/cm with the suppressor [38]. The suppressor allows the plating surface to be wetted more easily and enables high-aspect-ratio features to be wetted at the bottom via capillary action. This is why a lot of suppressors are also commonly referred to as "wetting agents."

5.3.5 Levelers

Levelers are a type of suppressor molecule that are added to plating chemistries to improve plating uniformity. Levelers adsorb and suppress the rapid plating at corners, edges, or other geometric irregularities (i.e., protrusions and bump-out areas) on a plated surface [40]. Their names stem from their ability to make plating across a surface more flat or "level." Levelers minimize surface roughness, reduce topographical variation, and reduce variation in the height of the elements being plated. Levelers are often charged, nitrogen-containing polymers [41–45] with molecular weights in the range of 1,000–15,000 [46] and are utilized at low levels in the plating bath, often near or under 1 ppm. Under these ultralow concentrations, the leveler coverage is diffusion-limited; hence coverage tends to be highest at corners, edges, and elements where convection is greatest. This contrasts with suppressors, which are usually present in excess and thus are not sensitive to mass transfer differences. Examples of typical leveler molecules are shown in Figure 5.10.

5.3.6 Additive Interaction

Chronoamperometry tests on a rotating disk electrode (RDE), where the individual additive is injected in the plating solution, are a good way to examine additive interactions. Such tests allow the potentiostat to vary the voltage to maintain the current setpoint. An injection experiment consisting of sequentially injecting (1) accelerator, (2) suppressor, and (3) leveler, for three different additive systems is provided in Figure 5.11. Upon injecting the accelerator additive at 20 s, a slight decrease in polarization is observed. After 120 s elapses, the suppressor is injected into the plating solution and an increase in potential can be observed for all three (3) suppressors, although to a varying degree. Each plating system depolarizes between 120 s and 220 s. This is due to the accelerator species desorbing the suppressor. At the 220 s mark, the leveler additive is injected

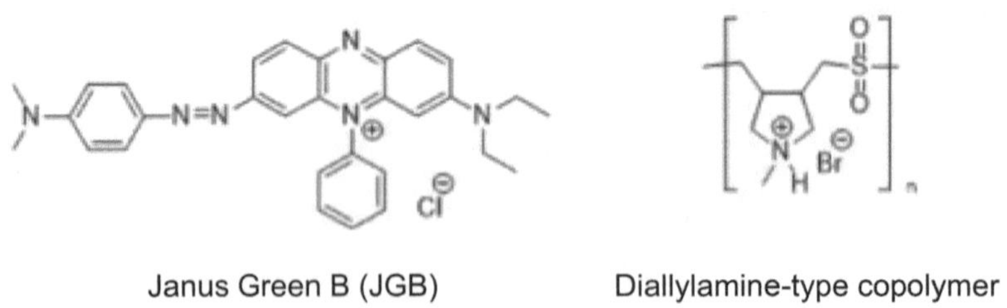

FIGURE 5.10 Example of molecules used as levelers [47].

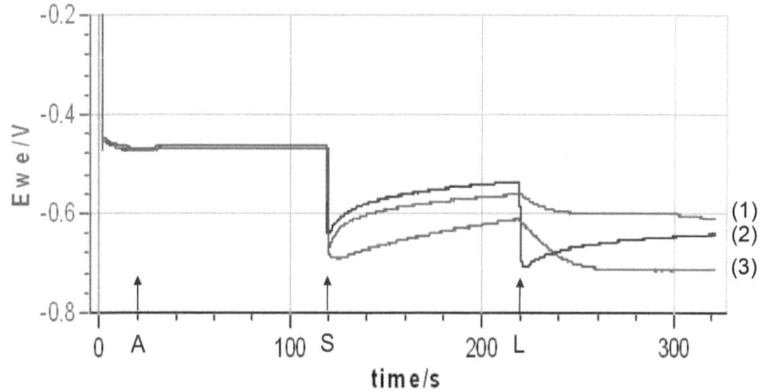

FIGURE 5.11 Injection experiment to characterize the additive (A = accelerator, S = suppressor, and L = leveler) function and the associated chronoamperometric response.

into the plating solution. The leveler corresponding to the curve 2 shows significant Add label (3) initial polarization of the electrode but then depolarization over time, while curve (1) shows only slight polarization of the electrode. The leveler corresponding to curve (3) shows very little initial polarization effects during the first 20 s but eventually increases in polarization which is indicative of accelerator displacement/deactivation.

5.4 HARDWARE AND PROCESS

5.4.1 Plating Reactor

The plating reactor is the core of the copper electrodeposition process. It contains several hardware and control elements which enable optimal current distribution and plating solution flow to the wafer surface [48]. Another key requirement of the plating reactor is defect-free plating; thus, the elimination of any particles or bubbles in the plating solution is very critical.

A typical horizontal plating reactor for a wafer-based system is shown in Figure 5.12. The wafer serves as the cathode. The wafer is placed into the plating cup assembly which seals electrical contacts around the periphery of the wafer [49, 50]. Most commercially available plating reactors incorporate phosphorus-containing copper anodes [51, 52]. These anodes are stored in the anode chamber which is isolated from the wafer by the incorporation of a membrane. The membrane material can vary depend on the platform, but state-of-the-art plating reactors utilize a cationic exchange membrane [53]. The membrane serves multiple purposes. First, the membrane isolates the plating additives from the anode, and thus reduces the degradation rate and saves the chemical cost. Second, the membrane isolates any particles which are formed on the copper anode, thereby ensuring defect-free plating. The flow and current distributor hardware helps maintain uniform current density and plating solution transport to the wafer surface [54].

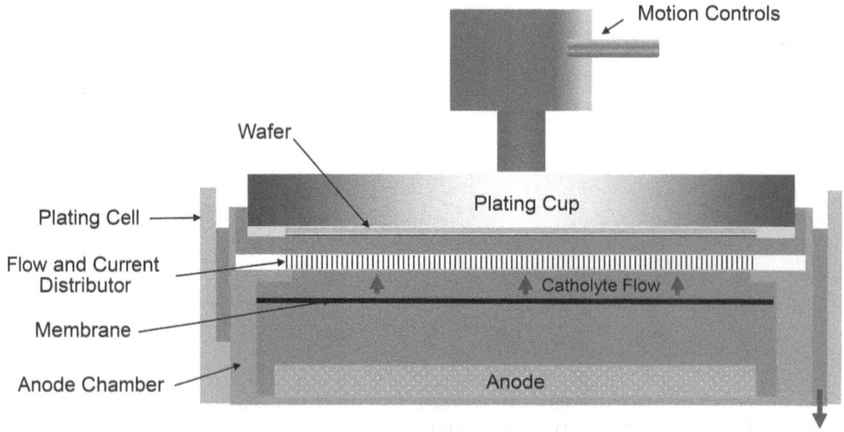

FIGURE 5.12 Plating reactor commonly used for damascene and TSV applications.

The wafer is rotated and precisely immersed into the plating solution by utilizing the motion control hardware. This ensures that bubbles are ejected from under the wafer and that an optimal current density is applied to the wafer while it is being immersed into the plating solution [55]. Figure 5.13a illustrates a two-dimensional model which shows the leading edge of the plating cup touching the plating solution and displacing liquid. Figure 5.13b depicts a three-dimensional model which shows the wafer entry sequence and the progression of the wetting front over the wafer surface.

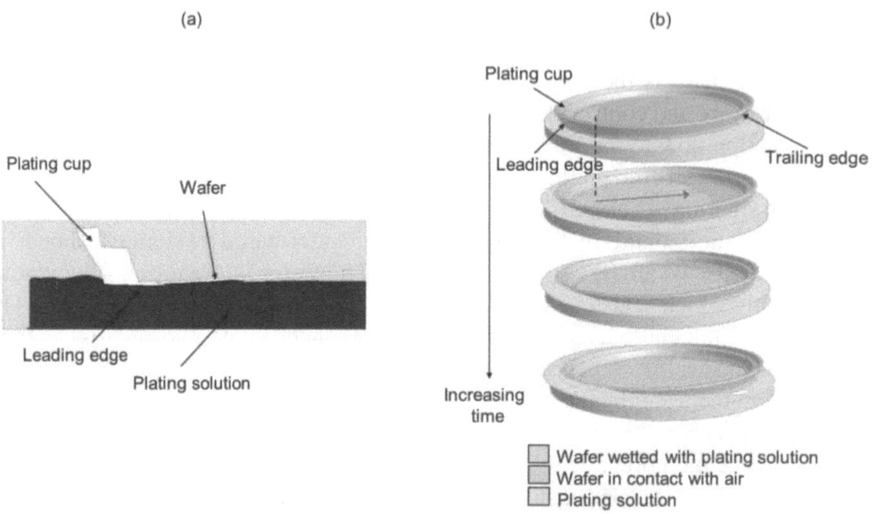

FIGURE 5.13 (a) 2D model showing wafer entry at the leading edge of the plating cup. (b) 3D model showing the wetting front propagation over the wafer surface during the entry process.

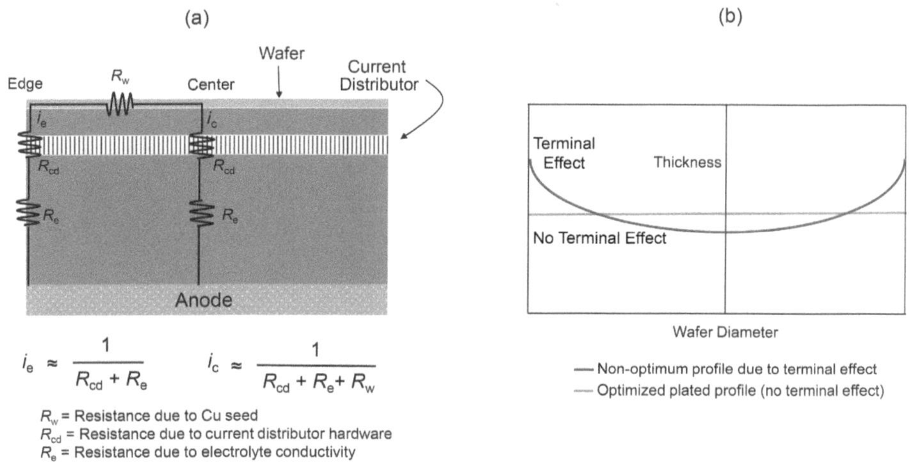

FIGURE 5.14 (a) Simplified plating circuit showing the critical resistances in the plating cell. (b) Copper-plated thickness distribution with and without the terminal effect.

One very key aspect of the flow and current distribution hardware is that it can address any "terminal effect" issues. Terminal effect is when the resistance of the copper seed (R_w) on the wafer starts to dominate the overall resistance of the circuit by creating an appreciable voltage drop across the wafer surface [56, 57]. When this condition arises, the current applied to the "terminals" (i.e., electrical contact points) at the wafer edge (i_e) becomes greater than the current applied to the wafer center (i_c). The net result is an increase in the copper deposition thickness at the wafer edge. The flow and current distribution hardware serves as a swamping resistor (R_{cd}) in the plating circuit and can prevent the terminal effect condition, thus resulting in a uniform plating thickness [54]. An illustration of the terminal effect is provided in Figure 5.14.

Applications involving copper electrodeposition through features formed in a photoresist require a specific plating reactor design [58]. In such applications, it is very important to deliver very uniform, high metal ion transport to the bottom of the features. For this design, the flow distributor is removed from the current distributor and located at the periphery of the wafer, and the plating solution is introduced in a crossflow-type manner [59–61] (Figure 5.15a). The plating solution enters the very narrow gap which is created by the wafer plane and the current distributor, and then penetrates through the photoresist openings on the wafer and transports the copper ions and plating additives to the feature bottom (Figure 5.15b). The wafer is rotated in this flow field to achieve averaging across the wafer and within the features.

5.4.2 Power Supply

A direct current (DC) power supply is used for the copper electrodeposition process. The power supply can apply a current or voltage to the wafer immediately as it enters the plating solution, or after several seconds of immersion in the plating solution. For

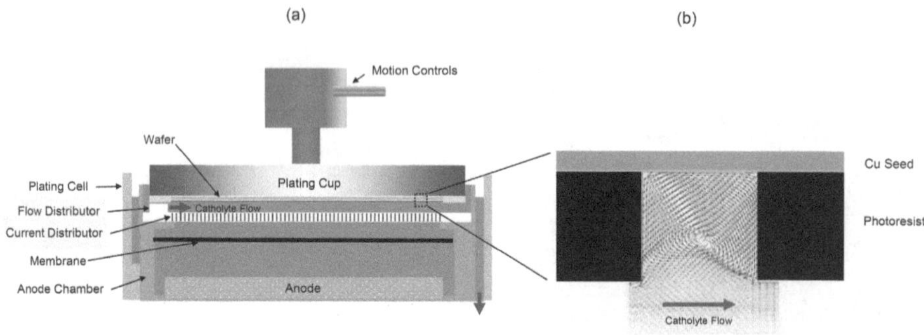

FIGURE 5.15 (a) High-convection plating reactor generally used for wafer-level packing (WLP) or through-resist plating applications. (b) 2D flow model shows that plating solution crossflow generates significant recirculation pattern within the photoresist features.

some plating applications, Cu seed dissolution may result in voiding or defects, so in such instances a current or voltage is applied to the wafer immediately upon contact with the plating solution.

After the wafer entry and plating initiation process, a plating current is applied to the wafer which is highly contingent on the type of wafer and plating process [62]. For damascene features which are typically used for HB, a current density of 3–15 mA/cm^2 (0.07–0.36 μm/min) is used for filling the feature and then a current density of 40–70 mA/cm^2 (1–1.75 μm/min) is used for plating the overburden, i.e., additional thickness over the top of the features.

Although most copper plating processes consist of a simple DC waveform, more advanced waveforms such as pulse [63, 64] and pulse-reverse [65] plating are used for some niche copper plating applications. Such waveforms have shown improved fill-performance in certain additives systems, as well as enhancing the copper metallurgy.

5.4.3 Fluid Distribution System

The fluid distribution system within the electroplating platform starts with a centralized plating bath reservoir. The VMS, DI water, and organic additives are dosed into the plating bath reservoir. A recirculation pump is connected to the plating bath reservoir and runs continuously to mix these components. A plating pump is also connected to the plating bath reservoir. It pumps catholyte through an in-line filter with media of 0.5–10 μm pore sizes to remove any particulates, and then delivers this fresh catholyte solution to the plating cell. The plating cell operates in an overflow manner; therefore, the spent catholyte flows by gravity back to the plating bath reservoir. There is significant hardware and controls used to create a quiescent flow to minimize aeration for this catholyte return to the plating bath reservoir [66, 67]. The anolyte solution is circulated through the anode chamber by a dedicated pump referred to as the separated anode chamber (SAC) pump [68]. The main purpose of this pump is to provide sufficient irrigation of the copper anode during plating to minimize $CuSO_4$ crystallization and passivation. See Figure 5.16 for the fluid distribution system.

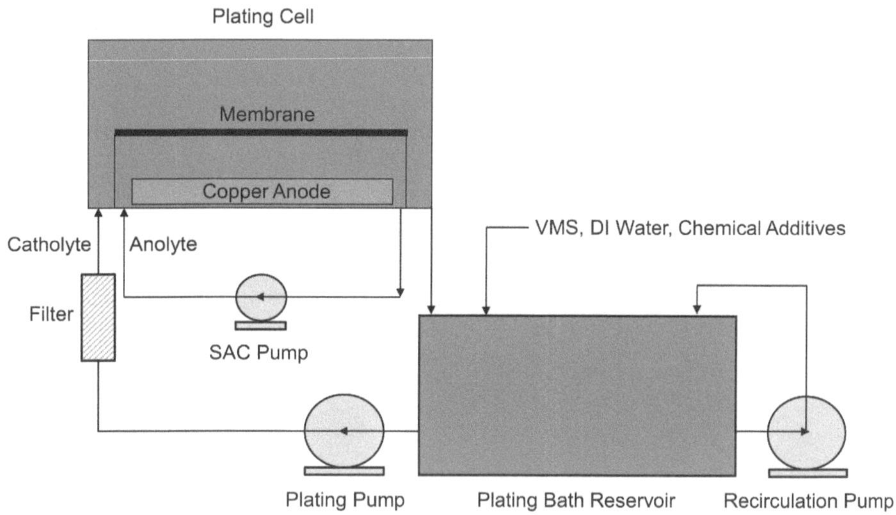

FIGURE 5.16 Fluid distribution system.

5.4.4 Dosing System

The dosing system is used to maintain the plating bath composition at target concentrations. The plating additives breakdown over time and need to be replenished so that the on-wafer plating metrics can be maintained [69–71]. There are dedicated additive bottles which allow for corrective dosing of each individual additive, while the VMS and DI water is dosed to the bath for both replenishment and evaporative control. Most plating baths are connected to a bath analyzer which is fed by the recirculation pump. The bath analyzer is used to measure each individual component within the bath and transmits this information to the plating bath control system (see Figure 5.17) [72].

The plating bath control system uses this information in a feedback mode and will dose any VMS, DI water, or additive if there is a deficit. If a bath measurement is out of specification, a warning may be displayed and in some instances the plating bath will be placed in an error state [73]. The bath analyzer is often multiplexed to several plating baths; therefore, it is possible that only one to two bath measurements are performed each day on a particular bath. For this reason, the plating bath control system also operates in a feedforward state. That is, the user defines dosing constants for the degradation rates of the plating additives, and the dosing system will automatically dose based on these settings.

As previously mentioned, the plating additives are known to degrade and form potentially harmful by-products which can alter deteriorate the plating performance. Wafers that contain a photoresist are known to leach from the wafer and interfere with the plating process. The bath replenishment, sometimes referred to as bleed and feed, helps maintain the by-products in the plating solution under control. The user specifies this replenishment value, and the system automatically doses VMS and additives to the plating bath. When the bath reaches a specified level, the dump valve opens, and a

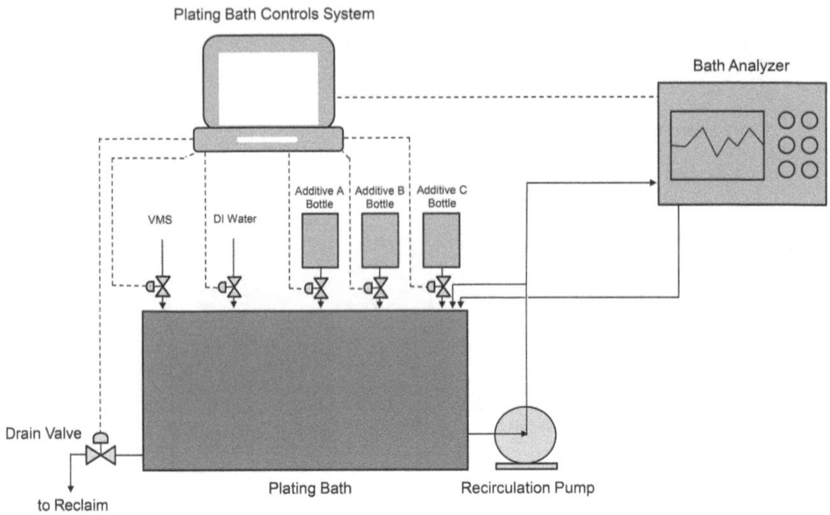

FIGURE 5.17 Copper plating bath dosing and control system.

small fraction of the plating solution is drained from the bath to the reclaim system and reclaimed in an off-board metal recovery system. This sequence is repeated over time and the by-products are managed as a result [74].

5.4.5 Wafer Processing Sequence

The plating modules and associated pretreatment and posttreatment modules are assembled in a manner which maximizes productivity (system throughput) while achieving the on-wafer specifications. The process sequence within a copper electroplating tool used for HB is shown in Figure 5.18. The wafer is (1) transferred from the

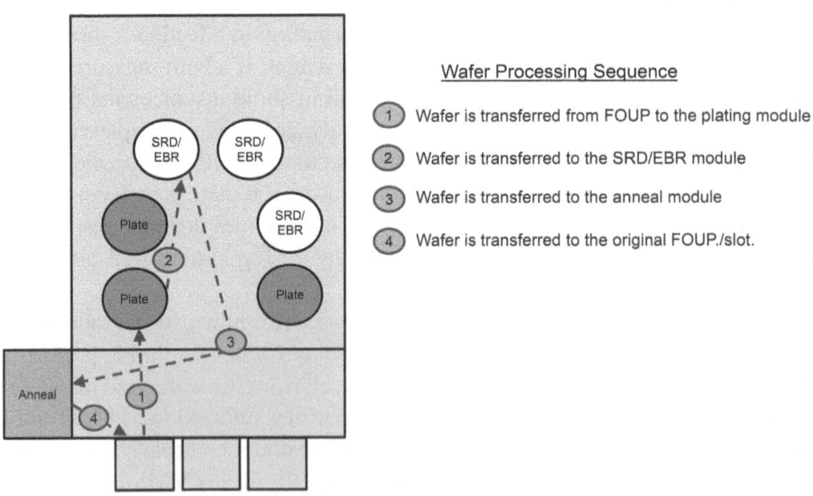

FIGURE 5.18 Wafer processing sequence [77].

FOUP (front-opening universal pod) to the plating module by robotics. The wafer is plated with copper to the desired thickness in the plating module. The wafer is then (2) transferred to the SRD (spin–rinse–dry)/EBR (edge-bevel removal) module. An EBR process is performed to etch the outer ~2 mm of copper at the far edge of the wafer in preparation for the downstream chemical mechanical planarization/polish (CMP) step [75]. The wafer is then spin rinsed and dried (SRD), and (3) transferred to the anneal chamber. In the anneal chamber, the wafer is heated to a specified temperature and time with ~3% H_2 forming gas [76]. The wafer is cooled and then (4) transferred to the original slot location within the FOUP. Once all wafers within a FOUP have been processed, the factory automation system will transfer the wafers to a downstream metrology inspection tool or the CMP tool.

5.5 COPPER INTERCONNECT APPLICATIONS

5.5.1 Damascene and SemiAdditive Process Flow

Copper interconnects are generally formed by one of two different process flows. The first such approach is the damascene process flow [78]. For this approach, features are etched in the dielectric (i.e., SiO_2) and then lined with a thin layer of barrier metal (Ti, Ta, TaN, etc.) and copper seed. The feature is then electrodeposited with copper to form a void-free deposit. The copper overburden (the overburden is actually necessary to drive grain growth in the film so we should not state that it is undesirable layer of Cu over top of the plated features) and underlying barrier metal are removed by the CMP process. The alternate interconnect approach is referred to as the semiadditive process (SAP) flow [79]. A photoresist material is dispensed on the surface and features are patterned by lithography and development. The features are subsequently plated by copper electrodeposition in the patterned regions where photoresist is not present. After plating, the photoresist is removed by a solvent-strip process, and the copper seed and barrier metal are removed from the surrounding features by a chemical etch step (see Figure 5.19).

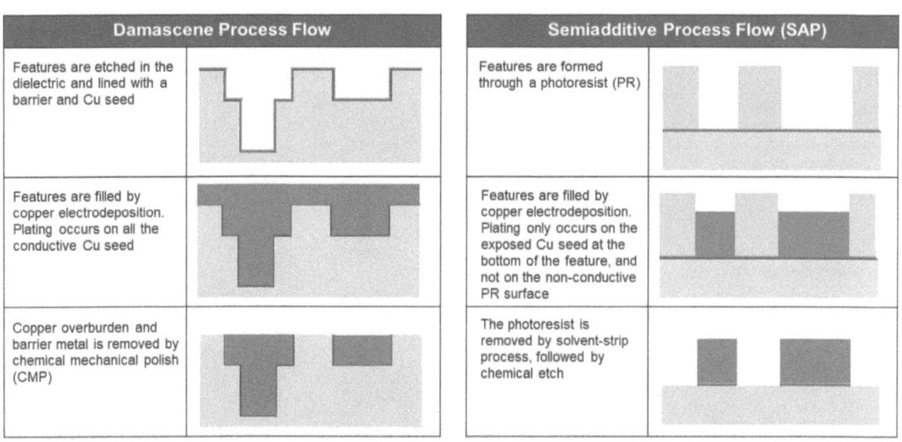

FIGURE 5.19 Damascene and semiadditive process flow.

5.5.2 Damascene Plating: Bottom-up Fill Mechanism

Copper electroplating of damascene features requires void-free plating. Most of these features are greater than 1:1 aspect ratio; so if plated conformally, severe voiding would be observed. To avoid this, the practice of additive-assisted bottom-up Cu filling is utilized for semiconductor device metallization and is the technology that enables void-free copper electrodeposition in a very complex interconnect circuitry. There are various mechanisms which are published on bottom-up filling [80–83]. One such very recognized mechanism is described by Reid et al. [80] and is based on bottom-up filling by accelerator accumulation at the feature bottom (see Figure 5.20).

The wafer is immersed in the plating bath (a), and additives begin to adsorb the Cu seed while conformal plating begins (b). This initial conformal growth leads to a decrease in the surface area within features, especially at the bottom corners. The suppressing species is quick to adsorb to the field region and slowly diffuses within the feature. This results in a concentration gradient of suppressing species between the feature base and the field. At the corner surface at the feature base, the local current density is believed to undergo transition to a higher current which initiates the bottom-up fill process (c). Once bottom-up fill begins, a growth front moves upward within the feature, while continuing to accumulate accelerator species (d). The growth rate within the feature during this stage of filling reaches a maximum and continues until the feature is filled. In the absence of a leveler, the accelerator remains concentrated over features even after filling and can result in significant overplating above the feature (f). This overplating can be minimized with the proper concentration of a leveler molecule, which deactivates or buries the locally accumulated accelerator molecule (g).

When the plating additives adsorb onto the copper surface, they modify the copper deposition kinetics. Plating from a VMS bath (or VMS plus accelerator) results

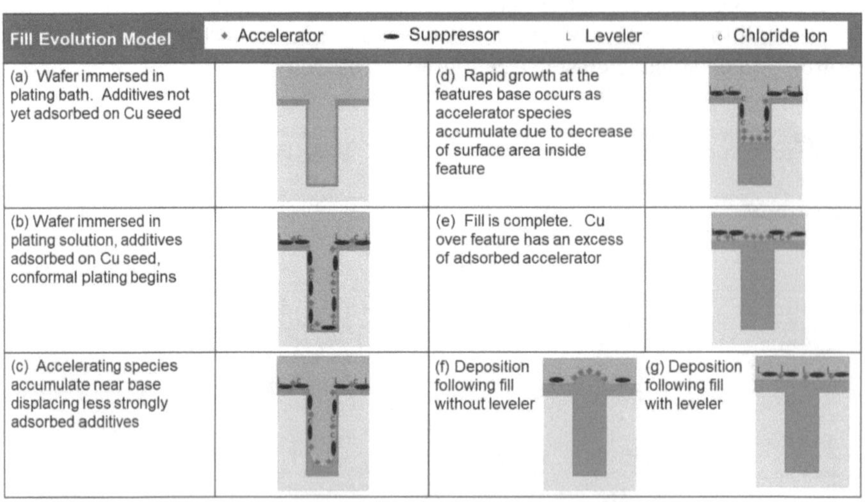

FIGURE 5.20 Fill evolution during bottom-up fill based on accelerator accumulation at the feature base [80].

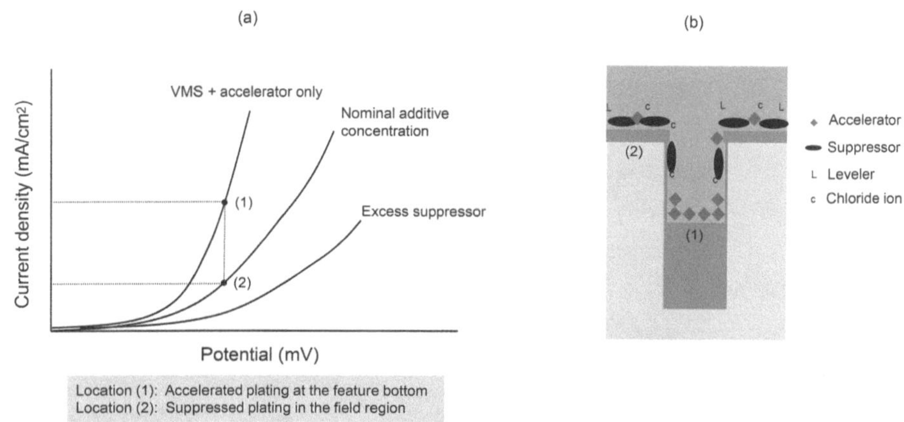

FIGURE 5.21 (a) Polarization characteristics of wafer field and feature base which lead to bottom-up fill based on depolarization within feature due to accelerator accumulation. (b) Partially filled feature showing accelerator accumulation on feature base and suppressor species on the field and upper sidewall.

in a significant deposition current with an applied voltage. See Figure 5.21a. The addition of the suppressor and leveler additives reduces the plating kinetics, while the addition of excess suppressor even further reduces the plating current. As shown in Figure 5.21b, the current near the feature base (Location 1) is estimated by measuring the current–potential behavior at an electrode immersed in an electrolyte containing an accelerator only. Similarly, the current–potential behavior for a plating solution containing all three organic additives best simulates the deposition on the field (Location 2). It is generally beneficial from a feature-fill perspective to center the copper plating process window in a region of maximum feature contrast (i.e., largest ratio between the field current and the feature base current) [84].

5.5.3 Damascene Plating: Fill Evolution for Different Feature Sizes

The fill evolution in 1.2 μm deep trenches ranging from 0.2 μm to 1.7 μm in width was electrodeposited for thicknesses ranging from 38 nm to 490 nm of copper on the field [85]. See Figure 5.22. The smallest feature (0.2 μm) is nearly fully filled after depositing only 38 nm of Cu, while the largest feature (1.7 μm) shows no bottom-up fill—only conformal deposition. As plating proceeds from 86 nm to 250 nm Cu thickness, the smaller features are fully filled, while the intermediate feature size (0.6–0.9 μm) transitions from partially filled to being fully filled. The largest feature size (1.7 μm) achieves full fill after depositing 490 nm thickness.

Contrasting the plating results of these various feature sizes reveals the bottom-up fill mechanism. That is, once sufficient corner growth and accelerator accumulation transpires, a rapid transition to accelerated bottom-up fill occurs for each trench size. There is also a clear feature size dependence of initiation of bottom-up fill acceleration. As the trench width increases, more time is required for corner growth to propagate and allow accelerator species to accumulate across the feature bottom.

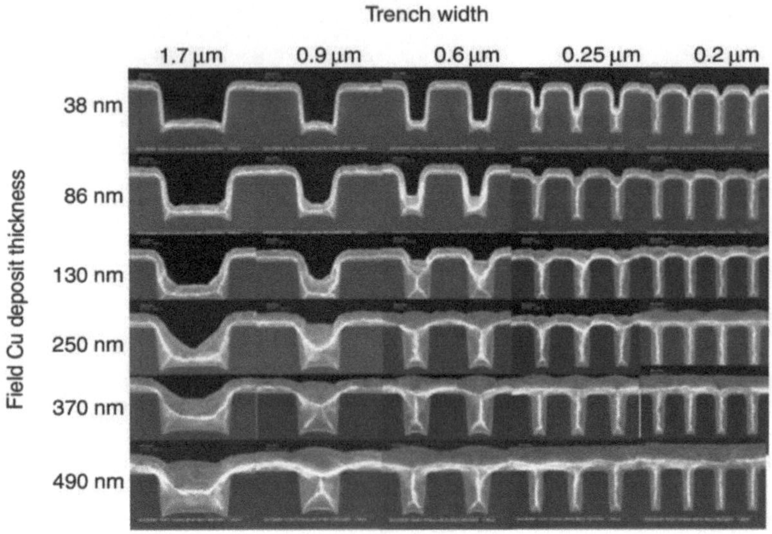

FIGURE 5.22 Damascene fill evolution in trenches as a function of trench width and copper thickness deposited on the wafer field [85].

5.5.4 Through-resist Plating Mechanism

Most copper plating baths that are designed for through-resist plating also contain accelerator, suppressor, and leveler additives, much like damascene plating baths [86]. However, the functional requirement is quite different from damascene plating, so specially designed additives are essential in meeting the on-wafer specifications [87, 88]. The accelerator acts as a copper grain refiner, and thus helps promote a smooth and bright deposit. Sufficient accelerator concentration is required to sustain the higher deposition rate which is typically associated with through-resist plating. The suppressor additive is largely responsible for wetting the photoresist but also polarizes the copper surface to enhance plating uniformity and regulate nodules or dendritic growth in through-mask plating. As discussed earlier, levelers are mass transport limited kinetic inhibitors. They are continuously incorporated into the plated copper film or deactivated at the electrode surface. Levelers play an important role in controlling the feature shape of copper pillars. This is achieved as the leveler transport is the highest to the center of the feature where it more strongly desorbs the accelerator species and minimizes the doming behavior. Another critical aspect is that the electroplating hardware needs to provide uniform high-rate convective mixing to ensure successful implementation of such chemistries [60].

5.5.5 Through-Resist Plating: Feature Shape and Die Coplanarity

Copper pillar shape can be affected by many factors (e.g., copper concentration, additive performance and concentrations, deposition rate, and diffusive and convective mass transfer conditions). At low deposition rates, the pillar shape is generally

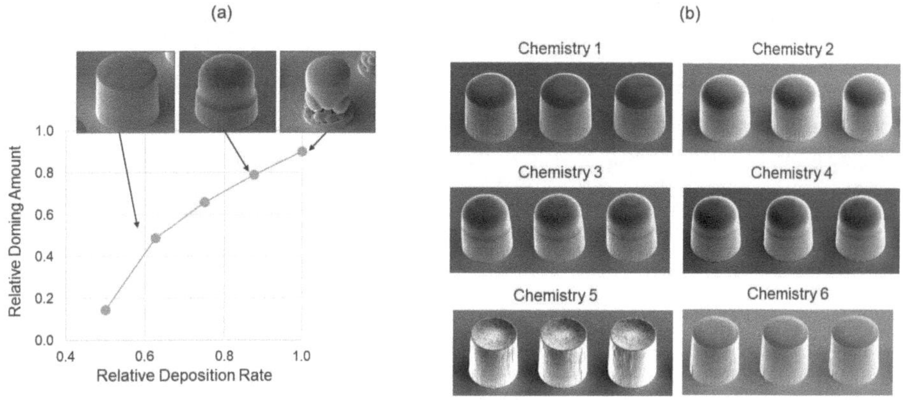

FIGURE 5.23 (a) Cu pillar shape as a function of relative deposition rate. (b) Effect of plating additives on feature shape at a relative deposition rate of 0.8 [90].

flat and uniform but deteriorate as the deposition rate increases. The features' relative center-to-edge deposition rate changes as the balance of copper and/or inhibiting transport varies in time or space. Any disparity in intra-feature microresistance translates to a non-ideal shape (e.g., domed or tilted) [89]. At very high deposition rates, the process becomes substantially copper mass transfer limited requiring delicate additive transport balance and control to avoid significant shape and thickness distribution deterioration (see Figure 5.23a).

Properly chosen plating additives are critical in controlling global and feature thickness distributions at higher deposition rates. This is demonstrated in Figure 5.23b. *Chemistry 1–4* resulted in varying levels of dome-shaped features, while *Chemistry 5* resulted in a dish-shaped feature which is indicative of an excessive-leveling chemistry. *Chemistry 6* resulted in the most optimal (flat) feature shape among all the plating chemicals.

In wafer-level packaging and through-resist plating applications, the within die coplanarity (CoP) of Cu pillar heights has always been critical for ensuring good electrical connections and low stresses in the final package. Unlike the damascene process flow, there are no downstream planarization steps after plating, so the Cu electrodeposition process serves a critical role for through-resist plating technology. Different die layouts and variable pillar dimensions present their own unique challenges for the copper plating process, but the design can be simplified into two categories. As illustrated in Figure 5.24, the coplanarity of Die Type 1 layouts are dominated by the locations of the Cu pillars within the die, and only slightly affected by concentration and convection of the electroplating solution. The coplanarity of Die Type 2 layouts are dominated by the size difference of the pillars rather than their locations within the die, and they are highly sensitive to both Cu concentration and convection. The primary strategy for optimizing coplanarity is to maximize the convection magnitude and uniformity within the plating reactor, in concert with the selection of an optimized plating chemistry [58].

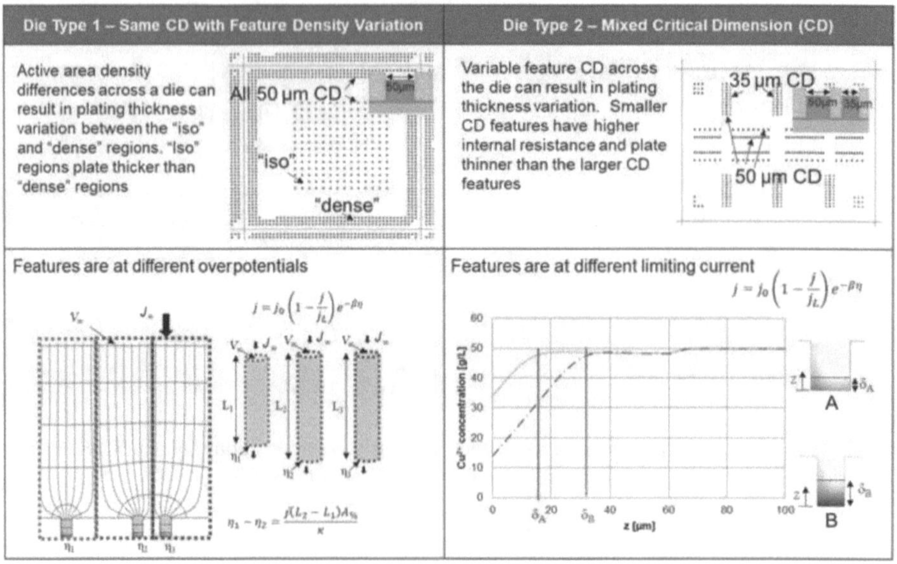

FIGURE 5.24 Die classification for coplanarity optimization [91].

5.6 FILM PROPERTIES

5.6.1 Grain Structure

The grain structure of electrodeposited copper is a critical film property. The traditional approach for damascene BEOL metallization is to engineer copper films with large grains to obtain optimal reliability and electrical performance. This is achieved by a complex combination of deposition current, plating additive selection, plating reactor convection, and film annealing requirement.

Grain growth is the process of grain boundary minimization. The effect is driven by the minimization of the excess free energies trapped in the film during electrodeposition, including that associated with lattice, elastic, interfacial, and grain boundaries [92]. Feature-scale reduction increases the interface-to-volume ratio, interfacial pinning at the sidewall, and trench/via bottoms [93]. Collectively, these items reduce the propensity for grain growth in the feature (see Figure 5.25).

Electrical and mechanical reliability are just as important for HB applications [95]. The fact that the copper-to-copper interconnect for HB is formed by diffusion bonding requires specific consideration to the copper grain structure, orientation, etc. As previously mentioned, the feature size tends to be much larger for HB interconnects (as compared to damascene features), and this enables an opportunity to tailor the copper film with specific grain characteristic. Recently, nanotwin (NT) copper and fine-grain (FG) copper films have shown significant promise for lowering the HB post anneal temperature required to produce high-quality bonding [96].

The nanotwin grain structure is quite unique with respect to a standard grain. In a metal with a face-centered cubic structure such as copper, a twin type of planar defect can be introduced where the normal stacking sequence of (111) planes is

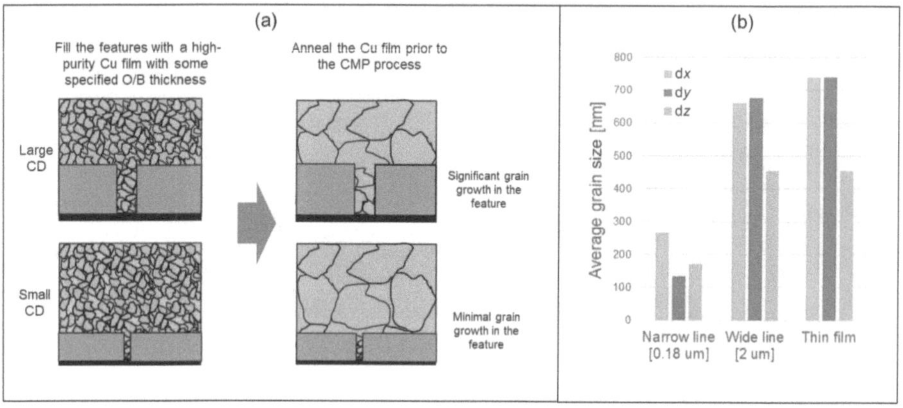

FIGURE 5.25 (a) Illustration of grain growth behavior in small and large CD (i.e., critical dimension) features. (b) Published Cu grain size study for narrow, wide, and thin (blanket) films [94].

mirrored across an internal boundary. The growth occurs in a layer-by-layer plane where the twin thickness is on the order of nanometers, hence the name "nanotwins" [97]. The (111) orientation is ideal for HB since the diffusivity for the (111) orientation is 1,000× higher than in the other directions (see Figure 5.26) [98].

The FG copper film is also unique versus a typical copper grain in that the grains remained pinned during thermal anneal. As described earlier, standard copper films undergo significant grain growth in coarse features and overburden regions during anneal. The FG copper film is engineered such that a high density of grain boundaries is maintained during the HB post-anneal step. The reason for this is because copper diffusion is much higher along the grain boundaries as compared to the lattice diffusivity [99].

As-plated copper grain comparison between a standard, FG, and NT film is shown in Figure 5.27. The average grain size for the standard copper film was 1.5 μm, whereas the FG film resulted in significantly smaller grains, an average of 0.18 μm. In the case of the NT film, an average grain size of 0.7 μm was achieved along with a twin spacing of 80 nm.

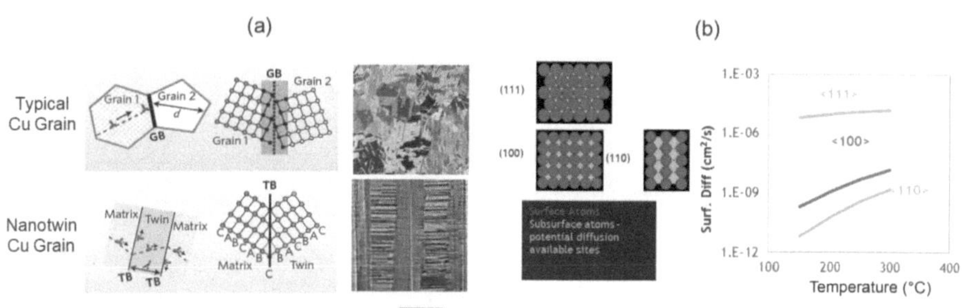

FIGURE 5.26 (a) Comparison of typical and nanotwin grains. (b) Atomic spacing and surface diffusivity for (111), (100), and (110) crystal structure [97].

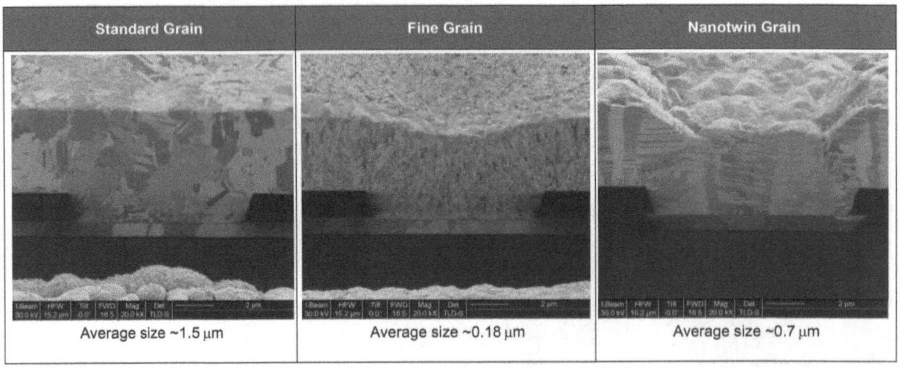

FIGURE 5.27 As-plated Cu grains structure in a HB pad. (HB pad test structure courtesy of Adeia.)

5.6.2 Film Stress

Film stress is rarely monitored in copper deposited for semiconductor interconnects, and its magnitude is highly sensitive to plating chemistry, mass transfer conditions, plating current, bath temperature, etc. This can lead to significant variabilities in the as-plated film stress for standard copper films [100]. The film stress for as-plated wafers and annealed wafers is provided in Figure 5.28. The as-plated stress for a suite of standard (STD) copper films was measured between −80 MPa (compressive) to +20 MPa (tensile). Annealing the STD copper films at 400°C for 120 seconds resulted in films

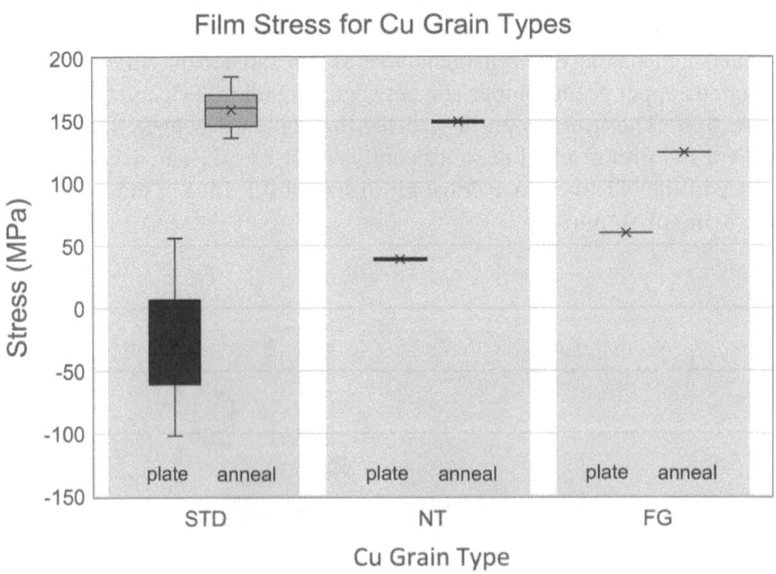

FIGURE 5.28 As-plated and post-anneal (400°C/120 s) film stress for standard (STD), nanotwin (NT), and fine-grain (FG) films. All films plated to 3.0 μm thickness.

under tensile stress (+140 MPa to +180 MPa). The NT and FG films both resulted in higher tensile films, +40 MPa and + 60 MPa, respectively, in comparison to the STD films. Both films became more tensile after annealing at 400°C for 120 seconds. The NT film achieved +150 MPa which was very similar to the STD films; however, the FG film had the lowest stress value of any film (post-anneal) at +125 MPa.

5.6.3 In-film Defects

One of the primary goals for the copper electroplating process is to minimize the occurrence of in-film defects [101, 102]. In-film Defects can be classified as pits, particles, protrusions, and stains (see Figure 5.29).

The most serious are pitting or missing metal defects, which are the result of plated copper failing to deposit on the copper seed layer. These defects can lead to electrical opens in the damascene circuitry following CMP, and are generally flagged during in-line metrology checks of monitor wafers. There are several potential drivers of pit-type defects: (1) incomplete wafer wetting due to air bubble entrapment; (2) non-uniform wetting due to suppressor–additive related polymer agglomerates, micelles, etc.; (3) preexisting contamination of the copper seed; and (4) severe oxidation of the incoming copper seed. Pit-type defects can be minimized through wafer-entry design, chemistry optimization, and queue time control.

Protrusion-type defects are normally associated with non-homogeneous additive adsorption. These regions can act as nucleation sites for excessive growth. This

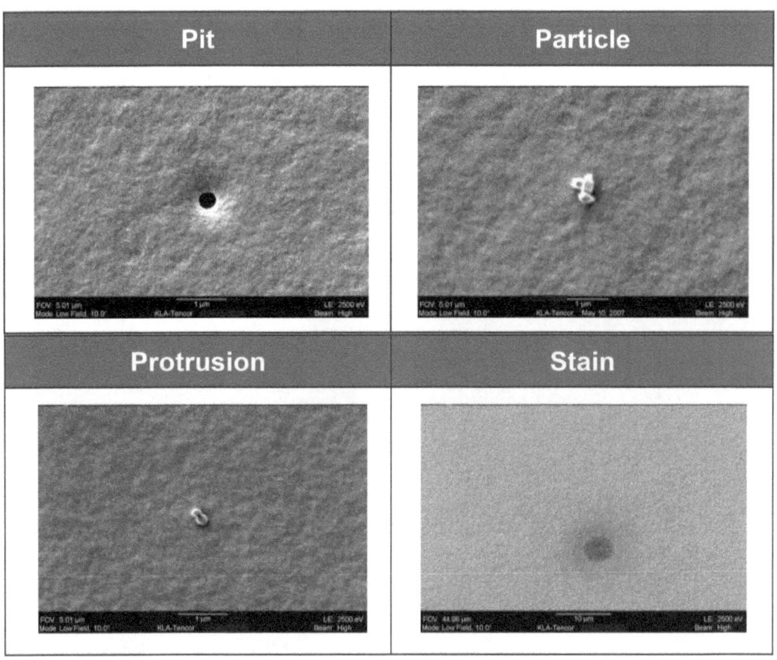

FIGURE 5.29 In-film defects for copper-plated films.

defect type can be overcome by the plating reactor and process set points, as well as proper chemistry selection.

Many sources within the copper electroplating tool can drive particle-type defects. Common sources include (1) small copper particles, which become dislodged from the plating cell hardware [103], and (2) airborne particulates, which arrive on the wafer surface. Mitigation strategies to eliminate particle defects include an anode membrane, optimized in-cell rinsing, plating solution filtration, and adequate clean-air flow within the plating tool mini-environment.

Stain-type defects can also be encountered. These are typically not yield-limiting in nature and driven by insufficient wafer rinse or residual misting within the plating tool mini-environment [104]. Stains can be prevented by careful design of the rinse and spray hardware, along with proper air balancing of the copper electroplating tool.

5.6.4 Integration and Reliability

Electrical characterization is the main method of assessing copper-to-copper interconnect integrity for HB applications [105]. This is done by conducting electrical resistance measurements of bonded daisy chain structures with a two-probe system. Such electrical characterization was recently evaluated [96] for the three microstructures—standard (STD), nanotwin (NT), and fine grain (FG)—and are discussed here.

The measured electrical median resistance of a daisy chain test structure having >31,000 copper interconnects is plotted as a function of temperature in Figure 5.30.

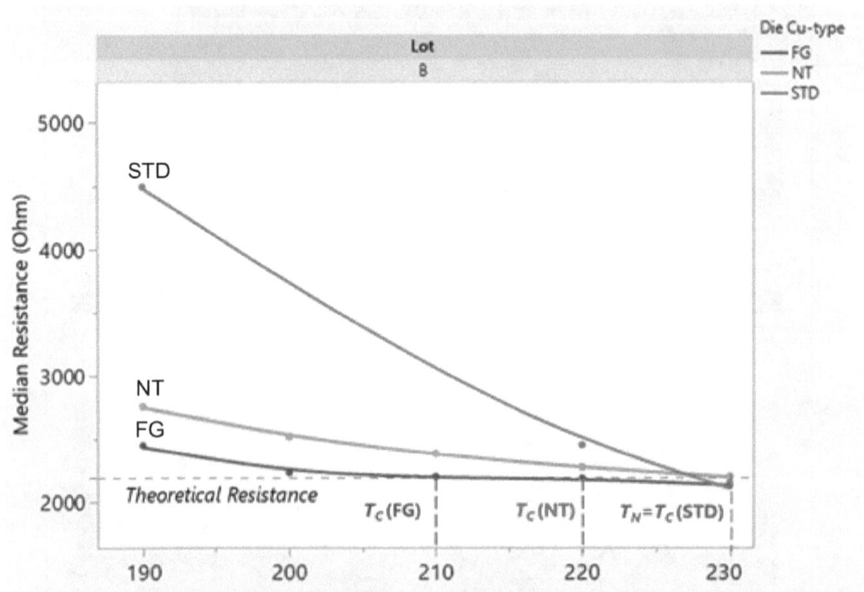

FIGURE 5.30 Via chain median resistance plotted as a function of the anneal temperature for three different microstructures: standard (STD), nanotwin (NT), and fine grain (FG) [96].

TABLE 5.3
Reliability Tests for HB Interconnects [105, 106]

Test	Standard	Test Condition
Autoclave	JESD22-A102D	121°C/100% RH, 15 psi, 168 hours
High-temperature storage	JESD22-A103	225°C, 2,000 hours
Moisture sensitivity test	J-STD-020D-1	MSL3 condition: 1. Bake at 125°C for 24 hours 2. Soak at 30°C/60% RH for 200 hours 3. Three times reflow at 260°C peak temperature
Temperature cycling with MSL3 preconditioning	JESD22-A104D Condition G	–40°C to 150°C, 2,000 cycles

Reliability tests for HB interconnects are listed. Autoclave test follows JESD22-A102D. High-temperature storage test follows JESD22-A103. Moisture sensitivity test follows J-STD-020D-1, and temperature cycling with MSL3 preconditioning follows JESD22-A104D condition G.

Each datapoint represents a sample size of 30–40 bonded die. The theoretical electrical resistance of the daisy chains was calculated based on the design and the actual fabricated dimensions in the test structures. Using the dimensions of the 31,000-link copper daisy chain, the theoretical value of the daisy chain was calculated to be 2,200 Ω. When the circuit resistance equals the theoretical resistance (to within 5% deviation), the joint formation is complete. This is referred to as the complete temperature, T_C, for that microstructure. The normalized temperature, T_N, is when all three samples have reached T_C. The T_C value is 210°C, 220°C, and 230°C for the FG-Cu, NT-Cu, and STD-Cu, respectively. Therefore, the T_N for this test is 230°C. The data indicates that the average FG-Cu and NT-Cu completes interconnect formation at a lower thermal budget than STD-Cu. The FG-Cu consistently reached the theoretical resistance at lower temperatures than the NT-Cu and STD-Cu in this study.

The reliability of HB interconnects has been well established for standard copper grain structures [106]. Typical reliability tests consist of (1) autoclave, (2) high-temperature storage, (3) moisture sensitivity, and (4) temperature cycling, and must adhere to specific global standards identified by JEDEC. Refer to Table 5.3 for the JEDEC reliability tests for HB interconnects.

5.7 SUMMARY

The semiconductor industry has recognized HB as an enabling technology to meet sub-10 μm interconnect pitch requirements for advanced packaging. The HB process flow includes a copper electrodeposition step which is very critical to the formation of the Cu–Cu interconnect part of the hybrid bond. In this chapter, copper electrodeposition fundamentals and the various components of the electrodeposition plating technology for semiconductor metallization are discussed. Organic additives are used in the copper plating bath and are responsible for achieving feature fill and the desired deposit characteristics. The plating reactor serves as the core of the plating technology, while the fluidic handling, dosing system, and tool architecture are essential for achieving high device yields and optimized productivity. Recent progress in copper grain engineering, which includes nanotwin and fine-grain microstructures, has shown significant promise in forming robust Cu–Cu interconnects relative to standard Cu grain structures.

ACKNOWLEDGMENTS

The data examples presented in this chapter were collected by several Lam Research employees in support of copper electroplating development for the SABRE® and SABRE 3D® product.

REFERENCES

1. P. Andricacos, "Copper on-chip interconnections: A breakthrough in electrodeposition to make better chips", Electrochem. Soc. Interface, 8, 32 (1999).
2. C. Beaudry and J. Dukovic. "Faraday in the fab: A look at copper plating for on-chip wiring", Electrochem. Soc. Interface, 13, 40 (2004).
3. R. Akolkar, "Current Status and Advances in Damascene Electrodeposition", Encyclopedia of Interface Chemistry: Surface Science and Electrochemistry, Elsevier (2008) 24–31.
4. S. Lee, Y. Guo and C. Ong, "Electromigration effect on Cu-pillars (Sn) bumps", Electronics Packaging Technology Conference, 135–139 (2005).
5. C. Hau-Riege, Y. Yau and K. Caffey, "The electromigration behavior of copper pillars for different current distribution directions and pillar shapes", 2015 IEEE International Reliability Physics Symposium, IEEE (2016) 5C2.1–5C.2.5.
6. R. Tummala, P. Raj, A. Aggarwal, G. Mehrotra, S. Koh and S. Bansal, "Copper interconnections for high performance and fine pitch flip chip digital applications and ultra-miniaturized RF module applications", Electronic Components and Technology Conference, 102–111 (2006).
7. Y. Park, G. Park, B. Lee, J. Kim and K. Son, "Solder volume effect on electromigration failure mechanisms of Cu/Ni/Sn-Ag microbumps", IEEE Trans. Compon. Packaging Manuf. Technol., 10, 1589–1593 (2020).
8. L. Wang, G. Fountain, B. Lee, G. Gao, C. Uzoh, S. McGrath, P. Enquist, S. Arkalgud and L. Mirkarimi, "Direct bond interconnect (DBI®) for fine-pitch bonding in 3D and 2.5D integrated circuits", 2017 Pan Pacific Microelectronics Symposium (Pan Pacific), Pan Pacific, HI, (2017) 1–6.
9. G. Gao, T. Workman, L. Mirkarimi, G. Fountain, J. Theil, G. Guevara, C. Uzoh, B. Lee, P. Liu and P. Mrozek, "Chip to wafer hybrid bonding with Cu interconnect: High volume manufacturing process compatibility study", Proceedings of the International Wafer-Level-Packaging Conference (2019).
10. E. Beyne, S. Kim, L. Peng, N. Heylen, J. Messemaeker, O. Okudur, A. Phommahaxay, T. Kim, M. Stucchi, D. Velenis, A. Miller and G. Beyer, "Scalable, sub 2 μm pitch, Cu/SiCN to Cu/SiCN hybrid wafer-to-wafer bonding technology", IEEE International Electron Devices Meeting (IEDM), San Francisco, CA, (2017) 32.4.1–32.4.4.
11. Y. Ouyang, S. Yang, D. Yin, X. Huang, Z. Wang, S. Yang, K. Han and Z. Xia, "Excellent reliability of Xtacking™ bonding interface", IEEE International Reliability Physics Symposium (2021).
12. M. Shamsuddin, "Electrometallurgy", Physical Chemistry of Metallurgical Processes, The Minerals, Metals and Materials Society (2016), 523–557.

13. Y. Gamburg and G. Zangari, Theory and Practice of Metal Electrodeposition, New York, NY: Springer (2011).
14. M. Ladd, "Electrochemistry", Introduction to Physical Chemistry, 3rd Edition, Cambridge University Press (1998) 371–412.
15. J. Newman and K. Thomas-Alyea, Electrochemical Systems, 3rd Edition, New York, NY: John Wiley & Sons, Inc. (2004).
16. K. Krischer and K. Schönleber, Physics of Energy Conversion. De Gruyter (2015).
17. A. J. Bard and L. R. Faulkner., Electrochemical Methods Fundamentals and Applications. New York: John Wiley & Sons, Inc. (1980).
18. C. Wagner, "Theoretical analysis of the current density distribution in electrolytic cells", J. Electrochem. Soc., 98 116–128 (1951).
19. K. Popov, S. Djokic and B. Grgur, "The Current Distribution in Electrochemical Cells", Fundamental Aspects of Electrometallurgy, New York, NY: Springer (2002) 101–143.
20. R. Akolkar and V. Dubin, "Pattern density effect on the bottom-up fill during damascene copper electrodeposition", Electrochem. Solid-State Lett., 10 (6) D55–D59 (2007).
21. B. Buckalew, T. Ponnuswamy, J. Reid and Y. Takada "Optimization of fill rate uniformity across dense features", 217th Electrochemical Society (ECS) Conference, Vancouver, BC (2010).
22. N. Perez, "Kinetics of Concentration Polarization", Electrochemistry and Corrosion Science, New York, NY: Springer (2004) 121–154.
23. Bagotskii, V.S., Fundamentals of Electrochemistry, Hoboken, N.J.: Wiley-Interscience, 2006, 2nd ed.
24. P. Vereecken, R. Binstead, H. Deligianni and P. Andricacos, "The chemistry of additives in damascene copper plating", IBM J. Res. Dev, 49 (1) 3–18 (2005).
25. T. Atanasova, L. Carbonell, R. Caluwaerts, Z. Tokei, K. Strubbe and P. Vereecken, "Ultra-low copper baths for sub-35 nm copper interconnects", J. Electrochem. Soc., 160 (12) D3255–D3259 (2013).
26. W. Shao, G. Pattanaik and G. Zangari, "Influence of chloride anions on the mechanism of copper electrodeposition from acid sulfate electrolytes", J. Electrochem. Soc., 154 (4) D201–D207 (2007).
27. K. Hebert, "Role of chloride ions in suppression of copper electrodeposition by polyethylene glycol", J. Electrochem. Soc., 152 (5) C283–C287 (2005).
28. J. Casas, F. Alvarez and L. Cifuentes, "Aqueous speciation of sulfuric acid–cupric sulfate solutions", Chem. Eng. Sci., 55, 6223–6234 (2000).
29. F. Wang and Y. Le, "Bis-(3-sulfopropyl) disulfide acceleration of copper electrodeposition via molecular dynamics and quantum chemical calculations", Int. J. Electrochem. Sci., 15, 4931–4943 (2020).
30. B. Bozzini, L. D'Urzo, C. Mele and V. Romanello, "Electrodeposition of Cu from acid sulphate solutions in presence of bis-(3-sulphopropyl)-disulphide (SPS)", Trans. Inst. Met. Finish., 84 (2) 83–93 (2006).
31. H. Shen, H. Kim, M. Sung, T. Lim and J. Kim, "Thermodynamics aspects of bis(3-sulfopropyl) disulfide and 3-mercapto-1-propanesulfonic acid in Cu electrodeposition", J. Electroanal. Chem., 816 (2018) 132–137.
32. J. Luo, Z. Li, M. Shi, J. Chen, Z. Hao and J. He, "Effects of accelerator alkyl chain length on the microvia filling performance in copper superconformal electroplating", J. Electrochem. Soc., 166 (4) D104–D112 (2019).
33. M. Tan, C. Guymon, D. Wheeler and J. Harb, "The role of SPS, MPSA and chloride in additive systems for copper electrodeposition", J. Electrochem. Soc., 154 (2) D78–D81 (2007).
34. M. Willey and E. McInerney, "Adsorption and desorption kinetics of a block copolymer wetting agent used in copper electroplating", J. Electrochem. Soc., 156 (3) D98–D103 (2009).

35. W. Dow, M. Yen, W. Lin and S. Ho, "Influence of molecular weight of polyethylene glycol on microvia filling by copper electroplating", J. Electrochem. Soc., 152 (11) C769–C775 (2005).
36. S. Kim, D. Josell and T. Moffat, "Electrodeposition of Cu in the PEI-PEG-Cl-SPS additive system", J. Electrochem. Soc., 153 (9) C616–C622 (2006).
37. M. Pasquale, L. Gassa and A. Arvia, "Copper electrodeposition from an acidic plating bath containing accelerating and inhibiting organic additives", Electrochim. Acta, 53, 5891–5904 (2008).
38. S. Chang, Y. Wang, C. Hung, W. Lee and G. Hwang, "Role of surface tension in copper electroplating", J. Vac. Sci. Technol. A, 25 (3) 566–569 (2007).
39. J. Mendez, R. Akolkar and U. Landau, "Polyether suppressors enabling copper metallization of high aspect ratio interconnects", J. Electrochem. Soc., 156 (11) d474–d479 (2009).
40. S. Kim, D. Josell and T. Moffat, "Cationic surfactants for the control of overfill bumps in Cu superfilling", J. Electrochem. Soc., 153 (12) C826–C833 (2006).
41. J. Hatch, M. Willey and A. Gewirth, "Influence of aromatic functionality on quaternary ammonium levelers for Cu plating", J. Electrochem. Soc., 158 (6) D323–D329 (2011).
42. A. Wang, B. Chen, L. Fang, J. Yu and L. Wang, "Influence of branched quaternary ammonium surfactant molecules as levelers for copper electroplating from acid sulfate bath", Electrochim. Acta, 108, 698–706 (2013).
43. Z. Li, B. Tan, J. Luo, J. Qin, G. Yang, C. Cui and L. Pan, "Structural influence of nitrogen-containing groups on triphenylmethane-based levelers in super-conformal copper electroplating", Electrochim. Acta, 401, 139445 (2022).
44. E. Eliadis and R. Alkire, "In situ studies of Cu deposition in the presence of quaternary ammonium salts", J. Electrochem. Soc., 145 (4) 1218–1226 (1998).
45. C. Chang, X. Lu, Z. Lei, Z. Wang and C. Zhao, "2-Mercaptopyridine as a new leveler for bottom-up filling of micro-vias in copper electroplating", Electrochim. Acta, 208, 33–38 (2016).
46. J. Zhou and J. Reid, "Impact of leveler molecular weight and concentration on damascene copper electroplating", ECS Trans., 2 (6) 77–92 (2007).
47. M. Lee, Y. Lee, M Sung, S. Cho, Y. Kim and J. Kim, "Structural influence of terminal functional groups on TEG-based leveler in microvia filling", J. Electrochem. Soc., 147 102505 (2020).
48. J. Reid, R. Contolini, E. Opocensky, E. Patton and E. Broadbent, "Method of electroplating semiconductor wafer using variable currents and mass transfer to obtain uniform plated layer", US Patent 6,074,544, issued on June 13, 2000.
49. R. Rash, S. Ghongadi, K. Ganesan, Z. He, T. Majid, J. Hawkins, S. Varadarajan and B. Buckalew, "Electroplating cup assembly", US Patent 8,377,268, issued on February 19, 2013.
50. S. Ghongadi, R. Rash, J. Hawkins, S. Varadarajan, T. Majid, K. Ganesan, B. Buckalew and B. Evans, "Rapidly cleanable electroplating cup seal", US Patent 8,398,831, issued on March 19, 2013.
51. V. Rashkov, G. Raichevski and L. Vuchkov, "Influence of anodic films on the kinetics and mechanisms of the dissolution of copper anodes in bright acid copper plating", Bulg. Acad. Sci. Comm. Dept. Chem., 11, 459 (1978).
52. X. Cheng and B. Hisky, "Fundamental studies of copper anode passivation during electrorefining: Aprt 2 surface morphology", Metall. Mater. Trans. B, 27B, 610 (1996).
53. J. Reid, R. Contolini and J. Dukovic, "Membrane partition system for plating of wafers", US Patent 6,59,299, issued on May 27, 2003.
54. S. Mayer and J. Reid, "High resistance ionic current source", US Patent 7,622,0224, issued on November 24, 2009.

55. M. Ranjan, S. Ghongadi, F. Wilmot, D. Hill and B. Buckalew, "Wetting wave front control for reduced air entrapment during wafer entry into electroplating bath", US Patent 9,028,666, issued on May 12, 2015.
56. S. Armini, Z. Tokei, H. Volders, Z. El-Mekki, A. Radisic, G. Beyer, W. Ruythooren and P. Vereecken, "Impact of 'terminal effect' on Cu electrodeposition: Filling capability for different metallization options", Microelectron. Eng., 88 754–759 (2011).
57. K. Chen, Y. Wang, L. Chang, F. Li and S. Chang, "Investigation of overpotential and seed thickness on damascene copper electroplating", Surf. Coat. Technol., 200, 3112–3116 (2006).
58. B. Buckalew, "Enhanced CuP plating process marginality for packaging", SEMI WLCSP Equipment Technology Forum, Hsinchu, Taiwan (June 2016).
59. R. Abraham, S. Mayer, B. Buckalew and R. Rash, "Cross flow manifold for electroplating apparatus", US Patent 9,624,592, issued on April 18, 2017.
60. S. Mayer, B. Buckalew, H. Fu, T. Ponnuswamy, H. Camilo, R. Rash and D. Porter, "Enhancement of electrolyte hydrodynamics for efficient mass transfer during electroplating", US Patent 9,834,852, issued on December 5, 2017.
61. B. Kagajwala, B. Buckalew, L. Chua, A. Berke, R. Rash and S. Mayer, "Apparatus and method for electrodeposition of metals with the use of an ionically resistive ionically permeable element having spatially tailored resistivity", US Patent 10,014,170, issued on July 3, 2018.
62. S. Mayer and J. Reid, "Semiconductor counter electrode for electrolytic current distribution control", US Patent 8,147,660, issued on April 3, 2012.
63. T. Ponnuswamy, B. Pennington, C. Berry, B. Buckalew and S. Mayer, "Pulse sequence for plating on thin seed layers", US Patent 8,500,983, issued on August 6, 2013.
64. C. Seah, S. Mridha and L. Chan, "DC/pulse plating of copper for trench/via filling", J. Mater. Process. Technol., 114, 233–239 (2001).
65. K. Ganesan, "Capable copper electrodeposition process for integrated circuit–substrate packaging manufacturing", Ph.D. thesis, Arizona State University (2018).
66. R. Abraham, R. Rash, D. Porter, S. Mayer and J. Ostrowski, "Bubble and foam solutions using a completely immersed air-free feedback flow control valve", US Patent 10,208,395, issued on February 19, 2019.
67. S. Banik, A. Berke, G. Graham, G. Kearns, L. Chua and B. Buckalew, "Cross flow conduit for foaming prevention in high convection plating cells", US Patent 11,585,007, issued on February 21, 2023.
68. D. Kim, S. Ghongadi, Y. Takada, L. Huang and T. Majid, "Membrane design for reducing defects in electroplating systems", US Patent 9,677,190, issued June 13, 2017.
69. P. Mocoteguy, C. Gabrielli, H. Perrot, Z. Zdunek and D. Sanz, "Influence of the anode and accelerator on copper bath aging in the damascene process", J. Electrochem. Soc., 153 (12) G1086–1098 (2006).
70. W. Lee, C. Hung, S. Chang and Y. Wang, "Bis-(3-sodiumsulfopropyl disulfide) decomposition with cathodic current flowing in a copper-electroplating bath", J. Electrochem. Soc., 157 (1) H131–H135 (2010).
71. M. West, M. Anderson, Q. Wang and T. Bailey, "Quantitative monitoring of copper electroplating additives and their breakdown products", Electrochemical Processes in ULSI and MEMS, Proceedings Volume 2004–17, Pennington, NJ (2004) 41–56.
72. R. Contolini, J. Reid, S. Mayer, E. Broadbent and R. Jackson "Use of on-line chemical analysis for copper electrodeposition", Advanced Metallization Conference 1999 (AMC 1999), edited by M. Gross, Warrenton PA: MRS Press (2000), 117–123.
73. J. Reid, "Electroplating chemistry on-line monitoring and control system", US Patent 6,458,262, issued on October 1, 2002.
74. P. Andricacos, J. Dukovic and L. Romanikw, "Method for controlling species concentrations", US Patent 5,352,350, issued on October 4, 1994.

75. S. Mayer, S. Varadarajan and D. Preston, "Edge bevel removal of copper from silicon wafers", US Patent 7,780,867, issued on August 24, 2010.
76. T. Ponnuswamy, J. Sukamto, J. Reid and S. Mayer, "Two step copper electroplating process with anneal for uniform across wafer deposition and void free filling on ruthenium coated wafers", US Patent 7,964,506, issued on June 21, 2011.
77. E. Patton, T. Cacouris, E. Broadbent and S. Mayer, "Sequential station tool for wet processing of semiconductor wafers", US Patent 8,026,174, issued on September 27, 2011.
78. J. Kriz, C. Angelkort, M. Czekalla, S. Huth, D. Meinhold, A. Pohl, S. Shulte, A. Thamm and S. Wallace, "Overview of dual damascene integration schemes in Cu BEOL integration", Microelectron. Eng., 85, 2128–2132 (2008).
79. Y. Lee, K. Harr, C. Lee, J. Kim, D. Kim and Y. Kweon, "The formation of ultra fine Cu metal line with sub-5 micron using photolithography and semi additive process", ECS Trans., 52 (1) 337–341 (2013).
80. J. Reid, S. Mayer, E. Broadbent, E. Klawuhn and K. Ashtiani, "Factors influencing damascene feature fill using copper PVD and electroplating", Solid State Technol., 43 (7) 86–99 (2000).
81. R. Akolkar and U. Landau, "Mechanistic analysis of the "bottom-up" fill in copper interconnect metallization", J. Electrochem. Soc., 156 (9) D351–D359 (2009).
82. T. Moffat, D. Wheeler and D. Josell, "Superconformal electrodeposition of copper", Electrochem. Solid-State Lett., 4, C26v.
83. A. C West, "Theory of filling high-aspect ratio trenches and vias in presence of additives", J. Electrochem. Soc., 147 227 (2000).
84. D. D. Josell and D. Wheeler, "A simple equation for predicting superconformal electrodeposition in submicron trenches", J. Electrochem. Soc., 148, 767 (2001).
85. J. Reid, "Damascene Copper Electroplating", Handbook of Semiconductor Manufacturing, CRC Press (2017) 16.1–16.47.
86. Y. Zhang, S. Chung, E. Walch, C. Rietman, C. Wang, T. Richardson, E. Kuo and M. Tseng, "A high-speed cu pillar bump plating process", 2008 3rd International Microsystems, Packaging, Assembly & Circuits Technology Conference, IEEE (2008) 28–31.
87. G. Poon, D. Williams and K. Chin, "Optimising the lithographic patterning effect in an acid copper electroplating process", Int. J. Adv. Manuf. Technol., 16, 881–888 (2000).
88. Y. Oh, S. Chung and M. Lee, "Optimization of thickness uniformity in electrodeposition onto a patterned substrate", Mater. Trans., 45 (10) 3005–3010 (2004).
89. K. Thorkelsson, S. Banik, S. Mayer and B. Buckalew, "Plating challenges associated with high-density fan-out (HDFO) technology", 18th Electronics Packaging Technology Conference, Singapore (November 2017).
90. B. Buckalew, T. Ponnuswamy, S. Mayer, K. Thorkelsson, J. Oberst and G. Graham, "Copper electrodeposition advancements for FOWLP", 18th Electronics Packaging Technology Conference, Singapore (November 2016).
91. G. Graham, L. Chua, B. Buckalew, T. Ponnuswamy and S. Mayer, "Within die coplanarity improvement strategies for electroplated Cu pillars", 20th Electronics Packaging Technology Center (2018), 600–607.
92. C. Seah, S. Mridha and L. Chan, "Annealing of copper electrodeposits", J. Vac. Sci. Technol. A, 17 (4) 1963–1976 (1999).
93. V. Vasko, I. Tabakovic and S. Riemer, "Structure and room-temperature recrystallization of electrodeposited copper", Electrochem. Solid-State Lett., 6 (7) C100–C102 (2003).
94. J. Harper, C. Cabral, P. Andricacos and L. Gignac, "Mechanisms for microstructure evolution in electroplated copper thin films", J. Appl. Phys., 86, 387–392 (1999).
95. S. Konishi, M. Moriyama and M. Murakami, "Effect of annealing atmosphere on void formation in copper interconnects", Mater. Trans., 43 (7) 1624–1628 (2002).

96. L. Mirkarimi, C. Uzoh, D. Suwito, G. Fountain, T. Workman, J. Theil, G. Gao, B. Lee, B. Buckalew, J. Oberst and T. Ponnuswamy, "The influence of Cu microstructure on thermal budget in hybrid bonding", IEEE 72nd Electronic Components and Technology Conference (June 2022).
97. K. Lu, L. Lu and S. Suresh, "Strengthening materials by engineering coherent internal boundaries at the nanoscale", Science, 324, 349–352 (2009).
98. C. Tseng, C. Liu, H. Lin, Y Chu, C. Chen, D. Lyu, K. Chen and K. Tu, "Low-temperature and low-pressure copper-to-copper direct bonding enabled by creep on highly (111)-orientated Cu surfaces", ICEP-IAAC 2015 Proceedings (2015), FA2-1.
99. B. Rebhan and K. Hingerl, "Physical mechanisms of copper–copper wafer bonding", J. Appl. Phys., 118 135301 (2015).
100. A. Engwall, Z. Rao and E. Chason, "Residual stress in electrodeposited Cu thin films: Understanding the combined effects of growth rate and grain size", J. Electrochem. Soc., 164 (13) D828–D834 (2017).
101. S Varadajaran, D. Kalakad and T. Cacouris, "Understanding and reducing copper defects." Semicond. Int., 25 125–130 (2002).
102. J. Lu, L. Chen, D. Gonzalex, H. Guo, D. Rose, M. Marudachalam, W. Hsu, H. Liu, F. Cataldi, B. Chatterjee, P. Smith, P. Holverson, R. Guldi, N. Russell, G. Shinn and J. Luttmer, "Understanding and eliminating defects in electroplated Cu films", Proceedings of the IEEE 2001 International Interconnect Technology Conference (Cat. No.01EX461), 280 (2001).
103. V. Prabhakar, B. Buckalew, K. Ganesan, S. Ghongadi, Z. He, S. Mayer, R. Rash, J. Reid, Y. Takada and J. Zibrida, "Wafer electroplating apparatus for reducing edge defects", US Patent 8,172,992, issued on May 8, 2012.
104. D. Lee, S. Pyo, C. Ko, M. Lee, S. Kim and J. Lee, "Effect of acidity on defectivity and film properties of electroplated copper films", J. Electrochem. Soc., 151 (3) C204–C209 (2004).
105. G. Gao, L. Mirkarimi, G. Fountain, D. Suwito, J. Theil, T. Workman, C. Uzoh, G. Guevara, B. Lee, M. Huyhn and P. Mrozek, "Low temperature hybrid bonding for die to wafer stacking applications", 2021 IEEE 71st Electronic Components and Technology Conference (ECTC), IEEE (June 2021) 383–389.
106. G. Gao, L. Mirkarimi, T. Workman, G. Guevara, J. Theil, C. Uzoh, G. Fountain, B. Lee, P. Mrozek, M. Huynh and R. Katkar, "Development of low temperature direct bond interconnect technology for die-to-wafer and die-to-die applications-stacking, yield improvement, reliability assessment", International Wafer Level Packaging Conference (IWLPC), IEEE (2018).

6 Planarization for Advanced Packaging and Hybrid Bonding

Brian J. Brown, Ekaterina Mikhaylichenko, Jianshe Tang, Jun Qian, and Yen-Chu Yang

6.1 INTRODUCTION: BACKGROUND AND DRIVING FORCES

IBM first developed chemical–mechanical planarization (CMP) around 1983 in its Base Technology Lab [1]. Modeled after polishing in jewelry-making, the early processes filled a trench in material A with material B, and then polished back material B to the material A surface. IBM termed CMP of metal trenches polished back to a dielectric mold "damascene" and metal patterns filled with dielectric and polished "cloisonne." Although initial reaction was negative due to fears of defectivity by the introduction of "dirty" slurries into the cleanroom, no other process then or now can deliver on the topography needed for multilevel metallization.

6.2 CMP FUNDAMENTALS

Production CMP systems used in semiconductor fabrication facilities today are sophisticated, integrated systems normally separated into three subsystems. The factory interface (FI) accepts front opening universal pods (FOUP) of wafers from the factory automation system. Robotics transfer wafers from the FI to the CMP module, then to the cleaner for post-CMP cleaning, and finally back to the FI for subsequent process steps. The FI configuration often contains dry, in-line metrology for pre- or post-thickness measurements.

The CMP module contains multiple platens (typically four) and multiple heads (four or more). Individual wafers can sequence to the platens in parallel (batch) or in sequence (multistep), depending upon the desired process sequence. The CMP ecosystem shown in Figure 6.1 illustrates the co-optimization of both polishing hardware and polishing consumables. The polishing head positions the wafer's device side in contact with a polymeric pad, typically polyurethane, laminated to the platen with pressure-sensitive adhesive. Both the platen and head rotate to create relative velocity between the wafer and the pad. The CMP head applies pressure to the back of the wafer against the pad and has a retaining ring to prevent the wafer from slipping out of the head during CMP. A dispense arm delivers polishing slurry onto the pad upstream of the head, and grooves in the pad transport the slurry under the wafer. To prevent glazing of the pad by buildup of abrasives and by-products and to maintain a controlled texture for stable removal rate and planarization,

DOI: 10.1201/9781003408673-6

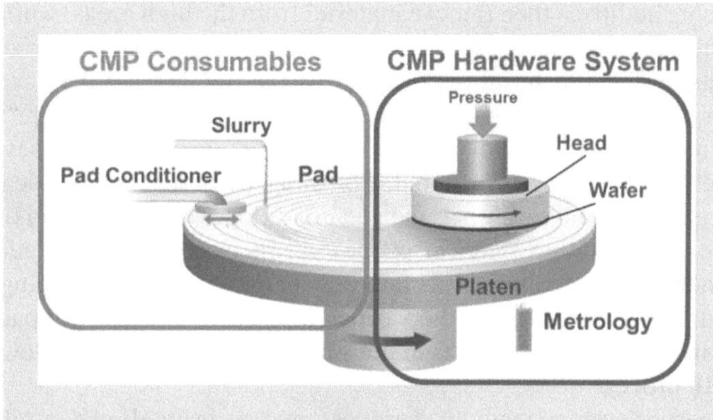

FIGURE 6.1 CMP module elements and ecosystem.

a rotating diamond disk or plastic brush mounted to a sweeping arm conditions the pad. Different processes employ conditioning of the pad ex situ in-between wafers, in situ during wafer polish, or with a combination of the two. Sensor(s) mounted in the platen can measure thickness across the wafer as they rotate under the head.

Material removal mechanisms involve three body interactions—wafer, pad, and slurry. Slurries are complex suspensions containing abrasive particles, stabilizing agents, pH buffers, rate-enhancing additives, inhibitors, and, in the case of metal CMP, corrosion inhibitors and oxidants. Figure 6.2 schematically shows the polish process. The inhibitors or oxidants coat or oxidize the surface, respectively, protecting it from chemical attack by rate-enhancing additives. When the pad moves across the wafer surface under pressure, asperities in the pad carry abrasives to remove inhibitor "a" or oxide from the protruding, high-pressure areas.

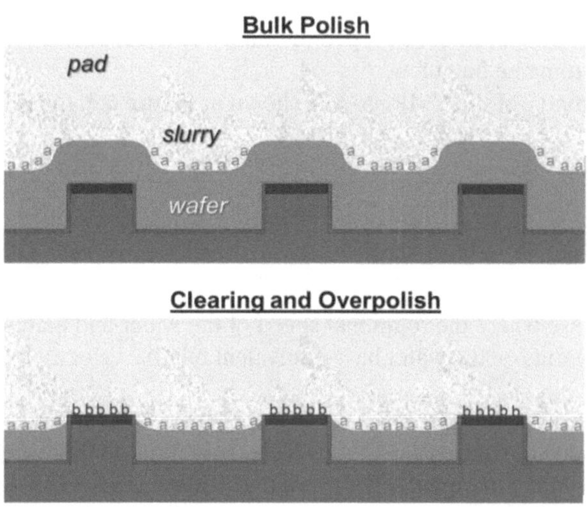

FIGURE 6.2 Progression of chemical–mechanical planarization process.

The enhancing additives then remove material from the high areas, while low areas remain protected, and this leads to a planarized surface. Once planarized, the surface exhibits uniform material removal across the die. During the clearing of the bulk polish film from the stop film "field" area, a different inhibitor "b" protects the stop film from being removed and inhibitor "a" or oxide represses dishing in trenches between the stop-layer mold. Polishing by-products (chemical reactants and products, agglomerated slurry, and pad/conditioner debris) remain on the wafer after polish, necessitating removal by post-CMP cleaning. While additives and abrasives in the slurry determine the interplay between chemical etching, mechanical removal, and suppression of the various films on the wafer, the pad asperities play a role in how slurry transports to the surface and penetrates the low areas during the CMP process.

Pad properties such as modulus, hardness, and porosity all play a role. The diamond-conditioning disk design (diamond geometry and layout) and its application (speed and downforce) will determine the density and morphology of the pad asperities and the local transport of slurry to the wafer surface. The pad grooving geometry, slurry arm dispense position, slurry flow rate, and retaining ring design all play a role in the macro-transport of slurry across the wafer.

One of the most critical output parameters for CMP is the polish rate, whether it be average, wafer to wafer (WTW), within wafer (WIW), or within die (WID). In 1927, Preston [2] described the fundamental equation for polishing rate, dz/dt, at any point on a workpiece which scales with the local polishing pressure p and the relative velocity v between that point and the polishing pad contacting it.

$$\frac{dz}{dt} = K_p \; pv \tag{6.1}$$

The scaling factor K_p, known as the Preston coefficient, is not necessarily a constant and can depend upon many factors such as temperature, slurry (constituents, concentration, distribution, thickness), pad (modulus, texture, porosity), and by-product concentration, to name but a few.

For the geometry of the CMP module shown in Figure 6.3, the relative velocity v_x at a given point x on the wafer is as follows:

$$v_x = r_{pw}\omega_p \sqrt{1 + 2\frac{r_{wx}}{r_{pw}}\left(1 - \frac{\omega_w}{\omega_p}\right)\cos\Theta + \left[\frac{r_{wx}}{r_{pw}}\left(1 - \frac{\omega_w}{\omega_p}\right)\right]^2} \tag{6.2}$$

which, in the case where the rotational speed of the wafer and platen are equivalent, shows that all points on the wafer have equivalent relative velocity in both magnitude and direction.

In practice, the platen and head normally rotate at a small speed offset for several reasons. Speed offset ensures the velocity direction on the wafer relative to the notch changes with time preventing a "wedge" in the thickness profile across certain die features. Pad grooving can lead to an interference thickness pattern on the wafer which speed offset mitigates. Finally, speed offset causes the in situ metrology

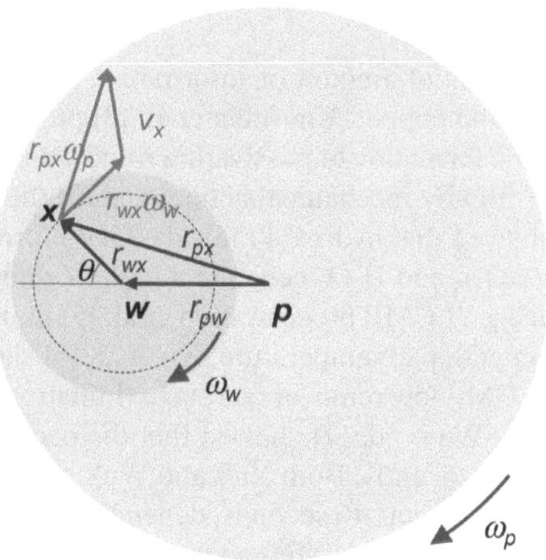

FIGURE 6.3 Kinematic diagram showing relative velocity at point X on wafer.

sensors to scan the wafer in different trajectories over time, allowing averaging of radial thickness for asymmetric incoming profiles. However, even with this considered, the polish rate uniformity seldom achieves that predicted from the kinematic model due to variations in the pressure and K_p across the wafer. One way to compensate for inherent non-uniformity is via design of the CMP head, for example, by using annular pressure zones in the CMP head behind the wafer to tune the removal rate profile. The discontinuity between the wafer and the retaining ring leads to the polishing pad bulging (termed "pad rebound"), and the wafer digging into the pad which leads to non-uniform pressure near the edge of the wafer (see Figure 6.4). Optimization of CMP head design in concert with the pad mechanical properties can help improve the pressure distribution at the wafer edge, but as non-uniformity requirements across a 300 mm fall below 10 nm, machine design and consumables design become even more critical.

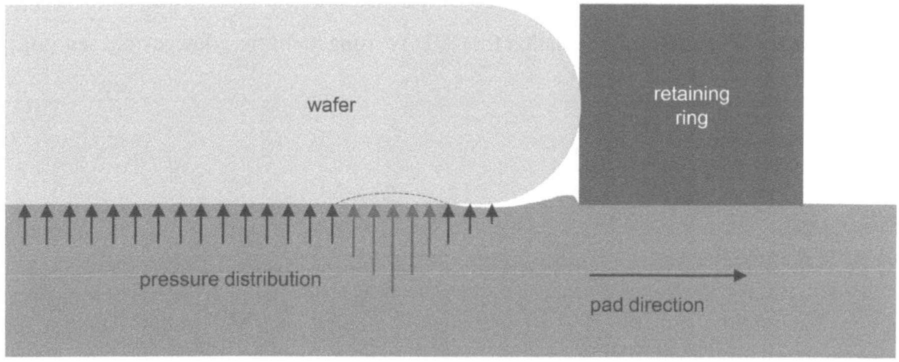

FIGURE 6.4 Pad rebound at the wafer edge leading to non-uniform pressure.

6.2.1 Metal CMP

The literature has a substantial amount of information about CMP of conventional metals such as tungsten and copper. Kaufman et al. [3] proposed the mechanism for W CMP in 1991 as (1) the formation of passivating film by chemical oxidation, (2) the removal of passivating film by mechanical action, and (3) the repeat of steps (1) and (2). Poddar et al. [4] showed that hydroxyl radicals (OH*) form a soft WO_3 film, and mixed oxidants of $Fe(NO_3)_3$ and H_2O_2 generate a high W removal rate due to in situ generation of OH* during W CMP process. Kim et al. [5] found that W removal rate has a strong correlation with pad temperature. Xu et al. [6] showed that W morphology correlates with the CMP low removal rate stage (initiation) by calculating instantaneous W removal rate. Wu et al. [7] showed that the pad temperature impacts W removal rate initiation significantly. Both Xu's and Wu's studies showed that removal rate initiation could last as long as 40 seconds, depending on the W film morphology, as shown in Figure 6.5. Heating the CMP pad from 25°C to 45°C before starting polishing could reduce the initiation time by about 50%, as seen in Figure 6.6.

Copper is the wiring material of choice in modern semiconductor devices due to its high electrical and thermal conductivity, electromigration resistance, and low

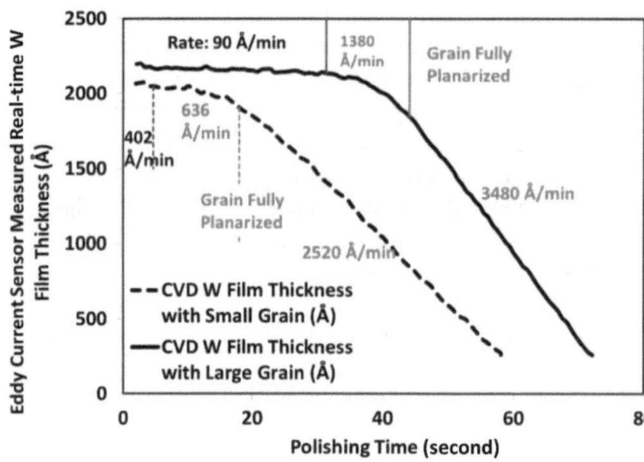

FIGURE 6.5 W removal rate evolution for CVD W using real-time eddy current sensor [6].

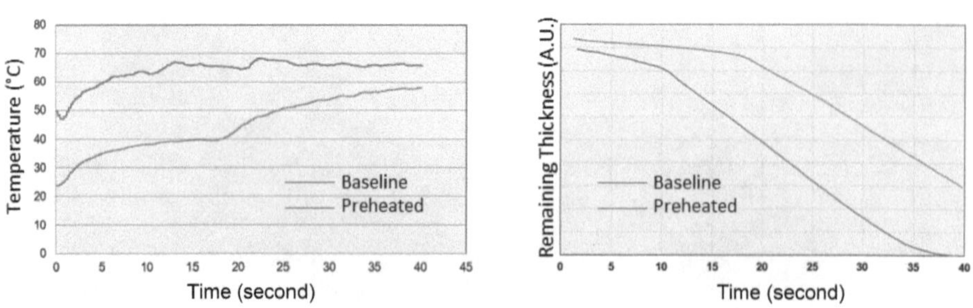

FIGURE 6.6 W CMP initiation time reduction by increasing pad temperature [7].

cost [8, 9]. Copper wiring structures included both single damascene (conductor) and dual damascene (conductor and via). Like W CMP, the Cu CMP mechanism involves both chemical passivation and mechanical abrasion by abrasive particles for planarization and removal [10]. Adding oxidizers, such as H_2O_2, to the slurry creates the passivation layer.

A typical Cu CMP process has the following three steps [11, 12]:

1. *Bulk Copper:* Removal of copper overburden at a high rate to planarize the copper prior to the next step. High pressures typically reduce process time. In situ WIW process control during bulk Cu CMP can ensure that the next step will uniformly clear the Cu. Bulk Cu removal targets a thin Cu remaining thickness with no barrier exposure, either by active endpoint control or by time.
2. *Copper Clearing:* Cu clearing at lower pressures and speeds to remove remaining copper film or residue, stopping on the barrier. These conditions lower the temperature to improve dishing and erosion and allow for a more precise endpoint control to minimize overpolish time. An active endpoint normally monitors clearing, and this step often involves an overpolish to ensure complete clearing of the entire wafer.
3. *Barrier:* Complete removal of the barrier along with dielectric (the amount depending upon integration considerations) electrically isolates the Cu lines. Careful selection of consumables and process optimization will ensure that the Cu and dielectric are as close to the same height as possible at the end of the CMP process. Active endpoint or in-line metrology with APC (advanced process control) monitors process completion (see below).

The Cu CMP removal rate follows the Preston equation, increasing with the increasing downforce and velocity. Because temperature modulates the reaction rate at the chemically active copper surface, temperature often plays a key role in materials removal, selectivity, topography, and defect control.

Increasing the polishing pad temperature increases the Cu removal rate; however, higher temperatures can also increase dishing and erosion [13, 14]. Using heated slurry, Liu et al. [15] showed that higher process temperatures both increased copper removal rate and degraded surface morphology. Wu et al. [16] developed a system to adjust the pad temperature with two modules which contact the pad surface during Cu CMP. The two modules circulated either heated or chilled water behind a plate contacting the pad to modulate the Cu removal rate. Optimizing the contact location improved wafer non-uniformity.

Chang et al. [17] developed a non-contact pad surface temperature control system with both heating and cooling capability. The system allows for rapid temperature change over a single wafer process to simultaneously improve both throughput and topography. During the bulk CMP step, heat boosts the removal rate, allowing for less time to reach the bulk endpoint. After the bulk endpoint, cooling rapidly drops the temperature to reduce dishing and erosion. During the barrier step, cooling can improve topography and control selectivity. Figure 6.7 schematically illustrates the

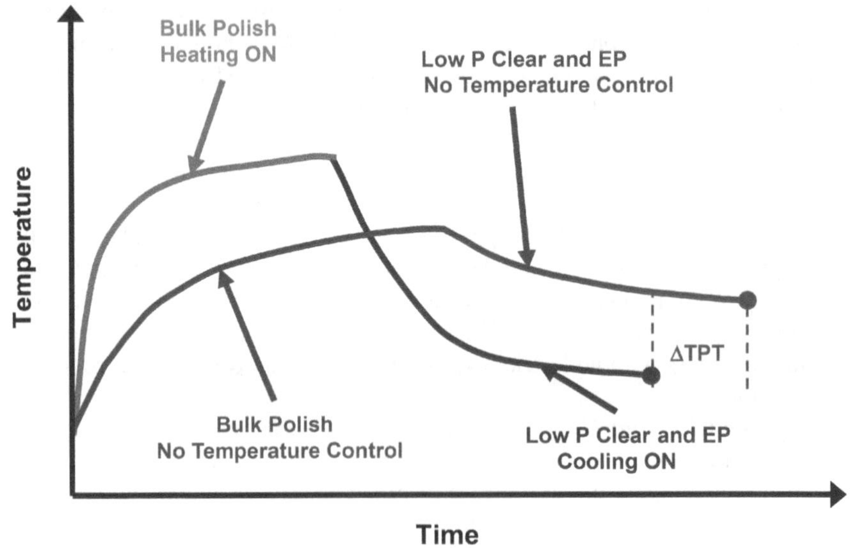

FIGURE 6.7 Temperature control concept for Cu CMP.

temperature profile during the Cu bulk and clearing steps on a single platen with and without temperature control.

During wet processes like CMP, copper chemical corrosion is always of concern because, unlike aluminum, copper oxidation is not self-limiting. Figure 6.8a shows a top-down SEM image of typical Cu line corrosion. Cu is also prone to galvanic corrosion when connected to another metal of a different electrode potential. Cobalt serves as a liner for copper seed deposition and, due to its low electrode potential compared to copper, can lead to etching of Cu at the Co–Cu interface, as seen in Figure 6.8b. Wang et al. [18] demonstrated that addition of specific amine inhibitors can suppress barrier/liner galvanic corrosion. Photons with energy above the bandgap of silicon illuminating a p–n junction connected to two Cu lines will generate voltage between the lines. When in a wet environment, this can lead to photocorrosion of Cu lines [19] and in the extreme case, growth of Cu dendrite shorts along the

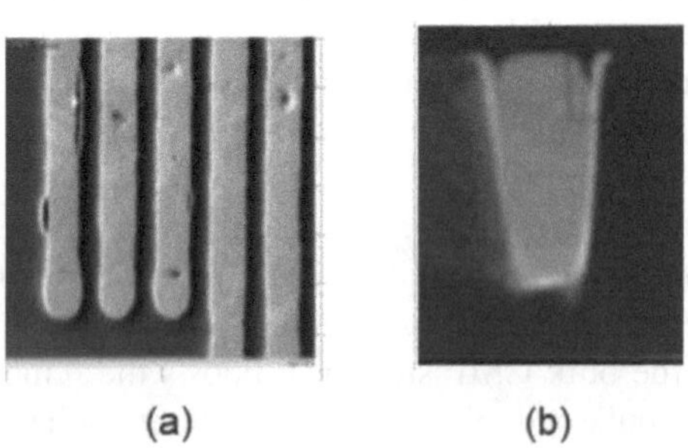

FIGURE 6.8 Copper corrosion. (a) Cu line corrosion. (b) Cu/Co galvanic corrosion [18].

dielectric surface between the lines. CMP tools commonly use opaque doors in the CMP module and a CMP cleaner to prevent photocorrosion.

Other defect modes prevalent in Cu CMP include particles (slurry abrasive), organic residue (slurry additives and pad debris), and scratches. The CMP system normally has an integrated post-CMP cleaner; however, reducing the number of defects on wafers leaving the CMP module will result in reduced defects after cleaning.

For example, if the cleaner has three sequential cleaning stations, each with a PRE (particle removal efficiency) of 90%, then if the number of particles at a given size threshold coming out of the polisher is 100,000, after cleaning the count will be 100. If we reduce the post-polish particle count to 50,000, the post-cleaning count reduces to 50.

Another factor to consider is brush loading. The majority of post-CMP processes employ contact cleaning methods with PVA brushes which can become loaded with particles over time. These particles can redeposit on the wafers, requiring a brush change and thereby machine downtime. Lowering the defects on wafers incoming to the cleaner can extend the mean time between brush changes. Finally, because no post-CMP cleaning technology can "repair" microscratches, prevention requires optimization within the CMP module.

Lu et al. [20] developed a process of using thermal energy to aid in pad cleaning. The process sequentially heats and cools the pad. Thermal energy application to the pad removes defects upstream of the wafer head followed by a pad cooling step, so that by the time the pad rotates to the head/wafer for buffing, the temperature is low. Using thermally assisted pad cleaning improves defect performance by ~60%, as shown in Figure 6.9. SEM inspection of the defects shows that thermally assisted cleaning reduces all defect modes significantly.

6.2.2 Dielectric CMP

A variety of suspensions of metal oxide abrasives will polish silicon oxide films. Modern dielectric CMP fall into two categories: silica-based slurry processes and ceria-based slurry processes. Early slurries for dielectric film polish, such as Cabot SemiSperse—SS12, had colloidal suspensions of silica particles (mean particle size of 160 nm) in an alkaline medium (KOH at pH ~11). The material removal

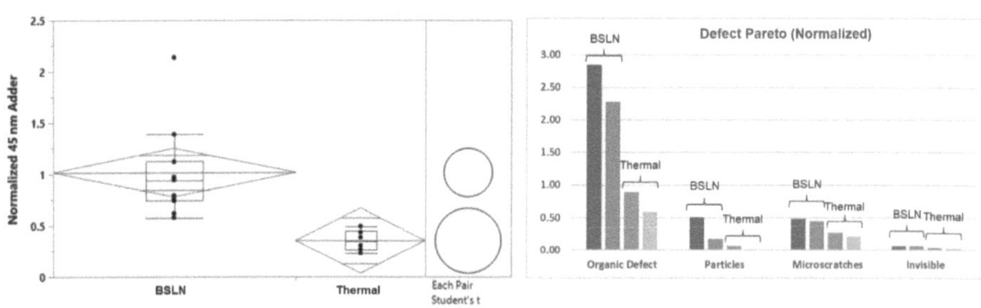

FIGURE 6.9 Thermally assisted pad cleaning post-CMP and cleaning.

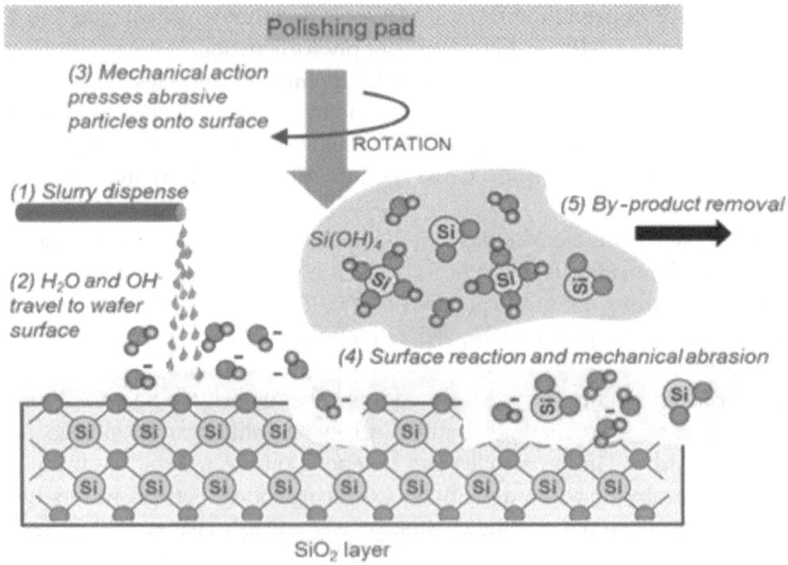

FIGURE 6.10 Alkaline silica abrasive slurry dielectric CMP mechanism.

mechanism in dielectric CMP includes both chemical and mechanical components [21], as shown in an example for alkaline silica-based slurries in Figure 6.10. A passivation layer forms through hydrolysis; hydroxyl ions travel to the surface, breaking the Si—O—Si bonds and leading to the formation of a softer surface passivation layer. Mechanical force then abrades the suspended particles across the passivated surface, removing the passivation layer [22]. The cycle of chemical reaction, silica dissolution, and material abrasion leads to material removal.

For silica slurry CMP, the average removal rate increases linearly with downforce (between 1 psi and 5 psi) following the Preston equation, as shown in Figure 6.11a. With uniform pressures applied to the wafer, a typical material removal profile

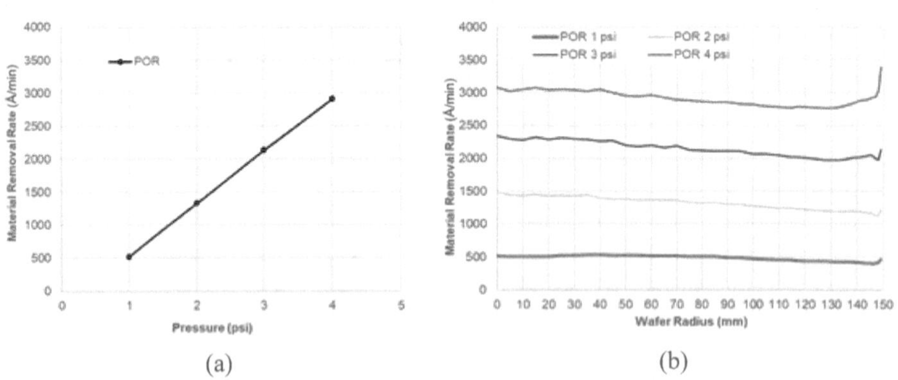

FIGURE 6.11 Oxide removal rate with alkaline silica slurry. (a) Removal rate versus polishing pressure. (b) Polishing profile at uniform pressure.

exhibits slightly edge fast behavior, as shown in Figure 6.11b, which becomes more pronounced at higher polishing pressures. Use of a multi-zone head can normally address the edge-fast profile.

Ceria abrasive-based slurries offer several advantages over silica-based slurries. Ceria offers higher oxide removal rates and better oxide-to-nitride selectivity (>200:1) resulting in low nitride loss (<2 nm) for oxide stop on nitride processes. Ceria slurry challenges include stability of the dispersion, vulnerability to surface scratching, and difficulty in removing ceria particles during subsequent post-CMP cleaning.

Srinivasan et al. [23] provided a review of the literature around the mechanism of ceria CMP. Ceria abrasives have a large surface chemical action during SiO_2 polishing. Ceria abrasive particles have a surface "shell" enriched in Cerium(III) ions. These Ce^{3+} surface sites, along with oxygen vacancies, form peroxo and superoxo complexes in the presence of O_2, critical for silicon dioxide removal, through the proposed reaction:

$$-\mathrm{Ce}-\mathrm{OH}+-\mathrm{Si}-\mathrm{O}^{-} \leftrightarrow -\mathrm{Si}-\mathrm{O}-\mathrm{Ce}-+\mathrm{OH}^{-} \tag{6.3}$$

As the ceria particles held by pad asperities abrade the oxide surface, the weaker Si—O—Si bonds break, allowing the stronger Ce—O—Si to carry SiO_2 with the abrasive, thereby making this material removal both chemical and mechanical in nature. While Cook et al. [24] suggested that ceria abrasives remove one SiO_2 molecule at a time, more recent studies [25] show that the SiO_2 film does not dissolve, rather it scrapes away in lumps, as shown in Figure 6.12.

Ceria slurry processes tend to exhibit non-linear (non-Prestonian) behavior with respect to polishing pressure, especially higher pressures above 3 psi, as seen in

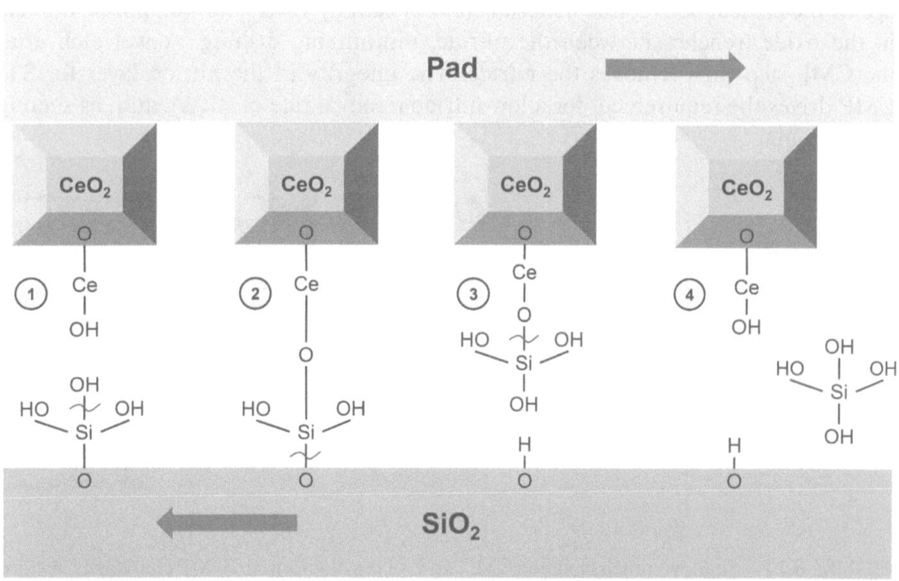

FIGURE 6.12 Silicon oxide polish mechanism with ceria abrasive-based slurry.

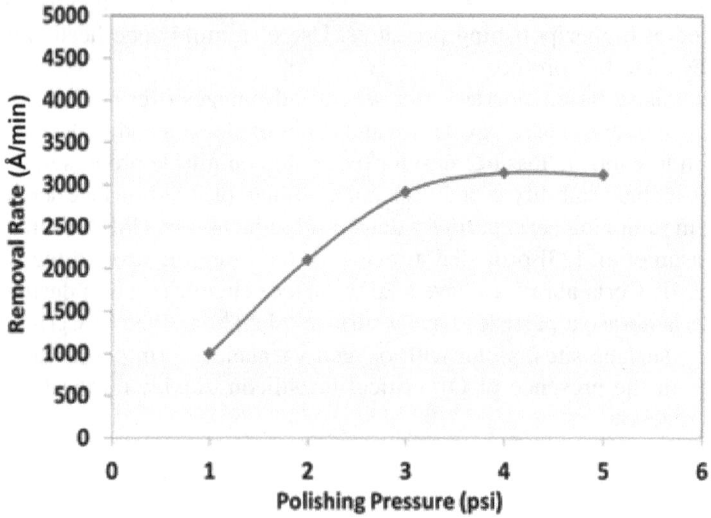

FIGURE 6.13 Removal rate versus polishing pressure for ceria slurry-based dielectric polish.

Figure 6.13. Typical removal profiles with ceria polish exhibit fast edge rate, making it challenging to tune flat profile without introducing a "checkmark" near the edge.

Stop on nitride processes (SON), such as shallow trench isolation (STI) CMP, commonly use ceria abrasive-based slurries. Depending on the additive type and concentration, material removal optimization can achieve highly selective silicon oxide removal against the silicon nitride hard mask. The nitride layer prevents damage of the critical active area materials underneath by CMP and "supports" the pad in the oxide trenches between the nitride, minimizing dishing. A wet etch after the CMP step then removes the nitride. The integrity of the nitride layer for STI CMP drives the requirement for a low nitride removal rate of <10 Å/min, as seen in Figure 6.14a.

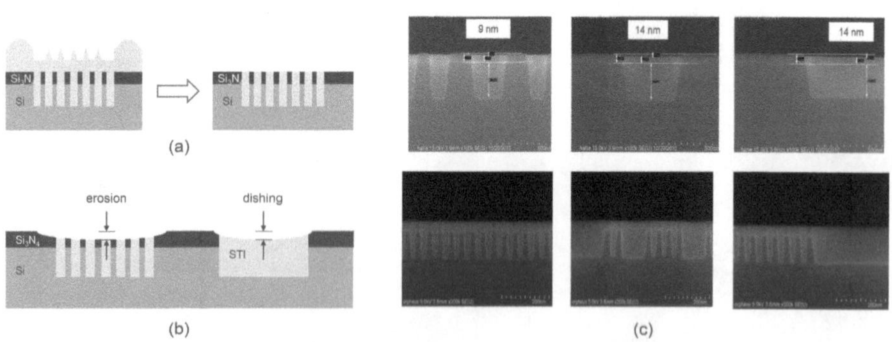

FIGURE 6.14 Shallow trench isolation CMP. (a) Process diagram. (b) CMP challenges: erosion of dense array, dishing in isolated features. (c) SEM image showing minimum oxide erosion and SiN loss across dense array, and minimal dishing in semi-isolated and open features.

Extensive studies of the SiN removal mechanism [26–28] aid in selecting the additives to suppress the nitride removal rate. The literature suggests silicon nitride oxidizes to silicon dioxide when exposed to air or water.

$$Si_3N_4 + 3O_2 \rightarrow 3SiO_2 + 2N_2 \tag{6.4}$$

$$Si_3N_4 + 6H_2O \rightarrow 3SiO_2 + 4NH_3 \tag{6.5}$$

The CMP process does not remove the nitride directly, rather it removes the chemically modified oxide surface layer and the oxidation/removal cycle continues. In highly selective slurries, additives preferentially adsorb on the surface of the nitride to prevent its conversion to oxide and thus suppress Si_3N_4 removal.

STI CMP challenges include control of the erosion in dense arrays and dishing in low-density areas, enhanced at high oxide removal rates (Figure 6.14b). On the other hand, any remaining oxide over the nitride will block the subsequent wet removal of the nitride. The requirement for complete oxide clearing while minimizing dishing, erosion, and nitride loss makes end point control critical. Figure 6.14c represents an example of optimized STI process with minimal oxide erosion across the dense array and minimal dishing in semi-isolated and open features.

CMP consumables play a critical role in the on-wafer performance metrics, such as topography, dishing, erosion, WID uniformity, WIW uniformity, and defectivity. As device dimensions continue to shrink, meeting these tight and often conflicting requirements becomes ever more challenging. Optimization of process parameters—polishing pressure, head and platen rpm, pad conditioning timing, downforce and rpm, and endpoint algorithms, among others—for each consumable set will dictate the final performance.

6.2.3 Post-CMP Cleaning

As semiconductor device technology approaches critical scaling limitations, not only do the number of CMP steps increase, but CMP defect reduction also becomes more critical for yield. Semiconductor device manufacturers traditionally have considered CMP a “dirty” process. The mechanical action of pad and slurry against the wafer surface generates a variety of non-removable defects (scratches, pits, corrosion) as well as removable defects—abrasive particles, chemical by-products, agglomerated slurry, and pad/conditioner debris, examples shown in Figure 6.15. After completing the polishing, post-CMP cleaning must address removable polishing defects, which play a critical role in the overall defect performance [29].

Additional post-CMP cleaning challenges include spontaneous corrosion of metal films, or galvanic corrosion, previously shown in Figure 6.8. Soft polishing films such as Cu can be easily damaged (scratched) during cleaning, as shown in Figure 6.16.

Cu cleaning requires precise control of the shear force during cleaning to maximize particle removal efficiency while ensuring no damage. Cleaning becomes more challenging with the time between polishing and cleaning as the bond between contaminants and surface becomes stronger; therefore, all commercial semiconductor

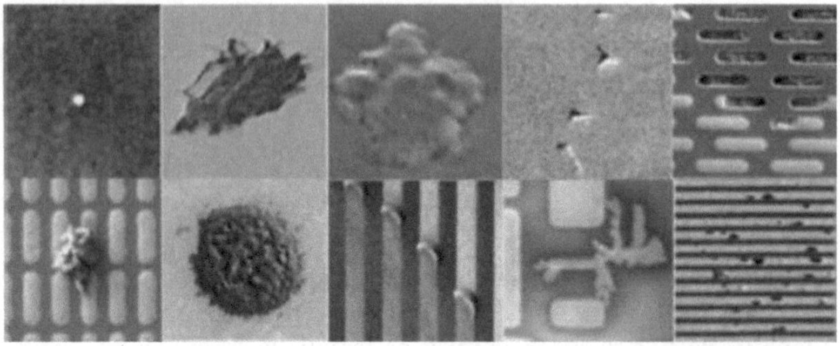

FIGURE 6.15 Examples of defects observed in CMP. *Upper images from left to right:* slurry ball; fiber-texture organic residue; smearing polishing by-product; chatter-mark scratch; missing metal. *Lower images from left to right:* agglomerated slurry; polishing organic by-product; watermark; Cu dendrites; Cu corrosion [29].

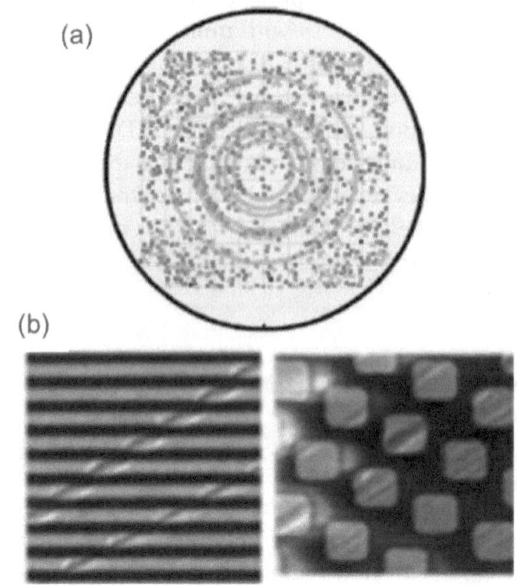

FIGURE 6.16 (a) Wafer scan map of copper circular scratches generated by brush scrubbing. (b) SEM images of the scratches [29].

CMP systems have integrated post-CMP cleaners. Drying is another critical step in post-CMP cleaning; the defect performance of the full system can never be better than the performance of its dryer. Since typical post-CMP surfaces have multiple films exposed, drying needs to simultaneously accommodate a combination of hydrophilic and hydrophobic surfaces. This drives the use of surface tension gradient based drying, also known as Marangoni dry, in post-CMP cleaners.

To remove a particle attached on a surface, the cleaning force must overcome the interaction forces between the particle and the substrate [30]. Typically, the cleaning

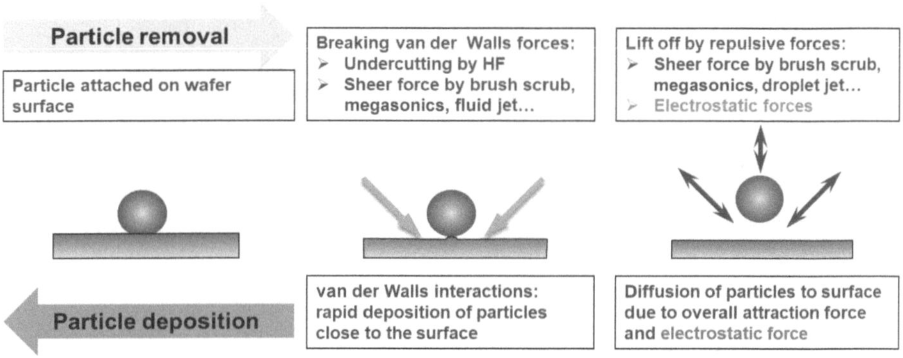

FIGURE 6.17 Schematic description of particle removal and particle deposition from a wafer surface [31].

method applies a mechanical shear force acting on the particle–surface interface and/or its chemical etches, for example, with hydrogen fluoride chemistry [31]. After overcoming the adhesion forces, repulsive forces must remove the particle from the surface to prevent redeposition, as shown in Figure 6.17. To prevent particle reattachment during post-CMP cleaning, one should select a cleaning chemistry that modifies the zeta potentials of the wafer surface and the particles to have the same charge, thereby repelling particles away from the surface. In practice, alkaline solutions and/or surfactants will lead to electrostatic repulsion. Once outside of the electrostatic region, Brownian motion will carry the particles across the flow boundary layer, where fluid flow can carry them away.

Typical post-CMP cleaners consist of several cleaning modules implementing several types of cleaning processes—chemical–mechanical buff, double-sided roller brush scrub, pencil brush scrub, megasonics (immersion or nozzle), high-velocity droplet jet cleaning, and surface tension gradient-based drier. Post-CMP cleaning widely employs double-sided PVA (polyvinyl acetal $[C_4H_6O_2]_n$) roller brush scrubbing because of its high efficiency of removing loosely attached slurry particles. Figure 6.18 shows a schematic representation of post-CMP cleaner modules.

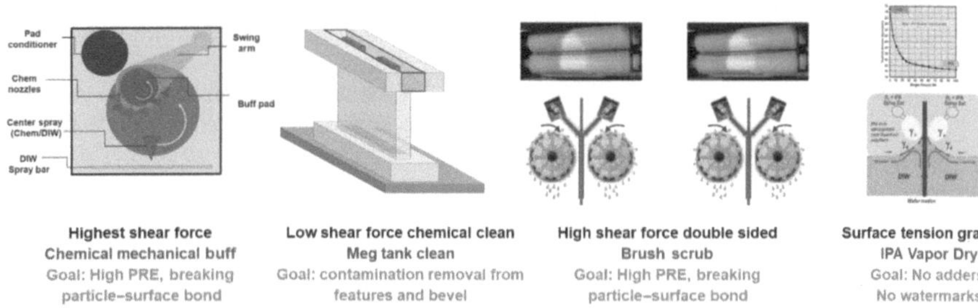

FIGURE 6.18 Schematic diagram of a post-CMP cleaning modules.

6.2.4 CMP Process Control

Developing a successful CMP packaging process requires mitigation of multiple sources of variation which can lead to variations in WIW and WTW uniformity. All modern CMP tools need to match the performance among multiple platens, polishing heads, and conditioners and accommodate for variations in pads, slurry, and conditioning disk consumables, normally changed multiple times per week in production. Figure 6.19 summarizes major sources of variation that can affect the performance of a CMP process. A variety of process control techniques address these variations in the increasingly challenging CMP applications for advanced technology nodes and advanced packaging applications [32, 33].

In addition to process drift from the CMP tool subcomponents and CMP consumables with time, the process must compensate for incoming film thickness or film property variations. As an example, a typical Cu layer may have a removal amount of 0.5 μm. A typical deposition uniformity of 1% translates to a 50 Å range across the wafer. If the polish removal uniformity is perfect, then when the thinnest part of the wafer clears, the thickest part still has an additional 50 Å of polishing. This leads to overpolish of 50 Å of the thinnest part, typically for less than 10 seconds. As one moves to thicker Cu films used in packaging, up to 10 μm, this non-uniformity becomes 1,000 Å, illustrating the more significant challenge. By measuring the film in situ during the CMP process and "flattening" profile across the wafer prior to clearing, one can address the incoming non-uniformity.

Dry reflectometers integrated into the CMP system's factory interface enable in-line thin film thickness measurements of a wafer's incoming and post-CMP thickness to allow an advanced process control (APC) feedforward/feedback approach to controlling post-CMP thickness. This approach provides more sampling and a

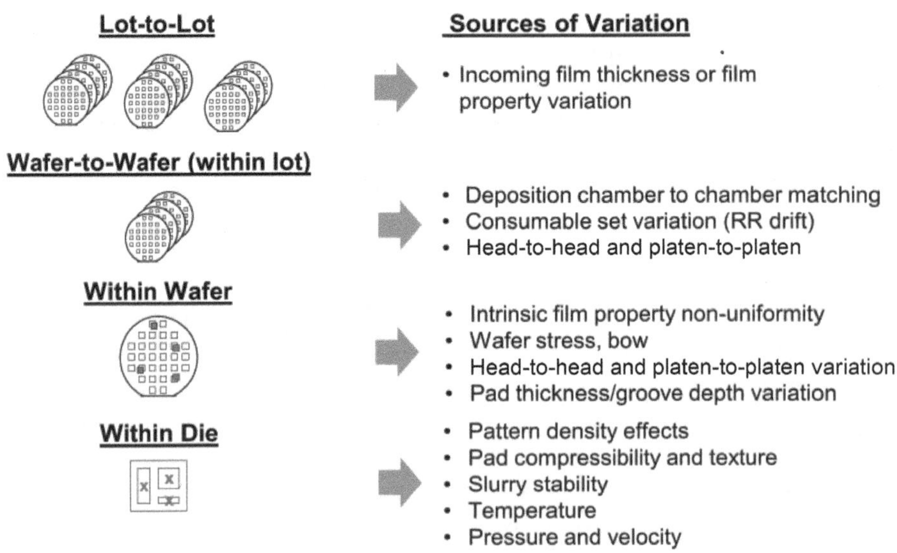

FIGURE 6.19 Variation effects on non-uniformity.

shorter feedback loop compared to using data from stand-alone metrology tools, albeit at a higher cost per wafer. For a CMP system with a stable removal rate, this method effectively corrects the rate and profile drift over time. Optimization of the number of sites per wafer and number of wafers measured per lot can minimize the throughput impact. In-line APC works best to correct for systematic variations, such as differences between duplicate hardware sets (platens, heads, or conditioners) or slow changes over time (pad or conditioning disk wear). When lot-to-lot incoming film thickness changes or an inherently unstable process leads to unpredictable removal rate variations, feedback delay between wafer polishing and wafer post-measurement can significantly deteriorate post-wafer-to-wafer and within wafer non-uniformity.

In situ sensors inserted inside the rotating platen enable real-time control of post-thickness for both dielectric and metal CMP applications. A controller converts the sensor signal into a thickness and then provides real-time changes of head zone pressures to create the desired WIW thickness profile and final thickness endpoint, as illustrated in Figure 6.20a. Figure 6.20b shows an example of how the method quickly corrects the removal profile of an inherently non-uniform polish process. In situ control of within-wafer profile during bulk polish achieves a more uniform clearing to the stop film, thus minimizing the risk of residue, dishing, and erosion. Multiple sensors strategically located in the polishing platen can efficiently provide complete wafer coverage enabling the resolution required for precise in situ within-wafer control.

For dielectric films, a spectrograph feeds spectral signals to the controller which converts them to thicknesses and provides real-time control of the polishing pressure to achieve the desired thickness profile after CMP. Depending on the application and thickness of the film, either a visible or near-infrared spectrograph detects film removal. Interpretation of the spectral signal has employed a variety of techniques over the years, from spectral feature tracking to modeling, to the use of machine learning techniques. For stop-on-film processes, the change in platen or head torque needed to maintain a constant rpm on the transition between films of differing friction coefficients can provide film clearing endpoint control.

For metal CMP processes, eddy current sensors typically detect in situ film thickness during polish. A variety of eddy current sensors can cover a wide range of thicknesses and metal films, such as Cu, Al, W, Co, Mo, Ru, Ta, TaN, Ti, TiN, and other alloys. In

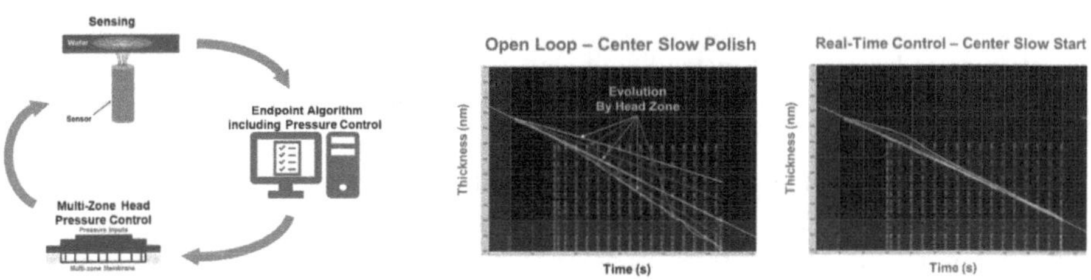

(a) (b)

FIGURE 6.20 (a) In situ process control methodology. (b) Comparison of real-time thickness for an open-loop versus closed-loop process.

most cases, one type of sensor measures thickness during bulk polishing, while another measures thickness and endpoint for clearing of the metal film to dielectric.

Imaging systems integrated onto the CMP tool can quickly detect residue, film stress induced cracking or other anomalies. They can also detect gross over- or under-polish so that the CMP system software can then automatically route under polished wafers back into the tool for rework.

Another major source of process variations comes from the CMP polishing pad. Over the course of pad life, non-uniform pad removal can induce wafer profile variations and the need for a premature pad change. To address this, closed-loop control of the pad conditioning recipe ensures the desired pad profile using an in situ pad thickness sensor.

CMP packaging applications benefit from a variety of the described process control techniques to improve performance while minimizing the cost and throughput impact. Different thickness (thick copper and silicon) and materials (polymers and epoxy) have driven extension of the existing sensor technologies. The evolution of CMP packaging applications will continue to drive new sensors, additional control loop algorithms, and enhancements in the ability to monitor components and consumables.

6.3 PACKAGING FLOWS USING CMP

In today's front-end semiconductor process manufacturing process flows, CMP is the state-of-the-art technology for planarization; however, until recently, manufacturers considered CMP as overkill to meet the traditional packaging requirements. The excessive cost of consumables (slurries, pads, and disks) has limited CMP adoption in the backend to a handful of steps such as TSV formation and reveal. However, as packaging begins to adopt hybrid bonding, CMP becomes the technology enabler. The extreme topography requirements at bonding interface to achieve acceptable product yield drives the need for CMP. Typical processes employing CMP in hybrid bonding include Cu bond pad formation, Si thinning, and TSV formation and reveal. For chip to wafer bonding, CMP can also planarize post-bonding, inter-die gapfill materials. Figure 6.21 illustrates some of the steps utilizing CMP.

6.3.1 CMP in Wafer-to-Wafer Hybrid Bonding

Hybrid bonding provides a solution for 10 μm pitches and below by completely avoiding the use of bumps, and instead connects dies in packages using small copper-to-copper connections. Hybrid bonding provides a permanent bond that combines a dielectric bond (SiO_2) with embedded metal bond (Cu) to form interconnections, which allows face-to-face connection of the wafers, often referred to as a direct bond interconnect (DBI). Figure 6.22 shows a typical wafer-to-wafer bond pad formation flow. After dielectric deposition, lithography, and etch create trenches and vias in the dielectric surfaces, a deposited metal barrier prevents the subsequent Cu from diffusing into the dielectric film, causing leakage. Next, Cu PVD seed and electroplating fill the trenches. CMP removes the thick Cu overburden deposited in the field, clearing the dielectric between the Cu bond pads. As in traditional Cu CMP, Cu bond

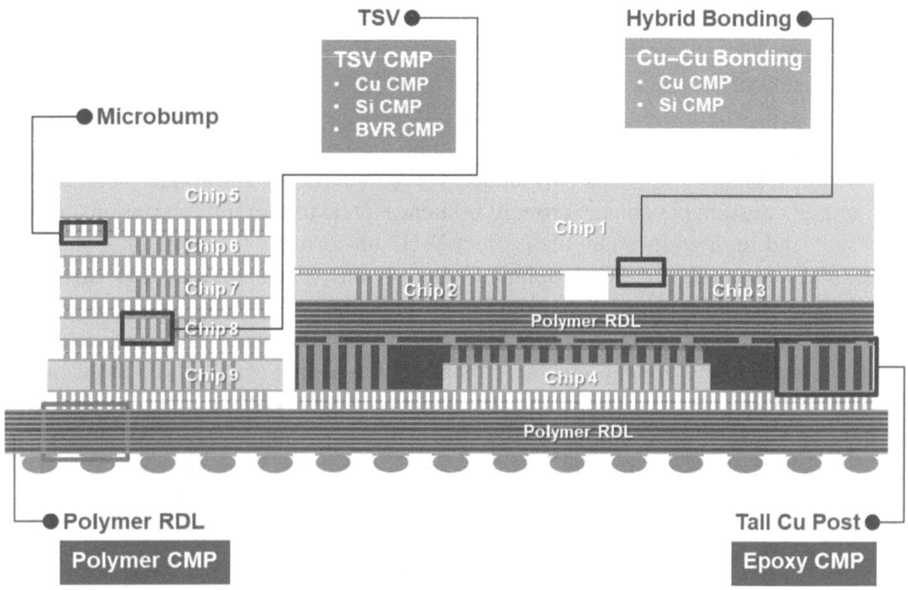

FIGURE 6.21 CMP applications in packaging.

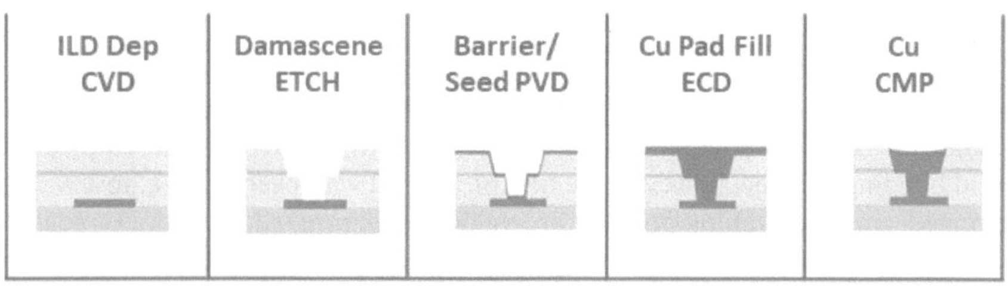

FIGURE 6.22 Copper bond formation process flow.

pad formation has three steps: Cu bulk to planarize the Cu, Cu clearing to clear the Cu from the field, and barrier removal to clear the barrier and some dielectric from the field.

In the ideal world, the surface of a Cu bond pad and its adjacent dielectric film should be at the same level. Figure 6.23 defines three parameters commonly used to describe the topography in Cu CMP. Cu dishing describes when the Cu bond

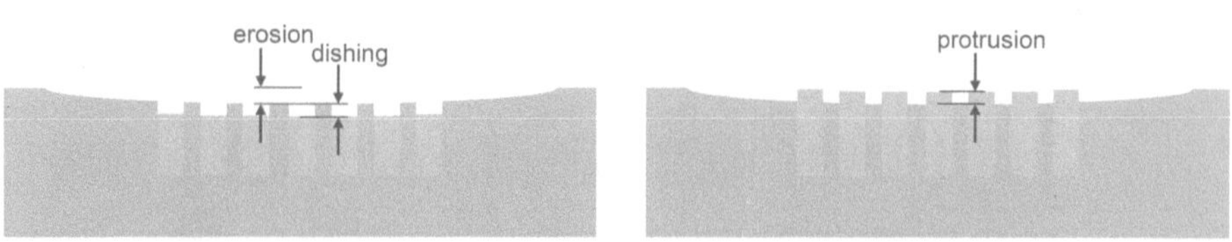

FIGURE 6.23 Cu dishing, Cu protrusion, and oxide erosion after Cu CMP.

pad recesses below the adjacent dielectric film, while Cu protrusion describes when the Cu protrudes above the dielectric. Oxide erosion describes loss in the dielectric over an array of trenches versus the oxide in an open field. Optimization of the Cu clearing step, which typically has dished Cu, and the selectivity of the barrier and dielectric to Cu in the barrier CMP step dictate the final topography. Controlling the WIW polish rates for all steps will reduce WIW topography variation. While the die layout and consumable choice strongly influence WID topography variations, polish pressure and temperature can also affect WID uniformity. Consumables (slurries, pads, disks, retaining ring) and tool matching (platens, heads) drive WTW variations in CMP, but a robust process control plan will also help to alleviate it.

After two wafers complete Cu bond pad formation, they move to surface preparation, bonding, and anneal. During annealing, Cu grains grow out of the trenches and the bond pads of the upper and lower wafers connect at the bonding interface. When the Cu dishing is too large, the Cu bond pads cannot form a robust electrical connection, leading to yield loss or reliability failures from "opens." When Cu protrudes from the bond pads, a void in the dielectric can form, leading to reliability failures ("shorts"). The final dishing and protrusion requirements will depend upon the anneal temperature, bond pad size and spacing, and integration; however, a typical dishing target can be as low as 2 +/– 1 nm. After successfully bonding the two wafers, bulk grinding thins the Si on one side, followed by Si CMP to planarize and improve the surface finish and to allow electrical connection to the backside.

6.3.2 CMP in Chip-to-Wafer Hybrid Bonding

Until Cu bond pad formation, chip-to-wafer hybrid bonding flow matches wafer-to-wafer bonding quite closely. After Cu CMP of two wafers in the chip-to-wafer hybrid bonding scheme, however, the first pretreated wafer serves as a host wafer for pretreated, singulated chips from the second wafer. Figure 6.24 shows a typical chip-to-wafer bonding flow. After bonding, a thick dielectric deposited film fills the gaps in-between chips, called inter-die gapfill, and this does not exist in wafer-to-wafer bonding. Next, CMP planarizes the dielectric and exposes the chip surface, forming a "chiplet." A through oxide via (TOV) can connect the chiplets for subsequent levels of packaging. Similar to TSV, TOV requires Cu CMP to create the interconnect.

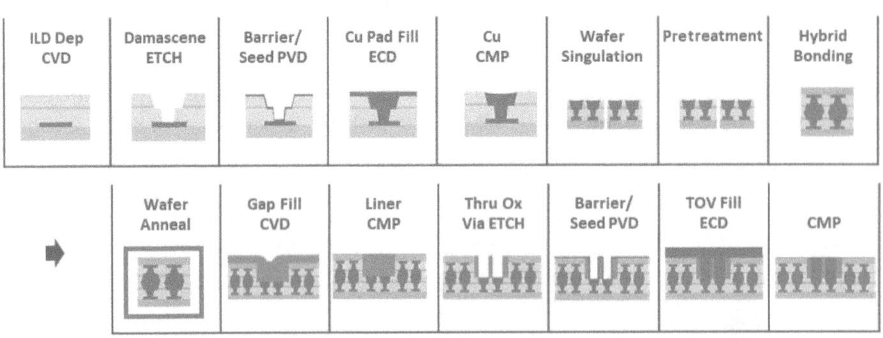

FIGURE 6.24 Chip-to-wafer hybrid bonding flow.

6.3.3 New Requirements for Through Silicon Vias (TSV)

TSV formation involves TSV etch, CVD, PVD, ECD, and Cu CMP, followed by temporary bonding to a carrier and backside via reveal (BVR) processing to expose the bottom of the TSV trenches. BVR typically involves Si grinding, CMP and etch, dielectric deposition, and dielectric CMP again for Cu reveal (BVR CMP). Figure 6.25 shows the TSV formation and reveal process flow.

As devices continue to scale, TSV scales toward smaller diameters and higher aspect ratios. In addition, the industry is driving toward replacement of microbumps with TSV direct bonding. This approach is especially valuable for high-bandwidth memory (HBM) as it allows for more stacking of DRAM chips in a limited space. To enable TSV direct bonding, traditional CMP consumables and equipment in use for TSV formation and reveal will not meet the stringent dishing requirement for hybrid bonding without improvements.

6.3.4 CMP for Redistribution Layer (RDL)

Packaging RDL, wafer molding, and gapfill commonly employ polymers and epoxies. Polymers' high bonding strength and low material cost compared to SiO_2 or SiCN make them a popular dielectric candidate for hybrid bonding. Traditionally, thinning of polymers and epoxies employed grinding; however, the more stringent planarization and surface quality requirements drive CMP as a replacement or supplement to grinding for advanced packaging applications.

For RDL formation, polymer CMP can be employed by one of two methods. The first method patterns and planarizes the polymer, followed by Cu RDL built on top of the polymer. This flow has the risk of slurry particles remaining within the vias, requiring high-performance post-CMP cleaning. The second method builds the Cu RDL followed by polymer coating and CMP to planarize and reveal the Cu before subsequent bonding. This flow has selection of a consumable set with the desired polymer/Cu selectivity as the main challenge.

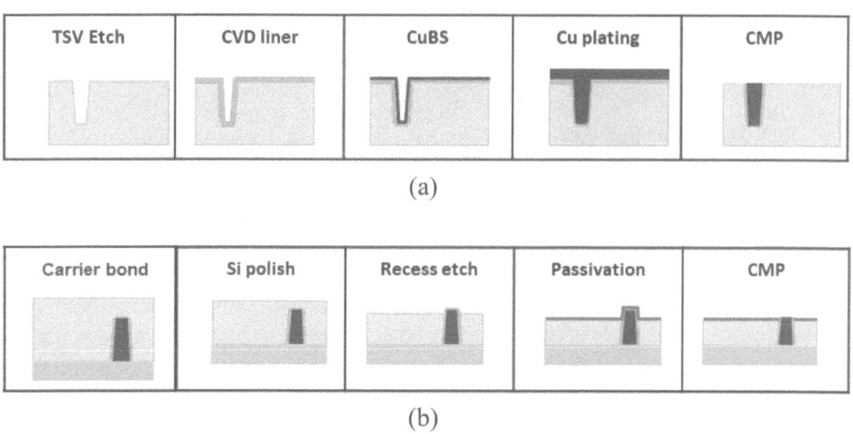

FIGURE 6.25 Through silicon via (a) formation and (b) reveal.

6.3.5 CMP for Fusion Bonding

The typical process flow for Backside Power Distribution Network (BSPDN, see 6.4.1.1), and Backside Illuminated (BSI) CMOS Image Sensor (CSI) [34] involve bonding the top dielectric of a device wafer to a carrier wafer, followed by grinding of the silicon to a thin layer, Dielectric to dielectric bonding is ofter referred to as fusion bonding. In order to ensure a high strength, void-free bond, the dielectrics are polished to form a smooth, flat surface prior to bonding. Often, a plasma treatment activates the surfaces prior to bonding. Although not metal CMP, oxide-to-oxide bonding topography requirements match those of copper-to-copper bonding to ensure a reliable, void-free bond.

6.4 CMP APPLICATIONS FOR HYBRID BONDING

6.4.1 Thick Cu CMP

The fundamentals of CMP developed over the last 20 years for Cu interconnects extend to the packaging arena for backside Cu interconnects, Cu TSV, and Cu bond pad formation. The packaging applications differ in that they tend to have much larger copper removal amounts, so Cu polish rates need to increase to achieve manufacturable tool throughputs. In the case of Cu bond pad formation prior to hybrid bonding, the topography requirements are as difficult as, if not more than, any required to produce for state-of-the-art logic copper interconnects.

6.4.1.1 Backside Cu Interconnects

Modern logic device wiring usually has interconnect signal lines and power delivery lines totaling more than 15 layers varying in thickness from fractions of a micrometer to 10 μm or more, as illustrated in Figure 6.26 [35]. Intel introduced the Buried Power Rail (BPR) and BSPDN technologies in 2021 [36]. By moving the power delivery

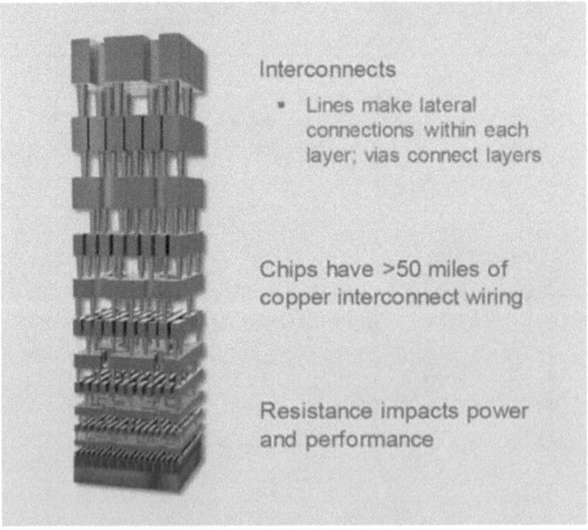

FIGURE 6.26 Frontside Cu interconnect illustration [35].

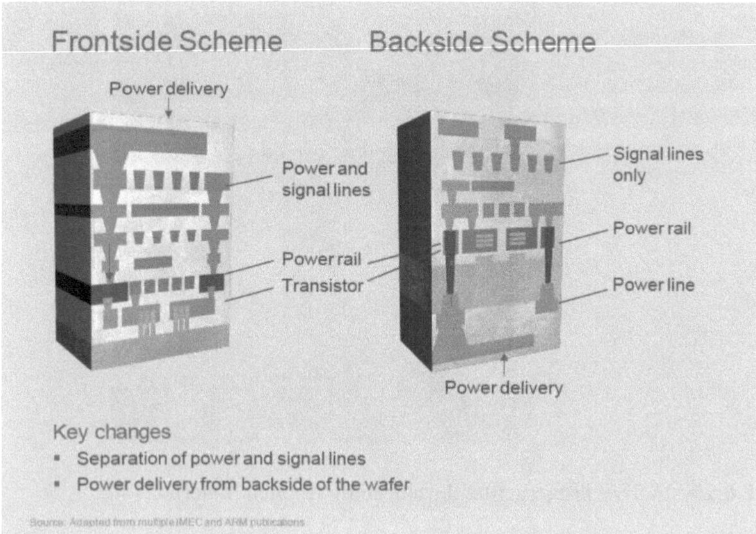

FIGURE 6.27 Backside power distribution network [35].

layers to the wafer backside, the package directly delivers power to the transistors, reducing both the voltage drop and package area. Figure 6.27 shows how to independently optimize the thick Cu backside power delivery layers and the signal layers [35]. The typical process flow for BSPDN involves fusion bonding the topside of the device wafer to a carrier wafer, thinning the device backside silicon with a combination of grinding and CMP (see 6.4.2), forming buried power rails to connect to the device from the backside, and building up a multi-level BSPDN on top of the BPR.

With a move to multi-micrometer Cu thicknesses, increasing the removal rate remains a significant challenge for high productivity without sacrificing the topography performance. Figure 6.28 shows Cu removal rate and pad temperature evolution

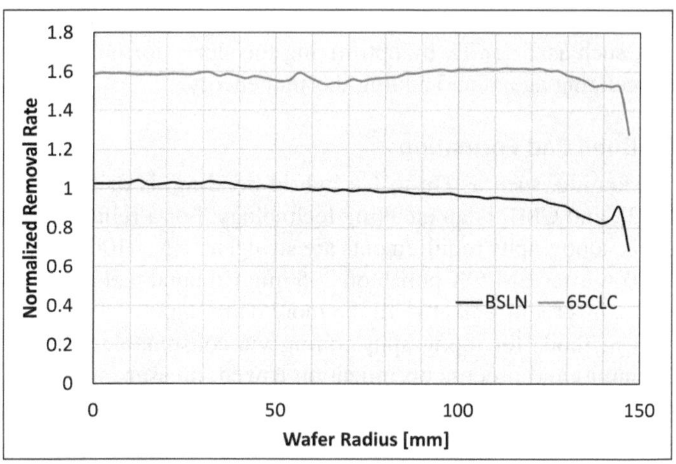

FIGURE 6.28 Heating effect on thick Cu CMP [37].

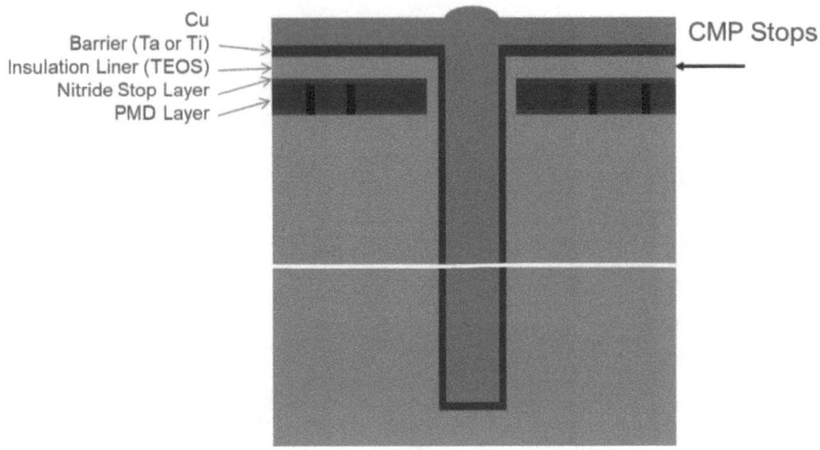

FIGURE 6.29 A TSV test structure developed by Applied Materials [40].

with and without heating on Cu blanket wafers. Rapidly heating to 65°C increases the removal rate by 60% [37].

6.4.1.2 TSV Cu

Interposers employ TSV for packaging in high-end logic, memory, and sensor devices for applications such as servers and Al [38]. TSV Cu CMP requires the removal of thick amounts of copper, barrier, and dielectric while still maintaining control over topography and defectivity [39].

Figure 6.29 shows a TSV test structure developed by Applied Materials [40], where the Cu thickness can vary from 1 μm to several micrometers, and the barrier thickness vary from a few thousand angstroms up to 1 μm, depending upon the via size and aspect ratio. CMP can stop on or within the TEOS, nitride, or PMD layers, depending upon the integration flow desired. While normally mechanically dominated, CMP can achieve removal rates of a few thousand angstroms per minute of barrier layers, such as Ta or Ti, by optimizing the slurry formulation, selecting the appropriate oxidizing agent, and adding thermal energy.

6.4.1.3 Cu Bond Pad Formation

Advanced packaging, such as Cu-to-Cu hybrid bonding, is driving semiconductor scaling [41–43], and CMP is an enabling technology. For a reliable contact without dielectric voids, topography requirements are stringent, e.g., <100 Å dishing after Cu polish and <20 Å after BM/OX polish on 2–5 mm Cu bond pads. With Cu removals in micrometers, this requirement is all the more daunting.

Traditional methods for topography tuning via consumable screening (slurries, pads, conditioning) and process optimization (speed, pressure, process control) are not sufficient to meet these targets. Temperature control, especially during Cu clearing and barrier/dielectric polishing, is one path to achieve CMP bond pad formation requirements.

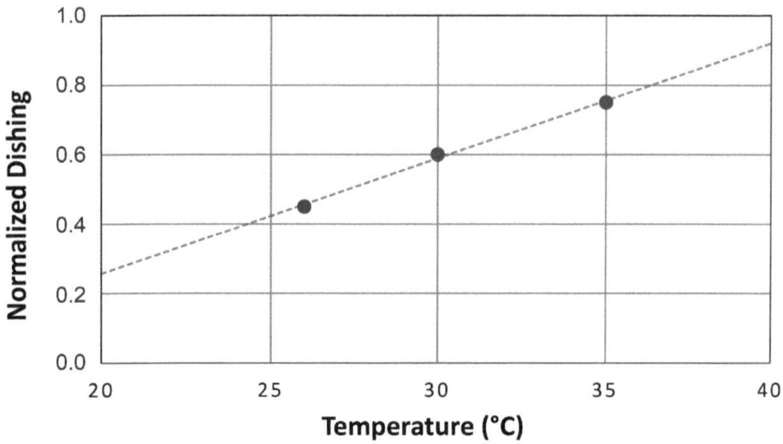

FIGURE 6.30 The relationship between process temperature and bond pad dishing.

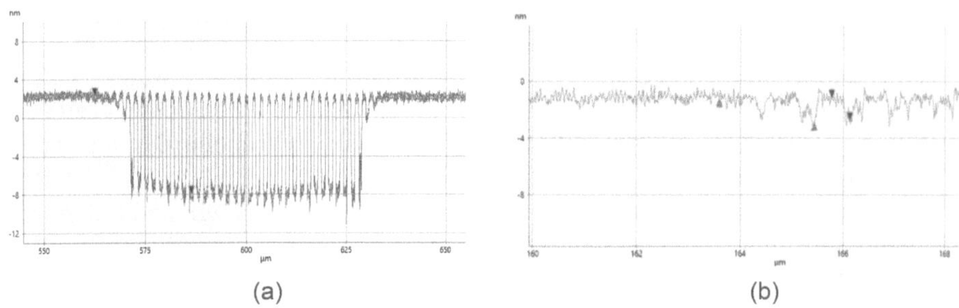

FIGURE 6.31 Typical post-CMP dishing of a 1 μm × 1 μm bond pad (a) post-Cu clear and (b) post-barrier/dielectric via AFM.

Figure 6.30 shows the relationship between post-Cu CMP bond pad dishing and process temperature. Figure 6.31 shows a typical post-Cu clear polishing dishing performance using AFM (atomic force microscopy) with cooling for a 1 × 1 μm bonding structure, achieving post-Cu clear and barrier/dielectric dishing of ~100 Å and ~20 Å, respectively.

6.4.2 Backside Silicon

Backside silicon wafer thinning is an integral part of the BPR and BSPDN process flows and silicon CMP enables backside thinning. Si CMP relies on several unique polishing performance metrics not used in other CMP applications.

1. *Global Metrics*
   a. *Total Thickness Variation (TTV):* The most critical requirement, calculated as the maximum thickness minus the minimum thickness

   b. *Bow:* Wafer deformation relative to a median surface
   c. *Warp:* Wafer deformation relative to the back surface
2. *Flatness Metrics*
   a. *Site Front Quotient Range (SFQR):* Difference between the maximum and minimum thickness value for each site relative to the least-squares best-fit reference plane over a fictitious die area of 26 mm × 28 mm
   b. *Site Backside Ideal Range (SBIR):* Difference between the maximum and minimum thickness for each site relative to the back surface over a fictitious die area of 26 mm × 28 mm
3. *Wafer Edge Metrics*
   a. *Edge Flatness:* Similar to SFQR but for the edge rather than a fictitious die
   b. *Edge Roll-off:* The rate of slope change of the average radial profile, measured in millimeters
4. *Metrics of Interest Common to Other CMP Processes:* nano-topography, defectivity, and surface metal contamination

IMEC successfully demonstrated the integration of buried power rails made of refractory metals—metallic elements like Ru or W [44–46]. The wafer-to-wafer bonding and thinning processes first bond the frontside of a device wafer to a carrier wafer followed by thinning of the device wafer backside by grinding, CMP, dry etch, and wet etch. Grinding thins the backside silicon layer to about 10 μm, and then CMP thins the remaining silicon from the backside (see Figure 6.32).

Thickness uniformity is a critical challenge for backside Si CMP and the incoming non-uniformity from backside grinding often transfers to the subsequent CMP step. The kinematics of the CMP system allows for radial non-uniformity corrections; however, the asymmetric character of grinding, as seen in Figure 6.33, is difficult to repair. One approach introduces a wet etch or dry etch process between the

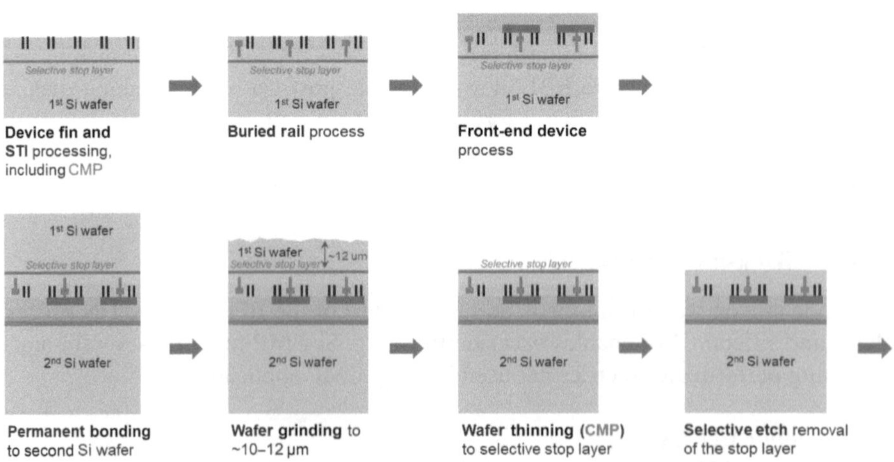

FIGURE 6.32 Si backside CMP after wafer bonding.

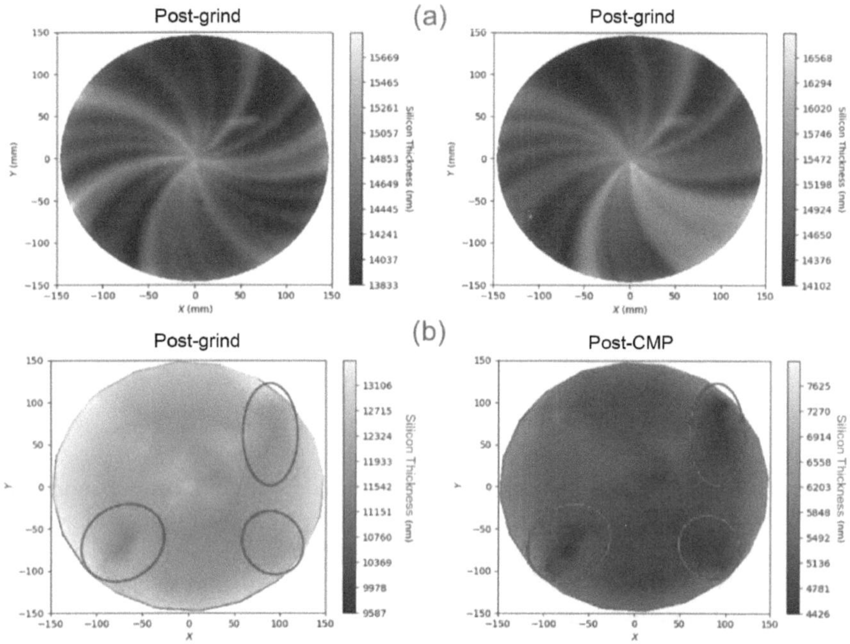

FIGURE 6.33 (a) High resolution (16,000 pts), Si thickness map after grinding showing machining marks. (b) Low resolution (1,000 pts), Si thickness maps after grinding and CMP showing transfer of grinding asymmetry to the CMP thickness profile.

grinding and the CMP in order to reduce non-uniformity and especially asymmetry (grinding marks) of the wafer incoming to CMP. Asymmetry notwithstanding, stopping in silicon at hundreds of angstroms below the device structures requires a well-controlled CMP process with non-uniformities of <1% for typical Si removals of >5 μm.

An alternative approach employs a stop layer in the silicon substrate below the device. In the case of the IMEC process flow shown in Figure 6.32, the first wafer had an embedded SiGe layer, which served as a stop layer for Si CMP. When the area of the wafer with the thinnest incoming Si reaches the SiGe, the rate there drops significantly while the Si in the thicker areas continue to polish, thereby improving the final global uniformity of the thinned wafer. A successful stop-on-film (SOF) scheme requires high selectivity to the stop layer and high silicon removal rates. Unfortunately, additives used to reduce the stop layer removal rate often reduce the silicon removal rate. Fine-tuning of the slurry composition and careful pad selection are essential in making this approach successful. Additionally, adding stop layers prior to device fabrication can present significant integration challenges, making this approach undesirable to many IC manufacturers.

Throughput presents another challenge. After silicon backside grinding, remaining silicon thicknesses are often several microns. Higher rate slurries needed to support these large removals often make profile control more difficult. Heating the pad

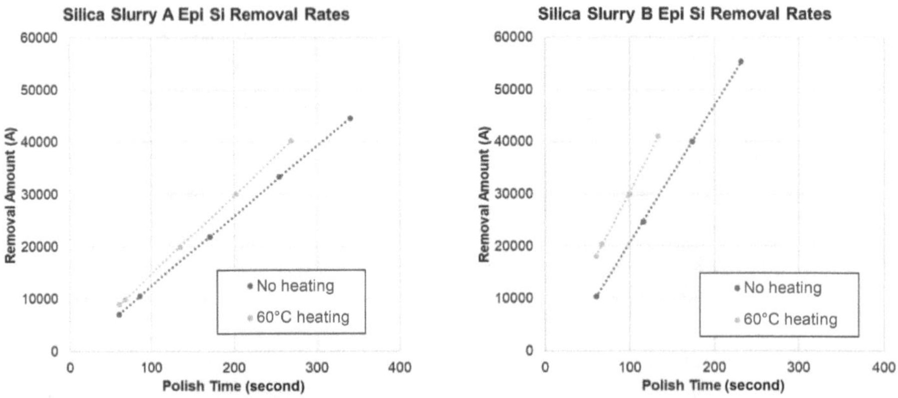

FIGURE 6.34 Silicon CMP removal rate enhancement with heating.

surface can increase the Si polish rate, the magnitude of rate boost dependent upon the consumable set. Figure 6.34 shows epitaxial silicon removal rates with and without heating for silica slurries A and B. Slurry A shows only a 12% rate boost with heating at 60°C. Slurry B's intrinsic rate is almost double that of slurry A's and heating increases its rate by an additional 33%.

Due to the wafer bevel, grinding and CMP of the silicon backside can lead to a fragile, re-entrant profile at the completion of CMP, as shown in Figure 6.35a. Figure 6.35b illustrates how "trimming" of the edge prior to wafer bonding with a dry etch or mechanical grinding can improve the edge integrity. For packaging schemes employing edge trim prior to prebonding CMP, increased contact pressure by the pad at the sharp corners of the trim can lead subsequently to rounding or edge roll-off (ERO) during CMP (see Figure 6.36). The ERO generated from CMP, in turn, can lead to water intrusion at the bonding interface, generation of bond voids, and reduced bond

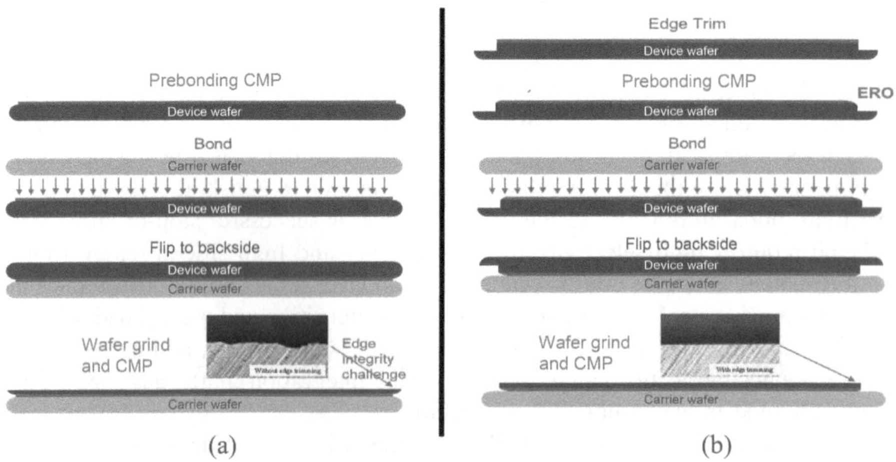

FIGURE 6.35 Backside silicon CMP (a) without edge trim and (b) with edge trim.

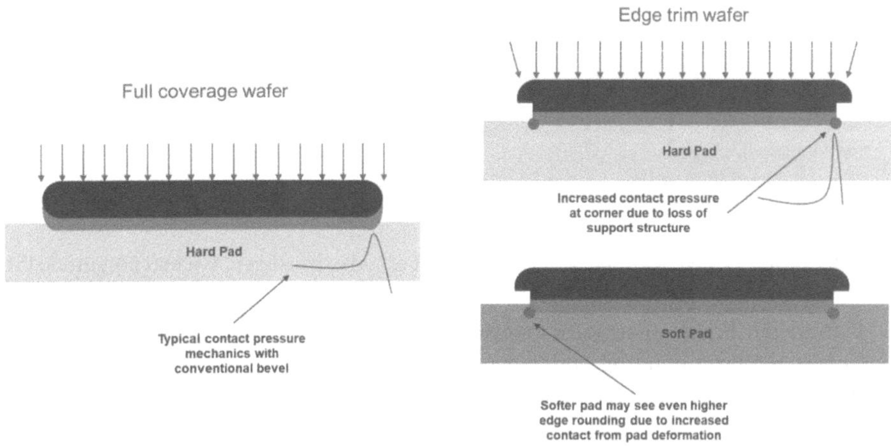

FIGURE 6.36 Edge roll-off mechanism.

strength. Pad hardness influences the magnitude of ERO, and significant efforts to minimize edge roll-off on the edge-trimmed wafers employ the most advanced CMP polishing head designs.

6.4.3 Polymer CMP

Although similar to oxide–Cu hybrid bond pad formation and TSV formation, polymer/Cu CMP presents specific challenges and benefits. During bonding, polymers can reflow, sometimes leading to air gaps around the metal. The lower modulus of the polymers however does provide for some stress relief and less sensitivity to surface particles or microscratches which could cause voids after bonding.

Chidambaram et al. [47] proposes a process flow using polymer CMP to realize hybrid bonding. Rather than a damascene approach, polymer/Cu CMP uses the cloisonne flow, as shown in Figure 6.37. The ability to spin-coat and thermal stability at the Cu–Cu diffusion bonding temperatures exceeding 300°C drive the selection of the polymer material. Additionally, the polymer must have enough mechanical toughness to survive CMP [47–49]. Polyimides have shown promise as candidates for polymer–metal hybrid bonding schemes [50].

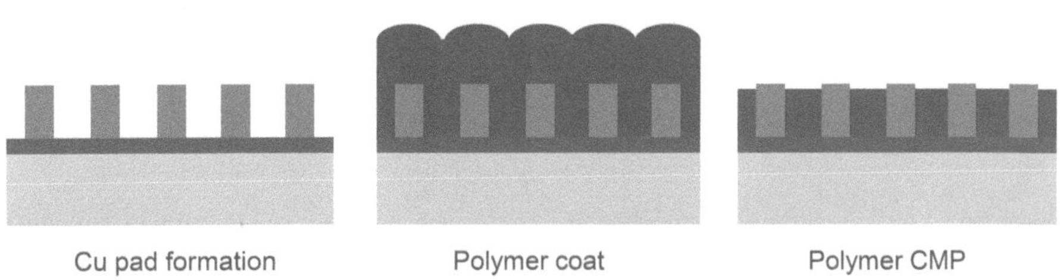

FIGURE 6.37 Use of polymer CMP step during wafer bonding and thinning.

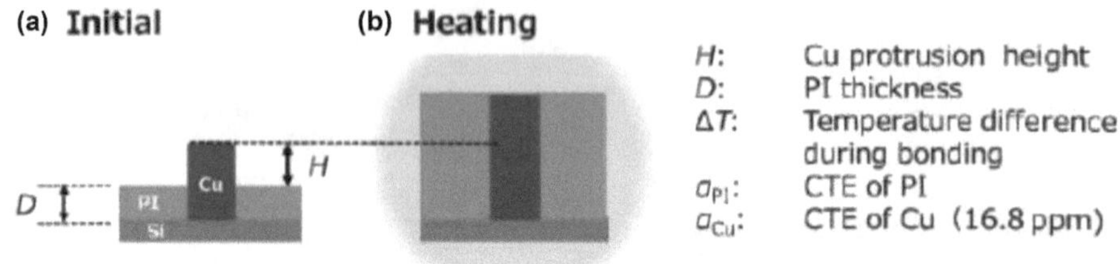

FIGURE 6.38 Schematic representation of the thermal expansion of Cu and polyimide [51].

Due to the higher thermal expansion coefficient of polymers compared to Cu, polymer/Cu hybrid bonding employs Cu protrusion rather than dishing required for oxide–Cu hybrid boding, as shown in Figure 6.38 [51]. Specific polyimide formulations require slurry customization and process optimization to achieve the desired Cu protrusion. Figure 6.39 illustrates how process conditions, such as CMP pressure, can modulate the polymer to Cu selectivity needed to achieve the required Cu protrusion.

Achievement of post-CMP polymer surface roughness with Ra (roughness average) below 1 nm requires polymer composition and CMP consumable co-optimization. Figure 6.40 shows the feasibility of achieving low post-CMP roughness values for the process developed by Showa Denko and HDMS [52]. Low roughness requires a soft CMP pad, and Cu protrusion prevention requires an accurate endpoint control method. Polymer delamination in the 8–10 mm region near the edge combined with high edge removal during the CMP step presents another challenge. Closed-loop thickness control using in situ thickness sensors and a multi-zone polishing head operating at low pressure can provide a solution to edge delamination.

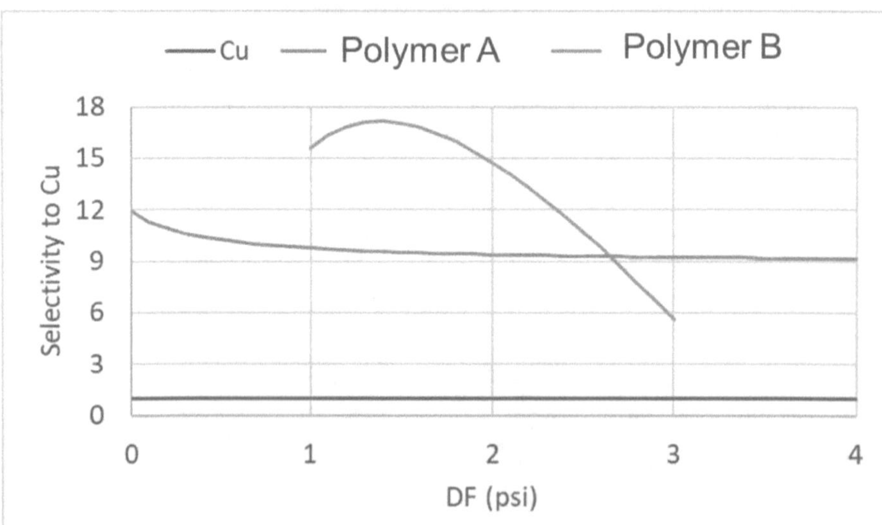

FIGURE 6.39 Polymer CMP removal selectivity to Cu for polymer A and polymer B as A function of downforce.

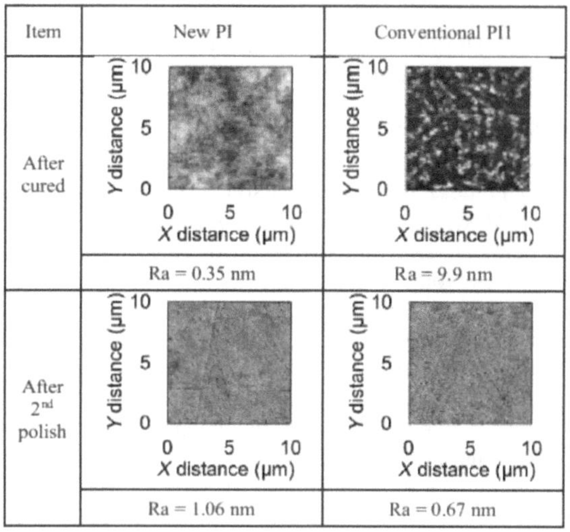

FIGURE 6.40 Surface roughness Ra values on 10 × 10 μm area for two polyimide type polymers: new PI and conventional PI [52].

6.5 CONCLUSION

Despite CMP as the planarization technique of choice for today's front-end semiconductor process manufacturing, backend semiconductor manufacturers have regarded CMP as overly complex to meet traditional packaging requirements. The excessive cost of consumables has limited CMP's adoption to only a few backend steps such as TSV formation and reveal. With the adoption of hybrid bonding, CMP will become a technology enabler in advanced semiconductor packaging. The extreme topography requirements at the copper-to-copper bonding interface drive the need for CMP to achieve acceptable product yield. With dishing requirements for Cu as low 2 nm +/– 1 nm, CMP for Cu hybrid bonding provides technical challenges in line with any state-of-the-art front-end CMP step. The additional challenge in hybrid bonding comes from the large removal amounts. The larger absolute incoming thickness variation will challenge the in situ process control to meet these tight topography requirements. Also, in order to meet reasonable machine throughput for high-volume manufacturing, real-time heating and cooling coupled with consumable selection are key. Hybrid bonding requires moving packaging CMP to state-of-the-art equipment, process control, and consumables.

REFERENCES

1. M. A. Fury, "The Early Days of CMP," Solid State Technology, vol. 40, no. 5, p. 81+, 1997.
2. F. W. Preston, "The Theory and Design of Plate Glass Polishing Machines," Journal of the Society of Glass Technology, vol. 11, pp. 214–256, 1927.
3. F. B. Kaufman, D. B. Thompson, R. E. Broadie, M. A. Jaso, W. L. Guthrie, D. J. Pearson, and M. B. Small, "Chemical–Mechanical Polishing for Fabricating Patterned W Metal Features as Chip Interconnects," Journal of the Electrochemical Society, vol. 138, no. 11, pp. 3460–3465, 1991.

4. M. K. Poddar, P. Jalalzai, S. Sahir, N. P. Yerriboina, T.-G. Kim, and J.-G. Park, "Tungsten Passivation Layer (WO_3) Formation Mechanisms during Chemical Mechanical Planarization in the Presence of Oxidizers," Applied Surface Science, vol. 537, pp. 147862–147892, 2021.
5. H. J. Kim, S.-G. Ahn, L. Qin, D. Koli, V. Govindarajulu, and Y. Moon, "Effects of Pad Temperature on the Chemical Mechanical Polishing of Tungsten," ECS Journal of Solid State Science and Technology, vol. 3, no. 10, pp. 310–314, 2014.
6. K. Xu, S.-H. Shen, J. Fung, H. Iravani, I. Carlsson, T.-Y. Liu, B. Swedek, S.-S. Chang, W.-C. Tu, T. Kitajima, K. Mikhaylich, B. Brown, S. Huey, and F. Redeker, "Impact of Film Morphology on Chemical Mechanical Polishing of Tungsten," ECS Journal of Solid State Science and Technology, vol. 5, no. 6, pp. 361–367, 2016.
7. H. Wu, P. Diep, T. Sekine, J. Tang, X. Wang, W. Shen, S. Cui, W. Lu, B. Brown, Y. Chen, K. Xu, H. Soundararajan, K. Chou, F. H. Chen, C. Pollard, S.-S. Chang, J. Domin, and S. Datar, "CMP In-situ Temperature Control Module: An Overview," Applied Materials Engineering and Technology Conference, 2021.
8. Y. Li and S.V. Babu, "Chemical–Mechanical Polishing of Copper and Tantalum with Silica Abrasives," Journal of Materials Research, vol. 16, no. 4, pp. 1067–1073, 2001.
9. M. C. Lin, M. Q. Wang, J. Lai, R. Huang, C. M. Weng, J. H. Liao, J. Tang, C. H. Weng, W. Lu, H. W. Chen, and J. T. Lee, "Metal Hard Mask Employed Cu/Low K Film Post Ash and Wet Clean Process Optimization and Integration into 65 nm Manufacturing Flow," Solid State Phenomena, vol. 134, pp. 359–362, 2008.
10. M. Sharma and C.-C. A. Chen, "Analytical Modelling of Material Removal in Copper Chemical Mechanical Polishing Incorporating the Scratch Hardness of the Passivated Layer on Copper Thin Film Wafer," ECS Journal of Solid State Science and Technology, vol. 11, no. 4, pp. 44007–44008, 2022.
11. K. Wijekoon, S. Tsai, M. Chandrachood, B. Brown, F. Redeker, S. Nanjangud, and G. Amico, "Multi-step Approach to Copper CMP Development," SEMI/Japan Technical Symposium, Tokyo, Japan, 1998.
12. L. Zhang, Y. Meng, Y. Xian, W. Zhang, H. Zhou, J. Fang, and Y. Zhang, "Copper Corrosion Issue Analysis and Study on Advanced CMP Process," China Semiconductor Technology International Conference, Shanghai, China, 2021.
13. S. Mudhivarthi, N. Gitis, S. Kuiry, M. Vinogradov, and A. Kumara, "Effects of Slurry Flow Rate and Pad Conditioning Temperature on Dishing, Erosion, and Metal Loss during Copper CMP," Journal of the Electrochemical Society, vol. 153, no. 5, p. 372, 2006.
14. V. Kakireddy, S. Mudhivarthi, and A. Kumara, "Effect of Temperature on Copper Damascene Chemical Mechanical Polishing Process," Journal of Vacuum Science & Technology B, vol. 26, pp. 141–150, 2008.
15. P. Liu, S. Bae, S. Hong, C. Bae, H. Seo, J. Lee, C. Tang and T. Kim, "Investigation of Thermal Effects in Copper Chemical Mechanical Polishing," Precision Engineering, vol. 73, pp. 195–202, 2022.
16. C. Wu, Y. Sampurno, X. Liao, Y. Jiao, S. Theng, Y. Zhuang, L. Borucki, and A. Philipossian, "Pad Surface Thermal Management during Copper Chemical Mechanical Planarization," ECS Journal of Solid State Science and Technology, vol. 4, no. 7, pp. 206–212, 2015.
17. S.-S. Chang, H. Soundararajan, C. Pollard, F. Chen, K. Chou, C. Lee, J. Domin, S. Datar, H. Wu, J. Tang, J. Oh, and S. Zuniga, "CMP Temperature Control," Applied Materials Engineering and Technology Conference, 2021.
18. Y. Wang, M. Gage, M. Yu, K. Xu, Z. Wang, and W.-C. Tu, "Managing Galvanic Corrosion of Advanced Barrier in CMP Process," Applied Materials Engineering and Technology Conference, 2013.
19. Z. Song, S. P. Neo, C. K. Oh, S. Redkar, and Y.-P. Lee, "Copper Corrosion Issue and Analysis on Copper Damascene Process," IEEE Transactions on Device and Materials Reliability, vol. 5, no. 2, pp. 206–211, 2005.

20. W. Lu, H. Wu, T. Sekine, H. Soundararajan, S.-S. Chang, C. Pollard, M.-P. Cai, J. Zhang, J. Tang, and B. Brown, "Defect Reduction in Metal CMP Applications," Applied Materials Engineering and Technology Conference, 2021.
21. M. Quirk and J. Serda, "Chemical Mechanical Planarization" in *Semiconductor Manufacturing Technology*, Columbus, OH: Prentice Hall, 2001, p. 523.
22. M. Krishnan, J. W. Nalaskowski, and L. M. Cook, "Chemical Mechanical Planarization: Slurry Chemistry, Materials, and Mechanisms," Chemical Reviews, vol. 110, no. 1, pp. 178–204, 2010.
23. R. Srinivasan, P. V. R. Dandu, and S. V. Babu, "Shallow Trench Isolation Chemical Mechanical Planarization: A Review," ECS J. Solid State Science and Technology, vol. 4, no. 11, pp. 5029–5039, 2015.
24. L. M. Cook, "Chemical Processes in Glass Polishing," Journal of Non-Crystalline Solids, vol. 120, pp. 152–171, 1990.
25. T. Hoshino, Y. Kurata, Y. Terasaki, and K. Susa, "Mechanism of Polishing of SiO_2 Films by CeO_2 Particles," Journal of Non-Crystalline Solids, vol. 383, no. 1–3, pp. 129–136, 2001.
26. P. R. Veera Dandu et al., "Silicon Nitride Film Removal during Chemical Mechanical Polishing Using Ceria-based Dispersions," Journal of the Electrochemical Society, vol. 158, no. 8, pp. H763–H767, 2011.
27. Y.-Z. Hu, R. J. Gutmann, and T. P. Chow, "Silicon Nitride Chemical Mechanical Polishing Mechanisms," Journal of the Electrochemical Society, vol. 145, pp. 3919–3925, 1998.
28. W. G. America and S. V. Babu, "Slurry Additive Effects on the Suppression of Silicon Nitride Removal during CMP," Electrochemical and Solid-State Letters, vol. 7, pp. G327–G330, 2004.
29. Y. Chen, K. Mikhaylichenko, B. Brown, and F. Redeker, "Post-CMP Cleaning," in *Handbook of Semiconductor Cleaning Technology*, K. Reinhardt and W. Kern (eds.), Cambridge, MA: Elsevier, 2018, pp. 253–301.
30. M. L. Zoeteweij, J. C. van der Donck, and R. Versluis, "Particle Removal in Linear Shear Flow: Model Prediction and Experimental Validation," Journal of Adhesion Science and Technology, vol. 23, no. 6, pp. 899–911, 2009.
31. R. Vos, M. Lux, K. Xu, W. Fyen, C. Kenens, T. Conard, P. Mertens, M. Heyns, Z. Hatcher, and M. Hoffman, "Removal of Submicrometer Particles from Silicon Wafer Surfaces Using HF-based Cleaning Mixtures," Journal of the Electrochemical Society, vol. 148, no. 12, pp. G683–G691, 2001.
32. S. Huey, B. Chandrasekaran, D. Bennett, S. Tsai, K. Xu, J. Qian, S. Dhandapani, J. David, B. Swedek, and L. Karuppiah, "CMP Process Control for Advanced CMOS Device Integration," ECS Transactions, vol. 44, no. 1, p. 543, 2012.
33. B. J. Brown, "Closing the Loop: CMP Process Control Methodology and Implementation," China Semiconductor Technology International Conference Symposium V: CMP and Post-Polish Cleaning, Shanghai, China, 2022.
34. A. Lahav, A. Fenigstein, A. Strum, and S. Rizzolo, "Backside Illuminated (BSI) Complementary Metal-Oxide-Semiconductor (CMOS) Image Sensors," in *High Performance Silicon Imaging Fundamentals and Applications of CMOS and CCD Sensors*, D. Durini (ed.), Cambridge, MA: Woodhead Publishing, 2020, pp. 95–117.
35. R. Freed, M. Sullivan, K. Moraes, M. Naik, S. Ramamurthy, and R. Achutharaman, "New Ways to Wire and Integrate Chips," Applied Materials Master Class, 2022.
36. Intel, "Intel Accelerated: Introducing New RibbonFET and PowerVia Technologies," *intel.com*, July 26, 2021. Available at: https://intel.ly/3GonMP1 (accessed April 25, 2023).
37. J. Tang, H. Wu, S.-S. Chang, M. Gage, T. Sekine, K. Xu, B. Brown, and S. Huey, "CMP Challenges and Solutions for Low-resistance Wiring and Backside Power Distribution Network Applications," Advanced Metallization Conference, Tokyo, Japan, 2022.

38. S. Nam, Y. Kim, A. Jang, I. Hwang, S. Park, S.-C. Lee, and D.-W Kim, "The Extremely Large 2.5D Molded Interposer on Substrate (MIoS) Package Integration: Warpage and Reliability," IEEE 71st Electronic Components and Technology Conference (ECTC), pp. 1998–2002, 2021.
39. M. Gage, F. Liu, and Y. Wang, "High Removal Rate Silicon CMP for Cu TSV Application," Applied Materials Engineering and Technology Conference, 2010.
40. W. Lu, M. Gage, S. Hu, T. Gartner, Y. Wang, Z. Wang, and W.-C. Tu, "Differentiators of CMP Reflexion/Reflexion LK Tools for Via-middle Cu TSV CMP Applications," Applied Materials Engineering and Technology Conference, 2012.
41. T.-S. Kim, S. Cho, S.-K. Hwang, K. Lee, Y. Hong, H. Lee, H. Cho, K. Moon, H. Na, and K. Hwang, "Multi-stack Wafer Bonding Demonstration Utilizing Cu to Cu Hybrid Bonding and TSV Enabling Diverse 3D Integration," IEEE 71st Electronic Components and Technology Conference, pp. 415–419, 2021.
42. A. Jourdain, F. Schleicher, J. De Vos, M. Stucchi, E. Chery, A. Miller, G. Beyer, G. van der Plas, E. Walsby, K. Roberts, H. Ashraf, and D. Thomas, "Extreme Wafer Thinning and Nano-TSV Processing for 3D Heterogeneous Integration," IEEE 70th Electronic Components and Technology Conference (ECTC), pp. 42–48, 2020.
43. J. Ryckaert, A. Gupta, A. Jourdain, B. Chava, G. van der Plas, D. Verkest, and E. Beyne, "Extending the Roadmap beyond 3 nm through System Scaling Boosters: A Case Study on Buried Power Rail and Backside Power Delivery," Electron Devices Technology and Manufacturing Conference (EDTM), pp. 50–52, 2019.
44. L. Peters, "Challenges in Backside Power Delivery," Semiconductor Engineering, November 17, 2022. Available at: https://semiengineering.com/challenges-in-backside-power-delivery/ (accessed May 8, 2023).
45. K. Vandersmissen, K. Devriendt, D. Pittevils, B. Ulenaers, P. Kalantzis, P. Ong, J. Vande Weeghde, and H. Struyf, "CMP Challenges for Buried Power Rail Integration," International Conference on Planarization Technology, Portland, OR, 2022.
46. N. Horiguchi and E. Beyne, "Powering Chips for the Backside," Chip Scale Review, vol. 26, no. 6, pp. 43–48, 2022.
47. V. Chidambaram, P. Lianto, X. Y. Wang, G. See, N. Wiswell, and M. Kawano, "Dielectric Materials Characterization for Hybrid Bonding," IEEE 71st Electronic Components and Technology Conference, pp. 426–431, 2021.
48. Y. Kwon, J. Seok, J.-Q. Lu, and T. S. Cale, "Critical Adhesion Energy of Benzocyclobutene-bonded Wafers," Journal of the Electrochemical Society, vol. 153, no. 4, pp. G347–G352, 2006.
49. C.-C. Hsiao, H.-C. Fu, C.-W. Chiang, O.-H. Lee, T.-Y. O. Yang, and H.-H Chang, "Feasibility Study of Nanotwinned Copper and Adhesive Hybrid Bonding for Heterogeneous Integration," IEEE 71st Electronic Components and Technology Conference, pp. 445–450, 2021.
50. K.-C. Shie, P.-S. He, Y.-H. Kuo, J. J. Ong, K. N. Tu, B. T.-H. Lin, C.-C. Chang, and C. Chen, "Hybrid Bonding of Nanotwinned Copper/Organic Dielectric with Low Thermal Budget," IEEE 71st Electronic Components and Technology Conference, pp. 432–437, 2021.
51. T. Shirasaka, T. Okuda, T. Shibata, S. Yoneda, D. Matsukawa, M. Mariappan, M. Koyanagi, and T. Fukushima, "Comprehensive Study on Advanced Chip on Wafer Hybrid Bonding with Copper/Polyimide Systems," IEEE 71st Electronic Components and Technology Conference, pp. 317–323, 2022.
52. S. Yoneda, K. Adachi, K. Kobayashi, D. Matsukawa, M. Sasaki, T. Itabashi, T. Shirasaka, and T. Shibata "A Novel Photosensitive Polyimide Adhesive Material for Hybrid Bonding Processing," Proceedings: IEEE 71st Electronic Components and Technology Conference, pp. 680–686, 2022.

7 Permanent and Temporary Wafer Bonding

Viorel Dragoi, Thomas Uhrmann, and Paul Lindner

7.1 INTRODUCTION

The adhesion between two clean flat surfaces placed in direct contact was observed since ancient times and started to attract the interest of the semiconductor industry already in the late 1970s. The spontaneous adhesion between two fused silica or glass surfaces (mostly known nowadays under the name "fusion bonding") was used by various companies for applications based on different substrates' combinations [1]. Another process of interest for industrial applications was the electric field–assisted bonding between glass–metal [2] or glass–silicon, mostly known nowadays under the name "anodic bonding." These early wafer bonding–based applications were initially using homemade fixtures and apparatus which were good enough to provide conditions for proving a concept or for manufacturing small volumes of devices. The relatively low complexity of these applications required the use of a combination of manually operated and semiautomated systems (even based on homemade solutions), providing both in situ wafer-to-wafer alignment and bonding.

With the increase in importance and the diversification of the applications foreseen for wafer bonding, the manufacturing of dedicated equipment started in early 1990s, to enable the specific process conditions for the various wafer bonding processes.

The evolution toward fully automated equipment had two main driving forces: technology and economics. The increasing number of applications emphasized the need for dedicated wafer-bonding equipment. Particularly, the development of highly specialized wafer-bonding equipment was the only path to ensure the levels of productivity, reliability, and reproducibility required by industrial manufacturing processes.

Probably one of the first fully automated bonders for volume production was designed to perform fusion bonding for silicon-on-insulator (SOI) substrates manufacturing: the process requires excellent contamination control, accurate handling of the two wafers when placed in contact, and high throughput. Such a system was able to ensure even front-end of line (FEOL) conditions for the fusion bonding process and had at that time the capability of performing 15–25 bonds/hour (for only the fusion bonding process, as the subsequent thermal annealing was performed in batches outside the bonding equipment).

The idea of using a mechanical carrier for the handling of thin or brittle substrates was used in the past for various processes (e.g., lapping, polishing, photolithography, etc.). One of the application areas pushing for the use of such techniques was the

DOI: 10.1201/9781003408673-7

handling of compound semiconductor substrates, which typically are very brittle and require mechanical support during handling, patterning, or thinning processes.

In the early times, various waxes and resins were used as adhesives for temporary bonding. The carrier material and carrier design were always chosen to match with the substrate to handle (particularly thermal matching) or with the final debonding method (e.g., carriers with through holes, for solvent debonding) [3]. The concept evolved in time to what is known today as temporary wafer bonding.

This chapter presents an overview of the main equipment concepts and how the main process requirements are fulfilled in semiconductor manufacturing equipment.

7.2 EQUIPMENT TECHNOLOGY FOR PERMANENT WAFER BONDING

The first generations of commercially available wafer bonding equipment were designed to cover the process specifications using an "universal" bond chamber able to run all wafer bonding processes. Such a system was very flexible by providing inside the bond chamber atmosphere control (from high/low vacuum to overpressure, and using different process gases as needed), two-sided heating with controlled heating and cooling ramps (temperature typically ranging from room temperature to 500–600°C), the possibility to apply contact force to the wafers, and even the possibility to apply an electric field for performing anodic wafer bonding.

With such capabilities, the equipment was very useful for applications, particularly for early micro- electromechanical systems (MEMS). However, the use of such universal bond chamber was sometimes limited, particularly in cases where performing a specific process would produce residual contamination which might impact the next processes to run in the same equipment. Examples can be given for metal contamination (e.g., after metal wafer bonding processes or anodic wafer bonding), particles contamination (e.g., after glass frit bonding processes), or organic contamination (after glass frit or adhesive wafer bonding using polymer layers) interfering with high cleanliness applications based on fusion bonding. For such reasons, the typical use of the universal bond chamber was limited by users to a single process type, depending on the applications envisioned.

The wafer bonding–based applications are always requiring the two substrates to be aligned. The type of alignment must be selected depending on the final alignment accuracy needed, on wafer types, and on the bonding process intended for use.

7.2.1 Process Flows for 3D Interconnect Fabrication

Once the fabrication of wafer-level 3D interconnects by aligned wafer bonding was considered, wafer bonding stepped into a new era of developments and penetrated new and very advanced applications areas. Three important processes flows used for such applications are shown in Figure 7.1.

Among all the wafer bonding processes [4], metal thermocompression (TC) wafer bonding was historically the first process of choice in a via-first approach due to its capability to directly form a metal interconnection between two wafer surfaces. As

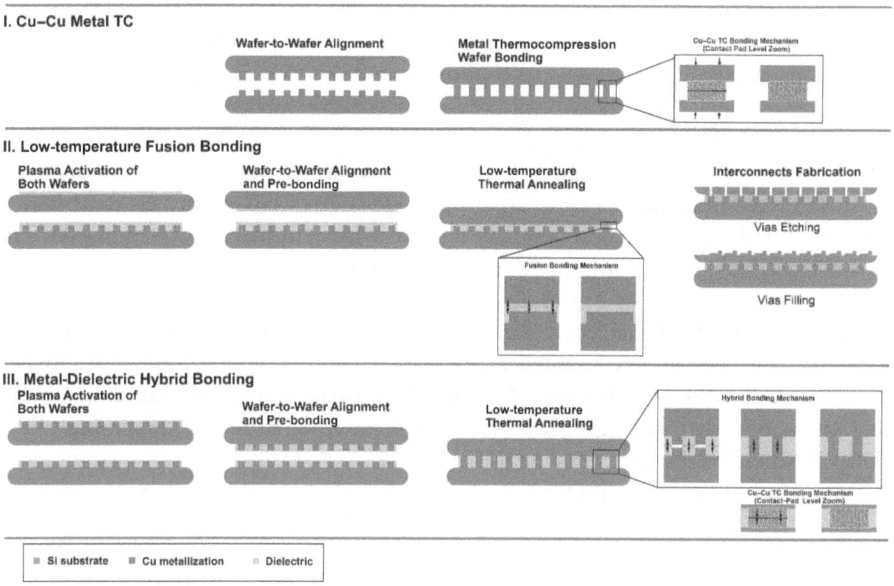

FIGURE 7.1 Schematic process flows used for 3D interconnects fabrication.

shown in Figure 7.1 (process flow I), the fully patterned wafers are optically aligned and bonded to each other (pad-to-pad) using metal TC bonding process. In the first stage of the metal TC bonding process, a high contact pressure is applied to the bonding area (typically with wafer size pistons pressing the two wafers against each other) to allow the metal atoms from the two surfaces to come into close proximity. In the second stage, the heating applied during the process (typically 300–400°C for Cu–Cu bonds) facilitates atomic diffusion, leading to bonding between the metal pads, with the total elimination of the initial interface.

The high contact force applied to the entire wafers, however, can lead to problems such as misalignment. Additionally, due to the very low metal bonding area (typically less than 10% of the total wafer area), the robustness of the bonded pairs was quite low during the post-bonding thinning steps (usually performed by grinding), causing a high risk of breakage. This was one of the main process drawbacks preventing this technology from really taking off as the main path for the fabrication of 3D interconnects. There were attempts to overcome this problem, e.g., by increasing the bonding area to 100% by filling the gaps between the Cu pads on the wafer surfaces with a polymer in a polymer–metal hybrid bonding. The material used was a flowable polymer (benzocyclobutene, or BCB), which was filling the gaps between the pads during the process due to its low viscosity during the bonding process [5]. In principle the method worked, but the added polymer processing (coating, optionally also lithography for directly patterning the polymer acting as a negative resist) and the added risk of polymer cross-contamination of the complementary equipment prevented this process from going mainstream. However, this can be considered an early attempt for hybrid bonding processes.

With the development of the low temperature fusion bonding based on plasma treatment of the surfaces in the early 2000s [6–10], new opportunities opened for the fabrication of 3D interconnects. Figure 7.1 (process flow II) introduces a process flow for a via-last approach: two patterned wafers were bonded using a fusion bonding process, then one wafer was thinned, vias were etched in both wafers and then filled with metal to produce the electrical connections between the two wafers [11].

In the low-temperature fusion bonding process, both wafers are covered with a high-quality dielectric layer (extremely flat and with extremely low roughness). The two surfaces are treated in a plasma for short time, optionally rinsed after plasma treatment using water for airborne particle removal, then aligned and pre-bonded (placed in contact) at ambient conditions. As a result of the pre-bonding, the two dielectric surfaces adhere to each other due to van der Waals forces established between molecules from the two surfaces. This weak bond produced is reversible and must be further strengthened using a low-temperature thermal annealing (200–400°C) which results in a permanent bond.

More recently, a very important process was developed based on a combination of low-temperature fusion bonding and metal TC bonding (without the need to apply an external force) performed simultaneously: hybrid wafer bonding [12]. In hybrid wafer bonding, the surfaces of both fully patterned wafers are covered with a dielectric layer using various chemical vapor deposition (CVD) methods, and then the metal contact pads are opened by chemical mechanical planarization (CMP), resulting in a metal pad surface recessed from the dielectric surface. The wafers are surface activated using a plasma treatment, optically aligned and bonded (process flow III). The plasma activation of the surfaces allows establishing a fusion bond between the two dielectric surfaces placed in contact. During the subsequent thermal annealing, the dielectric–dielectric fusion bond is strengthened, and simultaneously, the two recessed Cu pads are thermally expanding. When the two metal pads expand enough to come into contact, the dielectric bond is already at maximum strength, so the continuing thermal expansion of Cu pads applies implicitly a high contact pressure at the Cu–Cu interface, leading to bonding between the two Cu pads. Due to the absence of the external force, a very high alignment accuracy can be maintained.

Up to now in this chapter, the metal TC bonding process was mentioned for historical reasons. However, since this process is not used as much in 3D interconnect manufacturing and advanced packaging compared to other processes, only fusion and hybrid bonding processes will be further discussed, and these two processes are also using similar equipment platforms.

While fusion bonding is used mostly in a monolithic integration approach and for layer transfer, hybrid bonding is becoming a mature technique for fabrication of 3D interconnects and for advanced packaging. The two processes are based on the same general process flow: the two wafers to be bonded are activated in a plasma, then optionally cleaned for removing potential airborne particles on the surfaces, and then optically aligned and brought into contact under ambient conditions. After this first step, the adhesion developed between the two surfaces is strong enough to preserve the wafer-to-wafer alignment but the bonds are still reversible, and the two wafers can still be separated. In order to strengthen the bond, a subsequent thermal annealing (<400°C) is employed, resulting in the formation of a permanent connection based on covalent bonds between the dielectric layers from the two surfaces. In order to perform a successful bonding process, the following process steps are employed: substrates activation by plasma treatment, substrates cleaning, optical alignment and pre-bonding (namely, contacting the two surfaces at room temperature), followed by thermal annealing.

Based on the process flow for industrial manufacturing, equipment for hybrid bonding is split between the general equipment in the fab and the dedicated wafer bonding process equipment.

7.2.2 Fusion and Hybrid Wafer Bonding

Typically, the wafers intended for bonding are characterized after the manufacturing process, and finally undergo a wet bench cleaning, prior to being loaded onto the bonding equipment.

The fully automated bonding equipment integrates various functional modules performing different processes. First the wafer surface conditioning is performed (e.g., metal native oxide removal—optional for hybrid bonding, dielectric surface activation by plasma treatment, and final single wafer cleaning with deionized water). The activated wafers are then loaded onto the optical alignment module where they are aligned and placed into contact (pre-bonded), before going to an optical alignment verification module.

After unloading from the bonding equipment, the pre-bonded wafers are annealed in an external oven. After annealing, the wafers might be further inspected for bonding defects (typically using IR inspection or acoustic microscopy) and for bond strength measurement.

The first important aspect for fusion/hybrid bonding is to ensure the high quality of the incoming substrates. As the process depends on the contact between two rigid surfaces, the requirements in terms of flatness and micro-roughness are extremely strict in order to ensure high-quality bonds. Once the substrates are fabricated, they undergo a batch chemical cleaning process which fulfills the required incoming wafer contamination requirements before being loaded to the fully automated bonding equipment.

Inside the fully automated equipment, the wafers travel between the different process modules based on a robot handling system with the end effector not touching the front side of the wafers. The environment inside the fully automated equipment is well controlled by using a fan-filter assembly, ensuring particles-free, organics-free, and controlled moisture laminar airflow in the wafer-handling area, compliant with ISO 3 or FS 209 E—class 1 standards.

A typical manufacturing process flow is shown schematically in Figure 7.2.

In the first step, the wafers are loaded into a plasma chamber where they are surface treated. The gases used for the plasma treatment are typically nitrogen, oxygen, and some gas mixtures (e.g., O_2:N_2 mixture, H_2 diluted in an inert gas, etc.) [13].

After the plasma treatment, the wafers can be further cleaned to make sure that prior to bonding, there is no airborne particle on the wafer surfaces during the handling and plasma treatment. This cleaning step is designed as a "mild" chemicals-free

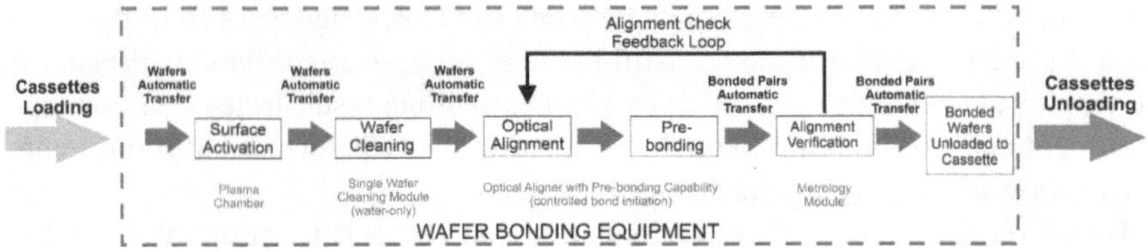

FIGURE 7.2 Schematic process flow of fusion and hybrid wafer bonding processes.

cleaning, for already high-cleanliness wafers, using only ultrapure deionized water. In order to increase the cleaning efficiency, additional features can be added to the cleaning process, e.g., megasonic nozzles [14], megasonic area transducers [15], or pressurized jet nozzles. In the early days, brush scrubbers were also used; however, due to the complexity of the substrates used for bonding (e.g., complex metal–dielectric surfaces with some surface topography), brushes are not much used anymore. Even if the standard cleaners use just water as the cleaning agent, mild chemistries can be also used (e.g., 2% NH_4OH or H_2O_2) depending on the application requirements. Water is typically the fluid of choice, as it simplifies the chemicals management inside the bonding equipment (no need for special fluid lines, special lids on cleaning modules, or special tanks) as well as in the fab (no need for additional drain lines for chemicals besides the standard water drains).

After the cleaning step, the two wafers are loaded sequentially into the aligner module, which performs two processes: first the two wafers are optically aligned using their predefined alignment keys on their surfaces, then they are placed into contact under ambient conditions in order to produce the first step of the fusion bonding process or the dielectric–dielectric fusion bonding part of the hybrid bonding process.

7.2.3 Wafer-to-Wafer Alignment

Wafer to wafer alignment offers device manufacturers different possibilities to meet their product requirements. The choice of the appropriate alignment process is very important and must be made already in the device design phase. The types of wafer-to-wafer alignment methods used in conjunction with wafer bonding are summarized in Figure 7.3.

As shown in Figure 7.3, two main categories of wafer-to-wafer alignment methods are generally used for wafer bonding:

1. *Mechanical Alignment:* The flats/notches of the two wafers are aligned to each other. A combination of pins and mechanical actuators is used to perform the process by pressing the two wafers' edges against edge pins. The alignment accuracy strongly depends on the tolerances of wafer dimensions, and the method cannot be used for wafers with different dimensions (e.g., one wafer is edge trimmed). For SEMI standard wafers, the alignment accuracy is typically ±50 μm. An alternative mechanical alignment process is based on the use of cameras to detect the edges of the two wafers' positions; for such a process, an alignment accuracy of less than ±15 μm and ±0.03° can be achieved for wafers with diameters up to 300 mm.

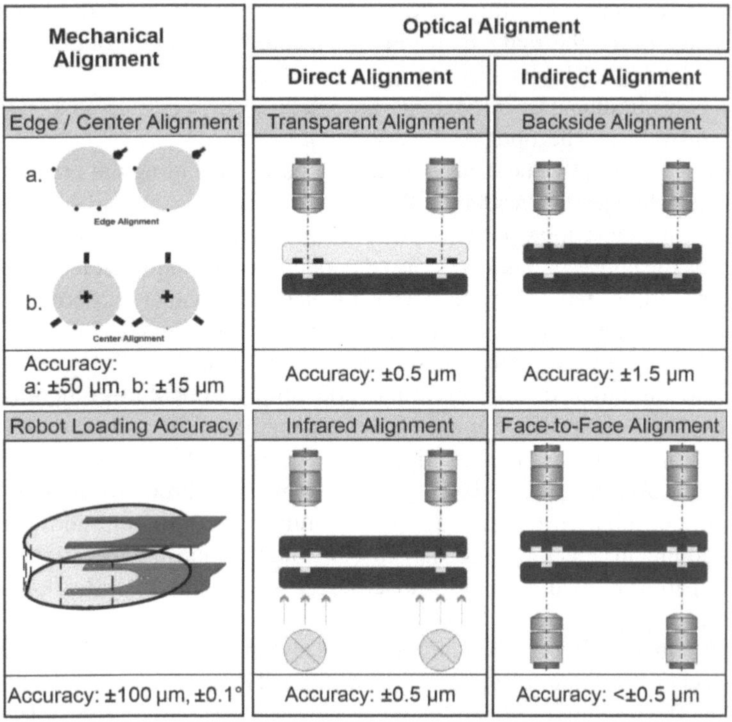

FIGURE 7.3 Summary of the alignment capabilities for aligned wafer bonding processes.

For applications requiring a lower alignment accuracy, robot handling systems with optical pre-aligners can be used to position the wafers. This results in lower alignment accuracy (about ±100 μm and ± 0.1°) but allows for a free programmable offset angle between the wafers, which can be an advantage for some applications.

2. *Optical Alignment:* The two wafers are precisely aligned using optical alignment equipment. In this case, the two wafers contain alignment keys which are located using advanced optical systems and thus a high alignment accuracy can be achieved, typically in the range of 0.05–10 μm. In the direct alignment process flow, the alignment keys are simultaneously in the view field and the alignment is performed live. The indirect alignment method performs the alignment by using the digitized images of the alignment keys on the two wafers: the two wafers are inspected sequentially, the positions of their alignment keys are stored, and then the system performs the alignment of the two wafers by restoring their positions.

The final post-bonding wafer-to-wafer alignment accuracy has two components: the optical alignment performance and the bonding process–induced misalignment. Among the bonding parameters with impact on alignment accuracy, the most significant are the compression of intermediate layers (for bonding of intermediate layers, for example, in eutectic bonding, adhesive bonding, or glass frit wafer bonding) and the difference in the thermal expansion coefficients of the two bonding partners or of the interface bond layers (e.g., in metal bonding). For such bonding processes, typically two alignment specifications are used: accuracy prior to bonding (provided exclusively by the optical alignment equipment performance) and accuracy post-bonding (the final value measured after bonding, including the process-induced misalignment budget). Considering these two factors, fusion and hybrid bonding methods are the most appropriate to provide extremely high post-bonding alignment accuracy, as the above-mentioned issues do not apply during the dielectric-to-dielectric fusion bonding step: the two surfaces are placed in contact at room temperature, no heating and compression forces are used, and the bonding interface remains rigid during the entire process, including the thermal annealing.

For fusion and hybrid wafer bonding applications, optical alignment is the process of choice. Among the optical alignment methods, face-to-face alignment is preferred [16, 17] as this method does not require the transmission of any light through the wafer (e.g., as in the infrared transmission direct alignment). The use of transmitted radiation induces significant limitations to the method as the wafer material doping affects the light transmission and the multiple metallization layers in a fully patterned wafer would impact negatively on the visibility of the alignment keys.

The alignment method relies on the use of two pairs of microscopes, with each pair observing two alignment sites on each wafer. The high accuracy of the method is based on the alignment principle: the system is self-calibrating in order to eliminate some optical errors and the gap between the two wafers is extremely small. The small gap results in a very small travel range for the top wafer at the pre-bonding step, which eliminates other potential mechanical errors. Additionally, very precise six-axis stages allow for very precise indirect alignment of the wafers.

A very important topic in optical alignment is related to the alignment keys (also named "alignment marks," "alignment targets," or "fiducials"), which play a crucial role in ensuring the final performance of the alignment process. (Two examples of alignment keys are shown in Figure 7.4.)

Some of the important aspects one should consider when selecting an optical alignment process and the respective alignment keys are as follows:

- *Material:* The alignment keys can be fabricated by etching the pattern into the surface layer or by fabricating a metal pattern. The metal pattern is usually preferred as it provides a better contrast and thus is better used by the image recognition system of the aligner.
- *Position:* The alignment keys must be positioned within the mechanical travel range of the objectives used for image observation.
- *Size and Shape:* These two factors are extremely important as they are related directly to the accuracy of the alignment process.
- *Fabrication Process:* The definition of alignment keys must result in alignment keys with the same dimensions, brightness, and contrast within a wafer batch; otherwise, the alignment results might vary within the batch—particularly in cases where very high alignment accuracy is required (<500 nm).

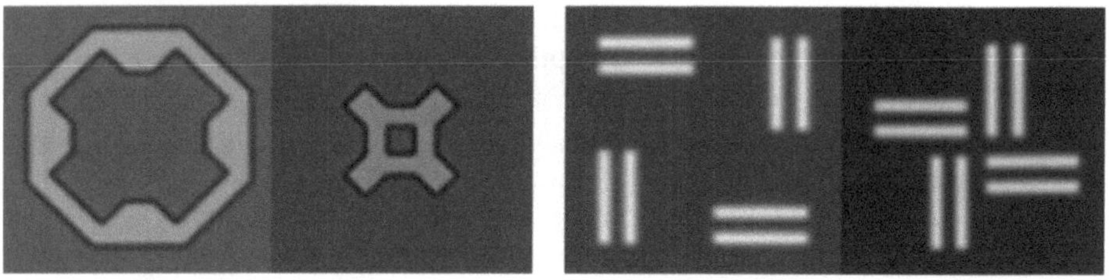

FIGURE 7.4 Examples of complementary alignment keys used for face-to-face wafer alignment: different size alignment key (*left*), and same size alignment key (*right*).

7.2.4 Overlay Analysis of the Bonded Wafers

After the optical alignment, the two wafers are pre-bonded: this process consists of bringing the two wafers into contact under ambient conditions for the dielectric–dielectric fusion bond. In order to preserve the highest post-bonding alignment accuracy, additional handling of the wafers is eliminated by using the optical aligner also for this step. The fusion bond (contact between surfaces) is initiated at one point using a bond initiation pin, followed by the wafers being released from the chucking system in a controlled manner. The fusion bond propagation also produces local deformation of the wafers and uncontrolled deformation might result in building stresses at the bonded interface. Such stresses have to be managed during pre-bonding as it might generate failures which cannot be detected until the final product qualification, when it can no longer be recovered and the bonded wafers have to be scrapped.

Traditionally, the post-bonding alignment accuracy was assessed by the measurement of the alignment keys at the same two positions used for the alignment process, and alignment accuracy range is typically ~1–5 μm in this case. The results are thus evaluated at the two locations measured, in terms of translation (in *x*- and *y*-directions), rotation and runout (also named "scaling," meaning a progressively increasing distance between two consecutive alignment keys along one axis).

With the increase in requirements for alignment accuracy better than 1 μm, this model to qualify alignment is not representative anymore and does not meet the requirements of certain applications.

For such applications, an overlay model was introduced to evaluate the alignment performance more precisely and globally across the wafer. For wafer-to-wafer bonding, the overlay accuracy between the wafers is usually evaluated using a box-in-box type measurement of the alignment keys located on the top and bottom wafers, respectively. The measurement results consist of sets of coordinates correlated with the corresponding misalignment values at each position, which are typically plotted as vector maps to visualize the results [17].

Such misalignment maps are usually analyzed using the overlay model. For wafer bonding, this model typically uses the same four fitting parameters: translations in *x* and in *y* directions, rotation of the two wafers with respect to each other, and runout (scaling), but measured at multiple sites across the bonded wafers pair.

After calculating the alignment correction factors (overlay model), the equipment feeds the information back to the optical aligner, and the alignment process is adjusted accordingly (per the feedback loop shown in Figure 7.2).

Sometimes, the above-described alignment verification is combined with an external measurement of each single wafer prior to bonding, and the results can be combined for developing the best process condition. Such measurements are performed with specific metrology equipment which is not integrated in the fully automated wafer bonding equipment.

Additionally, the tooling of the optical aligner can be customized in order to compensate for potential known failures related, for example, to the wafer shape or other wafer parameters (e.g., scaling effects) [16].

An important parameter of a successfully bonded wafer pair is distortion: this parameter represents the deviation of a pattern shape/size from its designed geometry. Distortion is typically observed after the thermal annealing of the bonded wafers, followed by thinning of one of the two wafers.

Patterns distortion can be influenced by the wafer's properties (e.g., shape, patterning accuracy, and wafer manufacturing processes) as well as by the bonding process (e.g., stress induced by deformation during pre-bonding, different wafers' temperatures when placed in contact, bond strength, etc.) and the post-bonding processes (e.g., thinning by grinding).

An overview of the factors contributing to the misalignment budget is presented in Table 7.1.

After unloading from the bonding equipment, the pre-bonded wafer pairs are thermally annealed at temperatures of 300–400°C for 1–3 hours. After the thermal

TABLE 7.1
Overview of the Parameters Influencing the Wafer-to-Wafer Alignment Accuracy

Manufacturing Step	Parameter
Substrates design	Surface layers Alignment keys: • Location • Shape/size • Contrast
Wafer fabrication	Wafer shape: • Bow • Warp Lithography: • Mask accuracy • Layer-to-layer registration Alignment keys fabrication: • Shape definition • Key edge definition

(Continued)

TABLE 7.1 *(Continued)*
Overview of the Parameters Influencing the Wafer-to-Wafer Alignment Accuracy

Manufacturing Step	Parameter
Alignment type	Alignment process chosen Aligner vision system • Camera • Optics (magnification, depth of field) Motion system (for indirect alignment): • Position measurement • Step resolution
Wafer bonding	Wafer bonding process of choice Process recipe optimization Equipment configuration
Accuracy assessment	Measurement method Measurement system accuracy

annealing, the wafers reach the maximum bond strength and they are ready for further processing. In the process of 3D interconnect fabrication, typically one of the bonded wafers is thinned down, and further wafers can be bonded on the thin wafer side of the stack for building complex 3D architectures.

7.3 EQUIPMENT TECHNOLOGY FOR TEMPORARY WAFER BONDING

With the increasing demand for processing of thin and ultrathin substrates, the temporary bonding technology concept has experienced a significant development over the past two decades [18]. Generally, in such a process, a thin wafer (usually <200 μm) is bonded to a second substrate (temporary carrier) by means of a temporary bonding layer, which produces a reversible bond between the two substrates. The use of a carrier is meant to provide the mechanical robustness and to support the subsequent processing on the back side of the device wafer after further thinning. At the end of the processing, the device wafer is debonded from the temporary carrier using various methods developed over the years based on the specific properties of the adhesive materials used [18, 19].

While the process is usually named "temporary bonding," the very important part is actually the final separation of the device wafer from the carrier (debonding), as it dictates the materials choice as well as the process flow for bonding/debonding, being the direct link to the final application. A process flow of a generic temporary bonding and debonding process is shown in Figure 7.5.

The three main categories of debonding processes are schematically illustrated in Figure 7.6 and will be further described in detail.

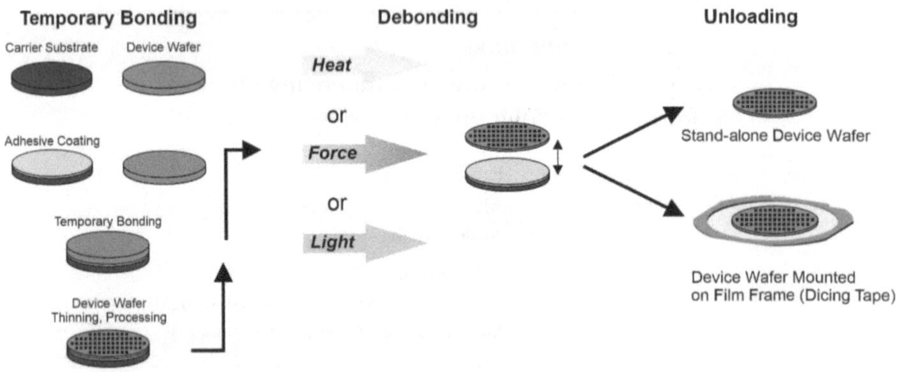

FIGURE 7.5 Schematic flow of a generic temporary bonding/debonding process.

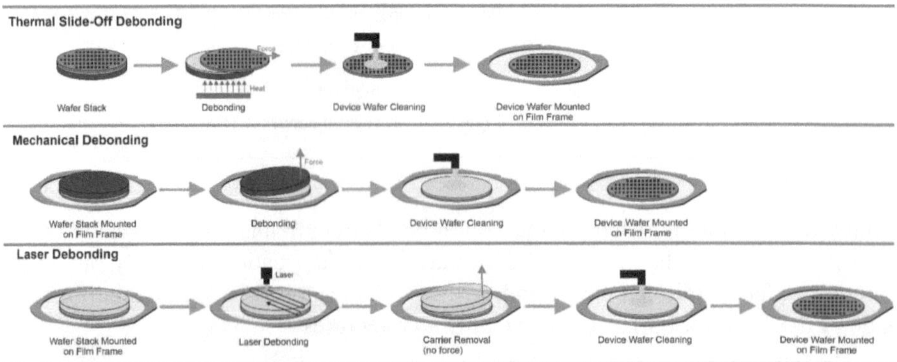

FIGURE 7.6 Schematic process flows for the three types of debonding processes: thermal slide-off, mechanical, and laser debonding.

7.3.1 Thermal Slide-off Debonding

Thermal slide-off is a method that employs the use of a thermoplastic material as the temporary adhesive interlayer between the device and the carrier wafer. The debonding is based on the reversible thermal behavior of the thermoplastic material: at elevated temperatures, the material exhibits a drop in viscosity, which enables debonding to be accomplished by simply sliding the wafers off each other. The main feature of thermal slide-off debonding is that the bonding and debonding are performed at temperatures which, depending on the thermoplastic material being used, can range between 150°C and 300°C. The thermal stability of the temporary bonded wafers depends significantly on the thermomechanical stress which can be observed during the process due to the thermoplastic's glass transition temperature (T_g) and its low viscosity at high temperatures [20].

Alternatively, a thermal lift-off process can be used for debonding. The difference between the working principles of thermal slide-off and thermal lift-off resides in

TABLE 7.2
Overview of the Main Types of Polymer Materials Used as Temporary Bonding Adhesives

Polymer Type	Debonding Process	Temperature Range	Carrier	Comments
Thermoplastic	Slide-off Mechanical	200–300°C	Any	• Remelts by reheating • Good topography coverage • Debonding temperature connected to thermal stability • Applications: memory, compound semiconductors
Thermoset	Mechanical Laser	300–450°C	Any UV or IR transparent	• Form 3D network by curing, cannot remelt • More rigid than thermoplastics • Applications: advanced packaging
Elastomers	Mechanical Laser	<260°C	Any UV or IR transparent	• High elasticity, hard to fracture • High peel strength
Hybrid polymers	Any (to be defined)	–	Any	Alloys and blends of elastomers, thermosetting and thermoplastic polymers designed for specific applications

the way the mechanical force is applied during debonding: in the slide-off approach, the device wafer is pulled parallel to the bonded interface; in the lift-off approach, the force is applied perpendicular to the bonded interface. Additionally, there might be differences between the adhesive types, and the choice of the debonding method is also made considering the particularities of the device wafer properties and process requirements.

Table 7.2 presents an overview of the main polymer material types used as temporary bonding adhesives.

The hybrid polymers mentioned in Table 7.2. can have different properties as they are designed for specific uses. These properties are triggering the choice of the temporary bonding process flow as well as of the materials used (e.g., carriers, and additional layers at the interface).

7.3.2 Mechanical Debonding

In the early times, temporary bonding and debonding processes were not usable for devices with strict temperature restrictions (e.g., having indium bumps), so the mechanical debonding process (mechanical peeling of the carrier) was introduced. Mechanical debonding enabled the use of release films mounted on frames for debonding and support of the device wafer, which allowed manufacturers to further reduce the thickness of the devices (≤50 μm).

The challenge for mechanical debonding resides in the requirement for adjusting the adhesive bond strength for the interface material and topology intended for use for each new device wafer.

Key features of the mechanical debonding include processing at room temperature and a strong dependence on the mechanical stress. Since mechanical debonding needs a low adhesion between the temporary bonding material and the wafer for a successful debonding process, its use for some applications might be limited as the high wafer stress associated with some processes can lead to spontaneous debonding [21]. The relatively narrow process window and the required adjustments for different device types limit the use of this technology.

7.3.3 Laser Debonding

Laser debonding is a technology that has been implemented with several variations [22–24]. The debonding mechanism depends on the type of the laser used as well as the temporary bonding adhesive or the specific additional layer added at the interface to enhance the debonding process (which allows for the release at the interaction with the laser). For this technical approach, the debonding initiation is triggered by laser exposure of the temporary bonded interface. Therefore, unintentional debonding initiation is very unlikely to happen in the case of laser debonding, in contrast to slide-off and mechanical debonding.

In this method, the laser scans the wafer following a path as shown in Figure 7.6, producing a network of dots at the interface. The dot size and pitch are designed in a way that the reaction of the release layer with the laser produces a full layer release.

One of the newest techniques developed for temporary bonding technology, ultraviolet (UV) laser debonding, is based on photochemical processes at the bonded interface: the light is absorbed into the release layer and the energy is used for breaking the chemical bonds in the material. The use of an UV laser for debonding imposes the use of a transparent carrier material (glass).

An alternative to using UV laser, infrared (IR) laser debonding works based on photothermal processes, where light is absorbed and transformed into heat, resulting in very localized high temperatures at the bonded interface. The breakage of the chemical bonds results in the production of fragments of the original polymer comprising gases, which increases the pressure around the bonded interface and subsequently supports the debonding process. This method is suitable for numerous applications due to the high adhesion of the temporary bonding adhesive to the wafers before the debonding process.

The reason to employ an IR laser is that in this case, a silicon wafer can be used as carrier due to its IR radiation transparence. This fact is of extreme practical importance, as in numerous semiconductor fabs, the use of glass wafers is not possible or desired. First of all, the handling of the glass wafers might create problems for standard equipment (e.g., some standard robot pre-aligners cannot handle glass substrates) or raise contamination concerns. The use of a silicon wafer eliminates all these problems and concerns as it allows for using standard equipment.

The laser debonding process is successful only with the accurate monitoring of the following parameters:

- *X/Y Beam Positioning:* The laser position is calibrated for the process.
- *Laser Focus:* This function allows for automated refocusing depending on low variations of the wafer shape.
- *Beam Shape:* Integrated micro-optics system shapes the beam for the highest process efficiency.
- *Laser Power.*
- *Overlap Accuracy:* The position of the scanning lines must be very accurate in order to ensure full layer transfer.
- *Debonding Force:* The force required to separate the carrier. (No high force is needed, but it should be uniform.)

7.3.4 Inline Metrology for Adhesive Layer Thickness Monitoring

Generally, the process flow is split between the bonding equipment and the complementary equipment in the fab. The temporary bonding/debonding equipment must serve several different processes, so typically a modular concept is used: process modules providing a single process are combined in fully automated configurations. The configuration is often made based on custom applications requiring specific process flows.

For the fully integrated temporary bonding equipment, the following processes can be included:

- Temporary bonding layer preparation: spin-coating or dry film lamination, baking.
- Inline metrology: as the total thickness variation (TTV) is a key parameter in various processes (particularly for wafer thinning), it is desirable to monitor it after processes such as coating, baking, bonding, and thinning.
- Film mounting on tape.
- Wafer mounting on the film frame.
- Wafer bonding in a thermal bonding chamber.

For the fully integrated debonding equipment, the following processes can be included:

- Debonding module: mechanical stage, thermal slide-off/lift-off module, laser module
- Single wafer cleaner for device wafer cleaning after debonding
- Special single wafer-on-film frame cleaner (optional)

Usually, the carriers are cleaned for reuse outside the bonding equipment in order to prevent equipment contamination.

The wafer stack TTV monitoring is of great importance particularly for wafer thinning processes. The spin-coating of the adhesive material is typically very accurate with respect to TTV. However, each additional process might affect this

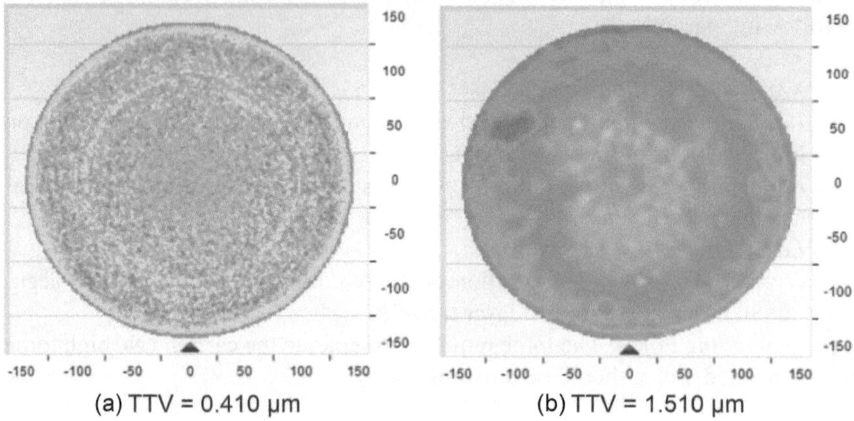

FIGURE 7.7 Temporary adhesive layer thickness uniformity measured. (a). After spin-coating and soft curing, and (b) after temporary bonding (Si wafers, 300 mm diameter, 5 mm edge exclusion, adhesive thickness of 20 μm measured at 72,000 sites).

important parameter. The implementation of an inline metrology inspection is able not only to provide the information on TTV but can also feedback the information to the integrated temporary bonding equipment, which is then able to sort the wafers that are out of specifications, allowing for correcting the process in order to minimize losses.

The metrology method is based on the use of two light interferometers measuring the wafer stack from two directions (top/bottom) in a predefined number of measurement points. The number of measurements per wafer is critical in providing accurate information on TTV; the higher the number of measurements, the better the resolution of the measurement.

An example of a wafer stack measured after two process steps is shown in Figure 7.7.

With the introduction of the newest laser-based debonding processes, the temporary bonding/debonding application areas have expanded significantly.

Figure 7.8 shows some of the application areas of the different types of temporary bonding processes.

7.4 EQUIPMENT TECHNOLOGY FOR DIE-TO-WAFER BONDING

Besides the wafer-to-wafer bonding processes, the combination of multiple functionalities onto a single die enables new packaging architectures and design concepts. The concept of die-to-wafer bonding allows bonding of multiple dies onto a wafer in a single process for heterogeneous integration applications. Such technological approach allows combining dies manufactured in different technologies and also combination of different materials (e.g., Si technology with compound semiconductors) and relies on the principle of maximizing the yield by combining the wafer manufacturing device yield with dies selected to be in specifications ("100% yield," or the so-called known good die) [25, 26].

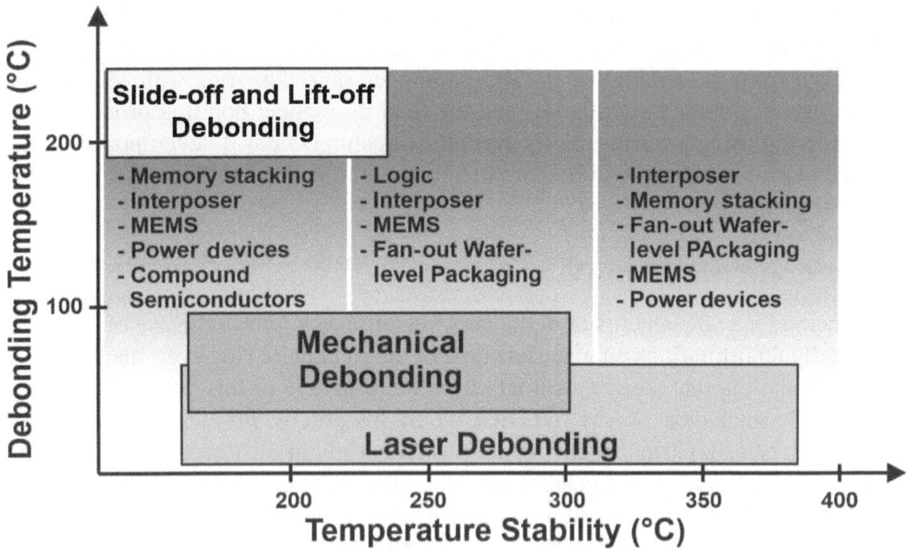

FIGURE 7.8 Application areas of the various types of temporary bonding/debonding techniques.

Based on the die dimensions and yield, wafer-to-wafer or die-to-wafer integration can ensure a cost-effective integration process.

The two base processes discussed in Sections 7.2 and 7.3 are also used for die-to-wafer bonding. Metal TC bonding, fusion bonding, and hybrid bonding can be used for permanent bonding, while temporary bonding/debonding processes can be employed as supporting processes, particularly for die handling.

Two different process flows can be employed: direct placement bonding and collective die-to-wafer bonding. The process flows will be briefly introduced for the same two final permanent bonding processes targeted: fusion bonding and hybrid bonding.

7.4.1 Direct Placement of Dies on Wafer

As shown in Figure 7.9, direct placement die-to-wafer bonding is a combination of temporary bonding of dies on a carrier and a plasma-activated low-temperature bonding process identical to wafer-to-wafer bonding.

Before the dies further go through plasma activation and cleaning, the dies are first temporary bonded onto a carrier in order to allow for their preparation for bonding

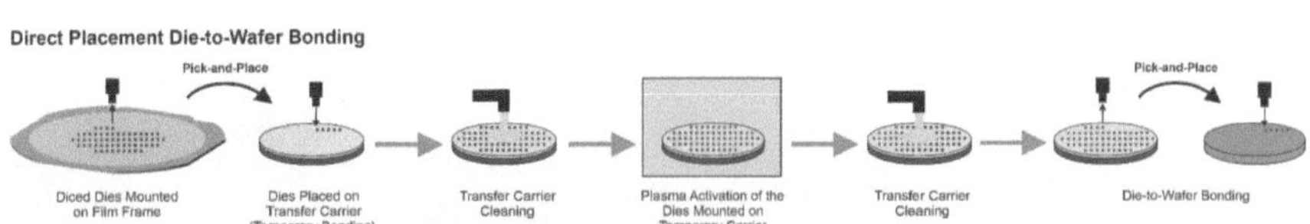

FIGURE 7.9 Direct placement die-to-wafer plasma-activated low-temperature bonding.

in a single process, with minimized risk of damaging the bonding face of the dies by individual die handling through the process. After activation and cleaning, the dies are picked from the temporary carrier and bonded onto the target wafer. Both temporary carrier population with dies and the final permanent bonding processes can be performed using a commercially available flip chip bonder. The temporary bonding and plasma-activated bonding are identical with the process flows introduced in previous sections.

7.4.2 Collective Die-to-Wafer Bonding

This method has the benefits of being very versatile as it allows the use of dies with practically no limitations on materials type or die thickness. However, the drawback of this method is that some dies might be hard to handle in this process flow (e.g., multiple die stacks for SRAM and DRAM). In this process flow, the contamination control (especially particles) during die placement might also be difficult, and this might affect the quality and yield of the fusion and hybrid wafer bonding processes.

Collective die-to-wafer bonding is an alternative process used for the final permanent bonding of a wafer-to-wafer process (the process flow is shown in Figure 7.10.).

In this process, the dies are first temporary bonded on a carrier wafer and plasma processed, similarly to the direct placement method described above. The target wafer is also plasma activated and cleaned. The two wafers are placed in a face-to-face aligner where they are aligned and pre-bonded. The pre-bonded wafers are thermally annealed at low temperatures (considering the temporary adhesive properties) and then the carrier is removed. Alternatively for this process, a laser debonding can be performed prior to thermal annealing.

The method has the advantage of providing a clean and reliable process similar to the wafer-to-wafer bonding. Bonding of dies of different sizes has been proven for various processes. Figure 7.11 shows an example of collective die-to-wafer bonded dies of various sizes.

However, there are some significant drawbacks which can limit the use of this method. First, the alignment accuracy of die-to-wafer bonding is affected by an error

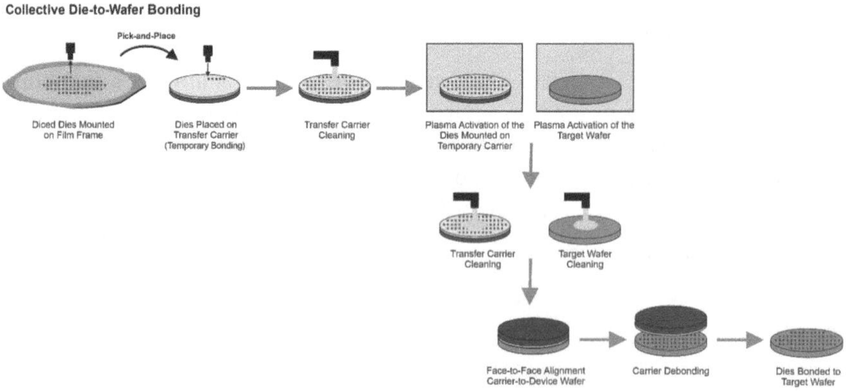

FIGURE 7.10 Collective die-to-wafer plasma-activated low-temperature bonding.

FIGURE 7.11 Example of collective die-to-wafer application using plasma-activated bonding of dies of various dimensions: photograph of the surface showing the dies bonded (*left*), and acoustic microscope image (*right*) showing high bonding quality of the dies (black contrast stands for bonded areas).

propagation from the initial placement of the die to the temporary carrier. Another significant disadvantage is that the die thicknesses for a bond must be in a very narrow range, as too high variations might result in no-bond for the significantly thinner wafers; therefore, in order to perform this process, the dies have to be sorted by thickness prior to the process. Despite these limiting factors, the method is quite mature and is used in volume production.

7.5 FINAL REMARKS

Equipment technology for fusion/hybrid wafer bonding as well as for temporary bonding/debonding directly reflects the evolution of the applications and the increased adoption of wafer bonding technology in semiconductor manufacturing. With the growing demand for heterogeneous integration and increased performance and decrease in size, wafer bonding technology has evolved significantly over the past decade. In this chapter, process flows and important aspects related to the equipment technology are introduced for permanent and temporary wafer bonding.

When configuring an equipment, the technology and the economics criteria must be implemented. It is also possible, when using modular equipment concepts, to start with a base configuration providing just the required functionality, and further upgrade it with additional redundant modules with the goal to reach the economic targets (i.e., throughput). Such an approach is very useful particularly for the development of a new product, and when successful, the same equipment can be moved from R&D to production. In this way, the processes developed in R&D phase can be optimized for production with only a minimum effort on process development.

The device manufacturers understand the importance of each factor involved in their processes (substrates, materials, processes, and process tolerances), so it is nowadays a common practice for them to team up with various equipment manufacturers to combine the optimization efforts in all areas to ensure a successful manufacturing process.

Additionally, a significant effort is spent on numerical simulation and modeling on both sides: device functionality and device manufacturing processes.

REFERENCES

1. J. Haisma, G. A. C. M. Spierings, "Contact Bonding, Including Direct Bonding in a Historical and Recent Context of Materials Science and Technology, Physics and Chemistry. Historical Review in a Broader Scope and Comparative Outlook," in Materials Science and Engineering, vol. 37, pp. 1–60, 2002.
2. G. Wallis, D. I. Pomerantz, "Field Assisted Glass–Metal Sealing," in Journal of Applied Physics, vol. 40, no. 10, p. 3946, 1969.
3. V. Dragoi, T. Glinsner, P. Lindner, C. Schaefer, T. Kazuta, D. Hornik, M. Reiche, "Temporary and Permanent Wafer Bonding for Reliable Backside Processing of Compound Semiconductor Wafers," in Proceedings of the 28th International Symposium on Compound Semiconductors, Institute of Physics Conference Series, vol. 170, no. 4, pp. 313–318, 2002.
4. B. Kim, T. Matthias, V. Dragoi, M. Wimplinger, P. Lindner, "Temporary and Permanent Bonding," in 3D IC Stacking Technology, B. Wu, A. Kumar, S. Ramaswami (eds.), McGraw-Hill, 2011, p. 409.
5. J. J. McMahon, J.-Q. Lu, R. J. Gutmann, "Wafer Bonding of Damascene-patterned Metal/Adhesive Redistribution Layers for Via-First Three-Dimensional (3D) Interconnect," in Proceedings of the IEEE 55th Electronic Components and Technology Conference, 2005, pp. 331–336.
6. V. Dragoi, S. Farrens, P. Lindner, "Plasma Activated Wafer Bonding for MEMS," in SPIE Smart Sensors, Actuators, and MEMS, vol. 5836, pp. 179–187, 2005.
7. V. Dragoi, P. Lindner, "Plasma Activated Wafer Bonding: In Situ and Ex Situ Processes," in ECS Transactions, vol. 3, no. 6, pp. 147–154, 2006.
8. M. Gabriel, V. Cetin, S. Hansen, M. Reiche, I. Radu, M. Eichler, "Capabilities of an Ambient Pressure Plasma for Activation in LT Wafer Bonding Processes," in Proceedings of the ECS Semiconductor Wafer Bonding VIII: Science, Technology, and Applications, PV2005-02, 2005 pp. 50–57.
9. V. Dragoi, G. Mittendorfer, A. Filbert, M. Wimplinger, "Wafer Bonding for Backside Illuminated CMOS Image Sensors Fabrication," in Proceedings of the MRS Spring Meeting, vol. 1249, 2010, p.1249-F08-06.
10. V. Dragoi, E. Pabo, J. Burggraf, G. Mittendorfer, "CMOS-compatible Wafer Bonding for MEMS and Wafer-level 3D Integration," in Journal of Microsystems Technology, vol. 18, no. 7–8, pp. 1065–1075, 2010.
11. K. N. Chen, C. S. Tan, "Integration Schemes and Enabling Technologies for Three-dimensional Integrated Circuits," in IET Computers & Digital Techniques, vol. 5, pp. 160–168, 2011.
12. P. Enquist, "High Density Bond Interconnect (DBI) Technology for Three Dimensional Integrated Circuit Applications," in MRS Proceedings, vol. 970, p. 0970-Y01-04, 2007.
13. T. Plach, V. Dragoi, G. Mittendorfer, M. Wimplinger, K. Hingerl, "Wafer Bonding of Plasma Activated Surfaces," in ECS Transactions, vol. 13, no. 25, pp. 39–45, 2008.
14. V. Dragoi, P. Lindner, S. Farrens, C. Schaefer, "Bonding 300 mm SOI Substrates," in European Semiconductors, vol. 25, no. 3, pp. 31–35, 2003.
15. F. Fournel, L. Bally, D. Dussault, V. Dragoi, "Innovative Megasonic Cleaning Technology Evaluated through Direct Wafer Bonding," in ECS Transactions, vol. 33, no. 4, pp. 495–500, 2010.
16. F. Kurz, T. Plach, J. Süss, T. Wagenleitner, D. Zinner, B. Rebhan, V. Dragoi, "High Precision Low Temperature Direct Wafer Bonding Technology for Wafer-Level 3D ICs Manufacturing," in ECS Transactions, vol. 75, no. 9, pp. 345–353, 2016.
17. T. Plach, B. Rebhan, V. Dragoi, T. Wagenleitner, M. Wimplinger, P. Lindner, "High Accuracy Aligned Wafer Bonding for Wafer-level Integration," in ECS Transactions, vol. 86, no. 5, pp. 145–158, 2018.

18. S. Pargfrieder, P. Kettner, M. Privett, J. Ting, "Temporary Bonding and Debonding Enabling TSV Formation and 3D Integration for Ultra-Thin Wafers," in Proceedings of the IEEE 10th Electronics Packaging Technology Conference, 2008, pp. 1301–1305.
19. V. Dragoi, C. Flötgen, J. Burggraf, L. Oggioni, T. Suga, "Emerging Wafer Bonding Technologies," in Handbook for Silicon Based Materials and Technologies, M. Tilli, T. Motooka, V.-M. Airaksinen, S. Franssila, M. Paulasto-Kröckel, V. Lindroos (eds.), Elsevier, Burlington, 2020, p. 651.
20. J. Azémar, "Fan-out and embedded die: Technology and market," in Yole Development Report, 2015.
21. T. Uhrmann, J. Burggraf, H. Wiesbauer, J. Bravin, T. Matthias, M. Wimplinger, P. Lindner, "Temporary Bonding on the Move towards High Volume: A Status Update on Cost-of-Ownership," in Proceedings of the IEEE 16th Electronics Packaging Technology Conference, 2014, pp. 378–382.
22. P. Andry, R. Budd, R. Polastre, C. Tsang, B. Dang, J. Knickerbocker, M. Glodde, "Advanced Wafer Bonding and Laser Debonding," in Proceedings of the IEEE 64th Electronic Components and Technology Conference, 2014, p. 883.
23. T. Uhrmann, M. Pichler, J. Bravin, D. Burgstaler, B. Povazay, "Laser Debonding Enabling Ultra-thin Fan-out WLP Devices," in Proceedings of the IEEE 7th Electronic System-Integration Technology Conference, 2018.
24. T. Uhrmann, P. Urban, B. Povazay, M. J. Gruber, J. Bravin, D. Burgstaller, M. Wimplinger, B. Thallner, "Novel IR Laser Cleaving for Ultra-thin Layer Transfer and 3D Stacked Devices," in Proceedings of the IEEE 73rd Electronic Components and Technology Conference, 2023, pp. 1749–1753.
25. T. Uhrmann, J. Burggraf, M. Eibelhuber, "Heterogeneous Integration by Collective Die-to-Wafer Bonding," in Proceedings of the IEEE International Wafer-Level Packaging Conference, 2018.
26. J. Burggraf, M. Pires, T. Uhrmann, "Collective Die Bonding: An Enabling Toolkit for Heterogenous Integration," in ECS Transactions, vol. 98, no. 4, pp. 173–180, 2020.

8 Die-to-Wafer Hybrid Bonding for Direct Copper Interconnections

Jonathan Abdilla and Stefan Schmid

8.1 INTRODUCTION

8.1.1 Packaging Technology Milestones

Back in the 1970s, wire bonding was the packaging technology which enabled the electrical communication from the die to the package, as seen in Figure 8.1 showing the die placement accuracy for each interconnect technology. Gold wires were typically used to connect the bond pads onto the leadframe or substrate. As devices became more complex, wire bonding also increased in intricacy, with multi-tier wire bonding and die-to-die (D2D) wire bonding to link different dies. In the 1990s, a new revolutionary technology emerged: flip chip bonding. This method enabled the population of the entire die area with I/O connections, unlike wire bonding which remains limited to bond pads at the edges of the die. Flip chip also enabled a reduction in pitch by employing solder balls, thereby enabling an even greater contact density. With limitations arising from solder balls and copper pillars with a solder cap, mainly due to the reflow effect whereby solder melts and could lead to bridging if the pitch is too small, thermocompression bonding (TCB) emerged. This packaging method consists of heating the die in place and then cooling while retaining the die, thereby controlling the lateral position as well as the z position while the solder melts and solidifies sequentially. This pushed the contact density even higher; however, the process is slow since each die must undergo heating and cooling. Fan-out alleviated this issue for some packages, whereby the substrate on which the die is bonded is larger than the die itself, and the solder balls extend beyond the die size. Of course, this meant that laterally the package is still larger; however, for those applications and modules that have no strict size limitations, it provides a great way to achieve higher contact density.

This brings us to more recent times whereby HB provides the highest increase in contact density in the history of semiconductor packaging.

8.1.2 Drivers for D2W/D2D HB

With the traditional Moore's law scaling slowing down, the semiconductor industry has turned to find other solutions on how to keep up with the technological pace. To build a fab that can process 5 nm nodes and smaller requires huge investment and experienced

 DOI: 10.1201/9781003408673-8

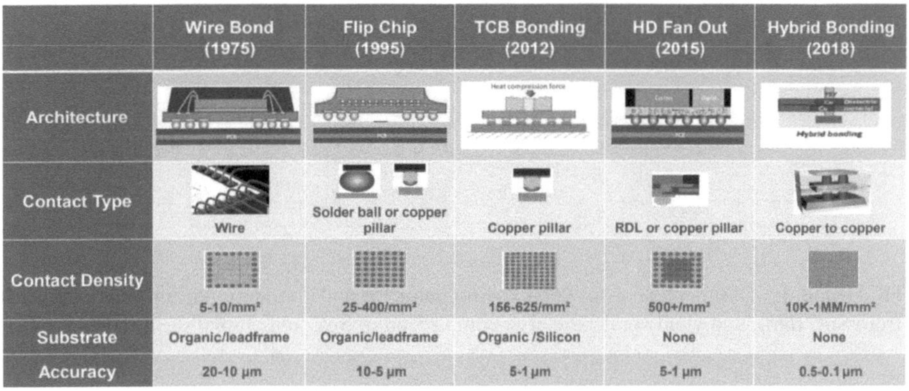

FIGURE 8.1 Timeline of major packaging technologies.

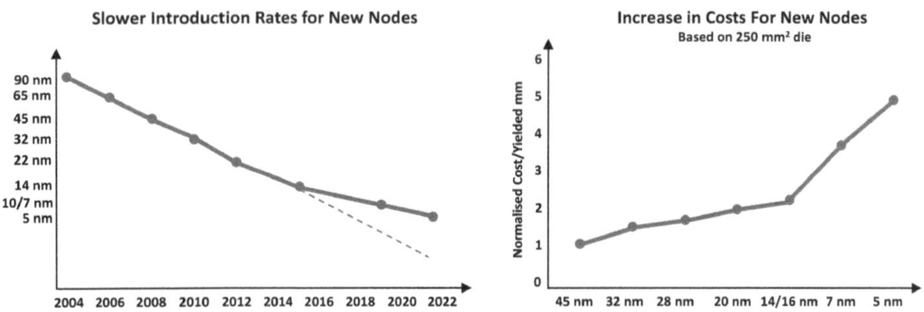

FIGURE 8.2 Moore's law slowing down (*left*) and drastic increase in costs with reducing tech nodes beyond 7 nm (*right*). (Adapted from Ref. [1].)

manpower, all of which are not easily available, as illustrated in Figure 8.2. Moreover, the gains from scaling down the tech node are no longer yielding the same advantages as the process becomes ever more complex. The demand for ever more powerful devices with lower power consumption, driven by Artificial Intelligence (AI), Augmented Reality (AR)/Virtual Reality (VR), cloud computing, and edge computing as well as Advanced Driver Assistance Systems (ADAS), is pushing for technology to keep up. System on chip (SoC), which has served the industry well for the past years, is reaching an inflection point where the cost of development, the lack of flexibility, reticle size limitations, as well as yield issues among other factors, are giving rise to heterogeneous integration, with the use of chiplets being at the forefront, as depicted in Figure 8.3.

Chiplets are relatively small chips which are designed to be specialized in their task. Typically, a chiplet only performs one specific task, and in doing so can do it at great efficiency. As shown in Figure 8.4, chiplets can be of different technological nodes, thereby only utilizing higher, more expensive, more complex tech node chiplets where needed, and they can also be of different materials (silicon, III–V, etc.). Heterogeneous integration aims at combining these chiplets into one package. This of course requires not only hardware interfacing but also software and communication protocols. Currently, there are products in high volume being manufactured using chiplets from the same suppliers. The aim, however, would be that different

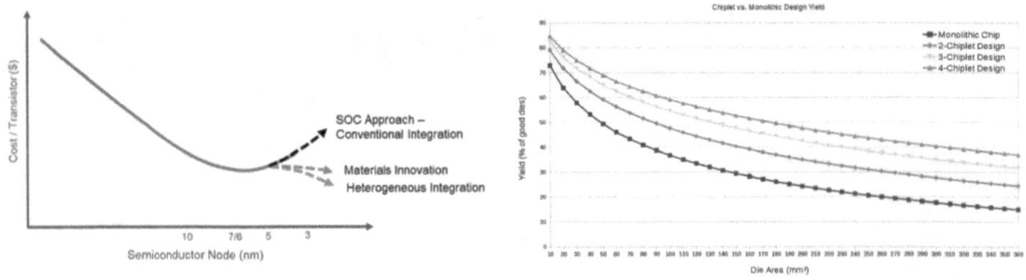

FIGURE 8.3 Cost per transistor versus semiconductor node: inflection point requiring shift from SoC (*left*), and yield versus die area comparison of SoC and chiplet design (*right*) [2].

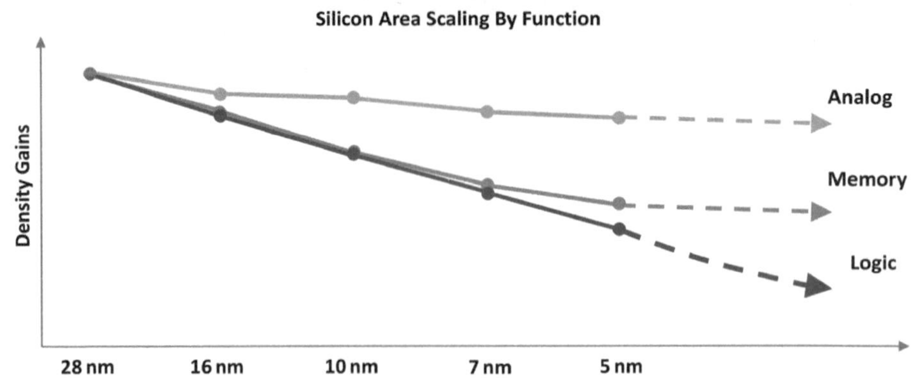

FIGURE 8.4 Not all silicon functions benefit at the same magnitude from semiconductor node scaling. (Adapted from Ref. [3].).

suppliers conform to common standards such that a huge library of chiplets can be used by anyone, reducing time to market and development costs.

Chiplets give rise to one important aspect – they can be stacked to leverage 3D integration. With miniaturization still a market driver, 3D integration offers an opportunity to fit all the package requirements without exceeding lateral constraints. HB is the packaging scheme which promises great benefits in this regard.

Figure 8.5 illustrates the drivers for HB for the four main product pillars. HB allows for higher interconnect densities which is the driver for high-performance computing (HPC), AI, and data center applications (DC). This is required since reducing the tech node results in smaller transistor sizes and thus the possibility to have more I/Os. However, to avoid solder bridging, flux cleaning, and underfill issues, HB is the interconnect technology of choice for bond pad pitches smaller than 10 μm.

Higher frequencies are also possible with HB for several reasons. Parasitic capacitance and inductance can limit the performance of high-frequency circuits. HB techniques can help minimize these effects by enabling precise control over the placement and configuration of different components. Moreover, the avoidance of solder as an interconnect and the surrounding underfill (or molded underfill) help to drive further

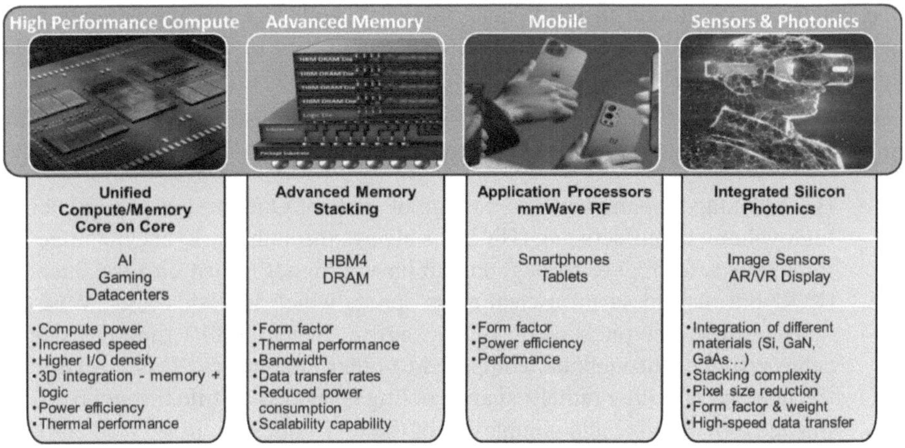

FIGURE 8.5 HB drivers for the four main pillars: HPC, advanced memory, mobile, and sensors and photonics [4].

down parasitic effects. Another factor is better materials matching: different semiconductor materials have different properties, such as electron mobility and thermal conductivity. By choosing the right combination of materials, one can optimize the overall performance of the circuit for high-frequency operations. Finally, HB allows for the direct integration of different materials with minimal interconnect lengths. Shorter interconnects lead to lower resistance, capacitance, and inductance, which collectively reduce signal propagation delays and losses. This is particularly important at higher frequencies where signal integrity becomes more critical.

Power efficiency is also a crucial aspect, not just for larger server farms that consume enormous amounts of energy, but also for mobile ad wearable devices where battery life is important. By having direct metal contact with no solder in-between and no underfill, parasitic effects are removed, thereby improving power efficiency.

With applications requiring more and more compute power, thermal dissipation is a critical topic to be addressed. This, coupled with the fact that miniaturization is pushing for 3D stacking, poses challenges as the heat generated needs to travel throughout the entire stack before being dissipated. For solder-based solutions like TCB which are employing microbumps, the low thermal conductivity of the gap-fill material leads to high vertical thermal resistance. HB has the advantage that the stack behaves like a single die and thermal gradients are much better than for those assembled with solder interconnect technologies.

The combination of no standoff from Cu pillars (or solder bumps) and smaller die sizes since more interconnects can be fitted in the same area by utilizing lower tech nodes, leads to smaller form factors in both x, y, and z. This addresses several applications, such as stacked memory z-restrictions and reticle limits for logic chips.

With the emergence of ADAS and AR/VR, the number of sensors being deployed in the market has increased tremendously. This leads to an increase in the use of III–V materials to leverage their specific properties in these areas. As such, the integration of silicon to III–V materials is crucial, and HB has already been proven to be successful and the interconnect technology of choice to achieve the highest possible benefits.

8.1.3 Advantages of D2D/D2W HB

- HB advantages span over a vast range of sectors. One important topic for high-volume manufacturing (HVM) is always throughput. Every manufacturer strives for higher throughputs to have a higher return on investment (ROI) but also to save on cleanroom space, which is very costly. When targeting the fine-pitch niche of packaging, typically ≤10 μm, HB can achieve higher throughputs compared to other contenders. This is tied to the fact that no solder remelting and cooling is needed. While throughput is impacted by the size and complexity of the die, values of circa 2,000 chips per hour have been reported [5].
- Unlike other processes, there is no intermediate material in-between the active surfaces. There are also no solder or copper pillars, which means that warpage is kept to a minimum, but also that there are fewer thermal stresses due to no CTE (coefficient of thermal expansion) mismatch. This plays a very important part since failure rates associated with these two criteria can be very high if not controlled. Thin dies are more easily bonded with HB as compared with other techniques for the aforementioned reasons.
- HB offers reduced multi-die assembly issues, especially for stacking. In comparison, TCB high-memory stacks offer challenges not just in throughput, but also in heat application and possible impact to the already bonded layers.
- Low forces are required for HB as compared with other interconnect technologies. Some requirements for high forces have been seen, but mostly these address a multi-interconnect scheme package, mixing HB with TCB or to compensate for other factors, such as warpage or voids. It has been shown that for both single die and stacking, low-force HB can achieve excellent electrical test results [6].
- As solder-based applications entered the sub-10 μm bond pad pitch regime, many constraints surfaced. The solder bumps which have transitioned to microbumps are very delicate and subject to damage very easily. Slipping is another common phenomenon whereby a slight lateral misalignment coupled with reduced bondhead or wafer table stiffness results in the copper interconnect resting in-between the two bottom ones, thus creating a short. The solder content itself becomes so minimal that controlling the z height over the entire die becomes a challenge and is aggravated with larger dies and issues of warpage as typically seen in interposers. Tilt issues are another factor to consider with solder-based interconnects. Methods do exist to overcome these issues; however, these come at an extra effort and cost.
- Oxidation is exacerbated with TCB due to high temperatures, although many suppliers are working on fluxless applications with inert or reducing gas atmospheres as well as in situ plasma cleaning.

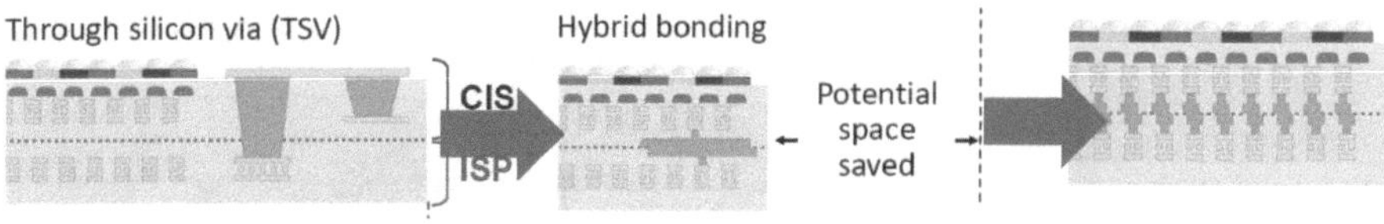

FIGURE 8.6 Gain in die area by leveraging HB 3D stacking. (Adapted from Ref. [7].)

- Gain in precious die real estate for additional modules can be achieved by switching to more advanced tech nodes, thereby having smaller transistors, and also by employing 3D stacking techniques (Figure 8.6).
- No flux or underfill is required for HB. With the reducing standoff from microbumps, it is becoming increasingly more difficult to fill the gap between the die and the substrate (or die and die). New materials are being developed but physical limitations are hard to overcome. Since no flux is needed for HB, no cleaning is required, which is also particularly challenging for microbumps. No-clean fluxes do exist, but some results have shown that these still contribute to delamination since it is so challenging to clean entirely all residues [8].
- When it comes to device performance, lower parasitic effects (no solder interconnect) have already been mentioned, as well as better thermal performance.
- There are fewer processing steps for HB as seen in Figure 8.7.

However, as with all interconnect technologies, HB also has its downsides. Cleanliness and particles play a huge role in process yield. Whereas in solder-based interconnect technologies, particles can be absorbed in the solder itself or fall in-between copper pillars, in HB a small particle will yield a large void, typically 10× the particle size (Figure 8.8). This will in the majority of cases result in electrical failures or reliability failures.

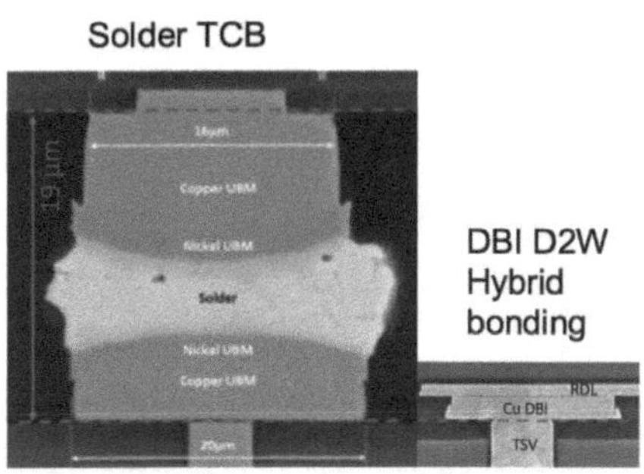

Device Side Process

Solder capped µbump process	DBI
Passivation opening	Yes
Lithography	Yes
Adhesion layer, Cu seed layer	Yes
Cu electroplating	Yes
Ni electroplating	
Solder electroplating	
Resist strip	
Wet etch (Cu + adhesion layer)	
Solder reflow	
Cu CMP: NO	Yes
THV across a wafer: µms	nms

5 steps removed

TSV Side Process

UBM & solder pad process	DBI
TSV reveal CMP	Yes
UBM (Adhesion layer, seed layer)	
Litho	
Cu electroplating	
Ni electroplating	
Resist strip	
Wet etch (Cu + adhesion layer)	
THV across a wafer: µms	nms

6 steps removed

FIGURE 8.7 Comparison of TCB versus HB: cross sections (*left*); device side process (*middle*); TSV side process (*right*) [9].

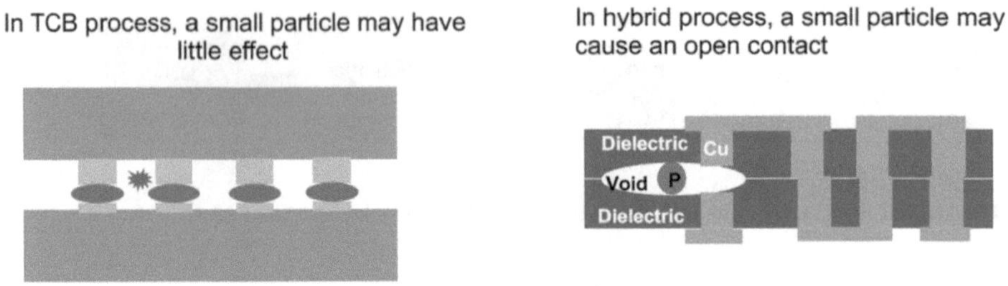

FIGURE 8.8 Impact of particles on TCB (*left*) versus on HB (*right*).

8.2 HB MECHANISM

8.2.1 HB Theory and Process Flows

To understand where HB stands in the overall scheme of stacked bonding technologies, Figure 8.9 gives an overall view in a hierarchical representation with typical examples. There are two main schemes – wafer-to-wafer (W2W) bonding and D2W or D2D bonding. W2W bonding is covered in other chapters in this book. D2W and D2D bonding are key enablers for heterogeneous integration.

Indirect bonding refers to a scheme whereby a secondary material exists in-between the dies or the die and the wafer. This could be in the form of epoxy in its various forms (conductive, insulative, silicon, etc.), or solder bumps, as well as pastes with high conductive content, usually silver, such as in sintered pastes. In the indirect bonding modes, this secondary material provides both the binding mechanism and the electrical connection where applicable.

For direct bonding, the focus will be on HB. In the context of this chapter, HB is understood as a bonding process where there is no material in-between the two dies or the die and the wafer. The bonding happens via van der Waals forces due to the highly polished and clean surfaces in contact with each other. The process occurs at ambient temperature without any heating of the bond stage or bond head. The forces involved are also low, in the range of 2–50 N. High forces may be used, but these are typically used to compensate for other issues which the manufacturer cannot control or chooses not to address, such as die warpage, for several reasons

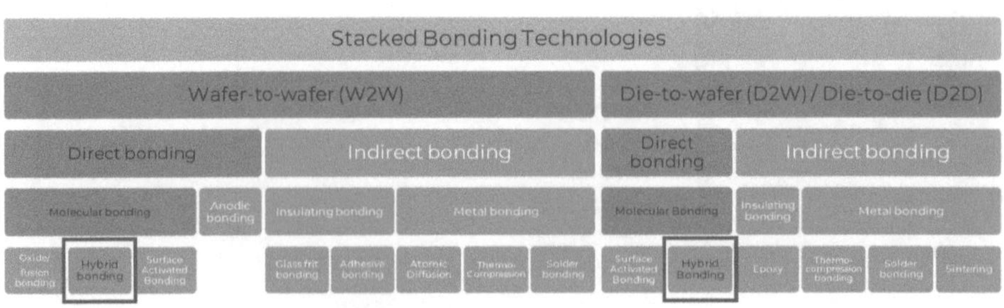

FIGURE 8.9 Interconnect technology schemes. (*Source:* AMD 3D V-CACHE with TSMC SoIC Packaging report, Yole SystemPlus, 2022 [10].)

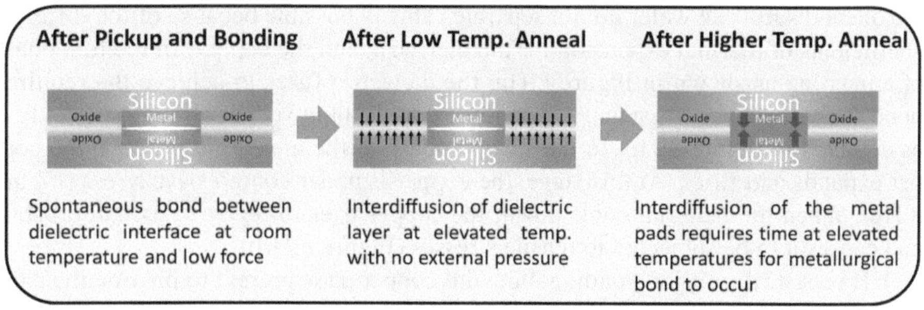

FIGURE 8.10 HB mechanism [11].

such as cost, time to market, reliability etc. The layers that come into contact are typically dielectric, such as SiO_2, SiCN, SiN, and others. The dielectric materials can be mixed and in some instances one of the layers can even have no dielectric; however silicon-to-silicon bonding without dielectric is not considered as HB in the context of this chapter.

Unlike TCB, D2W HB does not require high external pressure or temperature. As illustrated in Figure 8.10, in stage 1 (left), the dielectric layers fuse via the van der Waal's forces, causing a good degree of bond strength. At this stage, no electrical contact is present as the metal, typically copper, is recessed from the previous CMP process since copper is softer than the dielectric and erodes more quickly. (While there are methods to produce wafers with dielectric and protruding copper pads, these require specialized CMP and are not currently the norm for D2W HB.) In stage 2 (middle), a low-temperature process transforms the van der Waals bonds into covalent bonds, thus obtaining a stronger bond. In stage 3 (right), the temperature is elevated so that the copper pads expand, make contact, and fuse into each other, thereby creating an electrical path. During this process, the copper pads are pushing against each other trying to separate the two layers: this is where the importance for a strong dielectric bond strength comes into play. The populated wafer is then cooled down and singulated into individual chips.

To explore the process in more detail, the stage where the die is bonded to the substrate wafer at room temperature is considered as a stress-free state, as seen in Figure 8.11a. At this stage, there is no electrical contact. In order for the copper to expand, it requires annealing; and in order to fuse into each other, it requires a period of time in a state of pressure application. This pressure application is achieved by the copper itself as it expands and pushes against itself, while the dielectric ensures that

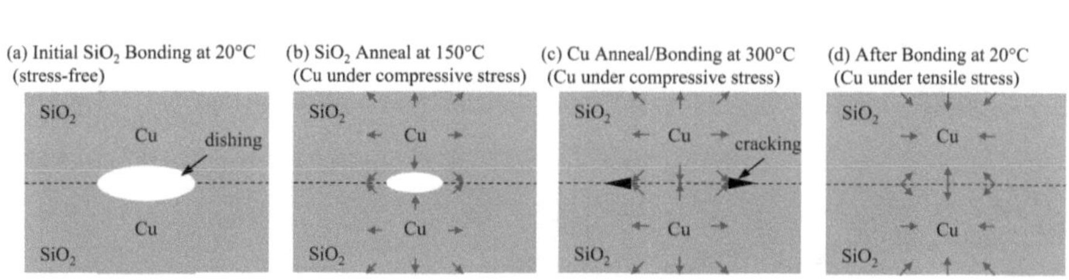

FIGURE 8.11 The stresses inside the HB bond pad area during annealing [12].

the die and substrate wafer do not separate. This is possible because of the different coefficients of thermal expansion for the dielectric and the copper. In the first phase of annealing as shown in Figure 8.11a, the dielectric fuses to achieve the required bond strength, but at this stage, it is possible that the bump edges already start to fuse as depicted in Figure 8.11b. In the second phase of the annealing process, the copper expands and fuses. At this stage, the copper is under compressive stress (Figure 8.11c). When the material cools down, the copper tries to revert to its natural state and causes it to be subjected to tensile stresses (Figure 8.11d).

HB is basically fusion bonding, but with copper pads present to provide the electrical interconnect after bond and anneal. As illustrated in Figure 8.12, in HB the surfaces of both dies (or, the die and the wafer) are Chemically and Mechanically Polished (CMP) to a very high degree, and then they are transferred to the packaging area. The source wafer is singulated into individual dies ready to be picked up and bonded. The next step is usually a wet clean to remove any particles or debris. This wet clean is not aimed to be aggressive as in the slurry removal wet cleaning after CMP; instead, it is designed to be a gentle clean such that it does not alter the surface roughness of the wafers, but at the same time, it removes any contamination present, especially for the singulated source wafer. Typically, deionized (DI) water is used with ultrasonic or megasonic heads, but other variations exist, also making use of chemicals. A degas step is highly recommended for the source wafers which are on the film frame. This helps to remove any excessive moisture and contamination since film frames are known to have issues. Among the diverse surface treatment methods, plasma treatment is widely utilized, and this surface treatment can easily functionalize the required functional group [13–15]. This plasma step is also aimed to be a gentle process so as not to alter the surface roughness, but at the same time removes the top layer of organic contamination just prior to bonding. Applying the plasma treatment to the oxide surface may strengthen the adhesion of the oxide interface by activating the surface with dense hydroxyl groups [16].

For fusion bonding, the wafers are then batch annealed to promote the diffusion of atoms and get a good bonding interface. For HB, a two-step annealing profile is usually employed (to be further discussed in the next section).

Variations to this flow exist where surface hydration can be skipped, plasma performed before DI water cleaning, degas skipped, and in some instances even plasma activation foregone. However, Figure 8.12 delineates a proven process flow as a point of reference.

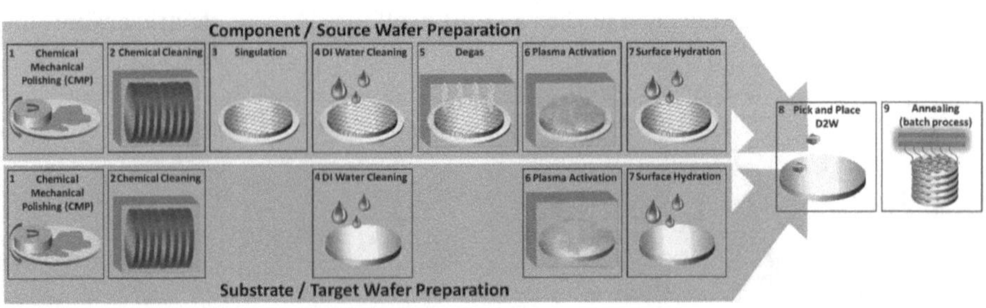

FIGURE 8.12 Typical flow for D2W fusion/HB.

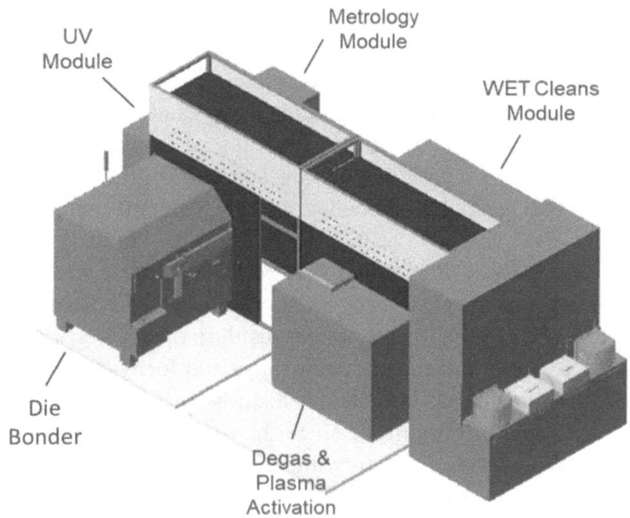

FIGURE 8.13 Example of an integrated tool [17].

For the aforementioned process, one can either opt to use standalone modules and rely on automation to transport the sealed containers from one equipment to another, or else one can choose to install an integrated line, which combines all of the steps from Step 4 to Step 8 (Figure 8.13). Advantages of such a system include full traceability of materials, no loading and unloading to cassette or front opening unified pod (FOUP) in-between process steps, queue time management, integrated metrology for post-bond feedback and correction, single point of contact for maintenance and repair, as well as the ability to map equipment parameters with post-bond results and build an intelligent database with specialized algorithms that can predict any excursions.

8.2.2 Factors Impacting D2W HB

Numerous factors can impact HB; however, it is worth mentioning that this interconnect technology is more susceptible to front-end and pre-preparation than any other technology. The reasons for these are several, but the fact that the bonding occurs purely through van der Waal's forces without any binding material in-between makes the bonding process heavily dependent on the material type, quality, and the pre-processing, which occurs on equipment that is not part of the bonder itself and which has a queue time.

i. *Dielectric Type:* Not only does the type of dielectric, but also the matching of dielectric types between different layers, impact the bond success and bond strength, be it a single die on the substrate wafer or even 3D stacking. The dielectric type also has influence on whether the bond will occur with only one side having dielectric and the other being silicon or III–V material.

Even if the same dielectric type is used, the behavior and end results are not ensured to be the same or even comparable, as the exact process of dielectric formation (TEOS, CVD, PVD, etc.) has a major impact.

ii. *Surface Roughness:* This is a critical factor controlled mainly by CMP and the slurry used. The dielectric also has an impact here. Low surface roughness ensures successful bonding through high-surface-area contact between the two layers being bonded. Ideally <0.5 nm surface roughness is required for HB.

iii. *Linked to the Surface Roughness Is Also the Copper Pad Dishing:* These are linked because the CMP process also impacts the amount of dishing of the copper pads, as high dishing might result in open electrical connections if the copper cannot expand enough to fuse and form an electrical connection. On the other hand, insufficient dishing will result in die separation from the substrate or bottom die since the expansion rate will be too large for the dielectric fusion force to withstand. The rule of thumb for optimal Cu pad dishing is about 3 nm.

iv. *Intrinsic Warpage:* The HB configuration features a complex 3D structure, with materials variations along the *z*-direction (layer structure) as well as in the *x–y*-plane (metals embedded in dielectrics). This complex layout generates intrinsic stresses in the material and eventually leads to a certain "non-planar" shape of the wafer and the dies, respectively. To avoid excess warpage, ideally the stress is already considered during the chip design phase, although it is impossible to completely avoid any waviness of the material. Typically, the die warpage can range from a few micrometers to 100 µm. With the advent of thin die HB in the near future, the die warpage will potentially be even higher, since for thin dies the same amount of stress leads to higher deformation/warpage (to be further discussed in Section 8.5).

v. *Cleanliness:* This is another parameter of paramount importance. Cleanliness requirements for HB include (a) no presence of particles (for prevention of voids, as illustrated in Figure 8.14); and (b) no contamination at the nano-level, which could lead to other issues (such as outgassing, no dangling bonds in places of contamination, etc.), even though it may not

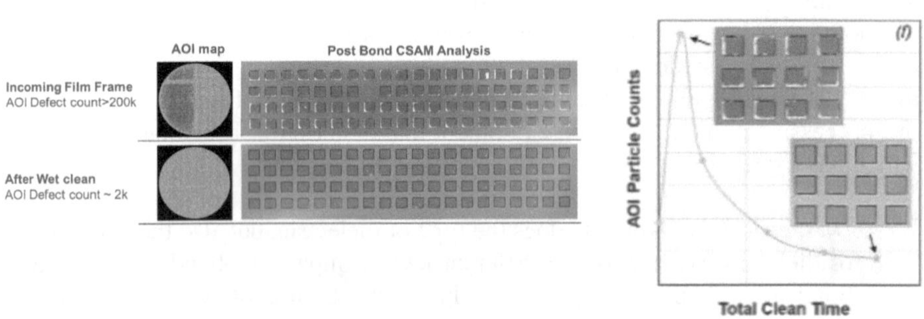

FIGURE 8.14 Impact on void formation from particles in HB [18].

physically hinder the surfaces making contact with each other. Cleanliness needs to be ensured in both the equipment itself and also in the bonding media. As using chemicals for wet cleaning could also alter the surface roughness, a balance needs to be found.

vi. *Plasma Activation:* Plasma activation ensures that organic contamination is removed to ensure successful HB and promotes the formation of dangling bonds and bonding between hydroxyl groups and water molecules. However, the plasma process can alter the surface roughness as well as alteration of the copper pad profile among other things, leading to issues in bonding. If the wrong gas is used in the process, creation of additional oxidation could ensue, which impacts negatively the bonding process.

vii. *Ratio of the Copper Area to the Dielectric Area in the Surface/Bonding Layer:* Enough dielectric surface area is required to ensure sufficient bond strength at time zero so that the initial bonding strength through the van der Waal's forces can withstand wafer movement through the rest of the process flow, especially before annealing. The dielectrics have a finite flatness, as shown in Figure 8.15, leading to a further reduction of the effective dielectric area contributing to the initial bond. The effective dielectric area has to be large enough to withstand the opposing forces applied by the expanding

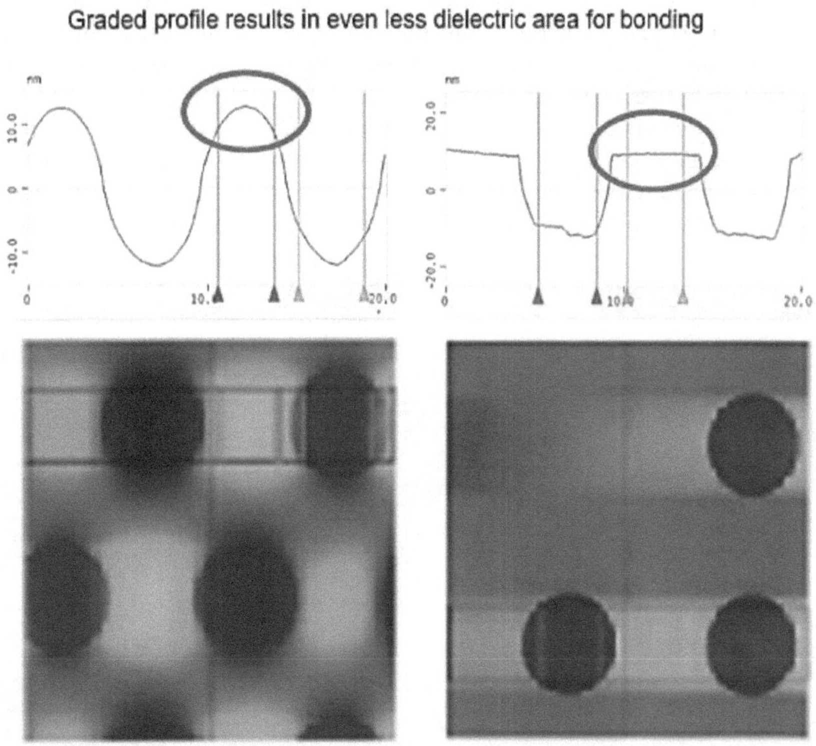

FIGURE 8.15 Profile for dielectric: non-optimal (*left*) and optimal (*right*) [18].

copper in the bond pads during annealing. Typically, for actual materials, a ratio of 20% copper is the upper limit for a successful process, as it has been reported that above 20%, delamination issues were seen, attributed to a lower dielectric surface area [19].

viii. *Distribution of Bond Pads:* Not only the aforementioned percentage of copper pads versus dielectric is important, but also the distribution of the bond pads throughout the wafer and respective dies is critical. If one part of the die has a higher concentration of bond pads, it could lead to warpage and complications during bonding and annealing. Other factors are pitch variations between bond pads and different bond pad sizes in the same die.

ix. *Fiducial or Alignment Marks Also Have an Impact:* The right fiducial design is needed to maximize placement accuracy. As shown in Figure 8.16, position uncertainty decreases with increasing fiducial size (pink line). Circular fiducials give best position results (green line). Overexposure or insufficient contrast increases uncertainty (third line from bottom and second line from bottom).

x. *CMP Needs to Ensure That Bond Pad Dishing Does Not Exhibit Any Issues:* For example, having a barrier which is higher than the surrounding dielectric, as depicted in Figure 8.17, would result in die separation during bonding as the already higher barrier expands more during annealing, resulting in die separation. The other factor is that dishing needs to be within spec throughout the entire wafer (center versus periphery). Too low dishing will result in copper expansion, pushing and separating the die from the bottom layer, and too high dishing will result in the copper not able to expand enough to fuse and form an electrical connection.

xi. *Annealing Process:* Typically, the annealing process occurs in two or more steps as depicted in Figure 8.18. A low-temperature anneal (100–200°C) is first carried out to promote bonding of the dielectric, and a second step at a higher temperature (200–350°C) is performed to expand the copper bond

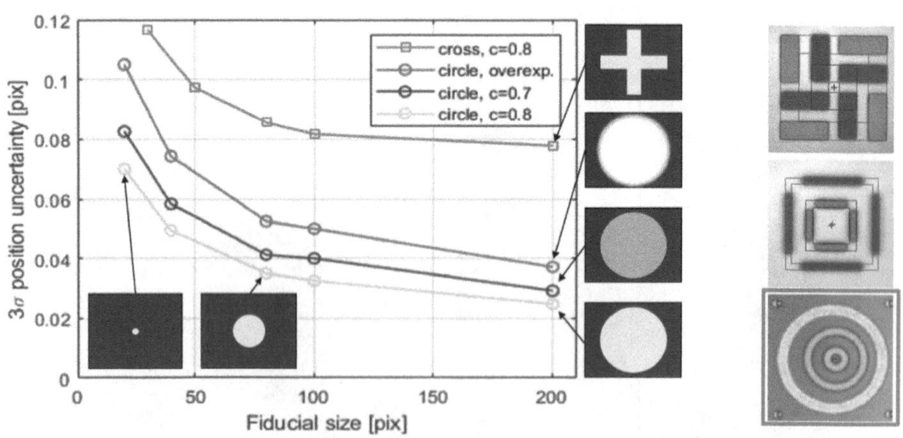

FIGURE 8.16 Fiducial properties and impact on the accuracy of the recognition (*left*), and optimized fiducial design in green (*right*) [20].

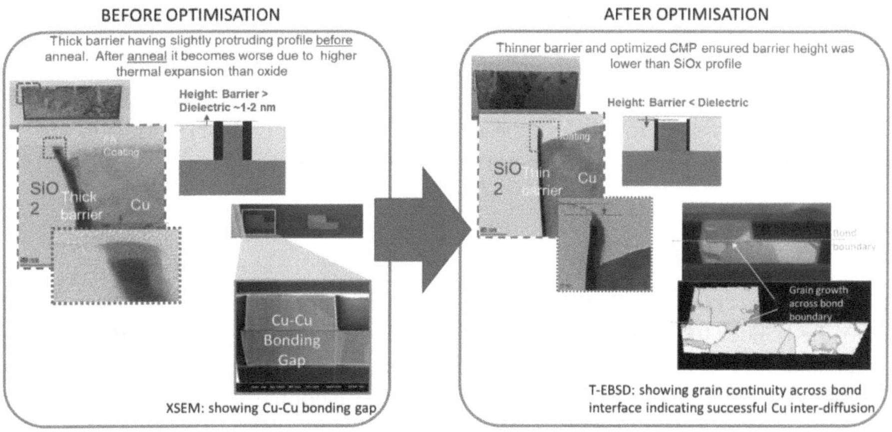

FIGURE 8.17 XSEM showing Cu–Cu bonding gap due to barrier–CMP interaction (*left*) and XSEM showing Cu–Cu diffusion and confirmation with T-EBSD showing grain growth across bond boundary after optimized thin PVD-CMP (*right*) [18].

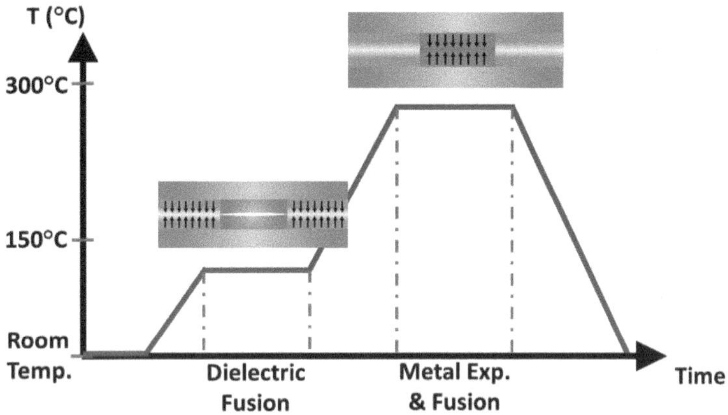

FIGURE 8.18 Two-step annealing process [11].

pads and promote diffusion and bond formation to obtain an electrical connection. The type of gas used is also critical. This is typically dependent on the type(s) of dielectric and bond pad metal used. Typical gases used are oxygen, nitrogen, and argon. Some profiles also involve multiple gases depending on the curing step or a mixture of gases.

xii. *ESD Is a Critical Topic in HB:* With HB addressing products having small tech nodes, the dies typically no longer have capacitive protection diodes and are thus more susceptible to damage from ESD. ESD is a core topic which must address the source and target dies, as well as all locations during the loading and unloading steps. Both the robot-handling and the wafer-holding modules as well as the pickup and bond tools need to address ESD topics. The ESD requirements for the D2W bonder are described in Section 8.3.6.

8.3 KEY PERFORMANCE FACTORS OF A D2W HYBRID BONDER

Flip chip bonders have evolved continuously over the last three decades. However, only with the latest developments of turning a die bonder into a front-end like tool did it become possible to address the needs of fine-pitch D2W HB. In this section, we present the key performance factors of a D2W hybrid bonder.

8.3.1 Cleanliness

The cleanliness requirements of a hybrid bonder are significantly higher than for any other die bonder today. First, particles locally hinder the dielectric bond to happen, which can then lead to a reduced strength of the bond. Second, particles at or in the vicinity of a copper pad can lead to a missing electric connection and thus to a failure of the device.

To keep both effects significantly low, a D2W hybrid bonder is typically placed in an ISO-4 to ISO-6 (class 10 to class 1000) cleanroom, together with the required pre-processing equipment. The maximally allowed particle count inside the machine where the material is bonded is even an order of magnitude smaller, with ISO-3 to ISO-5 conditions during operation.

To ensure such a high degree of cleanliness, the following should be performed:

1. Properly design the individual parts of the machine (e.g., no particle traps).
2. Properly manufacture the parts (e.g., low surface roughness).
3. Properly clean the parts before assembly.
4. Assemble the machine in a clean environment.
5. Suppress abrasion of moving parts.

For the cleanliness characterization, two types of measurements are commonly used:

- *Airborne Particle Count:* Number of airborne particles per volume and time within various size ranges
- *Witness Wafer Measurements:* Number of particles per time within various size ranges deposited on a witness wafer inside the bonder.

HB is extremely sensitive to particles and contamination, and surface preparation, one of the most crucial steps prior to bonding, is one of the biggest contributors to contamination if the process is not optimized and controlled. Figure 8.19 shows some issues and sources of contamination. For source wafers mounted on the film frame, outgassing needs to be addressed carefully. While W2W HB deals only with wafers, films give rise to additional issues which need to be addressed when dealing with D2W HB. Particularly during plasma cleaning, outgassing is known to be a phenomenon that occurs. The plasma also should not impact the surface roughness of the wafers, especially the copper pads and dishing, since this material has less hardness than the surrounding dielectric. In cases of excessive plasma or strong cleaning, the bond pad barrier which has a higher hardness than the dielectric becomes

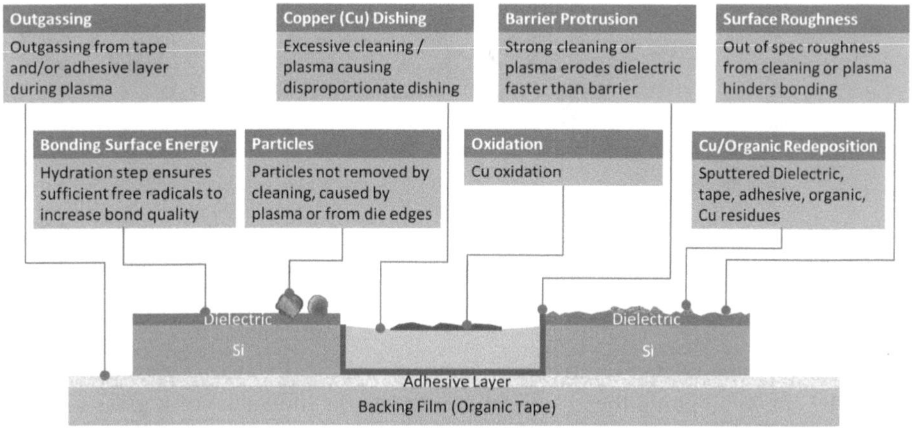

FIGURE 8.19 Surface preparation challenges for D2W HB [4].

exposed resulting in die delamination and voids. Any particles left over due to inefficient cleaning or generation of particles from the plasma process results in voids and electrical failure. When dealing with copper bond pads especially, oxidation is a factor to take into consideration. The surrounding atmosphere should be kept under temperature and moisture control, and ideally the queue time between plasma activation and bonding should be kept to a minimum. Finally, one needs to consider that all surrounding materials, such as the film adhesive, the copper pad, and the dielectric, could be sputtered and particles could end up on the active side.

8.3.2 Accuracy

As mentioned before, the HB technology is specifically interesting for state-of-the-art applications with the highest interconnection densities. The exact relationship between the bond pad pitch and the required accuracy eventually depends on the targeted yield. As a rule of thumb, it is common to demand a 3σ placement accuracy that is 1/10 of the bond pad pitch.

The overall placement accuracy typically has the following contributions:

a. *Equipment-related inaccuracy:*
   i. Inaccuracy due to the finite position stability of bond head and bond stage
   ii. Inaccuracy due to finite mechanical noise level and stability
   iii. Inaccuracy due to recognition of alignment marks (see also item c below)
b. *Process-related inaccuracy:*
   i. Inaccuracy due to variations in the touchdown process
   ii. Inaccuracy due to variations in the bond force trace
c. *Material-related inaccuracy:*
   i. Position inaccuracy of alignment marks, pad structure
   ii. Inaccuracy due to recognition of alignment marks (see also item a above)

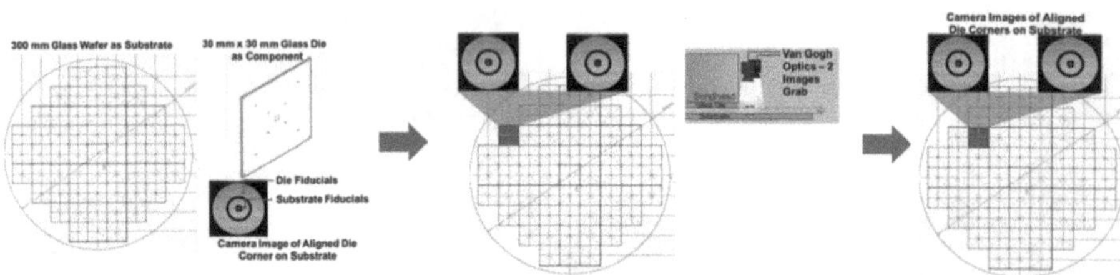

FIGURE 8.20 Sketch of the procedure for the determination of the alignment accuracy.

The equipment-related (in)accuracy (item a.) is typically characterized and measured by simply aligning the "ideal" alignment marks (fiducials) of a glass die with the alignment marks of a glass substrate, as illustrated in Figure 8.20. After the alignment is done, the glass chip is brought close to the surface of the glass substrate without touching. Then the alignment is verified by determining the actual position of both alignment marks (chip and substrate) again and calculating the standard deviation (σ) of this position difference for a representative sample size. The 3σ value described here is commonly called "alignment accuracy."

Multiple fiducials, often in the two upper corners of the glass die and the glass substrate, respectively, are used to perform the alignment. To verify the alignment, the positions of all the four fiducials in each corner of the die and substrate, respectively, are determined. The stated alignment accuracy is then the highest (worst) 3σ value of all four corners. Figure 8.21 shows typical values for this alignment accuracy procedure with state-of-the-art equipment reaching 55.81 nm @ ±3s (worst corner). Note that the size, shape, and contrast for the alignment marks on this glass material are typically chosen to have only a small/negligible contribution to the "alignment accuracy."

The process- and materials-related inaccuracy together with the resulting placement accuracy is further discussed in Section 8.6.

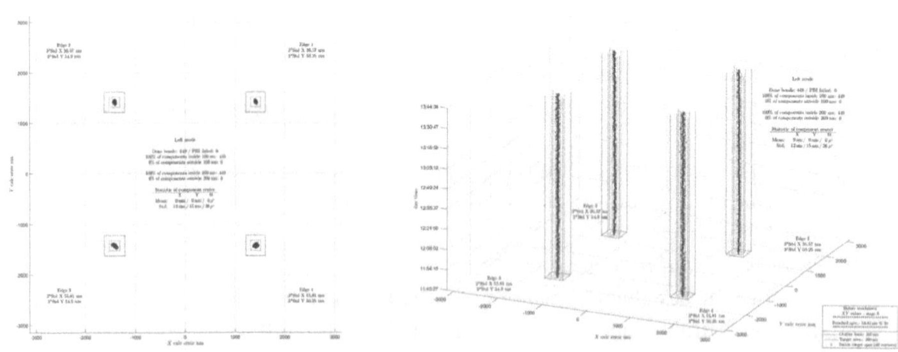

FIGURE 8.21 Alignment accuracy test results for all four corners of the die: 2D distribution (x and y) (*left*) and time dependency of the alignment (*right*).

8.3.3 Throughput

Throughput is essential for high profitability in a production environment. The throughput of a die bonder is typically defined by the number of bonded chips per hour (CPH) or units per hour (UPH). (In the case of a single unit consisting of N dies, the UPH number is N times lower than the CPH number.)

Since today's hybrid D2W bonder aims for the highest accuracy possible, the speed of these tools, typically 1,000–2,000 CPH, is much lower than the traditional flip chip bonders.

For profitability or cost-of-ownership calculations, it is important to also include the time needed to load and unload the source and the target wafers respectively (Figure 8.22), as defined by the so-called effective throughput (effective CPH, effective WPH), as follows:

$$\Delta t \quad \text{average time between two consecutive bonds (second)}$$

$$\Delta t_{\text{wc}} \quad \text{time needed to change the wafers (second)}$$

$$n_{\text{dpw}} \quad \text{number of dies per wafer}$$

$$(\text{pure})\,\text{CPH} = 3{,}600\,\text{seconds}/\Delta t$$

$$\text{effective CPH} = 3{,}600\,\text{seconds}/(\Delta t + (\Delta t_{\text{wc}}/n_{\text{dpw}}))$$

$$\text{effective WPH} = 3{,}600\,\text{seconds}/(\Delta t * n_{\text{dpw}} + \Delta t_{\text{wc}})$$

For large dies, the number of dies on a single wafer is smaller. Thus, the wafer change has a larger impact on the effective throughput.

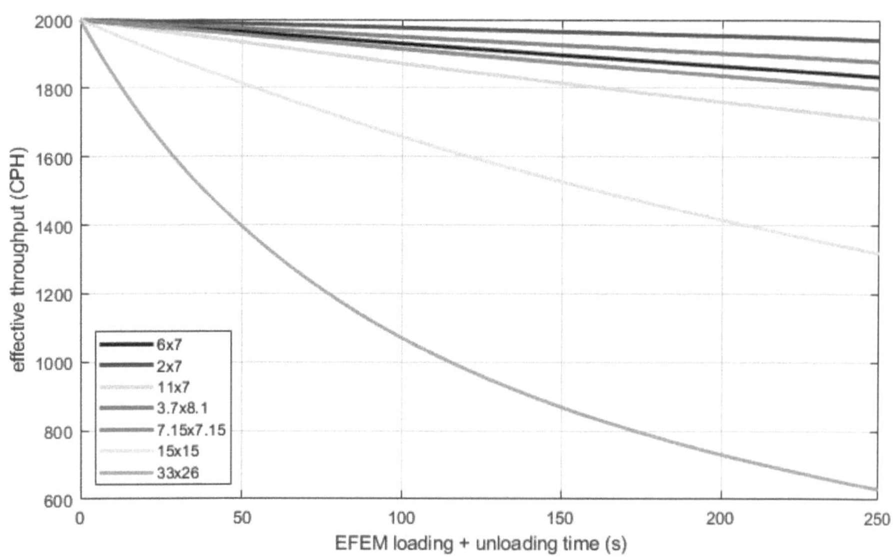

FIGURE 8.22 Impact of die bonder throughput versus loading and unloading times.

8.3.4 Bond Force

The bond force requirement for D2W HB differs from other established die attach technologies. For a stable process, it is important to have a good control over the applied force throughout the bond process. This typically starts with the so-called touchdown force, the force with which the touching between the component and the target wafer is detected. After touchdown, the force applied by the bond head system is then ramped up to the so-called bond force, which is then applied for a certain period of time (more details can be found in Section 8.4.4).

For a blank (non-patterned) material, the force simply needs to initiate the fusion bond process. In this case, the (minor) stress in the material can easily be overcome by the bond energy, assuming a proper activation of the material. For this case, the bond force of 10 N or less often turns out to be sufficient for successful bonding.

For a patterned material with low stress/warpage, the situation is expected to be similar. However, in case the deformation energy of the material exceeds the available bond energy, the force applied by the bond head system has to assist to overcome the stress in the material. For this purpose, a significantly higher bond forces (up to 50 N) might be needed.

8.3.5 Typical Material Dimensions

Typical values for state-of the-art D2W systems are given in Table 8.1 for reference.

8.3.6 Electrostatic Discharge (ESD)

The ESD requirements for fine-pitch HB are driven by the targeted CDM (charged-device-model) withstand voltage of 5 V, as seen in Figure 8.23.

TABLE 8.1
Typical Values for State-of-the-art D2W Systems

Dimension	Typical Values for State-of-the-art Systems
Target wafers (substrate wafers)	• Standard: 300 mm (12″) wafers • Upcoming: 200 mm (8″) and 150 mm (6″) wafers (e.g., for applications involving III–V materials, etc.)
Source wafers (component wafers)	• Standard: 300 mm (12″) diced wafers on film frames • Standard: 300 mm (12″) carrier wafers • Upcoming: 200 mm and 150 mm diced wafers on both film frames and carrier wafers (see above)
Component format	• Rectangular-shaped, constant height across the die • Aspect ratios typically range from 1:1 to 5:1
Die size	• Component size in x and y ranging from 1 mm up to the reticle size (approximately 30 mm) • Component sizes below 1 mm might possibly be relevant for future chiplet designs
Die thickness	• Dies are typically thinned to 50–300 μm prior to bonding • Dies with a thickness of 30–50 μm are considered thin dies and will become relevant for certain applications (e.g., memory)

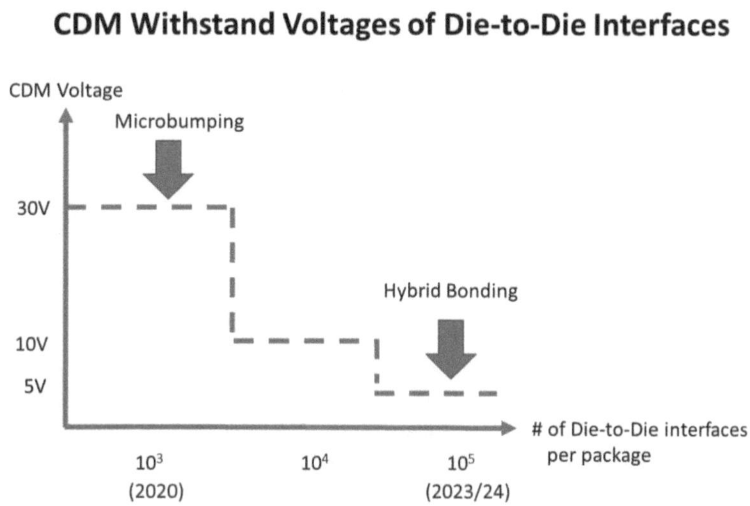

FIGURE 8.23 Industry roadmap for the number of die–die interfaces per 3D IC package and expected CDM withstand voltages [21].

This CDM withstand voltage roadmap results in specific measures when building a D2W hybrid bonder.

- *Grounding:* All surfaces inside the bonder which make contact with the HB material should not be insulating nor hard-grounded. Instead, the surface resistance as well as the resistance to ground should be *"dissipative conductive."* Typically, the term "dissipative conductive" refers to a resistance between 1e5 and 1e8 ohms.
- *Ionizers:* Ion sources are typically used to neutralize any charges which have not been drained off sufficiently. To assess the performance of the ionizers, a test plate is charged up to a voltage of ±1,000 V. Then, the time needed for the ionizer to reduce the voltage to ±100 V, the so-called decay time, is measured. In addition, the "balance voltage" (steady-state voltage) indicates how well the neutralization can be done. For HB, decay times well below 10 seconds and balance voltages comparable to the CDM withstand voltage are desired.

8.4 TYPICAL FEATURES OF A D2W HYBRID BONDER

To meet the key performance items mentioned above, a hybrid bonder typically features the following capabilities.

8.4.1 Automation Interface

Hybrid bonders (both W2W and D2W) will become a cornerstone for state-of-the-art advanced packaging fabs. These fabs are "front-end-like," featuring a strictly monitored cleanroom and a strong focus on automation. All tools in these fabs are

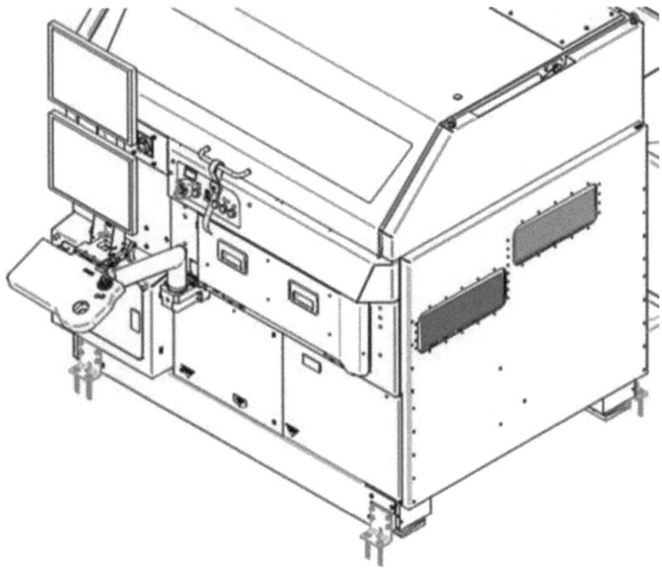

FIGURE 8.24 Example for the automation interface at a hybrid bonder.

expected to function without operator intervention (during "normal operation"). For this reason, all machines need to be connected to an overhead transport (OHT) via an "equipment front-end module" (EFEM).

Consequently, the automation interface of a D2W hybrid bonder has to be able to receive components and target wafers delivered by the robot end effectors of the EFEM. The wafers are put either directly onto their final position (needed to perform the bonding) or onto some kind of wafer transfer module. Such a wafer transfer module is needed, in case the reach of the end effector robot is not sufficient to place the respective wafer on the desired position inside the bonder.

The interface between the bonder and the EFEM may also feature a mechanical shutter. On the one hand, this adds additional complexity, but on the other hand, it allows separating the mini-environments from each other.

Figure 8.24 shows (in blue) the interface for the component/source wafer, which can be either a film frame wafer or a component carrier wafer, and (in green) the interface for the substrate/target wafer.

8.4.2 Component Wafer Handling and Die Picking

As stated in Section 8.3.5, the HB components are received on either film frames or carrier wafers (glass or Si). The handling of the component wafer as well as the subsequent die picking differs significantly between these two component wafer types.

- *Components on Film Frames*

  The metal frame is fixed to the film frame wafer table. The tape (for thermal release or UV release) is stretched such that the die–die distance on the tape increases to values sufficient for ejection. Ejection tools (e.g., needles, multistage tools, etc.) are used to release the dies from the tape.

- *Components on Carrier Wafers*
  For this type of component handling, the carrier wafers are coated with an adhesive film to ensure that the dies stay on the carrier wafer. To reduce the adhesion between the film and the components, the populated carrier wafer might get some thermal treatment ("temperature release"). Other films require a mechanical squeezing ("pressure release"), which can be achieved by applying a sufficiently high force using the die picking tool. For glass carrier wafers, a third possibility is to reduce the adhesion of the film by applying electromagnetic radiation (typically in the UV range) from the backside ("through the glass carrier"). However, in this case, the carrier wafer table has to feature the appropriate light source as well as special wafer chuck.

8.4.3 Flip Unit

In most cases, the components are provided on film frames with the bonding layer ("active side") facing upward, which requires flipping the die by 180° before bonding. Hence, a hybrid bonder features a flip unit, similar to what is used for other flip chip technologies. For a flip chip bonder, the flipper picks the die right during the ejection process, flips it by 180°, and presents it to the bond head with the proper orientation.

A crucial part for the hybrid bonding process is the flip tooling. Since it is touching the HB surface, it must be free of any particles. Non-contact methods exist and have been proven to work; however, these still need to be qualified in HB HVM.

8.4.4 Bond Head and Motion System

The bond head system is a crucial hardware of a hybrid bonder. During each bonding cycle, it needs to perform the following steps:

- *Take over the Die:* From the component supply system (typically from the flipper).
- *Transport the Die to the Uplooking Camera:* To determine the position of the die on the bond head.
- *Move to the Bond Position:* Hereby, the bond head has to move fast to enable high-throughput values and over a long distance (> target wafer diameter).
- *Perform Final Translational (x, y) and Rotational (θ) Corrections:* For high accuracy, precise spatial corrections need to be done as close as possible to the moment of the actual bonding.
- *Perform the Actual Bond Process:* To adapt to the specific process needs of the various applications and materials, during the actual bonding, the bond head needs to
  - move with different speeds along the z-direction (by finding the optimum between fast z-movements for the highest throughput and slow z-movements for precise and reliable bonding),

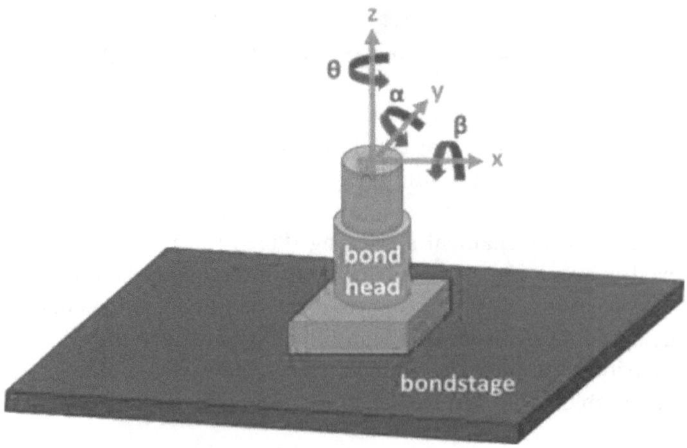

FIGURE 8.25 Typical definition for the three translational and the three rotational motion axes.

  - have some type of force measurement to detect the moment of touching/bonding and to manage well any potential "overtravel" in the z-direction (which can lead to materials damage and particle generation),
  - apply a proper force profile (touchdown and bond force along the z-direction for certain period of time),
  - have a mechanism to hold the dies on the bond head during transport until bonding (e.g., under-pressure, electromagnetic forces, etc.), and
  - feature switches to release the dies from the bond head after attachment (e.g., vacuum switches).
- *Have Sufficient Stiffness:* To avoid any spatial (translational, rotational) shift or tilting due to the reaction forces from the substrate surface.
- *Move Back to the Flipper:* To pick up the next die.

As indicated in the list above, motion axis for all three translational (x, y, z) as well as the rotational axes (θ) around the z-axis are required. In addition, the rotations around the x-axis and the y-axis (the angles β and α) of the bond head relative to the bond stage need to be adjusted (Figure 8.25), commonly referred to as the "bond head tilt adjust." The position stability as well as the repeatability of all movements is essential.

8.4.5 Camera and Optics System for High-accuracy Alignment

As mentioned in the previous section, hybrid bonding is particularly interesting with ultrahigh placement accuracy well below 1 µm towards 100 nm. In this regime, the position alignment of the die relative to the substrate needs to be done for each die individually. A rough sketch to illustrate this "local alignment" concept is shown in Figure 8.26. Such an alignment typically involves two camera systems. The "uplooking camera" is required to image the die fiducial, whereas the "downlooking camera"

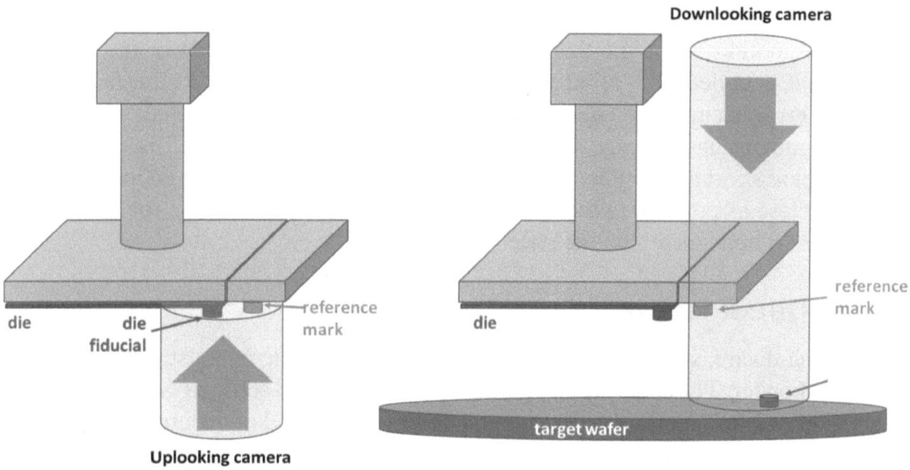

FIGURE 8.26 Vision alignment schemes.

takes images of the bond position fiducials on the target wafer. To determine the relative distance between the die and the bond position fiducial, the two camera systems need to be "linked." This connection is established by reference marks, which can be seen by both camera systems.

Typical challenges faced when integrating such a scheme into a die bonder are as follows:

- *Focal Height Difference between the Reference Mark and the Fiducials on the Material*
  During the alignment, both the die and the bond position fiducials are on different z-positions (heights) compared to the reference mark. Therefore, it is necessary to compensate for this spatial difference by smart optical solutions. One possibility is to introduce "focus shifting optical elements," which can shift the focal plane for parts of the camera's field-of-view (FoV). Another method is to split the imaging light into two different optical paths (with different path lengths). Eventually, the two optical paths are combined again and a single snap can be used to image the reference mark and the die/bond position fiducial. The latter method opens up the possibility to use a second camera (e.g., with a smaller FoV but higher resolution) to further minimize the alignment error.
- *Die Fiducial and Bond Position Fiducial Overlapping/at the Exact Same Position*
  In this case, as the die fiducial and the bond position fiducial are on the exact same position, the "downlooking camera system" cannot image the bond position fiducial right before the final bond move in the z-direction. In this case, an additional movement of the die relative to the downlooking camera system is needed. To enable this movement, typically a motion axis in either the x- or the y-direction is added.

- *Minimized Alignment Height*
  The motion of the die toward the substrate after finalization of the alignment is called the "blind move." The term "blind" refers to the fact that an error accumulated during this movement cannot be corrected, since the alignment has already finished. Therefore, the "blind move" is typically kept as short as possible. In other words, the alignment is desired to be done as close as possible to the substrate. Typical alignment heights for state-of-the-art bonders are on the order of 1 mm.

8.4.6 Defense Features

Besides hardware, software also plays a very important role in constructing the ultimate die bonder. The software should allow for control and setting of enough parameters to be able to set up adequate process flows for a myriad of different applications and materials. Traceability is also an important feature, with the bonding portion being paramount. A hybrid bonder should be able to trace the key machine parameters during the bonding process and record them for each die bonded (Figure 8.27). An easy GUI (Graphical User Interface) also helps the engineering team to analyze and solve issues in a fast manner so as not to impact production. The ability to integrate the bonder to the client host system is also important, and this typically requires customization for the different communication protocols.

8.4.7 Tool Change Concept

In heterogeneous integration, multi-die applications are the norm, and automatic tool change is a very important feature in die bonders, as it allows the bonder to be able to automatically change the ejection tool, the flip tool (if applicable), and the bond tool, to cater for different die sizes. This is especially important for HB since it allows a no-hands-in-bonder mode of operation, thereby reducing the impact of particles. Care

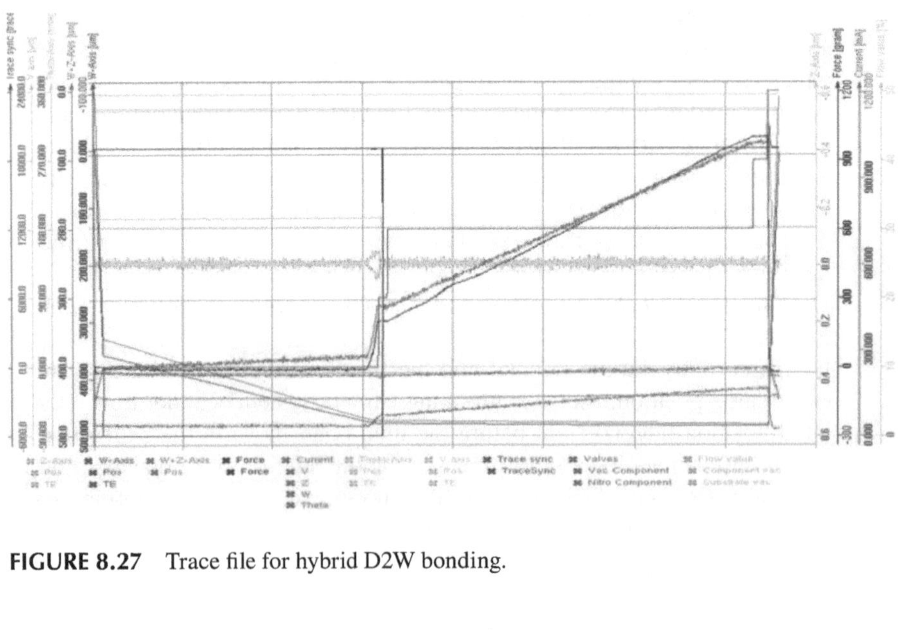

FIGURE 8.27 Trace file for hybrid D2W bonding.

needs to be given that the tool change does not create particles or disruption in the flows. Attention must also be given to the holding mechanism, be it vacuum or via magnets (or a hybrid) or other means, so that it does not generate contamination. The process should also ensure that there is no impact on the bond placement accuracy after tool change. This is usually achieved by having intelligent calibration routines in place.

8.5 D2W/D2D HB PROCESSES

8.5.1 Main Process Flows

In D2D/D2W HB, a singulated die is bonded onto an already bonded die (D2D) or on a wafer substrate (D2W). The standard process, as delineated in Figure 8.12, is that the source or component wafer is mounted on the film frame and singulated. Several singulation methods exist; however, laser or plasma singulation processes are preferred to mechanical dicing, for several reasons. The film frame wafer is then subjected to a wet clean, degas, plasma activation, hydration, and finally bonding. A mass anneal process is then carried out.

There exists a slightly different flavor of this process, whereby a reconstitution of the singulated wafer is carried out, as illustrated in Figure 8.28. This is done for several reasons:

- HB wafers are very expensive because they typically employ the latest state-of-the-art technology. As such, designers strive to make the most of the wafer surface area available to fit as many dies as possible. This means that the street width between dies is becoming smaller, in the range of 40 μm or lower. Employing laser and plasma singulation methods also means that the kerf width is kept small, in the range of 20 μm or less, unless intentionally widened by having multi-pass processing or custom-designed masks. The latter however is undesirable since every additional singulation alteration generates additional particles and increases the risk of damage to the components. The silicon debris and other particles in the kerfs are challenging to remove since the kerf is so narrow. Megasonic wet cleaning and other solutions try to remove the debris; however, it still remains a challenge. A reconstitution to a new, clean film frame or wafer carrier ensures that the majority of the particles are left behind.

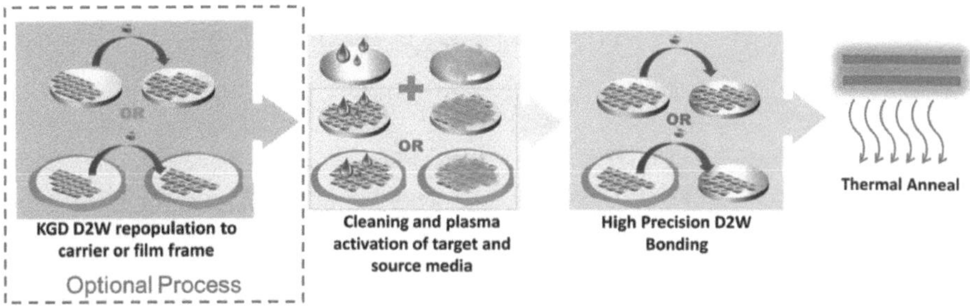

FIGURE 8.28 Simplified D2W process flow showing optional reconstitution [11].

- It also gives the opportunity to pick up only known good dies (KGD), meaning only the dies which have passed testing are processed. Although wafer maps exist to delineate this, yield might have an impact on throughput, as component wafers with a higher yield require less frequent wafer changes for the same number of dies placed. Processing only good dies through the pre-processing steps of cleaning, degas, plasma, and hydration also increases efficiency.
- The reconstitution can have larger die spacings to ensure that the wet cleaning modules can remove any particles from between the dies.
- The reconstitution can be done in such a way that the number of components matches the number of bond pads on the substrate or target wafer, making a 1:1 scenario. This however is unrealistic since every substrate wafer will also have a certain yield which therefore alters the amount of good bond positions, and therefore will never match the number of components on the reconstituted wafer.
- The reconstitution can be done in such a way that the number of components *does not* match the number of bond pads on the substrate or target wafer. This is specifically done so that the loading and unloading of both target and source wafers do not occur at the same time. Reasons could be to minimize particle generation by limiting movements in the bonder or to maintain throughput making changeovers faster, since a dual loading/unloading can lead to waiting times for operations which need to be done in series.
- Another flavor is to reconstitute on a carrier (silicon or glass) rather than another film frame. A temporary bonding material (TBM) will be needed to hold these dies in place. Rigid carriers are seen as being more clean than films and require no degas step. They are also widely used in the front-end areas and proven for many years to be of high cleanliness standards.

8.5.1.1 D2D HB

For the context of this chapter, D2D HB is a form of stacking. The bottom die needs to be in a form which can be accepted by the bonder and bonded on, as seen in Figure 8.29. The most reliable way to do this is to have the bottom dies bonded on a wafer or carrier with TBM. The source dies can be presented on a film frame or a

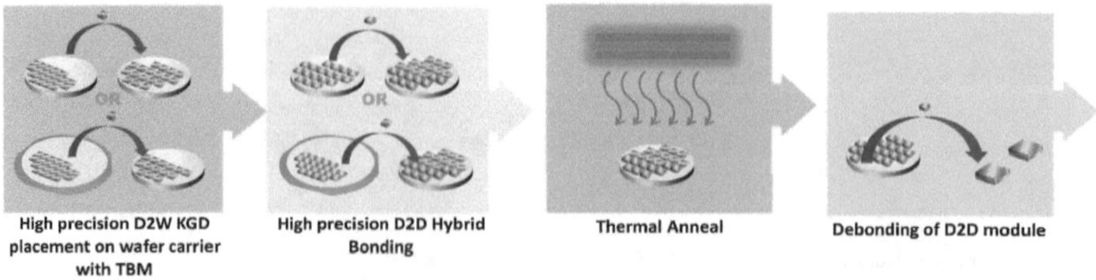

FIGURE 8.29 Simplified D2D process flow (wet cleaning and plasma activation are not shown for simplicity).

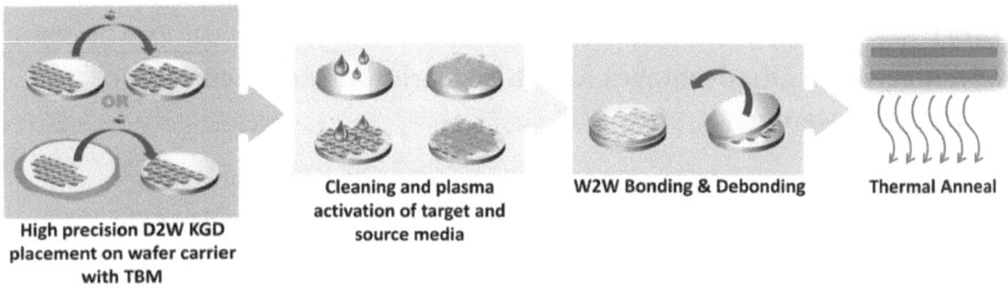

FIGURE 8.30 Simplified CoD2W process flow [11].

wafer carrier, and these are hybrid bonded onto the dies. An anneal process follows and then debonding of the D2D modules.

8.5.1.2 CoD2W HB

This approach involves bonding from a film frame (or a wafer carrier) to a wafer carrier, as seen in Figure 8.30. The dies are bonded face up at a high accuracy, and no plasma activation is required prior to this step since the dies are held in place via a TBM on the receiving wafer carrier. The populated wafer carrier and the matching substrate wafer now undergo wet cleaning, plasma activation, and hydration, followed by W2W bonding. The latter processes typically occur in a W2W bonder. After bonding, the wafer carrier with TBM is debonded. This process still involves the need of a high-accuracy bonder, since the dies on the wafer carrier need to match accurately the bond positions on the substrate wafer. Moreover, it requires additional equipment for W2W bonding. The final bonding accuracy is subject to a higher degree of error: one coming from the D2W bonder placement of the dies, the other from the W2W bonder final bonding.

8.5.1.3 W2W HB

In W2W HB, two wafers are bonded onto each other without TBM or another intermediate layer (Figure 8.31). The top and bottom dies need to match since the wafers are rigid and there is no possibility of lateral adjustment. The W2W bonder typically houses all modules necessary, such as plasma, wet cleaning, and the bonding chamber itself, which can also be under vacuum. The bonding itself takes place quickly, especially when compared to D2W bonding. There is no possibility of KGD, which means that the yield of the top and the bottom wafers will contribute to a lower

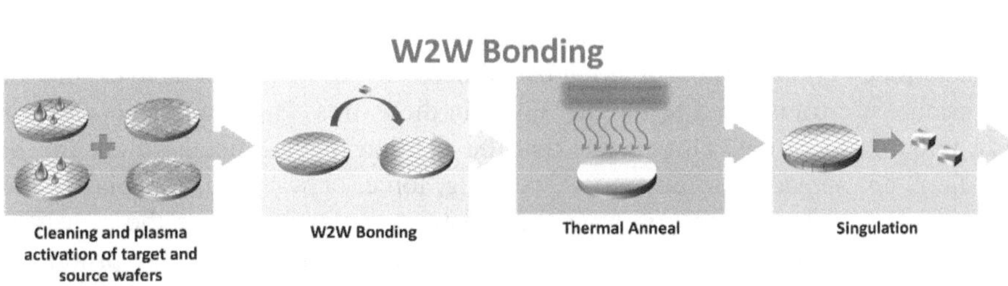

FIGURE 8.31 Simplified W2W process flow.

TABLE 8.2
Advantages and Disadvantages for HB Scenarios for Advanced Logic and Stacked Memory

	Logic		Stacked DRAM–HBM
D2D/D2W	✓ Simplified process ✓ Lower cost ✓ Optimum die alignment yield ✓ flexibility × Requires maintaining cleanliness through longer time period as wafer is populated	**D2D/D2W**	✓ Enables Complex Chiplet Integration ✓ Lowest cycle and queue times ✓ Highest accuracy and yield ✓ flexibility × Currently limited to >30 μm die thickness
CoD2W	✓ Lower risk of particle defects × High complexity and higher cost × Requires additional W2W bonding equipment × Error accumulation from 2 processes yields to lower placement accuracy × Tolerance on wafer thickness is critical	**CoD2W**	✓ Lower risk of particle defects × High complexity and higher cost × Requires additional W2W bonding equipment × Error accumulation from 2 processes yields to lower placement accuracy × Tolerance on wafer thickness is critical × Limited to less complex packages
W2W	✓ Ability to rework × Limited to 2-4 stack × Non KGD process results in yield loss × Dissimilar die size	**W2W**	× Not a viable flow

overall yield since it is very rare that both wafers' failed dies coincide at the same locations, or that the yields are 100% on both wafers. After bonding, the wafers are annealed to fuse the dielectric and achieve copper expansion, fusion, and electrical connection. Singulation follows to finally achieve the individual modules.

A summary comparing the advantages and disadvantages of these processes is shown in Table 8.2.

8.5.2 Ejection Processes

The ejection method needs to be clean and not produce any particles, a particular problem for ejection methods making use of needles, but it also needs to ensure that no residues are transferred to the die. For thin dies, stress is an important factor to take into consideration. This means that the ejection method needs to ensure that the die is not subjected to any stress, bending, force, or warpage, especially since the technical nodes are approaching the angstrom regions, making the transistors more sensitive to damage. Figure 8.32 shows issues encountered in the production floor when the bonder is not set up correctly, resulting in die damage or generation of particles. The complexity of the ejection method, however, extends itself further

FIGURE 8.32 Issues with ejection system setup: touching dies if ejection center not aligned with die center (*left*), touching dies if ejection system is not parallel to film frame (*right*).

beyond the ejection module, to the media used to supply the dies (film frame, carrier with film, carrier with spin coated TBM, etc.), any pre-processing needed (UV irradiation, IR exposure or thermal application), any stress relief applied to the wafer, the singulation method, and many more.

8.5.2.1 Needle Ejection

In needle ejection (Figure 8.33), the ejection module has a series of needles which can be plastic or metal tipped. These are spaced in such a way that they cover most of the die backside for support but are sufficiently spaced such that when vacuum is applied, the tape can be sucked in-between the needles to reduce the adhesive contact area with the die backside. This is the most common type of die ejection and not the preferred type for thin dies (<50 μm).

8.5.2.2 Multistage

In the multistage tool, the tool consists of up to three concentric stages which are individually controlled (Figure 8.34). All stages are raised in order to start the peeling process. The middle and inner stages are raised further sequentially so as to initiate die edge peeling. The inner stage is raised even further until only a small

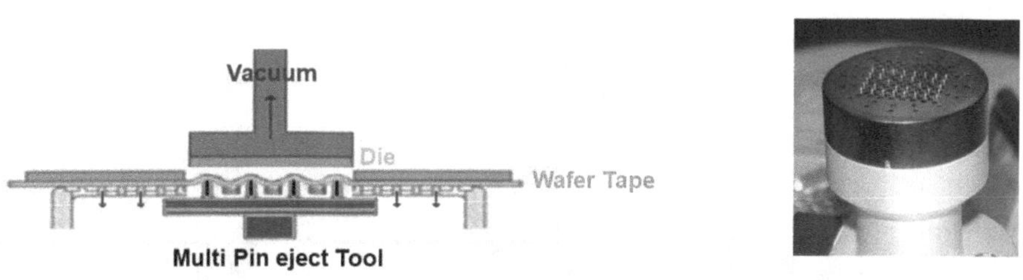

FIGURE 8.33 Multi-pin ejection module: schematic of die release (*left*); actual photo of ejector tool with needles raised (*right*).

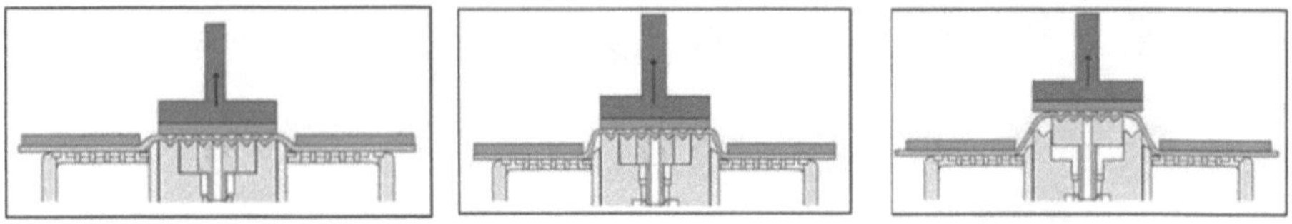

FIGURE 8.34 Multistage ejector process flow.

central area is making contact with the die backside. Since the contact area with the tape is reduced drastically, the residual adhesion is low enough such that die pickup can be carried out.

8.5.2.3 Multi-disk

In the multi-disk tool, the ejector module consists of a series of disks or blades which are individually controlled (Figure 8.35). The process starts with the disks fully retracted such that the die on the film frame is making contact with the flat part of the ejector module. The disks then all extend to the programmed top height, and symmetrically the disks at the outer edge start recessing, such that a peeling is initiated at the edge of the die. This disk recessing continues in pairs on opposite sides until only the middle disks remain, which are lowered as the last step before die detach and pickup.

8.5.2.4 UV Release

Special tapes drastically reduce the adhesion when exposed to UV radiation of a certain wavelength and intensity for a defined duration. This allows high adhesion for pre-processing steps, such as singulation and wet cleaning, and low adhesion prior to pickup to ensure low stress during the ejection process. This is one of the most preferred methods for ejection; however, care needs to be taken to ensure tape compatibility with wet clean (especially if chemicals are used), and compatibility with plasma activation. UV release also exists for glass carriers; however, this is typically laser based, not lamp or LED based.

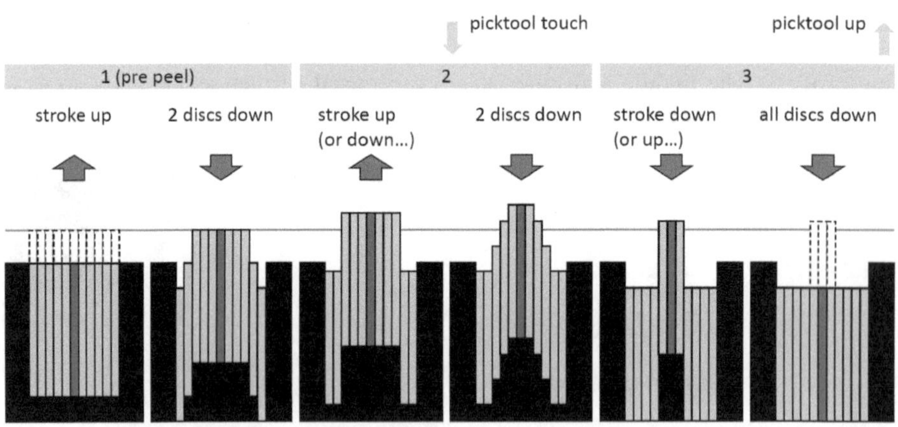

FIGURE 8.35 Multi-disk ejection tool process flow.

8.5.2.5 IR Release

This method is similar to UV release; however, an IR source is used. The advantage is that this method allows the use of silicon carriers as a medium carrying the source dies, since IR at certain wavelengths can pass through silicon.

8.5.2.6 Thermal Release

Thermal release is another method of ejection, whereby the ejection module has a flat head that can be heated. Once the film is heated, the tape forms bubbles and foams, such that the surface area in contact with the die backside decreases, thus decreasing the adhesion strength. A wafer-level thermal release solution also exists and is used in the market. Due to oxidation of the copper pads, this method is not the preferred approach for HB.

For peel energy measurement, as foil suppliers use different peel angles for their measurements, the below equation to convert peel forces to peel energies is ideally used:

$$G = \frac{F}{b}(1 - \cos\varphi)$$

where φ = peel angle; F/b = force per width; and G = peel energy. Peel energies of a few J/m^2 are state of the art. As a rule of thumb, the higher the required peel energy, and the thinner the die, the lower the process speed must be adjusted to.

8.5.2.7 Component Orientation

The process flow for HB follows two scenarios: face-up or face-down bonding. This means that one can either place the die active side up (face up) on the target wafer, or place the die active side down (face down) on the target wafer. The components are typically supplied to the bonder active side up on the film frame or carrier. Face-down is the approach used for D2W or D2D HB, whereby the die is picked, flipped by 180°, and then bonded onto the target wafer, as illustrated in Figure 8.36 (top). Face up is the approach employed for CoD2W, where the dies are picked and placed with the active side facing up, so that when the wafer carrier reaches the W2W bonding step, the wafer is flipped by 180° and the active side will come in contact with the

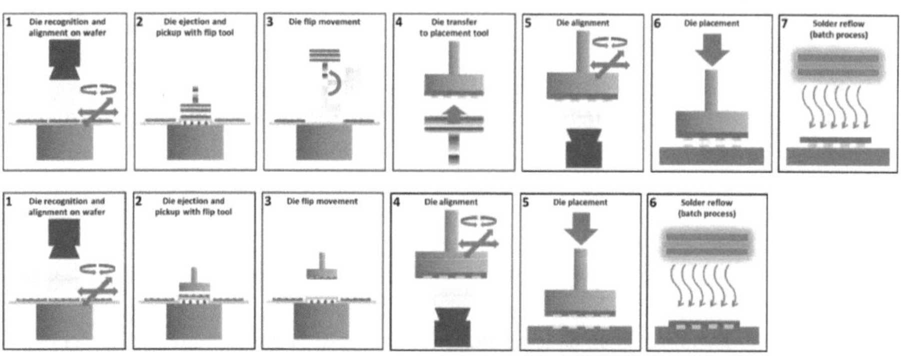

FIGURE 8.36 Process flow with flip (*top*) and process flow with no flip or direct pick (*bottom*).

target wafer. This means that the dies need to be picked directly and bonded, or else double flipped, a technique used in some bonders where direct pick is not possible because of physical constraints.

8.5.3 Actual Bonding Process

At the core of the HB process is the bond front propagation. This needs to be initiated in a single place; otherwise, having multiple bond fronts could lead to void entrapment, as seen in Figure 8.37. When bonding with a flat tool, the point of contact is uncertain, as tolerances in tool manufacturing, bond head tilt and die warpage all impact which point will make initial contact with the substrate wafer.

Theoretically, for the D2W HB scenario, one could obtain a successful void-free bond by initiating the bond front propagation in several ways, as illustrated in Figure 8.38: making contact with one edge of the die, making contact with one corner of the die, or making contact at the center of the die. The latter requires die shaping, but it is the most employed mode of operation for D2W HB, since it has advantages over the other two schemes. The tooling needs to ensure that the die is not damaged when shaped and that any stretching is not transferred to the die after bonding, resulting in loss of accuracy or bond pad overlay.

Tools exist which can actively shape the die to obtain different die shaping. The amount of die shaping is critical since it defines the area of contact during touch-down and to some extent also the speed of the bond front. Figure 8.39 (right) shows varying degrees of die shaping for 50 μm and 100 μm dies. For the 100 μm die, the maximum warpage is lower than the 50 μm die, which is expected due to the larger amount of bulk silicon. The important point is that the pseudo-colored images in

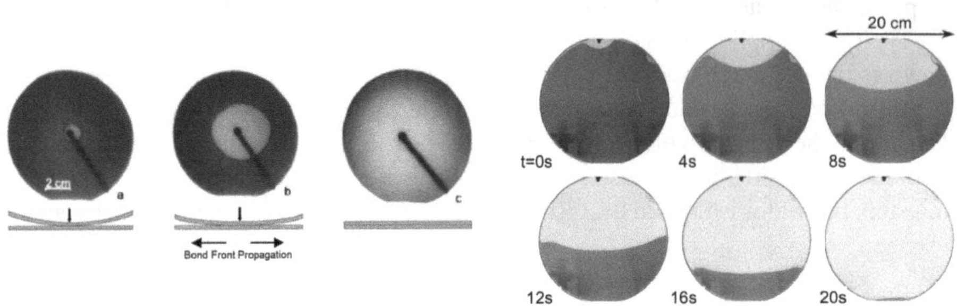

FIGURE 8.37 Initiation of direct bonding on wafer, showing bond front propagation from center toward wafer periphery (*left*) [22], and bond front propagation from edge toward opposite edge (*right*) [23].

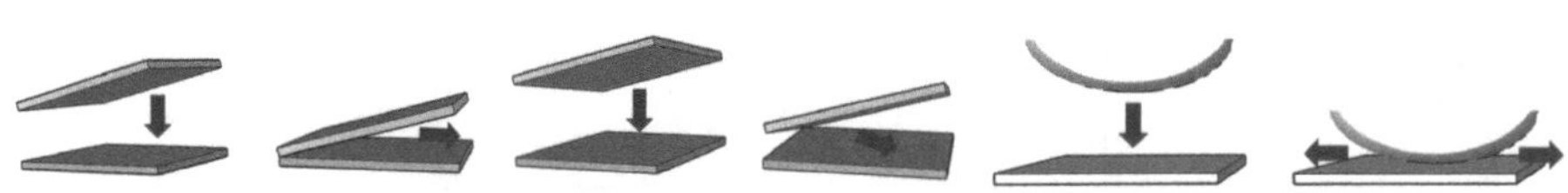

FIGURE 8.38 Bonding schemes for bond front propagation control: initial contact at one edge (*left*), initial contact at one corner (*middle*), initial contact in the middle of the die after shaping (*right*).

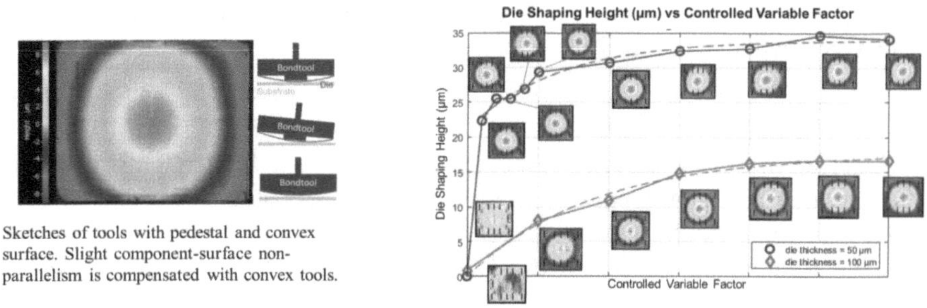

FIGURE 8.39 Sketches of tools with pedestal and convex surface (*left*), and maximum warpage height for 50 μm and 100 μm thick dies at different die shaping parameters (*right*) [18].

Figure 8.39 for each point show that the highest point of die shaping occurs at the center (range part), thus ensuring that the initial point of contact is occurring in the middle of the die.

8.6 CHARACTERIZATION

To determine whether bonding was successful so as to determine the best-known processes as well as to detect failure modes and understand the root causes, characterization is needed. This section aims to provide an overview of the main methods of characterization.

8.6.1 IR Microscope

For TCB, X-ray was the choice of inspection to determine placement accuracy and detect possible defects. For HB, the sub-10 μm bond pad pitches provide a challenge and infrared inspection is the method of choice due to the higher resolution offered.

Figure 8.40 shows a typical IR image of the die corner after placement. The outer ring is the fiducial on the substrate/target wafer, and the inner ring is the fiducial on

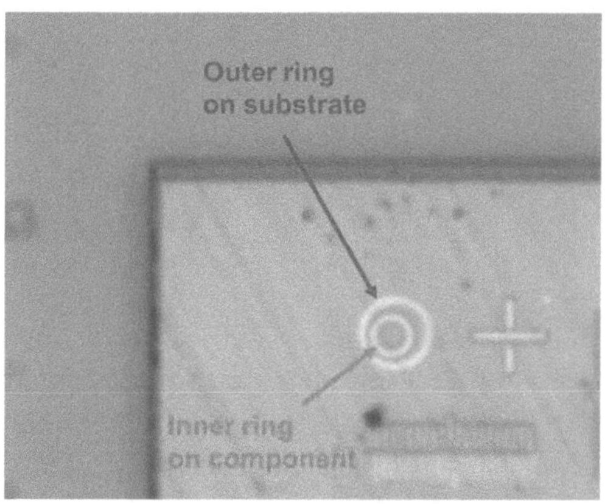

FIGURE 8.40 Post-bond infrared image of the die corner.

the component/die. The accuracy is determined by comparing the position of the two fiducials relative to each other.

8.6.1.1 Why Infrared Light?

The bandgap of Si is 1.14 eV, corresponding to a wavelength of roughly 1,100 nm. Thus, the material shows significant absorption in the visible wavelength regime (400—800 nm). Since Si is an indirect semiconductor, the absorption drops quickly with increasing wavelength beyond the bandgap. For this reason, IR microscopy can be used for post-bond analysis of Si-based materials. Figure 8.41 shows the absorption spectra for thicknesses between 100 μm and 2 mm.

In contrast to Si, copper (like most other metals) shows strong absorption in the infrared regime. Therefore, alignment marks should ideally be placed inside a so-called keep-out-zone (KOZ) without any patterning. However, materials real estate is very precious, thus requiring maximizing the number of transistors or RDLs. Outside the KOZs with metal patterning present, the IR microscopy has to be thoroughly tested and a study for finding the IR wavelength with the best performance needs to be done.

Typical state-of-the-art values for placement accuracy are depicted in Figure 8.42, showing ±108 nm @ 3σ for bonding on a temporary bonding material in the CoD2W approach, and ±263 nm @ 3σ for direct bonding of SiO_2 on SiO_2 dielectric, measured with an external, automated IR microscope.

The overlay between two wafers has a complex dependence on many parameters including the alignment accuracy, mismatch of exposed patterns, topology of each wafer, wafer distortion, and the bonding process conditions. The consistency of the overlay fingerprint is also dependent on several sources of variation from the incoming wafer, such as oxide properties and CMP strategies, as well as the repeatability of the bonding equipment [19].

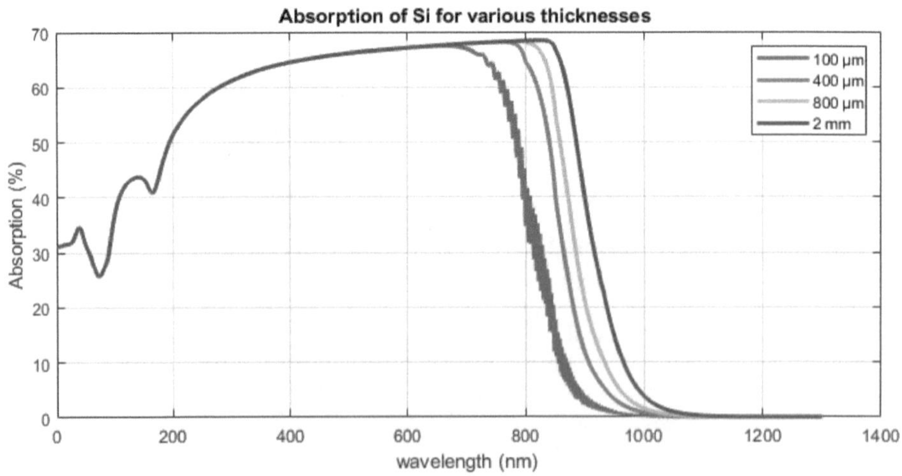

FIGURE 8.41 Absorption spectra of Si for various thicknesses of Si.

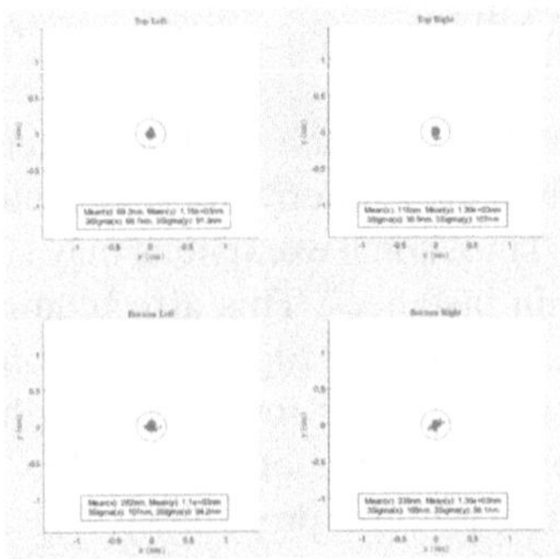

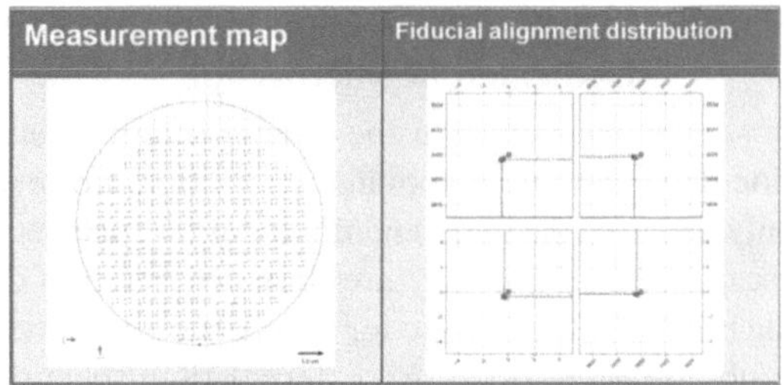

Parameter	Mean ± 3Sigma
Translation X (µm)	± 0.263
Translation Y (µm)	± 0.169

FIGURE 8.42 Worst corner placement accuracy: On a temporary bonding material (TBM) showing ±108 nm @ 3σ (*left*); using direct bonding showing ±263 nm @ 3σ (*right*).

8.6.1.2 Alignment Marks on the Substrate and on the Component in Different Focal Planes

Another aspect to consider is that the fiducial marks on the die and the fiducial marks on the substrate wafer are on different *z* heights and therefore the path traveled by the IR light is different, and this needs to be accounted for. Typically, a *z*-movement is required due to the depth of field (DOF) limitations, and this *z*-movement in the optics will induce some slight lateral movements, which contributes to the overall error of the system (Figure 8.43). Coupled to this issue is that the images will be taken sequentially rather than simultaneously. As such, special algorithms are needed to compensate for the latter two issues.

One other concern arises if the alignment fiducials on the substrate wafer are outside the die area. In this case, the IR light to detect the die fiducial will travel through air and then silicon, whereby for the substrate fiducial it will only travel through air. This will impact the data being measured and needs special algorithms to adjust for. Any non-orthogonality between the die surface and the IR camera will cause the IR light to travel at an angle and this will result in inaccurate readings. Finally, the dielectric layer itself also has an influence since its absorption capability will be different from that of silicon. Since hybrid bonding spans over various dielectric types, the IR system needs to be able to accommodate these scenarios.

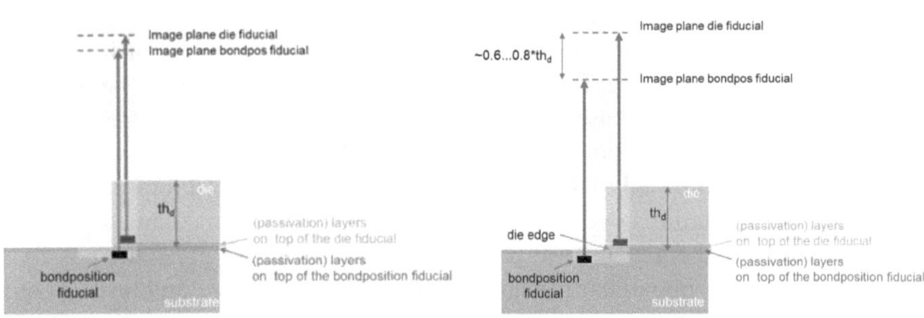

FIGURE 8.43 Visualization of the focal plane differences between bond position and die fiducial: when the bond position fiducial is underneath the die (*left*) and next to the die (*right*).

8.6.1.3 In Situ IR Inspection Inside the Bonder versus Dedicated IR Microscope

In situ IR inspection in the bonder offers several advantages such as timely excursion detection and automated correction; however, it is a CPH detractor and should only be used sparingly and not for 100% inspection. Thus, some excursions may still escape detection if they are only occurring in certain instances. This also requires the bonder to have space for such a camera, optics, and lighting setup and that proper calibrations are done. An external IR inspection scheme allows for a much higher sampling size for inspection to be carried out in parallel to the other activities of pre-preparation and loading/unloading, and this increases the probability that all defects are detected and corrective measures taken. The collection of this data could also be used to create a database which, when coupled to an AI system, can predict future excursions.

8.6.2 Confocal Scanning Acoustic Microscopy (CSAM)

One common method of determining if a bond is successful or not is by visualizing voids by CSAM, as seen in Figure 8.14. Factors addressed here are the size, the quantity, and the position of the voids. Voids are commonly classified as normal voids and micro-voids, and typical sources of voids are particles, air entrapment, outgassing during annealing (especially if the annealing temperature is higher than the dielectric deposition temperature), and also front-end defects (e.g., between the pad and the via).

Delamination between the copper pads can also manifest itself as voids, and this could stem from a number of reasons:

a. The annealing temperature is not high enough to promote grain growth.
b. The orientation of the copper microstructure is not optimal.
c. The microstructure of the copper pad (grain size, metastable, and nano-twinned) is not optimal [24].

8.6.3 Electrical Tests

Test vehicles typically employ daisy chain schemes in order to test the bonded dies and even subject them to reliability tests. For end products with fine pitches, traditional testing with probing methods have reached their limit in both sensitivity to the bond pads and the probe size. Figure 8.44 illustrates damage to Cu probe pads after testing. When using solder-based bonding, this damage can be overcome, as long as it does not affect the actual bond pads [25].

Some methods proposed to deal with testing of hybrid-bonded fine-pitch devices include the use of dummy bond pads, and this ensures that if some damage is caused during touchdown, the actual bond pads remain intact. It also gives the possibility to have the test pads manufactured larger to meet the probe size and testing equipment accuracy specifications in the market. Assembly of KGD, although expensive, becomes paramount and outweighs the cost of this extra step since the final product will have a higher yield and potentially fewer testing steps.

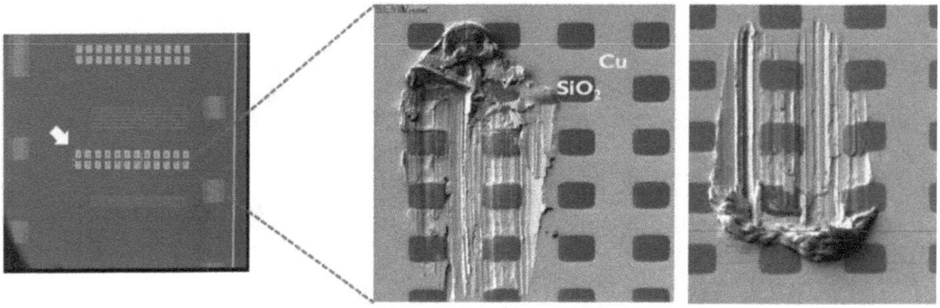

FIGURE 8.44 Smearing on copper pad after probe testing [26].

In the case of D2W hybrid bonding, this type of damages causes bond failures, even if these pads are for testing purposes only. The protruding metal will cause large hybrid bond voids, several orders of magnitude larger than the damage size. For hybrid bonding, currently the only option is to perform testing *before* applying the final hybrid bond pad layer. This is actually a challenging topic when it comes to D2W hybrid bonding.

8.6.4 Bond Strength

In standard die attach packaging involving glues, epoxies or solder, the industry standard test for bond strength is the die shear. This consists of having a die shear ram placed parallel to the die edge with a slight elevation of a couple of micrometers, and then moved laterally until a fracture occurs (Figure 8.45). The problem with HB samples is that even at time zero, before annealing, the bond strength can be strong enough that silicon fracture occurs rather than interface separation.

Some other methods have also been proposed. One such method involves a pull test, as seen in Figure 8.46. The bottom silicon coupon singulated from the substrate

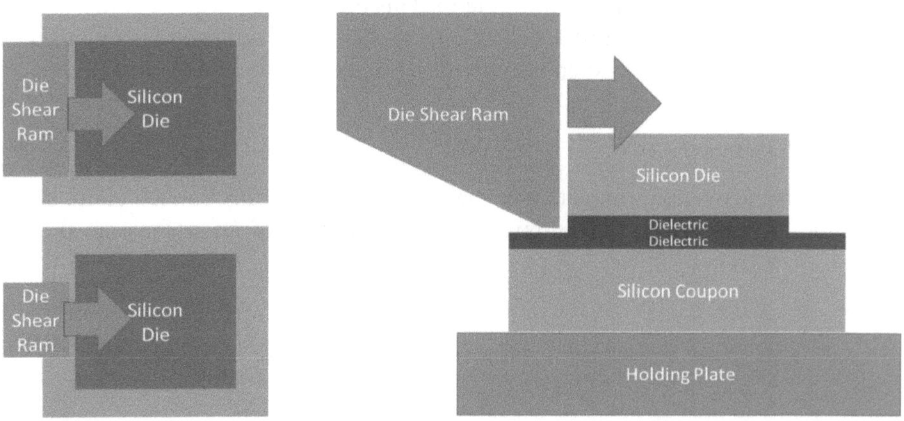

FIGURE 8.45 Die shear schematic: correct setup (*top left*), incorrect setup (*bottom left*), and side view (*right*).

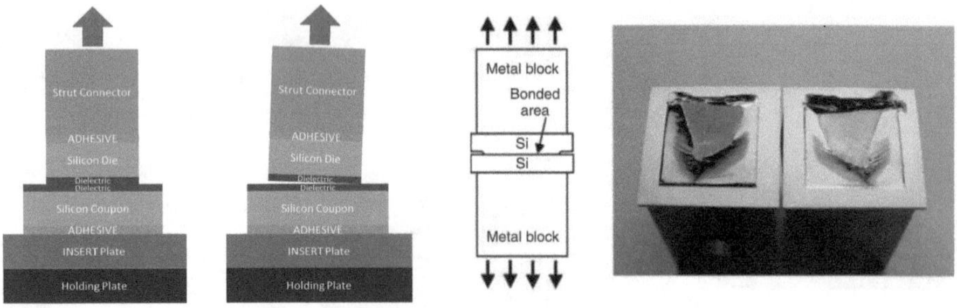

FIGURE 8.46 Pull test for bond strength characterization: schematic (*left*), and fracture surface of Si/Si bonding prepared at room temperature by surface activated bonding (SAB) (*right*) [27].

is glued to an insert plate, which in turn is held to the testing equipment via the holding plate. The top die backside is glued to a strut connector, which in turn is connected to the testing equipment's pulling axis. By pulling and monitoring the force required, one can get an indication of the bond strength. One particular issue with this method is that if there is slight non-parallelism, which could easily arise if the equipment is not rigid enough or if the glue applied is not even, then a peeling effect is achieved rather than a pulling effect. This method also requires singulating the bonded wafer and processing the adhesive to fully cure.

Another method proposed is a bend test (called the M3PB test). A coupon is singulated from a bonded wafer consisting of three dies. This coupon is then subjected to a three-point bend test, as seen in Figure 8.47, and the results are used to characterize the bond strength. As with the previous method, this test requires coupon singulation.

Another method consists of hybrid bonding two dies at an offset, mounting them vertically on a die shear equipment, and then performing the die shear on one die as depicted in Figure 8.48. This method leverages the single cantilever beam method and gives an indication of the bond strength. This method may not be appropriate for dies with thickness below 150 μm, as the test specimen is not rigid enough to bear the bonding strength, thus potentially resulting in die fracture during test [29].

8.6.5 Plasma Efficacy

After plasma activation, contact angle measurement is typically carried out to determine the process efficacy. This involves placing a drop of DI water onto the die

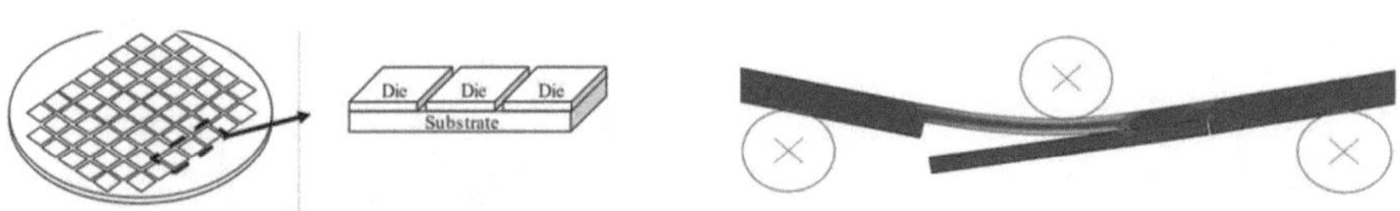

FIGURE 8.47 Sample preparation by singulating coupons from a bonded substrate wafer (*left*), and bend test showing delamination between the center die and the substrate (*right*) [28].

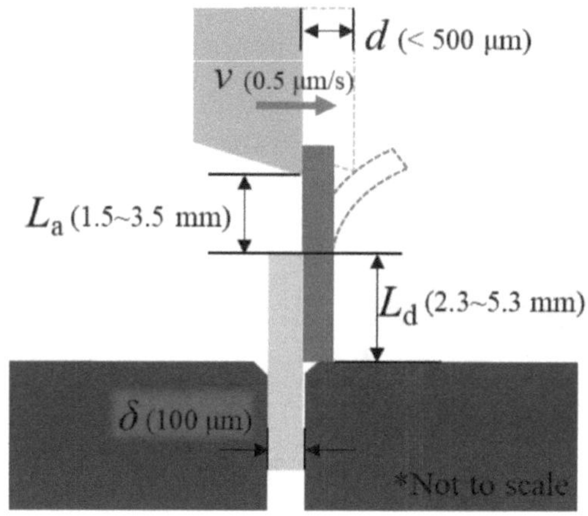

FIGURE 8.48 Schematic of D2D bonding strength [29].

or wafer, and then measuring the angle it makes with the substrate wafer surface (Figure 8.49). The smaller the angle, i.e., the flatter the water droplet shape, the better, since it means that there is a lower surface tension.

8.6.6 Scanning Electron Microscopy (SEM)

One way to analyze the bond quality is by using SEM. The material needs to be ground and polished to the desired section, and then subjected to an SEM analysis (Figure 8.50). This high-magnification equipment is capable of providing very detailed images. SEM images can be used to verify if fusion has occurred, if voids are due to particles, and to check placement accuracy, among other things. For the latter, both *x* and *y* planes need to be checked. This is a destructive method and takes ample time to prepare the samples and get results.

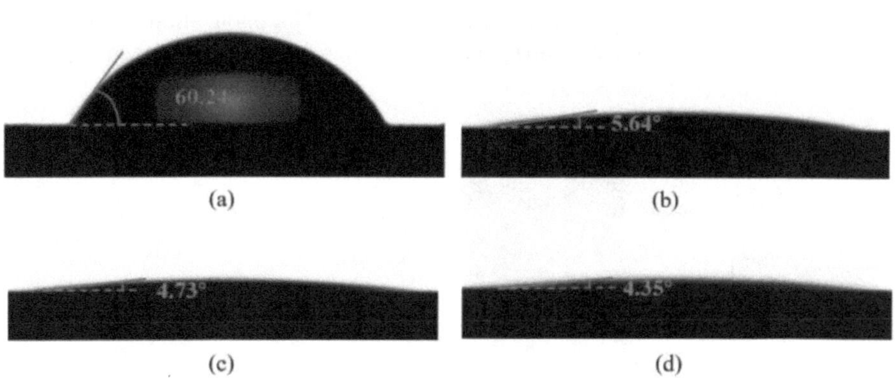

FIGURE 8.49 Contact angle test showing high surface tension (a), and low surface tension (b–d) [30].

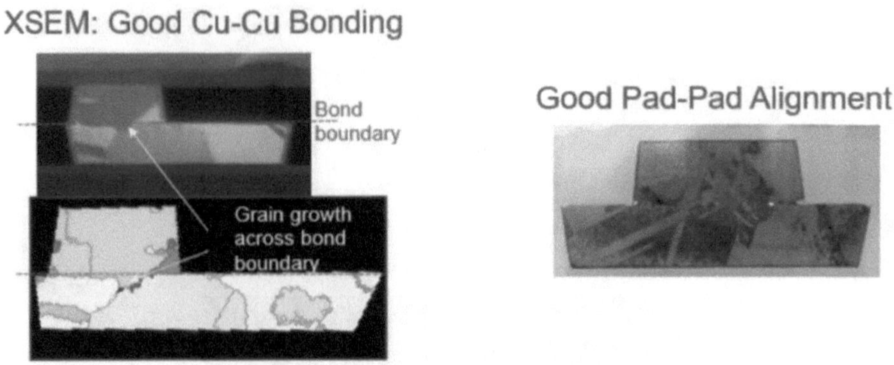

FIGURE 8.50 SEM image showing successful fusion and grain growth across boundary (*left*); SEM image showing cross section of high accuracy placement (*right*) [18].

8.6.7 Atomic Force Microscopy (AFM)

AFM, with a very small FOV, offers unprecedented levels of detail. AFM is typically used to measure surface roughness, surface profile, and bond pad dishing, among other elements.

8.7 END PRODUCT APPLICATIONS

HB has been around in HVM for several years, predominantly using W2W. In this section, some end products which are being mass produced and available on the market will be shown.

8.7.1 CMOS Image Sensor (CIS)

The earliest applications of HB include CIS stacked sensors, and the latest stacked CIS is a three-layer stack combining a circuit section at the bottom, a memory layer in the middle, and a top pixel section (Figure 8.51). The photo diode is at the top level rather than below the circuitry, and this enables faster shutter speeds. Quicker

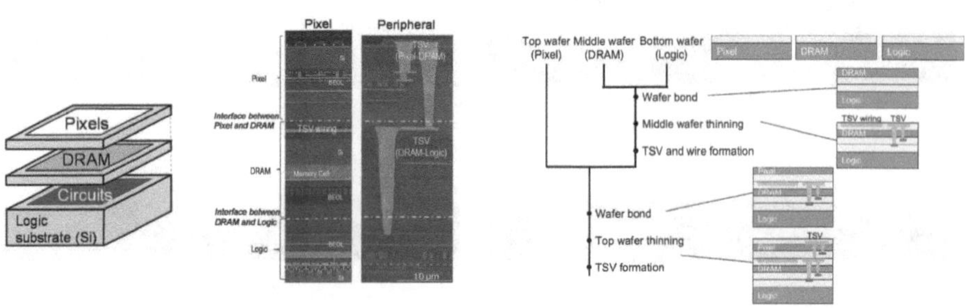

FIGURE 8.51 Details of three-layer CIS sensor (*left*); process flow for assembly of the CIS sensor (*right*) [31].

readout speed also allows for faster autofocus response for higher burst rates with autofocus. An additional DRAM layer is stacked to speed up image processing and data transfer. This stacking procedure makes the W2W bonding scheme more challenging since one needs to match the yields of three wafers, and therefore could benefit from a D2W or CoD2W process flow.

8.7.2 NAND Memory

Another application is stacked NAND memory, which employs W2W bonding as well [32]. This process flow bonds the peripheral circuitry face-to-face with the memory array (instead of alongside as in conventional schemes). The peripheral circuitry for the memory cell operation and I/O uses CMOS logic technology whose node matches the I/O speed and functions requirements and is manufactured on a separate wafer. The memory array wafer is then bonded to the periphery wafer by billions of metal Vertical Interconnect Accesses (VIAs), as shown in Figure 8.52. The advantages are minimization of the die area as well as improved latency. As the NAND product moves to smaller tech nodes and yields decrease, especially in the initial phases until the node matures, D2W or CoD2W bonding would definitely offer advantages.

8.7.3 Computing

W2W HB bonding of DRAM to logic has been employed in a product [33] which promises to revolutionize the crypto mining market by offering a more efficient and cost-effective solution as seen in Figure 8.53. By integrating the data storage unit

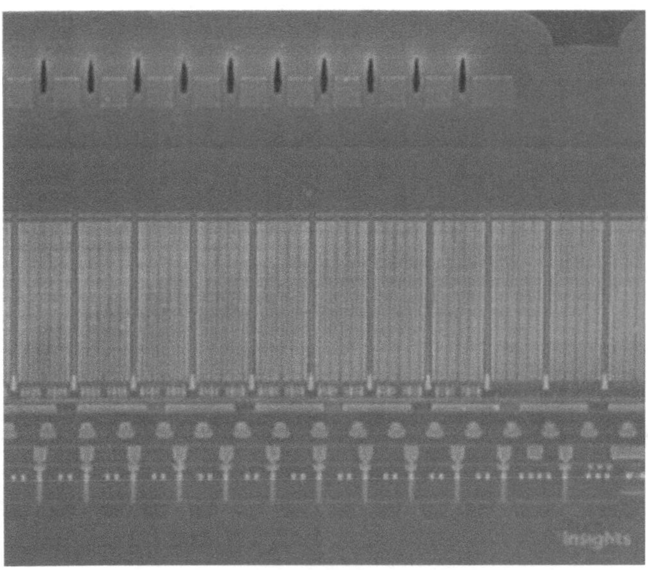

FIGURE 8.52 Xtacking SEM image of cross section [32].

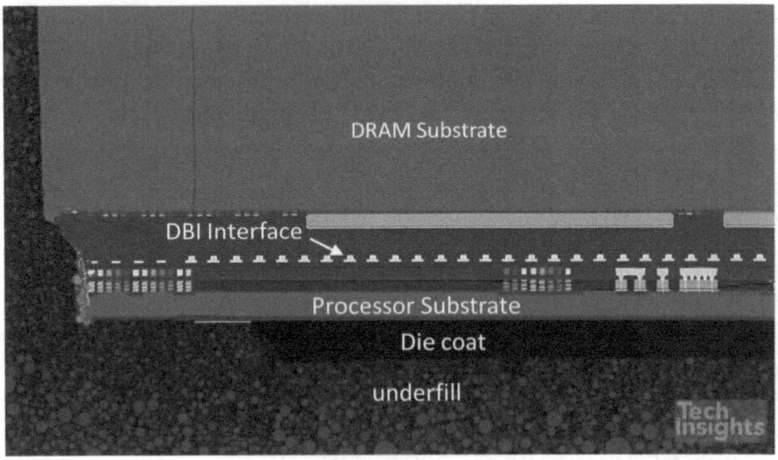

FIGURE 8.53 Application showing DRAM hybrid bonded onto logic. (Adapted from Ref. [33].).

and the computing unit on the same chip, a dramatic improvement in energy efficiency and performance can be achieved. The choice of utilizing HB as the interconnect technology is based on several facts, such as the lack of solder bumps, and no underfill is needed which simplifies the packaging process and provides better performance. The die thicknesses could also be reduced offering possibility for more stacked DRAM layers. And finally, the DRAM employed could also have smaller pitches resulting in a higher bandwidth.

Separately, in a product for gaming applications as well as servers, AI, and data centers, a cache memory die is hybrid bonded (D2W) onto a compute die (Figure 8.54) [32, 34, 35], utilizing fusion-bonded structural dies for heat dissipation and warpage control, and hybrid bonded memory to logic with 2.1 μm bond pads.

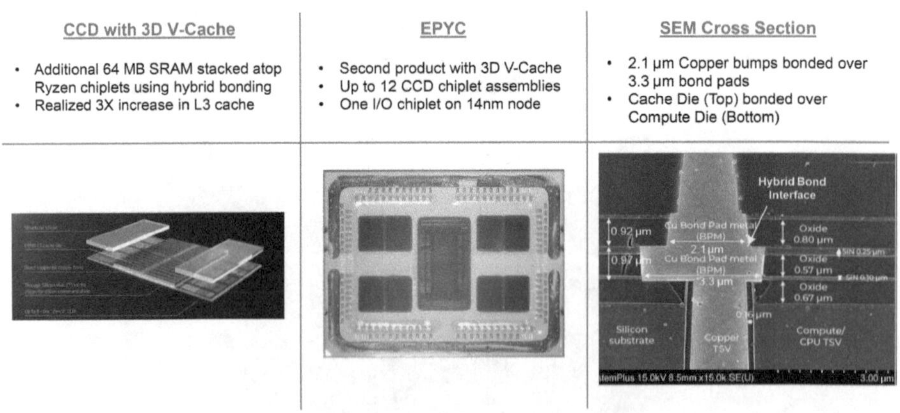

FIGURE 8.54 Cache die on compute die: exploded view (*left*) [32]; server edition with eight dies (middle) [34]; SEM cross section (*right*). (Adapted from Ref. [35].)

8.8 OUTLOOK

With the ever-increasing market demand for higher compute power and bandwidth, the market is looking at HB to be able to provide an interconnect technology capable of reliable sub-10 μm pitch wafers. The capability of combining different materials such as III–V to silicon or glass opens up a whole new field of applications to be explored with limitless potential.

Solder-based bumping in the industry is approaching the sub-20 μm pitch level today. Research at Imec has shown the possibility to extend the solder-based micro-bump roadmap further, down to about 5 μm pitch. This requires thermocompression bonders with a bonding accuracy of 1 μm [36, 37].

The emerging D2W HB technology has the potential to go significantly below the solder-based interconnect pitches. Already, interconnect pitches below 10 μm, down to 3 μm, have been reported [38].

As direct D2W bonding is performed at room temperature without large forces, very high placement accuracy can be obtained. This opens the option to reach down to 1 μm D2W bonding when reaching an absolute placement accuracy of 200 nm. This will allow for a much deeper 3D interconnect solutions as compared to the currently discussed 2.5D chiplet approaches [39].

Although D2W HB is still in its infancy, there are already products in the market in high volume in the top-level advanced packaging which are employing D2W HB, and it is only a matter of time before more products hit the market. The emergence of HB cluster tools in the market that integrate the bonder with wet clean, plasma, degas, and UV film frame irradiation modules also makes it easier for OSATs (outsource semiconductor assembly and test) to adopt such a technology, making it easier to proliferate the technology throughout the market.

ACKNOWLEDGEMENT

We would like to extend our sincere thanks to Applied Materials for their help and input, particularly the numerous joint trials and analysis that were carried out at the Singapore Centre of Excellence.

REFERENCES

1. M. Fuselier, “Design Leadership Imperative,” in The Growing Challenge of Semiconductor Design Leadership, 2022.
2. Chiplet, 28 February 2021. https://en.wikichip.org/wiki/chiplet?utm_content=cmp-true#google_vignette.
3. R. Swaminathan, “The Future of High Performance & Adaptive Computing,” in *ISES USA 2023*, 2023.
4. J. Abdilla, “An Insight into Die to Wafer Hybrid Bonding,” in Semi 3D & Systems Summit 2023, 2023.
5. C. Scanlan, “Interconnecting Chiplets,” in *NAVAT Chiplet Symposium*, 2023.
6. J.-J. Ong et al., “Low-temperature Cu/SiO_2 Hybrid Bonding with Low Contact Resistance Using (111)-Oriented Cu Surfaces,” *Materials*, vol. 5, no. 1888, p. 15, 2022.
7. F. Guyader, “3D Integrations for Smart Image Sensors,” in Semi 3D & Systems Summit, 2020.

8. S. Katoh et al., "Development of Fluxless Micro-bonding and Narrow Gap Filling Process," in *IEEE 73rd Electronic Components and Technology Conference (ECTC)*, 2023.
9. G. Gao et al., "Die to Wafer Hybrid Bonding: Multi-die Stacking with TSV Integration," in *2020 International Wafer Level Packaging Conference (IWLPC)*, 2020.
10. B. Hachemi et al., "Status of the Advanced Packaging 2023," Yole, 2023.
11. J. Abdilla, "Die to Wafer Hybrid Bonding: The Next Generation of Advanced Packaging," in Semi 3D & Systems Summit, 2022.
12. I. Pachenko et al., "Microstructure Development of Cu/SiO_2 Hybrid Bond Interconnects after Reliability Tests," *IEEE Transactions on Components, Packaging and Manufacturing Technology*, vol. 12, no. 3, p. 410–421, 2022.
13. S. Son et al., "Characteristics of Plasma-activated Dielectric Film Surfaces for Direct Wafer Bonding," in *IEEE 70th Electronic Components and Technology Conference (ECTC)*, 2020.
14. I. Fumihiro et al., "Influence of Composition of SiCN as Interfacial Layer on Plasma Activated Direct Bonding," *ECS Journal of Solid State Science and Technology*, vol. 8, no. 6, p. 346, 2019.
15. H.-S. Park et al., "Effect of Ar-N_2 Plasma Treatment on Copper Surface for Cu-to-Cu Wafer Bonding," in *IEEE 20th Electronics Packaging Technology Conference (EPTC)*, 2018.
16. B. Hwang et al., "A Study on the Surface Activation of Cu and Oxide for Hybrid Bonding Joint Interface," in *IEEE 73rd Electronic Components and Technology Conference (ECTC)*, 2023.
17. A. Shantaram, "Die-to-Wafer Hybrid Bonding to Address Next-Gen Electronics Packaging Challenges," in Semicon West 2023, 2023.
18. J. Abdilla, "Chip to Wafer Hybrid Bonding Development for High Volume Manufacturing," in IMAPS Device Packaging Conference, 2023.
19. T. Sherwood et al., "Aggressive Pitch Scaling (sub-0.5 µm) of W2W HB through Process Innovations," in *IEEE 73rd Electronic Components and Technology Conference (ECTC)*, 2023.
20. B. Brandstaetter et al., "High-speed Ultra-accurate Direct C2W Bonding," in *2020 IEEE 70th Electronic Components and Technology Conference (ECTC)*, 2020.
21. R. Ashton et al., "White Paper 2: A Case for Lowering Component-level CDM ESD Specifications and Requirements," 2021.
22. A. Plößl et al., "Wafer Direct Bonding: Tailoring Adhesion between Brittle Materials," *Materials Science and Engineering*, vol. 25, pp. 1–88, 1999.
23. E. Navarro, "Direct Wafer Bonding Dynamics," Université de Grenoble, 2014.
24. R. Schmidt, "Optimisation of Cu Microstructure to Improve Cu-to-Cu Direct Bonding," in Semi 3D & Systems Summit, 2023.
25. E.J. Marinissen et al., "Direct Probing on Large-array Fine-pitch Micro-bumps of a Wide-I/O Logic–Memory Interface," in *International Test Conference*, 2014.
26. Y. Lin, "Collective Die-to-Wafer HB Integration," in H2 IMEC Partner Technical Week, 2022.
27. H. Tagaki, "Direct Bonding," in *3D and Circuit Integration of MEMS*, Wiley, 2021.
28. K. Baek et al., "Chip Level Evaluation of Wafer-to-Wafer Direct Bonding Strength with Bending Test," in *IEEE 73rd Electronic Components and Technology Conference (ECTC)*, 2023.
29. J. Kim et al., "Novel Characterization Method of Chip Level Hybrid Bonding Strength," in *IEEE 73rd Electronic Components and Technology Conference (ECTC)*, 2021.
30. D. Li et al., "Effect of Combined Hydrophilic Activation on Interface Characteristics of Si/Si Wafer Direct Bonding," MDPI, 2021.

31. H. Tsugawa et al., "Pixel/DRAM/Logic 3-Layer Stacked CMOS Image Sensor Technology," in *International Electron Devices Meeting (IEDM)*, 2017.
32. J. H. Lau, "Recent Advances and Trends in Cu–Cu Hybrid Bonding," *IEEE Transactions on Components, Packaging and Manufacturing Technology*, vol. 13, no. 3, p. 420, 2023.
33. TechInsights, "Samsung's 3 nm GAA Process in Whatsminer ASIC," 2023.
34. Epyc, Wikipedia, June 2023. https://en.m.wikipedia.org/wiki/Epyc (accessed June 2023).
35. B. Dube et al., "AMD 3D V-Cache with TSMC SoIC Packaging," Yole SystemPlus, 2022.
36. J. Derakhshandeh et al., "10 and 7 μm Pitch Thermo-compression Solder Joint, Using a Novel Solder Pillar and Metal Spacer Process," in *IEEE 70th Electronic Components and Technology Conference (ECTC)*, 2020.
37. J. Derakhshandeh et al., "Low Temperature Backside Damascene Processing on Temporary Carrier Wafer Targeting 7 μm and 5 μm Pitch Microbumps for N Equal and Greater Than 2 Die to Wafer TCB Stacking," in *IEEE 72nd Electronic Components and Technology Conference (ECTC)*, 2022.
38. C.-T. Wang et al., "SoIC_H Technology for Heterogenous System Integration," in *IEEE Symposium on VLSI Technology and Circuits (VLSI Technology and Circuits)*, 2022.
39. E. Beyne et al., "3D SoC Integration, beyond 2.5D Chiplets," in *IEEE International Electron Devices Meeting (IEDM)*, 2021.

9 Design for Hybrid Bonding and Chiplets

Brandon Wang and Rong Bao

9.1 HYBRID BONDING: ENABLING ADVANCED CHIPLET-BASED DESIGNS

System-of-Chiplets design is a trend in modern electronics that involves creating a system by combining smaller, simpler "chiplets" that each performs a specific task, envisioned as system-of-chiplets building blocks akin to Legos. Compared to traditional monolithic system-on-chip (SoC), chiplet-based architectures provide numerous advantages, such as enhanced performance, decreased power consumption, and greater design adaptability. The emerging hybrid bonding technology, which is an advanced chip-to-chip interconnection method that combines the benefits of both copper and dielectric bonding, makes chiplet-based architecture design even more attractive. It provides promising solution for chiplet-based designs and 3DIC with essential benefits, including high interconnect density, fine-pitch capability, and improved signal integrity.

This section delves deeper into the advantages of hybrid bonding–enabled chiplet-based designs, examining their various benefits and implications for the electronics industry.

9.1.1 Overcome Physical Limits

As the semiconductor technology continues to advance, we are approaching the physical limits of transistor miniaturization. Moore's law, which drives the entire semiconductor industry for more than past half century, is slowing down recently, as shown in Figure 9.1.

On the other hand, the size of a single die is constrained by the lithography reticle limit, which affects the maximum area that can be exposed during the semiconductor manufacturing process. It also lowers the yields while the die size is at or close to the largest allowed size. As more computing power is required in modern chips, the size of advanced chips is hitting the reticle limit now, as shown in Figure 9.1.

Chiplets provide an alternative, allowing manufacturers to leverage the benefits of advanced nodes while circumventing size constraints. Chiplets overcome this limitation by dividing a large system-on-chip into smaller, interconnected units that can be manufactured separately and combined later. Hybrid bonding technology plays a pivotal role in enabling the assembly of these chiplets into a cohesive system. It not only improves the interconnect density and signal integrity but also reduces the pitch size between the chiplets, allowing for the creation of larger and more complex systems, referred to as System of Chiplets (SoC).

 DOI: 10.1201/9781003408673-9

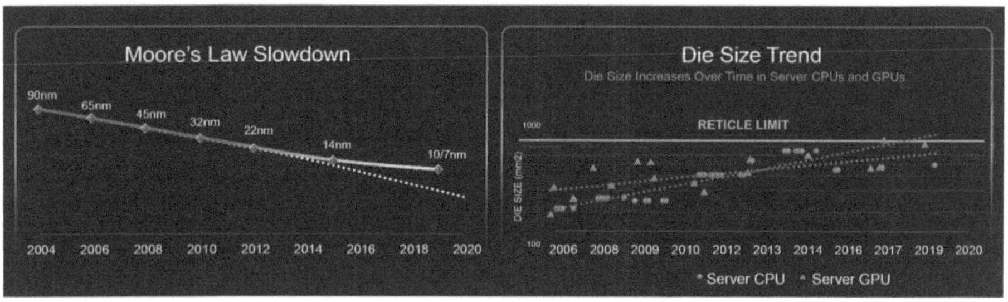

FIGURE 9.1 Physical limits on transistor size and die size [1].

9.1.2 Cost Reduction

Cost reduction is always the key and main driver of semiconductor industry moving forward. Chiplets can reduce costs in several ways throughout the design, manufacturing, and integration processes. Hybrid bonding technology plays a significant role in enabling these cost reductions by providing an efficient and reliable method for assembling chiplets into a complete system. By using standardized chiplets, manufacturers can reduce production costs by leveraging economies of scale and sharing development expenses across multiple products.

9.1.2.1 Lower Wafer and Manufacturing Cost

Chiplets allow system designers to choose the most cost-effective technology node for each component, optimizing for cost without sacrificing performance or power efficiency. Hybrid bonding technology further enhances cost reduction by providing a more reliable and efficient interconnect method. Hybrid bonding technology can simplify the overall manufacturing process due to its direct Cu–Cu bonding and reduced need for additional materials or complex processes. This simplification can lead to lower manufacturing costs and increased manufacturing throughput. Additionally, chiplet-based designs can potentially reduce packaging costs by using more cost-effective, multi-chip packaging techniques, such as advanced 2.5D and 3D packaging. By using standardized chiplets, manufacturers can maintain a smaller inventory of components, reducing inventory carrying costs and obsolescence risks.

9.1.2.2 Lower Development Cost

Chiplets simplify the design process by breaking it down into smaller, more manageable components. This can lead to reduced design costs, as well as shorter development cycles and faster time-to-market. Hybrid bonding technology can further streamline the design process by enabling tighter integration between chiplets. Pre-validated chiplets can be assembled more quickly than developing an entirely new monolithic chip, which reduces testing and validation costs. By using this modular approach, it shares development expenses, such as research, design, and validation, across multiple products.

9.1.2.3 Lower Re-spin Cost and IP Reuse

When a design error is discovered in a monolithic chip, the entire chip may need to be re-spun. With chiplets, however, only the affected component must be redesigned and remanufactured, resulting in significant cost- and time-savings. Hybrid bonding technology can contribute to further cost-savings by enabling a more precise and reliable interconnection between chiplets. Additionally, the modular nature of chiplets allows easy IP reuse across different designs. Chiplet-based designs can leverage preexisting and pre-verified intellectual property (IP) blocks, reducing the need for costly and time-consuming custom development. Chiplets enable the mass production of standardized components, which can be used across multiple products. This allows manufacturers to leverage economies of scale and reduce per-unit costs.

9.1.2.4 Improved Yield

Smaller chips generally have better manufacturing yields than larger, monolithic chips. Chiplets can be manufactured and tested individually, which improves yield and lowers costs by identifying and discarding defective components before assembly. Hybrid bonding technology can improve yield compared to microbumps or C4 bumps due to its fine-pitch capability, better mechanical stability, reduced thermal stress, and enhanced reliability. The combination of copper and dielectric bonding potentially offers superior performance and robustness, leading to fewer defects and higher overall yield rates. Higher yield rates mean fewer defects, less waste, and lower overall production costs.

9.1.3 Performance Improvement

9.1.3.1 Heterogeneous Integration

Chiplet-based designs enable heterogeneous integration, allowing designers to combine different types of processing elements optimized for specific tasks. For example, a high-performance CPU can be combined with an energy-efficient GPU, artificial intelligence (AI) accelerators, or memory controllers, each fabricated on its most suitable process technology. Hybrid bonding enhances the benefits of heterogeneous integration by providing more reliable and higher density connections between chiplets, further improving overall performance and efficiency compared to a monolithic chip design.

9.1.3.2 Improved Latency and Interconnect

Chiplet architectures often employ advanced interconnect technologies, such as high-bandwidth, low-latency interfaces, which facilitate faster communication between the chiplets. Hybrid bonding and TSV technologies enable silicon stacking in the *Z* direction, resulting in shorter total wire length, as illustrated in Figure 9.2.

9.1.3.3 Increased Compute Density

As hybrid bonding technology enables more efficient data communication in the *Z* direction, more silicon chips can be stacked, which means packing more functional units into a smaller area, providing greater processing power in a compact footprint.

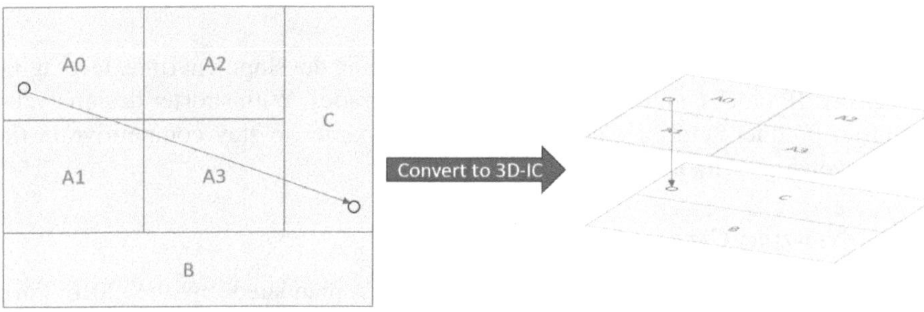

FIGURE 9.2 Hybrid bonding–enabled die stacking helps reduce total wirelength.

9.1.3.4 Reduced Power Consumption and Lower Thermal Cooling Cost

By enabling the use of specialized processing elements and more energy-efficient technologies, chiplet-based designs can achieve lower power consumption. Hybrid bonding further contributes to this by providing low-resistance and high-density interconnections, improving overall energy efficiency. This results in better performance per watt, a crucial metric in many modern computing applications, such as mobile devices and data centers. Chiplet-based designs also offer better thermal management due to their modular nature, as high-heat-generating dies can be split and heat can be more effectively dissipated across the chiplets, as only the thermal critical components need to be addressed. This helps maintain optimal operating temperatures, preventing thermal throttling and ensuring sustained high performance.

9.1.4 Scalability and Flexibility

Chiplet-based designs are inherently scalable and flexible, as they allow manufacturers to mix and match components to meet varying customer requirements. Individual components can be easily swapped or added without redesigning the entire system. This modularity, supported by hybrid bonding, makes it easier to adapt existing designs for new applications or to scale performance up or down as needed. It ensures consistent high performance with rapid adaption to changing market demands and technology advancements.

9.1.5 Smaller Size and Form Factor

By stacking smaller, interconnected chiplets instead of a single large system-on-chip, designers can create more compact, power-efficient systems that fit within tight form factor constraints. For example, stacking chiplets with hybrid bonding offers architectural and performance benefits for scaling augmented/virtual reality (AR/VR) models on highly resource-constrained edge devices. In contrast, two-dimensional off-chip memory interfaces are prohibitively energy intensive and bandwidth (BW) limited for AR/VR devices. To solve this, Meta proposes using advanced 3D stacking technology for high-density vertical integration to local memory and compute, increasing memory capacity within the same footprint at iso-BW with improvements in energy and latency [2].

9.1.6 Faster Time-to-Market

Chiplets help to accelerate time-to-market by reducing development time, leveraging pre-verified IP blocks, and simplifying system upgrades. With shorter design cycles and faster product iterations, chiplets enable companies to stay competitive in the rapidly evolving technology landscape.

9.1.7 Net-zero Carbon Emissions

The drive toward net-zero carbon emissions, as mandated by the 2016 Paris Agreement, is a global imperative that is shaping the future of various industries, including the semiconductor industry. The semiconductor supply chain, in particular, is under significant pressure to reduce its carbon footprint and contribute to global sustainability efforts. This is where multi-die chiplet designs can play a pivotal role.

Multi-die chiplet designs involve the use of multiple chiplets or dies linked together in a sophisticated package. This design approach provides several benefits that contribute toward sustainability and carbon neutrality. It allows for better silicon utilization, thus reducing waste during chip manufacturing. With the rise of AI/machine learning (ML) and the Internet of Things (IoT), the demand for high-performance, energy-efficient computation is growing. Multi-die designs can better cater to these demands as they allow for the integration of heterogeneous components, each optimized for specific tasks. This facilitates better energy efficiency as each task can be processed by the most suitable component, minimizing energy wastage. In conclusion, the shift toward multi-die chiplet designs is not only a response to the technical limitations of monolithic designs, but also a strategic move toward a more sustainable and carbon-neutral semiconductor industry.

In summary, hybrid bonding and chiplet-based designs offer transformative benefits. They significantly improve performance by enabling heterogeneous integration and employing advanced interconnects. They also reduce power consumption, enhance scalability, and improve manufacturing yield. Additionally, these designs offer rapid customization and upgradability, meeting the dynamic needs of various applications. Importantly, multi-die chiplet designs are a strategic response to sustainability challenges. By optimizing silicon utilization, facilitating energy-efficient computing, and supporting the shift toward edge computing, they play a crucial role in the semiconductor industry's push toward net-zero carbon emissions.

9.2 CHALLENGES AND NEW CONSIDERATIONS: FROM DESIGN PERSPECTIVES

9.2.1 Design Methodology and Additional Consideration

9.2.1.1 Readiness of Enablement Items and EDA Tools

To design integrated circuits (ICs), IC designers rely on two critical components: process design kits (PDKs) and electronic design automation (EDA) tools. As chiplet-based designs gain popularity and hybrid bonding technology emerges, specialized PDKs and EDA tools for 3D integrated circuits (3DICs) become necessary, as they play crucial roles in the design process.

PDK is a set of design tools, libraries, and files that are essential for designing ICs in a specific semiconductor fabrication process. PDKs are provided by semiconductor foundries and serve as the bridge between IC designers and the manufacturing process. They contain critical information about the process technology, design rules, device models, and libraries of predesigned components that can be used in the design process.

In the context of hybrid bonding–enabled 3DIC and chiplet-based design, there are many additional design rules and considerations that need to be addressed, compared to traditional chip design. The following are two typical examples:

1. *Warpage and Alignment in Hybrid Bonding:* 3DIC and chiplet designs involve stacking and connecting multiple layers or chiplets using hybrid bonding. This technology requires precise alignment between layers or chiplets due to the small bond size and high density of interconnections. Careful consideration of warpage and alignment is crucial to ensure proper electrical connections, signal integrity, and mechanical reliability. Enhanced electrical rule checks (ERC) are often required to verify alignment and maintain the high level of precision needed for successful hybrid bonding implementation.
2. *Hybrid bonding design rules:* Hybrid bonding is a critical interconnect technology for 3DIC and chiplet-based designs. Design rules related to dimensions, placement, and routing of hybrid bonds must be followed to ensure reliable connections between the layers or chiplets.

Foundries and outsourced semiconductor assembly and test (OSAT) companies are working to provide a complete set of PDKs that enable 3DIC designs with hybrid bonding. As the field of chiplet-based design continues to evolve, new concepts have emerged, such as the assemble design kit (ADK) and chiplet design kit (CDK), specific to chiplet-based design. These kits provide designers with the necessary information, tools, and resources to successfully design and integrate chiplets into larger systems using hybrid bonding.

EDA tools are the other fundamental components in the chip design process, as they automate various tasks, including circuit design and simulation, layout, and verification. EDA tools enable designers to create complex ICs more efficiently, accurately, and quickly than using manual methods. These tools facilitate the implementation of advanced design techniques and help optimize chip designs for performance, power consumption, and area.

However, current commercial EDA tools are mainly developed for traditional 2D chip designs, including their algorithms and user interfaces. To ensure the success of 3DIC and chiplet-based designs with hybrid bonding, EDA tools must be enhanced or overhauled for advanced 3DIC technologies. This can be achieved by the following:

1. Developing new algorithms and methods to handle the unique challenges posed by 3DIC and hybrid bonding, such as increased variability, reliability concerns, disaggregation, and integration
2. Enhancing the performance and scalability of EDA tools to handle the growing complexity and size of modern chiplet-based systems

3. Collaborating with foundries, IP providers, chiplet providers, and packaging houses to obtain accurate and up-to-date information about process technology, device models, design rules, and hybrid bonding techniques
4. Implementing support for emerging industry standards and methodologies for 3DIC design, such as new interface standards, new chiplet-based architectures, and new packaging technologies
5. Enhancing and customizing user interfaces and methodologies for 3DIC design with hybrid bonding

By addressing these challenges, EDA tools can better support the design of chiplet-based systems and 3DIC technologies with hybrid bonding, ultimately helping designers to realize the potential benefits of these advanced designs.

9.2.1.2 Interconnect

Interconnect technologies are a pivotal factor in shaping the overall performance and functionality of 3DIC systems. Figure 9.3 illustrates the progression of interconnect technology from the perspective of I/O density [3]. This evolution is largely motivated by the demands for enhanced performance, heightened integration, and superior reliability. Microbumps, through-silicon vias (TSV), and hybrid bonding stand at the forefront of contemporary interconnect and packaging technologies.

Microbumps: Microbumps are small solder bumps used to connect the layers or chiplets in a 3DIC or chiplet-based design. They provide a relatively simple and low-cost interconnect solution, but have limitations in terms of pitch, density, and scalability, which can impact the overall system performance.

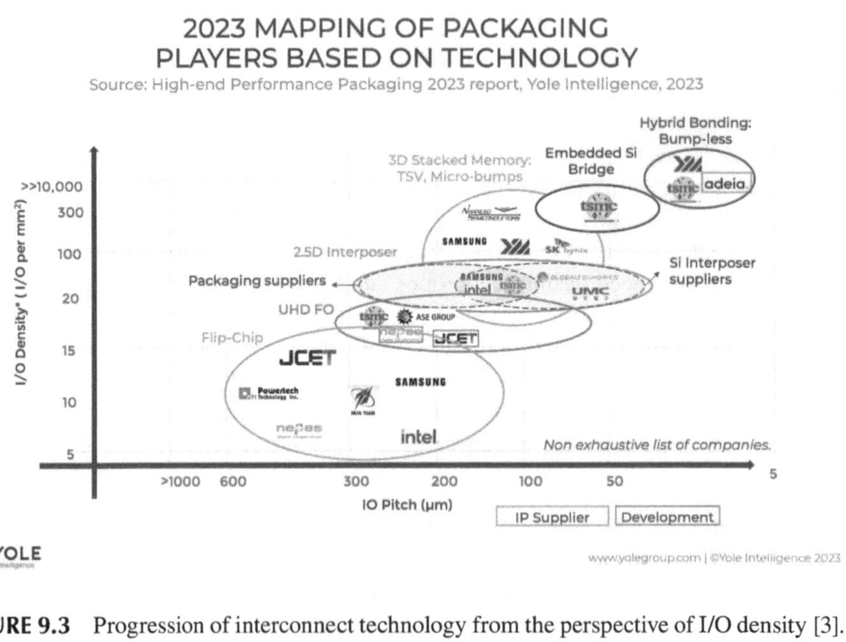

FIGURE 9.3 Progression of interconnect technology from the perspective of I/O density [3].

TSVs: TSVs are vertical interconnects that pass through the silicon layers in a 3DIC design, enabling a more direct and efficient connection between the layers. TSVs offer improved electrical performance and allow higher density interconnects compared to microbumps. However, they also introduce design challenges such as increased complexity, potential stress on the silicon layers, and the need for precise alignment during assembly.

Hybrid Bonding: Hybrid bonding is an advanced interconnect technology that combines the advantages of both direct and indirect bonding techniques. It enables ultrafine pitch interconnects and offers excellent electrical and thermal performance, making it suitable for high-performance 3DIC designs. Hybrid bonding involves the use of both metallic and dielectric materials to create strong, reliable, and low-resistance connections between chiplets or layers.

Hybrid bonding also introduces several design challenges for 3DIC designs and chiplet-based designs:

1. *Alignment Accuracy:* Due to the ultrafine pitch of hybrid bonding, achieving accurate alignment between the chiplets or layers is critical. Misalignment can result in poor electrical performance, increased stress on the components, and potential failure of the interconnects. Designers need to consider alignment tolerances and develop strategies to ensure accurate alignment during the bonding process.
2. *Materials and Process Compatibility:* The choice of materials and their compatibility with the bonding process is essential for achieving reliable hybrid bonding. Designers must consider factors such as thermal expansion coefficients, adhesion, and mechanical strength when selecting materials for the interconnects.
3. *Reliability:* Ensuring the long-term reliability of hybrid-bonded interconnects is crucial for the success of 3DIC designs. Designers need to perform thorough reliability testing, including mechanical stress tests, thermal cycling, and accelerated aging tests, to validate the robustness of the interconnects.
4. *Design Rules and Constraints:* Hybrid bonding introduces additional design rules and constraints that designers must consider when developing 3DIC systems. These include rules related to the pitch, layout, and materials used for the interconnects, as well as constraints related to the bonding process and assembly.
5. *Cost:* While hybrid bonding offers significant performance benefits, it can also increase the overall cost of the 3DIC design due to the additional process steps, materials, and equipment required. Designers need to carefully balance the performance benefits with the cost implications when considering hybrid bonding for their designs.

The advancements in hybrid bonding and advanced interconnect technology have resulted in significant improvements in performance, bandwidth, and latency for die-to-die and chiplet-to-chiplet communication. However, for the widespread adoption and success of 3DIC and chiplet-based designs, the development of interconnect standards is crucial. These standards facilitate seamless communication and compatibility between different components and systems, enabling designers to integrate chiplets from different manufacturers and use them in a variety of applications. Several organizations and industry consortia have been working on developing and promoting interconnect standards for 3DIC and chiplet designs. Some of the industry efforts and proposed standards include the following:

1. *UCIe (Universal Chiplet Interconnect Express):* UCIe is a consortium-driven standard aimed at providing a common and scalable interface for chiplet-based designs. It is designed to support a wide range of applications and use cases, from high-performance computing (HPC) to low-power edge devices. UCIe enables high-speed, low-latency communication between chiplets, simplifying the process of integrating heterogeneous components into a single system.
2. *AIB (Advanced Interface Bus):* Developed by Intel, AIB is an open-source, high-speed, and low-power interconnect standard for chiplet-based designs. It provides a simple and efficient way to connect multiple chiplets within a package, offering low-latency communication and scalable bandwidth. AIB has been widely adopted by several companies and organizations, which contributes to its growing ecosystem.
3. *Open HBI (Open High-bandwidth Interconnect):* Open HBI is another open-source interconnect standard that aims to provide a flexible and scalable interface for connecting chiplets in a heterogeneous system. It supports a wide range of data rates and offers low-latency communication between chiplets, making it suitable for various applications and use cases.

The importance of interconnect standards lies in their ability to facilitate seamless integration and communication between different components in a 3DIC or chiplet-based system. They provide a common language for components to communicate and enable easier interoperability between designs from different vendors.

Developing, aligning, and adopting interconnect standards, including those for hybrid bonding, present several challenges:

1. *Industry Consensus:* Achieving consensus among various industry players on a single standard can be difficult, as different companies may have vested interests in promoting their proprietary solutions.
2. *Technical Complexity:* Developing an interconnect standard that meets the requirements of diverse applications and use cases is a complex task, requiring collaboration and input from multiple stakeholders, including chip manufacturers, system integrators, and end users.
3. *Adoption and Ecosystem Development:* Promoting the adoption of a new interconnect standard requires significant effort to build a robust ecosystem around it, including the development of IP cores, EDA tools, test and measurement equipment, and other supporting infrastructure. This can be a time-consuming and resource-intensive process.
4. *Compatibility and Interoperability:* Ensuring compatibility and interoperability between different interconnect standards can be a challenge, as it may

require the development of bridges or translation layers to enable seamless communication between components using different standards.
5. *Standard Evolution and Updates:* As technology advances and new requirements emerge, interconnect standards need to evolve and adapt to remain relevant and effective. This can be challenging, as it requires ongoing collaboration and coordination among industry players to align on updates and revisions to the standards.

Despite these challenges, the development and adoption of interconnect standards, including those for hybrid bonding, are essential for the future growth and success of 3DIC and chiplet-based designs. By providing a common framework for communication and integration, these standards enable designers to develop more efficient, high-performance systems that can meet the increasing demands of various applications and use cases.

9.2.1.3 Power and Thermal Management

Stacking layers or integrating chiplets in close proximity, particularly with the use of hybrid bonding, can lead to increased power density and localized hot spots. Hybrid bonding enables ultrafine pitch interconnects and excellent electrical and thermal performance, but it also contributes to a more densely packed design, exacerbating power density and hot spot issues. Proper thermal management, including heat dissipation techniques and thermal-aware design methodologies, is crucial in 3DIC and chiplet designs, especially for high-performance applications. Designers must carefully balance power delivery, heat dissipation, and overall system performance while considering the unique constraints and challenges of chiplet-based systems:

1. *Higher Power Density:* As components are more densely packed in 3DIC and chiplet designs using hybrid bonding, power density increases, potentially leading to higher temperatures and thermal issues. Designers must employ power optimization techniques for 3DIC design, such as voltage scaling, clock gating, and dynamic power management, to minimize power consumption and alleviate thermal challenges.
2. *Heat Dissipation:* Effectively removing heat from 3DIC and chiplet designs with hybrid bonding can be challenging due to increased power density and stacked layers or closely placed chiplets. Designers need to adopt thermal-aware design methodologies and utilize advanced cooling techniques, including micro-channel cooling, thermal vias, and liquid cooling, to efficiently dissipate heat. Detailed discussions of thermal modeling and simulation for advanced 3DIC systems can be found in another chapter in this book.
3. *Temperature-aware Design:* The performance and reliability of the semiconductor devices are affected by temperature variations. Designers must perform temperature-aware simulations and analysis to ensure that the design can operate reliably under varying thermal conditions. This may involve characterizing devices at multiple temperatures, analyzing thermal-induced stress, and evaluating the impact of temperature on signal integrity and power delivery.

4. *Thermal Gradients and Hot Spots:* In 3DIC and chiplet designs, localized hot spots can emerge due to uneven power distribution and heat generation. These hot spots can cause thermal gradients that may impact the performance, reliability, and lifespan of the chip. Identifying and mitigating hot spots early in the design process is crucial to ensure a reliable and robust system.
5. *Thermal Coupling:* In 3DIC and chiplet designs, heat can be transferred between chiplets or packages due to their close proximity. The use of hybrid bonding may improve the thermal path for the stack but also increase thermal coupling between the die, adding complexity to the partitioning, floor planning, and simulation stages of the design process. Designers must consider thermal management when developing power and thermal strategies to minimize the impact of heat transfer between components. This includes thermal simulations to evaluate the thermal behavior of the system and optimize the placement and size of thermal interfaces such as thermal vias, heatsinks, and fans.
6. *Power Delivery Network (PDN) Design:* Providing stable power to all components in 3DIC and chiplet designs can be challenging due to the increased routing complexity and potential voltage drops. Designers must carefully design the PDN to minimize voltage fluctuations and ensure that all components receive stable power, taking into account the unique requirements of 3DIC and chiplet architectures. This may involve using multiple power planes, optimizing power delivery paths, and employing decoupling capacitors to minimize voltage noise.

By addressing these power and thermal management challenges, designers can create 3DIC and chiplet designs that achieve high performance, efficiency, and reliability, while minimizing the impact of temperature on the system.

9.2.1.4 Disaggregation and Integration

The process of deciding how to split, partition, or disaggregate complex system-on-chip into individual chiplets, especially with the use of hybrid bonding, involves several challenges from a design perspective:

1. *Functionality and Performance:* One of the primary challenges in chip partitioning is identifying the best way to divide the system-on-chip functions into separate chiplets without compromising system performance. The partitioning decision should take into account the communication latency and bandwidth requirements between different functional blocks to ensure that the disaggregation does not negatively impact the overall system performance.
2. *Modular versus Integrated Design Trade-offs:* Designers must carefully consider the trade-offs between the modularity benefits of chiplet-based designs and the potential downsides of increased complexity and communication overhead. Striking the right balance between these factors can be challenging but is essential for optimizing performance, cost, and time-to-market.

3. *Chiplet Size and Granularity:* Determining the appropriate level of granularity for chiplets can be challenging. Smaller chiplets may offer more flexibility for system configuration but could also lead to increased manufacturing complexity and costs. Designers need to find the right balance between chiplet size, manufacturing considerations, and system performance requirements.
4. *Heterogeneous Technology Integration:* When disaggregating a system-on-chip into chiplets, designers need to consider how different technologies can be combined effectively. This may involve addressing challenges related to integrating chiplets fabricated on different process nodes or leveraging different materials, which can impact performance, power consumption, and thermal characteristics.
5. *Design for Testability and Manufacturability:* Partitioning a system-on-chip into chiplets can introduce additional challenges related to testability and manufacturability. Designers need to ensure that each chiplet can be tested and verified independently, while also considering the implications of assembling and integrating the chiplets at the package level. This may require the implementation of specific test structures, additional design margins, or custom assembly techniques.
6. *Design Iterations and Optimization:* As the chiplet-based design evolves, it may be necessary to iterate on the partitioning and integration decisions to optimize performance, power, or cost. Designers need to establish a flexible and efficient design flow that allows rapid iteration and optimization, while also minimizing the impact of changes on the overall development timeline.

In summary, chip split/partition/disaggregation decisions in chiplet-based designs present several challenges from a design perspective. To overcome these challenges, designers must carefully consider a wide range of factors, including functionality, modularity trade-offs, granularity, heterogeneous technology integration, testability, IP reuse, design complexity, and design iteration.

9.2.1.5 Co-design and Co-optimization

Co-design and co-optimization are essential in 3DIC and chiplet-based designs, especially when using hybrid bonding, as they address the complex interactions and dependencies that arise in these systems and improve system-level power, performance, and area (PPA). Co-optimization can also help reduce the design loop for changes and iterations. However, it also poses many new challenges from the design perspective:

1. *Multi-domain Expertise:* Co-design requires expertise across various domains, including chip design, packaging, and system-level design. This can be challenging to achieve, as it requires close collaboration between teams with different backgrounds and skillsets.
2. *Tool and Methodology Development:* Co-design and co-optimization necessitate the development of advanced EDA tools and methodologies that

can handle the complexity of 3DIC and chiplet-based designs using hybrid bonding. These tools must be capable of addressing issues related to capacity limitation, heterogeneous integration, different databases, and seamless inter-tool combination, among others.

3. *Managing Design Complexity:* Co-design and co-optimization involve managing a higher degree of design complexity compared to traditional system-on-chip designs. This requires efficient design methodologies and tools that can handle large-scale, multi-domain optimizations while maintaining design flexibility and minimizing development time.
4. *Trade-off Analysis:* Co-optimization involves analyzing various trade-offs between performance, power consumption, cost, and other factors across different domains. Designers must develop efficient methods to explore these trade-offs and identify the best solution that meets the overall system requirements.
5. *IP Integration and Reusability:* Co-design and co-optimization must take into account the integration of different IP blocks and ensure their reusability across various chiplet-based designs. This can be challenging due to the need for standardization, IP customization, and compatibility considerations.

9.2.1.6 Security Consideration and Requirements

Traditionally, security for computational systems has focused on software aspects, and efforts in this area must remain ongoing. However, security becomes a critical concern for hardware, especially in chiplet-based designs using hybrid bonding, due to the integration of multiple chiplets that can introduce potential security vulnerabilities. Designers need to address these challenges to protect sensitive data and intellectual property. The semiconductor industry recognizes the importance of security in chiplet-based designs, and researchers and companies are developing new methodologies and techniques to ensure that these systems are secure and resilient against attacks.

Industry consortia and organizations are working to develop standards and guidelines for chiplet-based designs with hybrid bonding to address security concerns. These efforts aim to establish best practices and common frameworks for secure chiplet integration and communication. Additionally, there is a growing interest in developing security-focused IP blocks/chiplets and EDA tools that can be incorporated into chiplet-based designs to enhance security features. Some major security considerations for 3DIC and chiplet-based designs from design perspectives are as follows:

1. *Heterogeneous Integration:* Integrating chiplets with different technologies and IP blocks can introduce potential security vulnerabilities. Designers need to ensure that each chiplet's security features are compatible and effective when integrated into the final system.
2. *Supply Chain Security:* As chiplets may be sourced from different suppliers and manufacturers, ensuring supply chain security becomes a critical challenge. Designers must establish secure sourcing practices and maintain

traceability to mitigate the risk of counterfeit or tampered chiplets being introduced into the system.

3. *Inter-chiplet Communication Security:* Secure communication between chiplets is essential to prevent data leakage or unauthorized access. Designers need to implement encryption, authentication, and other security measures to protect inter-chiplet communication channels.
4. *Side-channel Attacks:* Chiplet-based designs can be susceptible to side-channel attacks, where an attacker could extract sensitive information by analyzing power consumption, electromagnetic radiation, or other physical characteristics of the system. Designers need to develop countermeasures to minimize the risk of such attacks.
5. *Hardware-based Security:* Incorporating hardware-based security features, such as secure enclaves or physically unclonable functions (PUFs), can help enhance the overall security of chiplet-based designs. However, designing and integrating these features can be complex and resource-intensive.
6. *Collaboration and Standardization:* To ensure security in chiplet-based designs, collaboration and standardization across the industry are vital. Sharing best practices, aligning on common security standards, and developing interoperable security solutions can help overcome the challenges associated with securing chiplet-based systems.

Overall, ensuring the security of chiplet-based designs with hybrid bonding requires a holistic approach that includes secure design methodologies, robust cryptographic solutions, and effective physical security measures. By addressing these challenges, designers can help protect against potential security threats and ensure the success of chiplet-based designs in a wide range of applications.

9.2.2 Design Verification and Test

9.2.2.1 Additional System-level Analysis and Verification

3DIC and chiplet-based designs using hybrid bonding require additional system-level analysis and verification compared to traditional chip design. These designs introduce new challenges and considerations that designers must address to ensure the correct functionality of interconnects and interfaces between chiplets, and the overall system performance. Here are some of the additional analysis and verification considerations designers need to address:

1. *Interconnect and Interface Verification:* Verifying the correct functionality of interconnects and interfaces between chiplets is essential in 3DIC and chiplet-based designs using hybrid bonding. Designers must ensure that interface connections are both logically and physically correct. Second, designers should also verify interconnect communication protocols, data integrity, and signal integrity. This may require the development of new verification methodologies and tools that can handle chiplet-to-chiplet interactions effectively.

2. *Multi-physics Analysis:* 3DIC and chiplet-based designs with hybrid bonding require comprehensive multi-physics analysis to account for the complex interactions between chiplets and the package. This includes thermal, electrical, and mechanical simulations to ensure that the system operates within acceptable limits and meets performance and reliability requirements. Designers must develop efficient methods for handling these multi-physics simulations and integrating them into the overall design flow.
3. *Signoff Corner Combinations:* The presence of multiple chiplets in a system can lead to an exponential increase in signoff corner combinations. Designers must find ways to reduce signoff corner combinations and find out the uncorrelated corners. This may involve adopting statistical or ML-based techniques to reduce the number of corners that need to be analyzed or developing methods for identifying the most critical corner combinations.
4. *Cross-die Coupling:* Due to the close proximity of dies in hybrid bonding technology, non-negligible RC couplings may occur between them. These couplings can impact signal integrity, power distribution, and system performance. Designers must analyze cross-die coupling effects and develop methods to mitigate potential issues. This may involve utilizing shielding techniques, or employing decoupling strategies to minimize the impact of cross-die coupling on the overall system performance.

9.2.2.2 3DIC Testability

Testing is an essential aspect of the semiconductor design process, ensuring the correct functionality, reliability, and performance of a chip. With the advent of 3DIC and chiplet-based designs, as these designs involve the integration of multiple chiplets using hybrid bonding, the potential for defects and errors increases.

3DIC and chiplet designs pose unique testability challenges due to the increased complexity and potential for defects in inter-chiplet connections. The integration of multiple chiplets can limit test access to individual components and inter-chiplet connections, and ensuring comprehensive test coverage for 3DIC and chiplet designs using hybrid bonding can be difficult due to the increased number of components and interactions. Ultimately, it leads to increased test costs.

9.2.2.3 3DIC Testing Methods

Several approaches and research directions are being explored to improve 3DIC and chiplet testing using hybrid bonding:

1. *Pre-bond Testing:* It involves verifying the functionality and performance of individual dies or chiplets before they are integrated into the 3DIC stack. This test ensures that only defect-free dies are used in the final assembly, which helps improve the overall system yield and reduce the cost of rework.
2. *Post-bond Testing:* It is performed after the dies or chiplets have been assembled and bonded together in the 3DIC stack or system. This test verifies the overall functionality, performance, and integrity of the assembled system, as well as the interconnections between the individual dies.

3. *3D Built-in Self-test (BIST):* This technique extends the traditional BIST approach to accommodate the stacked layers and interconnections in a 3DIC using hybrid bonding. It involves incorporating test circuitry within each layer and providing test access mechanisms through TSVs, micro-bumps, and hybrid bonds. The test circuitry can then generate test patterns and evaluate the responses for each layer independently or in a combined manner, depending on the specific test requirements.
4. *Design-for-Test (DFT) Techniques:* These techniques are being researched to enhance the testability of 3DIC and chiplet designs using hybrid bonding. These methods aim to embed test structures and capabilities within the design, allowing for efficient and effective testing throughout the design process.
5. *Redundancy and Fault Tolerance:* Integrating redundant connections or components in the design can improve the system's fault tolerance and minimize the impact of connection or packaging defects with hybrid bonding. By leveraging redundancy, a system can continue to operate even in the presence of certain defects, reducing the need for costly repairs or replacements.

9.2.2.4 3DIC Testing Standards

The development of a standardized test process for 3DIC and chiplet designs is crucial for streamlining testing and ensuring consistent and reliable results. Organizations such as IEEE are working on defining test standards, which will provide guidelines for test methodologies, test access mechanisms, and data formats. Some of the notable ones include the following:

1. *IEEE 1838—Standard for Test Access Architecture and Methods for 3D Stacked ICs:* This standard focuses on providing a unified test access architecture for 3DICs, with particular emphasis on TSV-based designs.
2. *IEEE 1149.1—Standard Test Access Port and Boundary-scan Architecture (JTAG):* Although not specifically tailored for 3DICs, the JTAG standard can be applied to 3DIC testing, especially when considering chiplet-to-chiplet interconnects and the testing of the digital portions within a chiplet-based design.
3. *IEEE 1500—Standard for Embedded Core Test (SECT):* This standard addresses the testability of embedded cores within a system-on-chip design. While initially intended for 2DICs, it can be adapted to suit the needs of 3DICs and chiplet-based designs with multiple embedded cores.
4. *IEEE P1839—Standard for Test Interface and Protocol for 2.5D/3D Stacked Integrated Circuits:* This project is currently under development by the IEEE Standards Association, aiming to provide a test interface and protocol for 2.5D and 3D stacked ICs.

It is worth noting that the development of 3DIC test standards is an ongoing process, and researchers and industry experts are continually working to refine and expand upon the existing standards to address the unique challenges posed by 3DIC and chiplet-based designs, especially with hybrid bonding. As the field progresses, we can expect the introduction of more specialized standards tailored to the specific requirements of 3DIC testing.

9.2.2.5 Chiplet Yield versus Package Yield versus System Yield

When dealing with 3DIC and chiplet designs, especially those utilizing hybrid bonding, it is essential to consider the different yield aspects that can impact the final product's cost. A hierarchical test methodology is used that tests the chiplet-based design at different levels of abstraction, starting from individual chiplets, moving to chiplet-to-chiplet interfaces, and finally, to the entire system.

Known-Good-Die (KGD) is a term used in the semiconductor industry to describe a die that has passed various tests and is confirmed to be fully functional and defect-free. These tests are performed at the wafer level or on individual dies or chiplets after they have been separated from the wafer. The identification of KGDs is crucial in 3DIC and chiplet-based system assembly because it helps to ensure that only fully functional dies are integrated into the final system, and KGD identification helps to maximize the system yield and minimize the likelihood of defective systems.

Even when assembling KGDs in a chiplet-based system, there is still the possibility of connection or packaging defects that could impact the overall system-level yield. Designing the chiplet-based system with modularity in mind can enable more straightforward repair and rework processes. When a package or connection defect is identified, a modular design allows for the replacement of individual chiplets or components rather than the entire system. This approach also facilitates the reuse of KGDs, further reducing the overall cost.

9.2.3 Business Model and Ecosystems: from the Design Perspective

9.2.3.1 Chiplet Marketplace and Availability-sourcing

The success of chiplet-based designs relies on a thriving ecosystem with a wide range of available and compatible chiplets, making it easier for designers to source and integrate the most suitable chiplets for their applications. The development of a vibrant chiplet marketplace, including those using hybrid bonding, has the potential to revolutionize design methodologies and enhance collaboration among various stakeholders in the semiconductor industry.

A healthy chiplet marketplace should offer diverse chiplets that cater to various functionalities, performance levels, and power requirements. This diversity allows designers to choose and integrate the most appropriate chiplets to meet their system's specific needs, enabling a higher level of customization and optimization.

A flourishing chiplet marketplace can lead to new design methodologies that focus on modularity, reuse, and customization. Designers can rapidly create advanced systems by selecting and integrating pre-built chiplets, reducing design complexity and shortening development cycles.

The chiplet marketplace necessitates robust supply chain management to ensure the availability and timely delivery of chiplets from various suppliers. Efficient supply chain management can help minimize sourcing risks and ensure seamless integration of chiplets into the final system.

Currently, the chiplet marketplace is still in its early stage, with several key developments and initiatives underway to create a more vibrant ecosystem.

9.2.3.2 Chiplet Abstract Model and Standard

Establishing standardized chiplet abstract models and configurable design parameters is critical for promoting interoperability and seamless integration of chiplets, including those employing hybrid bonding technology, from different vendors. These abstractions can also enable designing and analyzing chiplet-based designs in a hierarchical way.

Establishing standardized chiplet abstract models and configurable design parameters promotes interoperability, enabling designers to mix and match chiplets from different vendors without encountering compatibility issues. Standardized chiplet models and design parameters can help designers quickly evaluate and compare different chiplets, reducing the time and effort spent on design iterations and adjustments. This accelerates the overall design process, helping to shorten the time-to-market for chiplet-based products. By leveraging standardized chiplet models and configurable parameters, designers can choose from a wider variety of chiplets, often at a lower cost. This promotes cost-effective designs, especially for smaller companies that may not have the resources to develop custom chiplets in-house. Ultimately, standardized chiplet models can stimulate innovation within the chiplet ecosystem. With a common framework in place, chiplet vendors can focus on developing cutting-edge solutions and pushing the boundaries of chiplet technology.

The semiconductor industry has made several efforts to establish chiplet abstract models and standards, and groups such as the Open Domain-Specific Architecture (ODSA) under the Open Compute Project (OCP) and the UCIe consortium have been working on developing standards and interfaces for chiplets, ensuring better interoperability and seamless integration. EDA tool vendors are also working on enhancing their tools to support standardized chiplet models and design parameters, further facilitating the adoption of chiplet-based designs.

9.2.3.3 Chiplet Design Cycle

The chiplet design cycle must be optimized to reduce time-to-market and development costs. This requires close collaboration between design teams, IP providers, and foundries to streamline the development process and address potential challenges early in the design cycle.

9.2.3.4 Chiplet Life Cycle

Chiplet-based designs must consider the life cycle of each chiplet, including its weakest points, and align the life cycles of all chiplets in the system to ensure reliability and longevity. Designers must evaluate the cost trade-offs associated with using different chiplets to strike a balance between performance and cost.

9.2.3.5 Ownership and Responsibility

Defining ownership and responsibility in chiplet-based designs can be challenging, as it involves multiple stakeholders, such as integrators and component owners. Clear agreements and guidelines must be established to ensure that each party understands its responsibilities related to design, manufacturing, testing, and support throughout the product life cycle.

9.3 ARCHITECTURES AND APPLICATIONS OF SYSTEM-OF-CHIPLETS

The advent of hybrid bonding technology has given rise to a new era of heterogeneous integration, providing a promising solution for overcoming various limitations associated with traditional monolithic system-on-chip designs. System-of-chiplets is an innovative approach that leverages the advantages of hybrid bonding technology to create more efficient, high-performance, and customizable systems.

Hybrid bonding enables vertical stacking of dies, significantly enhancing integration density and communication efficiency between various components. This technology has a profound impact on the architecture and application of system-of-chiplets, providing a platform for the seamless integration of multiple chiplets, each handling different functions and processes. By combining the strengths of various chiplets in a single system, system-of-chiplets designs can achieve higher performance, better power efficiency, and increased versatility.

In this section, we will first explore the typical applications of chiplet-based designs, such as HPC, data centers, AI and ML, 5G and telecommunications, and consumer electronics. We will discuss why these applications are well-suited for chiplet-based designs, provide examples of industry-leading chiplet-based products, and examine the current industry status and trends.

Next, we will delve into an extensive system-of-chiplets architecture which integrates various chiplets, such as processor, memory, flash memory, accelerator, analog component, modem and RF module, optical and photonics chiplets, interconnect IP, security IP, and MEMS. For each type of chiplet, we will explore its role, challenges, and benefits in the context of chiplet-based systems, as well as the impact of hybrid bonding technology on its integration, communication, and performance.

By examining the various chiplet components and their interdependencies, along with their typical applications, we aim to provide a comprehensive understanding of the architectures and applications of hybrid bonding–enabled system-of-chiplets, highlighting the advantages, challenges, and future trends in this emerging paradigm.

9.3.1 Applications Using System-of-Chiplets

Let us first briefly explore the various applications that utilize hybrid bonding–enabled system-of-chiplets, demonstrating the potential of this technology in shaping the future of ICs. Hybrid bonding plays a crucial role in enabling the integration of chiplets and enhancing their communication efficiency, which is particularly beneficial in the following application domains.

9.3.1.1 HPC Systems

HPC systems require high levels of performance and energy efficiency, making chiplet-based designs an ideal fit. By combining multiple high-performance chiplets, such as CPUs, GPUs, and accelerators, with hybrid bonding technology, chiplet-based designs can achieve higher performance and better power efficiency compared to traditional monolithic designs. A prominent example is AMD's Ryzen processors, which leverage chiplet-based designs to deliver exceptional performance in HPC applications. By utilizing multiple CPU chiplets connected through their Infinity Fabric interconnect and taking advantage of hybrid bonding, AMD is able to optimize performance and power efficiency. For instance, in AMD ZEN3 with 3D V-CACHE technology, I/O die and one or more Core Complex Dies Flip-chip on organic substrate and V-cache is stacked on CCD through hybrid bonding.

9.3.1.2 Data Centers

Chiplet-based designs are well-suited for data center applications due to their modularity, customization, and power efficiency. Chiplets can be combined to optimize performance for specific data center workloads, such as AI and ML, networking, and storage. Intel's 4th Generation Xeon processors, also known as Sapphire Rapids (scalable processors), Sapphire Rapids HBM (CPU with high-bandwidth memory [HBM]), and Ponte Vecchio (data center GPU), leverage hybrid bonding–enabled chiplets, allowing for improved performance and customization for specific data center workloads. The industry is moving toward more customized and power-efficient chiplet-based designs for data centers to cater to diverse workload requirements.

9.3.1.3 AI and ML Applications

The growing demand for AI and ML applications has pushed the boundaries of traditional monolithic chip designs. Chiplet-based architectures offer unique advantages in addressing the specific requirements of AI and ML workloads. AI and ML algorithms are continually evolving, leading to a wide range of workload requirements. Chiplet-based designs offer flexibility to combine different types of processing elements, such as CPUs, GPUs, and AI-specific accelerators, in a single package. This allows system designers to create highly customized and scalable solutions that can adapt to various AI and ML workloads. Those applications require high computational capabilities, often leading to significant power consumption. Chiplet-based designs can help optimize energy efficiency and performance by enabling closer integration of specialized accelerators with other processing elements, such as CPUs and memory. By leveraging the hybrid bonding technology, chiplets can be vertically stacked, resulting in denser interconnects, reduced latency, and lower power consumption. This can lead to better overall system performance and energy efficiency for AI and ML workloads. By decoupling the design and manufacturing of different chiplets, specialized accelerators can be developed and improved independently from other components. This modular approach enables more rapid advancements in AI-specific processing capabilities, ultimately benefiting the development of AI and ML applications.

9.3.1.4 Consumer Electronics

Chiplet-based designs can be found in consumer electronics, offering improved performance and power efficiency for devices such as smartphones and gaming consoles. Hybrid bonding technology enables the integration of chiplets for CPU, GPU, and other components, providing better performance and efficiency. Apple's M1

ULTRA chip, used in their latest MacBooks, employs a chiplet-based design to connect two M1 Max dies through a silicon bridge (LSI, called UltraFusion by Apple) to make a larger system-of-chiplets with 2X physical metrics (CPU cores, GPU cores, Cache size, etc.).

9.3.1.5 Automotive

Chiplet-based designs for automotive applications provide a flexible, scalable, and efficient approach to meeting the diverse requirements of modern vehicles. By leveraging chiplet architectures, automotive systems can be customized to include a variety of specialized components, such as processors, memory, accelerators, and sensors, in a single package. This modular approach enables rapid innovation and efficient integration of new technologies, such as advanced driver assistance systems (ADAS), infotainment, and vehicle-to-everything (V2X) communication. Since chiplet-based designs can address the unique automotive requirements such as heterogeneous integration, thermal management, space constraints, and high levels of reliability and safety, chiplet adoption is expected to accelerate in the automotive markets in recent future years.

In summary, chiplet-based designs with the hybrid bonding technology are becoming increasingly popular in various applications due to their modularity, customization, and power efficiency. These examples demonstrate the widespread adoption of hybrid bonding–enabled system-of-chiplets across various applications and industries. The trend toward more integrated and optimized chiplet-based solutions, facilitated by the hybrid bonding technology, offers the potential for improved performance and cost-effectiveness in a diverse range of application domains. As the industry continues to innovate and develop new chiplet-based products, we can expect to see further advancements in performance, power efficiency, and customization, further solidifying the role of hybrid bonding–enabled system-of-chiplets in revolutionizing the semiconductor landscape. By 2032, as shown in Figure 9.4, the YOLE group expects that chiplet adoption will accelerate in consumer and automotive markets, and gain a foothold in defense, aerospace, industrial, and medical, leading to a total addressable market (TAM) of chiplet-based ICs in excess of $205 billion [3].

9.3.2 An Extensive System-of-Chiplets Architecture

Here, we present a hypothetical and extensive system-of-chiplets architecture, as shown in Figure 9.5, integrating a wide variety of chiplets in a single system. Hybrid bonding technology plays a crucial role in enabling this advanced architecture, providing a more efficient, compact, and high-performance solution.

9.3.2.1 Processor

Processors are the computational engines in chiplet-based systems. They are responsible for executing instructions and processing data. Hybrid bonding enables vertical stacking of processor dies, such as CPUs, GPUs, ASICs, and FPGAs, in the vertical direction. This allows for more efficient use of available space and reduced latency between processor dies. With hybrid bonding, the electrical connections between

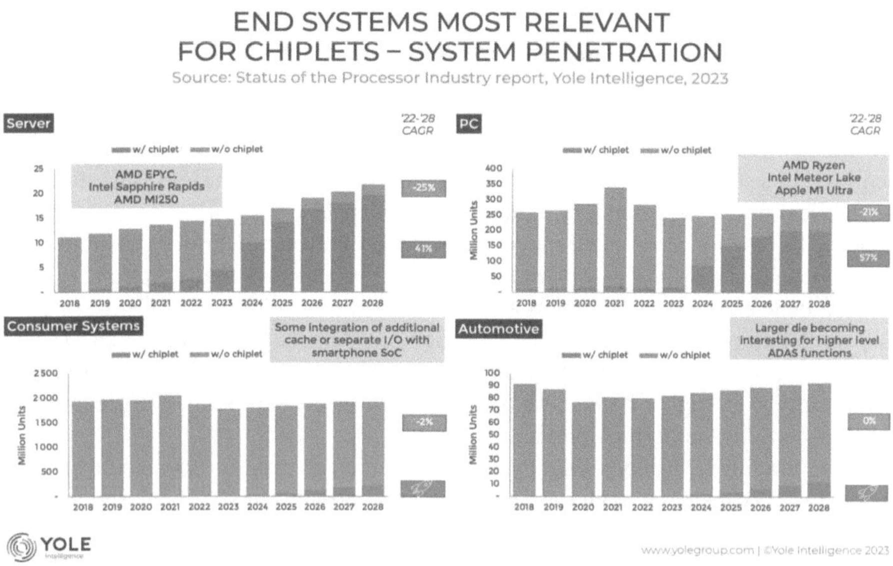

FIGURE 9.4 End systems most relevant for chiplets [3].

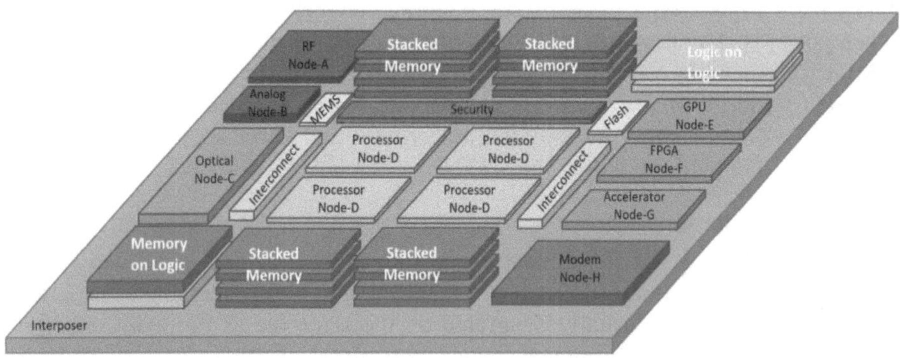

FIGURE 9.5 A hypothetical and extensive system-of-chiplets architecture.

processor dies can be denser, leading to improved communication and lower power consumption. This is especially beneficial for heterogeneous processor architectures, which combine different types of processors to optimize performance and power efficiency.

For example, AMD's Infinity Fabric utilizes chiplet-based design to connect multiple processor dies, such as CPUs and GPUs, in a single package [4]. Hybrid bonding can further enhance the performance and efficiency of this approach by enabling vertical stacking of the processor dies. Similarly, Intel's Foveros 3D packaging technology allows vertical stacking of various processor components, including CPU, GPU, and memory dies, to create highly integrated and compact chip solutions [5].

9.3.2.2 Memory

Memory chiplets store and manage data within the chiplet-based system. They are crucial for maintaining high-performance communication between processing units and other system components. Memory dies can also benefit from hybrid bonding technology. By vertically stacking memory dies, higher memory capacities can be achieved in a smaller footprint. This is particularly important for HPC and data center applications, where memory bandwidth and capacity are critical factors. Hybrid bonding allows for more efficient communication between memory dies and other components, such as processors, reducing latency, and power consumption.

A notable example of this approach is the HBM technology. HBM utilizes TSVs and microbumps to create a 3D stack of memory dies. Hybrid bonding can further improve HBM by providing higher density interconnects between the memory dies, resulting in even higher bandwidth and lower power consumption. This technology is widely used in modern chiplet-based products.

By leveraging hybrid bonding, the industry can create more compact and efficient processor and memory solutions, enabling new possibilities in performance and power optimization for various applications. Vertical stacking of dies through hybrid bonding can lead to better communication between components, reduced latency, and improved overall system performance.

9.3.2.3 Flash Memory

Flash memory is a type of non-volatile storage technology that retains data even when power is disconnected. It is widely used in various electronic devices, such as USB drives, memory cards, solid-state drives (SSDs), and smartphones, for data storage and transfer. With the improved interconnect density enabled by hybrid bonding, these SSDs could achieve faster data transfer rates and lower latency. For example, YMTC is using hybrid bonding in its 232-layer 3D NAND products [6].

9.3.2.4 Accelerator

Accelerator chiplets are specialized processing units designed to offload specific tasks from the main processor, such as AI, ML, or graphics processing. Accelerators in traditional chips are usually integrated on the same die as the main processor, which can limit their performance and customization. In chiplet systems, accelerators can be designed and manufactured separately, enabling better customization and optimization for specific tasks. Hybrid bonding can be used to improve communication between the accelerator chiplets and other components in the system, such as the processor and memory. This could lead to higher AI inference and training performance, lower latency, and increased energy efficiency.

9.3.2.5 Analog

Analog chiplets perform functions related to signal processing, power management, and other analog tasks. Analog components in traditional chips are typically integrated on the same die as the digital components, which can result in increased design complexity and noise interference. In chiplet systems, analog chiplets can be designed and manufactured separately, potentially reducing noise interference and design complexity. The challenge lies in achieving seamless integration of the analog chiplets with the rest of the system. The industry is focusing on developing high-performance and low-power analog chiplets and advanced packaging techniques for chiplet systems.

9.3.2.6 Modem and RF

Modem and RF chiplets are essential components in devices that require wireless connectivity, as they handle cellular and radio frequency communication, respectively. By combining these components in a chiplet-based design, the system can achieve high-performance wireless connectivity in a more compact and efficient solution.

Hybrid bonding technology can enable vertical stacking of modem and RF components, leading to improved communication between them and other chiplets in the system. This results in lower latency and enhanced overall system performance. In the case of RF components, hybrid bonding can potentially enable higher integration of elements such as power amplifiers, filters, and switches, leading to better RF performance, lower power consumption, and reduced overall system complexity.

The communication between modem and RF chiplets and other components like processors can be facilitated through specialized interfaces and interconnects, with hybrid bonding offering the possibility of more efficient and compact solutions.

9.3.2.7 Optical and Photonics

Optical and photonic chiplets are responsible for processing and transmitting optical signals. Optical and photonic components in traditional chips are often integrated on separate packages, resulting in relatively high latency and power consumption. In chiplet systems, optical and photonic chiplets can be integrated closer to the processing elements, improving communication bandwidth and reducing latency. The challenge lies in managing power consumption and heat dissipation in optical and photonic chiplets, and the industry is focusing on developing energy-efficient optical and photonic architectures and cooling solutions for chiplet systems. The AyarLabs TeraPHY optical I/O chiplet is an example of a chiplet-based product that integrates photonics with electronics [7]. This approach enables high-speed, low-power optical interconnects that can replace traditional electrical I/O, providing significant improvements in bandwidth, latency, and energy efficiency for data center and HPC applications. Hybrid bonding can facilitate communication between optical chiplets and other components in the system through dense, high-speed interconnects.

9.3.2.8 Interconnect IP

Interconnect IP chiplets, such as memory I/O controllers and network-on-chip (NoC) solutions, manage communication between different chiplets within the system. In traditional chips, interconnect IP is usually integrated on the same die as the main processor, which can limit scalability and customization. In chiplet systems, interconnect IP can be designed and manufactured separately, enabling better scaling and design flexibility. The Arteris FlexNoC5 is an example of an interconnect IP that can be used in chiplet-based systems [8]. Hybrid bonding can potentially improve communication between interconnect IP chiplets and other components, such as processors and memory, by providing a more efficient and compact solution. This can lead to better overall system performance and reduced latency. The challenge lies in maintaining low latency and high-bandwidth communication between the interconnect IP chiplets and other components, and the industry is working on advanced interconnect standards and interfaces to address this challenge and to support the integration of various chiplets in a seamless manner.

9.3.2.9 Security IP

Security IP chiplets handle encryption, authentication, and other security-related tasks in chiplet-based systems. In chiplet systems, security IP can be designed and manufactured separately, enabling better customization, isolation, and protection of sensitive data. The industry is focusing on developing advanced security architectures and secure interconnect solutions for chiplet systems.

9.3.2.10 MEMS

MEMS chiplets, such as accelerometers and magnetic sensors, are responsible for sensing and processing physical information in chiplet-based systems. MEMS components in traditional chips are usually integrated on separate packages or substrates, which can result in increased design complexity, form factor, and power consumption. In chiplet systems, MEMS chiplets can be integrated closer to the processing elements, reducing the overall form factor and potentially improving performance. The challenge lies in achieving seamless integration of the MEMS chiplets with the rest of the system and managing their unique mechanical and electrical properties. The industry is focusing on developing advanced packaging techniques and MEMS-specific interconnect solutions for chiplet systems.

In summary, we have explored an extensive system-of-chiplets architecture that incorporates various types of chiplets, including processors, memories, flash memories, accelerators, analog components, modem and RF, optical and photonics, interconnect IPs, security IPs, and MEMS. Hybrid bonding technology plays a critical role in enabling this advanced architecture, providing a more efficient, compact, and high-performance solution. By leveraging hybrid bonding, the industry can create more compact and efficient processor and memory solutions, unlocking new possibilities in performance and power optimization for various applications. The industry is also focused on overcoming the challenges associated with integrating these chiplets seamlessly, while maintaining low-latency, high-bandwidth communication, and addressing specific requirements, such as noise interference, heat dissipation, and mechanical properties.

9.4 DESIGN METHODOLOGIES AND TOOL FLOWS

In the previous sections, we delved into the fundamentals and the paradigm shift brought about by hybrid bonding and chiplet-based designs in the semiconductor industry. Let us now focus on the evolution of design flows and methodologies in response to this new era.

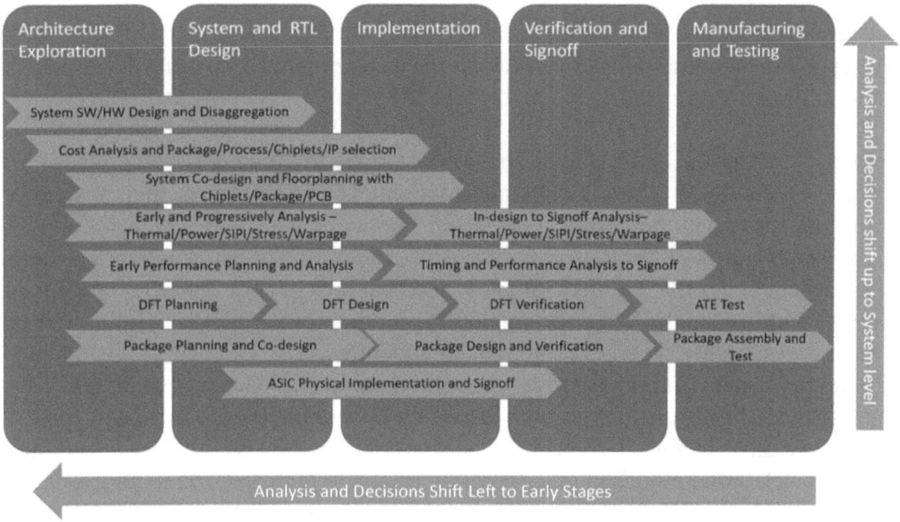

FIGURE 9.6 An overview of a typical heterogeneous integration design flow.

Hybrid bonding, with its unprecedented ability to deliver high-density, low-latency die-to-die (D2D) interconnections, has reshaped the landscape of chip design and manufacture. It has breathed new life into the pursuit of performance gains, power savings, and area efficiency, and it has also heralded the onset of a more diverse, modular, and customizable semiconductor ecosystem.

However, the impact of hybrid bonding extends beyond just the physical production and operation of chips. It has necessitated profound changes in how we approach the design of these systems. The transition from monolithic designs to chiplet-based systems implies a shift in design focus, moving away from the traditional model of chip-centric optimization to a more holistic system-centric viewpoint.

In Figure 9.6, we provide an overview of a typical heterogeneous integration design flowchart, tailored for a chiplet-based system. As we progress through this chapter, we will delve into how hybrid bonding and chiplet-based designs have triggered notable shifts in four main areas of the chip design process.

1. The shift of design focus up to the system level, driven by the need for optimal system partitioning and cost-effectiveness.
2. The need to bring analyses forward in the design process, allowing for better-informed decisions and higher quality outcomes.
3. The advent of co-design strategies to find the optimal balance between performance, power, area, and cost (PPAC) at the system level.
4. The enabling of high-bandwidth and low-latency D2D data communication, courtesy of hybrid bonding.

We will elaborate on these changes in the sections that follow, emphasizing how the industry is adapting its design strategies to harness the full potential of hybrid bonding and chiplet-based architectures.

9.4.1 Design Shifted up to the System Level

The evolution in the semiconductor landscape has prompted a shift from component-focused design to a system-level perspective. System companies and hyperscalers, with their vast resources and distinct application demands, are entering the chip industry, driving this shift. This transition is largely facilitated by the chiplet technology, which allows system companies to architect and customize their system-of-chiplets to align with their unique needs and applications.

Hybrid bonding technology plays a crucial role in this transformation, providing unprecedented density and performance, and enabling integration of heterogeneous chiplets within a single system-of-chiplets. This breakthrough allows system companies to create highly specialized system architectures.

One key aspect of this system-level approach is system partitioning, where a system is divided into distinct chiplets based on functionality, performance requirements, and other factors. This critical decision-making process involves analyzing if certain blocks can be stacked in a 3D configuration to optimize connectivity, performance, and power efficiency. The cost-benefit analysis of such decisions must also be considered, for instance, whether the increased performance and power efficiency justify the additional cost of 3D stacking.

System partitioning has a profound impact on the ability to build a functioning system with the desired performance at an optimal cost. When considering a chiplet-based design, it is crucial to look at the system as a whole, not just at individual functional blocks. Each partition's performance, its communication with other partitions, power and thermal management, among other factors, must be considered.

Typically, there are several viable partitioning solutions, each with its trade-offs concerning performance, cost, power efficiency, and time-to-market. The challenge lies in identifying the optimal solution that balances these factors. Tools and methodologies that can accurately model system performance and power consumption across different partitioning strategies are invaluable in guiding these decisions.

The industry's progression from design technology co-optimization (DTCO) to system technology co-optimization (STCO) also underscores the growing importance of system-level considerations. The DTCO approach, traditionally used in the semiconductor industry, emphasizes synergies between technology development and design at the individual chip level. It aims to optimize PPAC parameters within a single die or chip. STCO extends the concept of DTCO by incorporating system-level aspects, including the interaction between different chiplets or dies, the packaging technology (such as hybrid bonding), and system-level partitioning decisions. STCO takes into account the performance of the entire system and optimizes across multiple dimensions such as power, performance, thermal considerations, reliability, and cost. This shift to STCO is especially crucial for complex systems that include heterogeneous components, each potentially manufactured using different process technologies. For example, a system-of-chiplets for an AI application might contain high-performance CPU cores, a specialized AI accelerator, memory chiplets, and analog components, all of which have unique performance, power, and cost characteristics.

In conclusion, the shift from the chip-level to the system-level design methodology is revolutionizing the semiconductor industry, enabling faster innovation cycles, greater product diversification, and more comprehensive hardware optimization. Modern designers must thus equip themselves with a deep understanding of system-level design considerations, hybrid bonding technologies, and chiplet-based system-of-chiplets.

9.4.2 Analysis Is Shifted Early to Make Better Decisions

The transformation brought about by hybrid bonding and chiplet-based designs has necessitated a new approach in the design process. This approach demands a shift in analysis to an early stage in the design flow, allowing for more informed and strategic decisions. Additionally, this new paradigm requires that multi-physical analysis is integrated throughout the entire flow. Such integration helps to proactively identify potential challenges and devise effective solutions in the initial stages, thereby preventing future issues and enhancing the overall performance and reliability of the chiplet-based system.

One critical analysis that has been brought into the limelight is power analysis. In chiplet-based designs, chiplets may be sourced from different vendors, each with their unique power requirements and characteristics. This necessitates a detailed understanding of power consumption patterns across the system, including power distribution, potential voltage drops, and power grid integrity. By incorporating power analysis early in the design process, high-power consumption areas can be identified, energy-saving strategies can be established, and the specific power requirements of each chiplet can be effectively met.

Thermal analysis also holds a vital role in the chiplet-based design flow. With chiplets packed closely together, thermal hot spots can become a serious issue, leading to performance degradation, reliability problems, or, in the worst case, system failure. Comprehensive early-stage thermal analysis helps predict these hot spots, taking into consideration the thermal profile of each chiplet and the resulting thermal dynamics of the overall system. It is important to note that hybrid bonding, with its smaller interconnects and reduced distance between dies, can introduce additional thermal challenges, making early thermal analysis even more critical.

In chiplet-based designs, signal integrity and power integrity (SIPI) analysis is another fundamental element to be considered from the very beginning. Given the diverse signal and power characteristics of chiplets from different vendors, an accurate SIPI analysis ensures optimal system performance by preventing signal noise, power supply noise, and cross talk from impacting the system's operation. Early-stage SIPI analysis also contributes to efficient inter-chiplet communication, a key factor for the overall performance of a chiplet-based system.

Finally, stress analysis becomes particularly relevant in the context of chiplet-based designs using hybrid bonding. Mechanical stress factors, such as thermomechanical stress and stress induced by the hybrid bonding process, can negatively impact device performance and reliability. Incorporating stress analysis at an early design stage can guide strategic design decisions to mitigate potential risks associated with these stress factors.

In conclusion, incorporating these comprehensive early-stage analyses into the design flow enables designers to make more informed decisions, optimize the design process, and create robust, reliable chiplet-based systems. These practices allow designers to fully harness the benefits of hybrid bonding while managing and minimizing its associated challenges.

9.4.3 Co-design for the Optimal System PPAC

The era of hybrid bonding and chiplet-based designs has given rise to the need for a paradigm shift in design methodology. This shift has fostered the development of a new integrated design strategy known as co-design. Co-design encompasses a holistic methodology where physical and logical components, and importantly, chip and package elements, are designed in tandem to achieve optimal system PPAC.

Traditional system-on-chip design often treated chip and package as separate entities, with each being designed independently, creating a gap in optimization. This approach is no longer effective in the age of hybrid bonding and chiplet-based designs. With hybrid bonding providing high-density, low-latency interconnections, and chiplet designs creating systems out of diverse and reusable modules, the relationship between the chip and the package has never been more intertwined.

Chip-package co-design is now integral to achieving the system-level optimization necessary for modern designs. This methodology allows for considerations like power delivery, signal integrity, thermal management, and mechanical stress to be addressed concurrently in the design of both the chip and its package. Notably, it enables a designer to deal with the challenges and trade-offs that arise due to the complex interconnects and diverse components in chiplet-based systems.

The principle of co-design also encourages a level of collaboration between system architects, logic designers, and physical designers that is critical for the successful realization of a chiplet-based system. Such effective communication enables a shared understanding of the system-level constraints and targets, leading to a more cohesive and efficient design process.

As we continue to progress through the era of hybrid bonding and chiplet-based designs, chip-package co-design will play a pivotal role in harnessing the full potential of these technologies. By considering the chip and package as a co-optimized entity, designers can balance and meet the desired PPAC targets, delivering innovative and efficient systems that exceed the capabilities of traditional system-on-chip designs.

9.4.4 Hybrid Bonding Enables High Bandwidth and Low Latency Data Communication

Hybrid bonding is redefining semiconductor manufacturing and assembly, providing a robust solution for high-bandwidth and low-latency device-to-device (D2D) data communication, particularly in chiplet-based designs. This technological advance not only facilitates efficient inter-chiplet communication, but it also reshapes design flows and methodologies.

In traditional flip-chip technologies, interconnection pitch size has constrained data transfer rates. However, hybrid bonding, with its ultrafine pitch and high-density interconnects, empowers design methodologies to transcend these boundaries and substantially elevate data transfer rates. Consequently, this greatly augments system-level performance of chiplet-based designs. Additionally, by reducing signal path lengths, hybrid bonding minimizes latency and power consumption, further challenging conventional design approaches.

Moreover, hybrid bonding proves essential in enabling 2.5D and 3D stacking of chiplets. The vertical integration strategy not only optimizes space and improves integration density but also optimizes performance by decreasing the communication path between chiplets, thereby reducing signal propagation delay. This advancement necessitates a rethinking of design flows to encompass vertical integration strategies.

Hybrid bonding's significant contribution to design methodologies is particularly evident in the integration of HBM stacks. This technique allows more dense HBM stacking to be directly incorporated with the logic die or neighboring chiplets via hybrid bonding, enhancing memory bandwidth and reducing access latency. This highlights the importance of a comprehensive co-design and co-optimization approach, where both HBM and logic chiplets are considered as a unified system, and decisions are made at this holistic level to ensure optimal system performance.

Similarly, hybrid bonding enables the stacking of memory dies directly onto logic chiplets in a 3D configuration, amplifying data access speed and overall system performance. With this design approach, the flows must incorporate strategies to manage thermal and power factors and adopt advanced testing methodologies to ensure reliability.

Furthermore, hybrid bonding also allows for the design of high-speed I/O interfaces for chip-to-chip communication. This breakthrough boosts data exchange rates between chiplets and, subsequently, system performance. Nevertheless, design flows need to be restructured to handle signal integrity issues, power consumption management, and potential electromagnetic interference (EMI).

9.5 SUMMARY

Hybrid bonding technology stands as a key catalyst in the evolution of design methodologies for chiplet-based designs. By proficiently navigating the complexities that facets such as HBM, 3D memory stacking, and high-speed I/O interfaces present, chiplet-based designs utilizing hybrid bonding technology pave the way for next-generation, efficient, HPC solutions, revolutionizing design flows and methodologies in the process.

In this chapter, we have explored the profound transformation that hybrid bonding and chiplet-based designs have brought to design flows and methodologies in the semiconductor industry. However, it is crucial to note that these advancements also bring new challenges, particularly to commercial EDA tools.

Commercial EDA tools, used for designing and developing complex electronic systems, need to be significantly rethought and adapted to cope with the unique requirements and intricacies of chiplet-based designs. Traditionally, EDA tools have focused on monolithic designs, where all components are integrated into a single silicon die. However, with chiplet-based designs, we are now dealing with multiple individual dies that need to work together cohesively. This necessitates a systemic view that can span multiple dies and multiple physical domains, something current EDA tools may struggle with. Moreover, chiplet-based designs raise new considerations in terms of the physical layout, power distribution, signal integrity, thermal management, and the like. The co-design of chips and packaging, which is critical to the successful implementation of chiplet designs, further adds to the complexity. It requires EDA tools that can simultaneously consider and optimize both the chip and package designs.

In conclusion, the progression toward chiplet-based design, powered by the hybrid bonding technology, has radically altered the design landscape. It has triggered a shift toward system-level considerations, early multi-physical analysis, co-design, and high-bandwidth, low-latency data communication. These changes, while offering a promising approach for the next generation of efficient, HPC solutions, present new challenges that will require significant innovation and development in the EDA tool sector. This change is not just an evolution—it is a revolution in how we design and conceive of computing systems. The industry is at the threshold of a new era, and we can expect to see significant evolution in the EDA tool landscape to meet these new demands.

REFERENCES

1. L. Su, "Delivering the Future of High-performance Computing," in 2019 IEEE Hot Chips 31 Symposium (HCS), 2019.
2. L. Yang et al., "Three-Dimensional Stacked Neural Network Accelerator Architectures for AR/VR Applications," 2022.
3. YOLE Group, "Chiplet Market Update," Chiplet Summit, 2023.
4. AMD, "AMD Infinity Architecture." Available at: https://www.amd.com/en/technologies/infinity-architecture (accessed August 6, 2023).
5. Intel, "Intel Foveros Technology Explained." Available at: https://www.intel.com/content/www/us/en/corporate/usa-chipmaking/news-and-resources/video-intel-foveros-technology-explained.html (accessed August 6, 2023).
6. YOLE Group, "YMTC 232-Layer 3D NAND Memory." Available at: https://www.yolegroup.com/technology-insights/ymtc-232-layer-3d-nand-memory-an-unexpected-technological-breakthrough-the-chronicles-by-yole-systemplus/ (accessed August 6, 2023).
7. AyarLabs, "TeraPHY Optical I/O Chiplet." Available at: https://ayarlabs.com/teraphy/. (accessed August 6, 2023).
8. ARTERIS IP, "FlexNoC5 Interconnect IP." Available at: https://www.arteris.com/products/non-coherent-noc-ip/flexnoc (accessed August 6, 2023).

10 Thermal Modeling and Simulation for Advanced 3DIC Systems

Norman Chang, David Geb, Akhilesh Kumar, C.T. Wu, and Bo Ren

10.1 INTRODUCTION: CHALLENGES AND OPPORTUNITIES

With the recently introduced wafer-level front-end 3DIC chip stacking technology, such as TSMC's system on integrated chips (SoIC), chip-on-wafer (CoW) or wafer-on-wafer (WoW) stackings can also be assembled with backend 3DIC technologies such as chip on wafer on substrate (CoWoS) or integrated fan-out (InFO), as shown in Figure 10.1 [1]. Other foundries offer similar technology roadmaps [2].

Several emerging design optimization techniques on thermal considerations are needed for the new generation of 3DIC with closely packed chiplets and strong cross-die thermal coupling. The most important is the use of dynamic thermal management (DTM) with thermal throttling via dynamic voltage frequency scaling (DVFS) techniques. Traditionally, very few on-chip thermal sensors are placed and are usually not on optimal locations, as described in Refs. [3–6] for SoC designs due to the use of coarse mesh grid on-chip in chip-package system (CPS) static/transient thermal simulation. However, advanced 7 nm/5 nm designs such as the Vega 2.0 3DIC from AMD have 64 on-chip thermal sensors in place and therefore need more fine-grained thermal simulation, as shown in Figure 10.2. This implies that an accurate static and transient thermal simulation is needed with fine mesh grid to optimize the placement of on-chip thermal sensors.

Another emerging need of advanced 3DIC is the early estimation of static and transient thermal hot spots with transient-based partitions of power in discrete chips in a 3DIC. The thermal interaction is particularly significant in vertically stacked dies such as in SoIC with chip-on-wafer or wafer-on-wafer stack-up. Therefore, the architecture-level static and transient thermal simulation is required with the transient power profile of different chips computed from the vectors generated from emulation, as shown in Figure 10.3. Co-optimization of photonic ICs and digital ICs in upcoming co-packaged silicon photonics system implemented in 3DIC is another emerging challenge for advanced 3DIC, as shown in Figure 10.4.

It is well understood that thermal issues can severely degrade the performance and reliability of a chip and thermal runaway may occur if the temperature-dependent leakage power increases exponentially with increasing local temperature at certain areas in a 3DIC. A large value of peak temperature and thermal gradient on-chip caused by localized hot spot and cross-die thermal coupling can have a

DOI: 10.1201/9781003408673-10

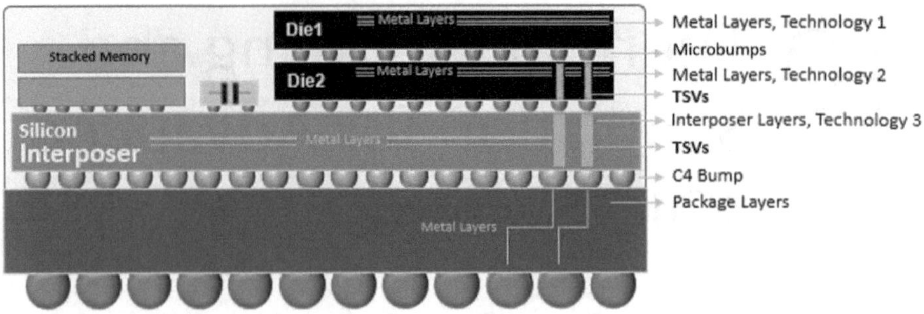

FIGURE 10.1 Industry trend of integrated front-end and backend holistic 3D heterogeneous integration.

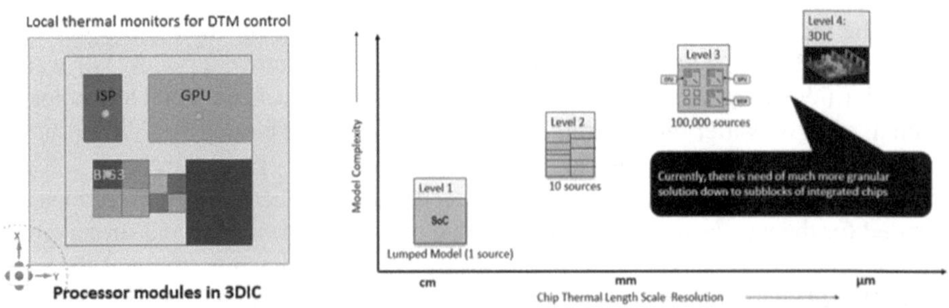

FIGURE 10.2 Many more on-chip thermal sensors are placed in advanced CPU/GPU 3DIC such as the Vega2.0 from AMD for DTM through fine-grain DVFS control and therefore need more fine-grained thermal simulation on-chip.

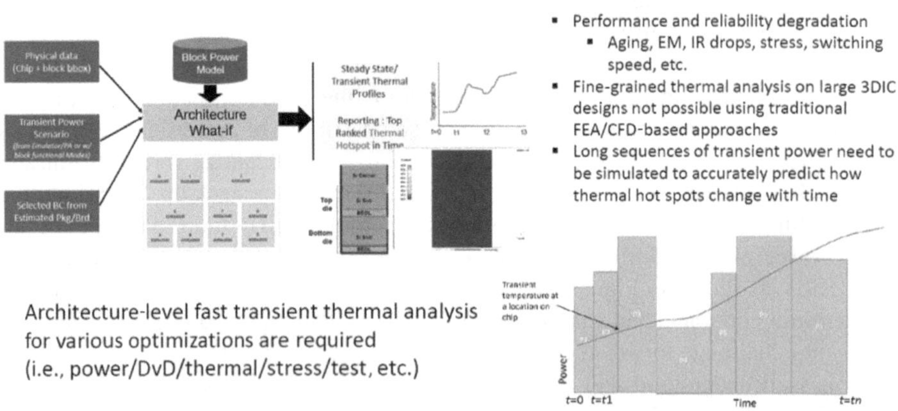

FIGURE 10.3 Architecture-level fast static and transient thermal simulation is required to help optimize the transient power partition of chips in 3DIC during boot-up sequence or for DTM purpose.

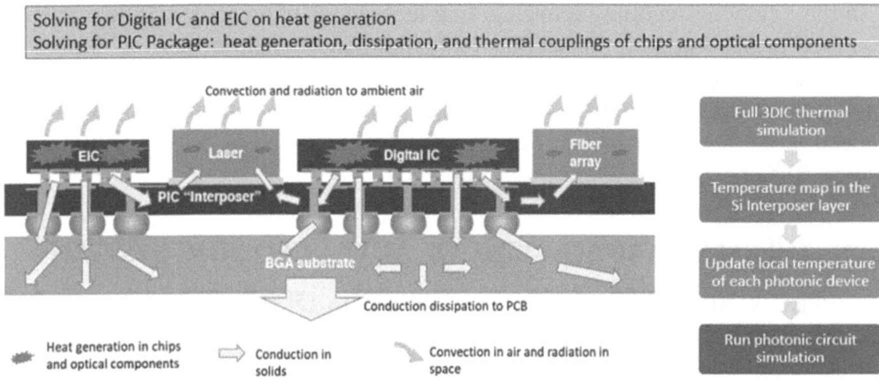

FIGURE 10.4 Co-optimization of photonic ICs and digital ICs in upcoming co-packaged silicon photonics system implemented in 3DIC is another emerging challenge for advanced 3DIC.

severely negative impact on transistor performance, stress, aging, electromigration (EM), voltage drops, and timing. Therefore, the following are the challenges and opportunities on thermal modeling and simulation for advanced 3DIC systems:

- Fine-grained static and transient thermal analysis on large 3DIC designs is required and demands adaptive meshing or machine learning (ML) technology to overcome the limitation using traditional CFD/FEA-based solvers [7–10].
- Architecture-level thermal and thermal-induced stress analysis is required due to the thermal coupling from cross-die horizontally and vertically with transient-based power profile among chiplets in 3DIC. Thermal-induced stress and warpage can impact the performance of circuits such as in C4 bumps, microbumps, TSVs, extreme low-K dielectric, etc., particularly for large 3DICs.
- Heterogeneous integration 3DICs may consist of analog/mixed-signal and digital designs which have very different thermal and stress requirements that need to be co-optimized among chiplets and packages in 3DIC.
- For silicon photonics 3DICs, accurate thermal gradient analysis is required for the co-optimization of the 3DIC package and the required thermal heater for PIC design.
- Testing of large 3DICs consisting of CPU/GPUs, etc. presents a major challenge due to multiple localized thermal hot spots and dynamic voltage drop affecting yield. Co-optimization of test techniques and localized thermal hot spots and Vdroop on 3DIC should be considered.
- Hybrid bonding is a major technique for Cu-to-Cu direct bonding in 3DIC and it also relies on a robust and consistent CMP process. Currently, there is no hybrid bonding modeling and simulation tool on thermal/stress for the whole Cu-to-Cu interface which is required for a successful 3DIC implementation. We would need the actual process measurement and correlation with simulation on Cu-to-Cu hybrid bonding.

In the following sections, active cooling techniques of advanced 3DIC system is described in Section 10.2. In Section 10.3, static and dynamic thermal simulation techniques and speed-up with ML are discussed in detail. Finally, Cu–Cu direct bonding modeling and numerical simulation techniques on thermal compression-induced stress are described in Section 10.4 and is followed with a summary in Section 10.5.

10.2 THERMAL MANAGEMENT OF ADVANCED 3DIC SYSTEMS

The subject of three-dimensional integrated circuit (3DIC) thermal management has increasingly been a topic of interest to researchers and the industry over the past decade, and review papers on the subject have become available. For example, in Ref. [11], the author assesses integrated cooling systems for 3DICs, considering challenges from temperature non-uniformity, localized hot spots, complex fluidic connections, and mechanical design. The author points to the importance of a codesign approach with circuit, software, and thermal designers working together. In Ref. [12], the authors provide an overview of the thermal management landscape with focus on heat dissipation from three-dimensional chip stacks with single-side, dual-side, and volumetric heat removal. The authors noted that the interlayer cooling approach scales with the number of dies in the chip stack, enabling extreme 3D integration; however, the microchannels and through-silicon vias (TSVs) must be carefully designed to avoid interference with each other. In Ref. [13], the authors analyzed the effect of several key attributes of the 3DIC structure on thermal hot spots, including geometry and material properties of the 3D stack, TSVs, thermal interface material (TIM), heat sink, and the number of layers, as well as the power distribution within the device layers. In Ref. [14], the authors provide an overview of electronics thermal challenges noting that 3DICs are one of two examples of the extreme limits for electronics cooling (with the other example being high-electron-mobility wide-bandgap transistor electronics). The authors observe a shift toward liquid cooling and note that a similar shift may occur with 3DICs, with embedded cooling within the package likely being required. In Ref. [15], the authors provide an overview of the intrachip enhanced cooling (ICECool) effort, launched by the Defense Advanced Research Projects Agency (DARPA) during 2012–2015 to target an order of magnitude improvement in chip level and hot spot heat fluxes for 3DICs, compared to the then state of the art (SOA). This program enabled fundamental investigation into embedded cooling strategies utilizing a combination of high thermal conductivity vias/substrates, thermoelectrics, and convective/evaporative microfluidics. Building on the ICECool effort, DARPA announced in 2023 the Miniature Thermal Management Systems for Three-Dimensional Heterogeneous Integration (Minitherms3D) program [16] calling for novel thermal management technologies to address aggressive targets in 3D HI thermal management.

There is a variety of techniques reported for thermal management of 3DICs. For example, liquid cooling involves circulating a liquid coolant through a heat sink (e.g., with microchannels). Liquid cooling can be an effective method for removing heat from high-power devices due to the high convective heat transfer coefficients and low thermal resistance to ambient it provides [17]. Liquid cooling can be classified into indirect and direct approaches, with the latter allowing for immersion cooling where the electronics (e.g., 3DIC) is directly submerged in a dielectric fluid, removing any thermal resistance between the package and the coolant [18]. Another option is embedded liquid cooling, which integrates microchannels into the 3DIC package [19, 20]. These microchannels can transport a dielectric liquid coolant through the 3D die stack to remove heat. Embedded cooling configurations allow for flow within (intrachip) or between (interchip) the chips in the stack. In any of the liquid cooling approaches, phase change (such as boiling and evaporation) can aid in increasing the convective heat transfer coefficient and reducing thermal resistance [21–24], aiding in high heat flux removal. Various liquid dielectric coolants have been assessed [25]. Heat pipes and vapor chambers are another effective thermal management solutions for 3DICs due to their potential for extremely high effective thermal conductivities and their ability to passively transfer or spread heat efficiently over long distances [26].

In addition to using working fluids for thermal management, a number of other techniques can be added to the thermal management solution, such as thermoelectric coolers (TEC), which effectively serve as solid-state refrigerators. TECs have been demonstrated as a sub-millimeter hot spot mitigation solution for wide-bandgap power amplifier transistors by integration directly on the electronic substrate [27]. Another technique is the use of heat spreaders, which spread the heat over a larger area to dissipate it more efficiently. Heat spreaders can be made from materials such as copper, graphite, or diamond. TIMs are used to improve the thermal contact between surfaces while being mechanically compliant [28]. In each of the cases, it is important to note that a combination of these techniques should be considered and that the 3DIC heat must ultimately be rejected to ambient which is typically air.

With increasing thermal challenges, more attention has been paid to addressing heat at its source. Thermal management techniques for 3DICs can be integrated directly into the 3D die stack or package. A number of such techniques have been reported for improving both conduction and convection heat transfer. For example, microchannel liquid cooling has been demonstrated with etched channels on the die [29, 30], holding promise to enable high-power 3D heterogeneous packaging with embedded cooling [31]. Adopting high-thermal-conductivity electrical insulators for the interlayer dielectric material can provide effective passive thermal management for 3DICs. In particular, it was shown that highly anisotropic insulators such as hexagonal boron nitride (hBN) are effective at mitigating both local hot spots and thermal coupling between dies [32]. Diamond is known for having extremely high thermal conductivity, which makes it an excellent material for thermal management. In diamond-based thermal management techniques, diamond structures can be integrated directly into the chip to provide efficient heat transfer [33, 34]. Laser-micromachined polycrystalline diamond forming extended surfaces for heat spreading has been demonstrated in power electronics [35]. Such embedded thermal management techniques can provide more efficient cooling for 3DICs than traditional cooling methods. However, they may also require more complex manufacturing processes and design considerations.

Compared to 2DICs, 3DICs can have higher power densities and present larger heat dissipation challenges due to the stacking of multiple layers of chips. Additionally, the interconnect used for the 3DIC chip stack can have a significant impact on thermal performance, as a high thermal resistance at the interconnect interface can lead to hot spots and reduced device reliability. With different materials being used in the interconnect in heterogeneous integrations, different coefficients of thermal expansion (CTE) can result in thermal mismatch and thermal-induced stress, ultimately leading to device failure. Various interconnect techniques exist for 3DICs. In Ref. [36], simulations were performed to compare thermocompression bonding (TCB) to direct bond Cu–Cu interconnects for a 3D stack. Results showed that direct bond Cu–Cu interconnect has better thermal (and electrical) performance due to reduced thermal resistance across the interconnect. Direct copper interconnect (DCI) is a promising technology for improving thermal–structural reliability in 3DICs. In Refs. [1, 37], authors demonstrate a new interconnect method that overcomes previous limitations.

Recently, backside PDN (BSPDN) has received attention. Although BSPDN has advantages, it comes with significant thermal challenges. In Ref. [38], the authors performed a thermal analysis and observed that BSPDN has higher vertical thermal resistance compared to frontside PDN (FSPDN) (which has only silicon substrate), and higher horizontal thermal resistance due to the thinned substrate in BSPDN reducing lateral heat spreading. The thermal effects become worse for stacked 3DIC with BSPDN on the bottom die. Due to the negative thermal effects, and special routing, accurate thermal analysis becomes important. Additionally, BSPDN, with its thinned substrate, requires a careful thermal/stress analysis as well.

Thermal management is therefore particularly critical for 3DICs to ensure reliability and prevent performance degradation. Some of the biggest challenges to thermal management for 3DICs are summarized as follows: (1) *Non-uniform heating:* This can lead to temperature gradients and localized hot spots, like the one shown in Figure 10.5, which can affect the performance and reliability of the device. Managing these temperature gradients is essential to prevent damage to the device. (2) *Thermal coupling between the stacked layers in 3DICs:* The heat generated by one layer can impact the layers above or below it. Therefore, designing thermal isolation between layers is an important consideration. (3) *Limited space due to the smaller form factor:* This limitation makes it challenging to implement traditional cooling techniques

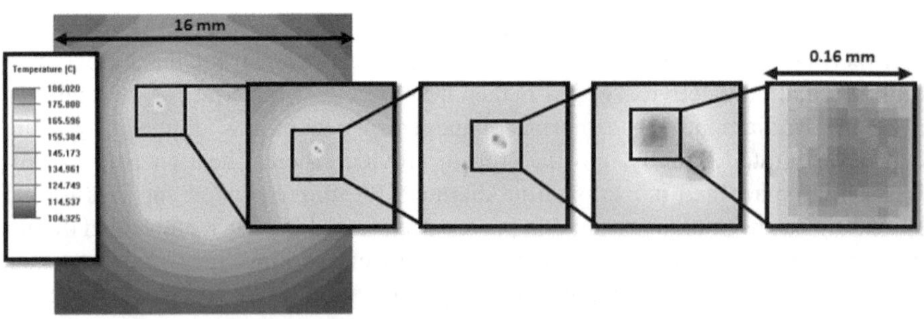

FIGURE 10.5 Hot spot on a chip due to localized high heat flux for a chip package system model.

effectively. Therefore, designing innovative cooling solutions that can fit in the limited space available is essential. (4) *Integration of heterogeneous components with different thermal properties:* The different thermal expansion coefficients of the different components can lead to thermal stress and reliability issues. (5) *Interconnect thermal management:* High current densities and thermal resistance can cause hot spots and CTE mismatches, leading to device failure. Electrical, thermal, and structural co-optimization of the geometry and materials, along with incorporating thermal management is required. (6) *Costs of solutions:* Thermal management solutions for 3DICs can be costly due to the increased demands and constraints of 3DICs. With significant challenges in 3DIC thermal management, thermal simulation becomes increasingly important, and efforts have been reported in this area [39, 40].

10.2.1 Simulation Approaches: System-level Thermal Modeling for 3DIC

With increased interest in thermal management for 3DICs, the significant thermal challenge that 3DICs present, and the variety of thermal management techniques applied in practice, effective simulation tools are needed in this space. Moreover, guidelines for comprehensive and efficient modeling of 3DIC CPS are needed. A 3DIC CPS model can be setup in a computational fluid dynamics (CFD) tool that is specifically designed for electronics thermal analysis [41]. Heat from a chip is ultimately dissipated to the surrounding fluid and CFD models how a device or system is being cooled by including necessary fluid and thermal radiation effects.

CFD tools can provide an integrated electronics cooling solution for IC packages, printed circuit boards, and complete electronic systems, along with multi-physics linkages to other physics solvers for electrical–thermal–mechanical co-simulation. Using a CFD approach, one can simulate fluid flow in both laminar and turbulent flow regimes, along with conjugate heat transfer that includes conduction, convection (both natural and forced), radiation (both thermal and solar), and Joule heating. CFD methods can provide detailed system-aware thermal boundary conditions on the 3DIC package. Advanced flow physics may be modeled in high detail. For example, fluid phase change, two phase flow, and heat transfer may be modeled with commercially available CFD software [42, 43]. Advancements in reduced order modeling (ROM) allow a complex (e.g., CFD) model to be reduced to an equivalent simple fast running model [44–46].

For thermal modeling accuracy, CAD designs of the electronics (e.g., 3DIC, PCB) and the mechanical structures (e.g., enclosures, fans, etc.) are imported into the CFD system-level tool. A trace layout file may be used to create a detailed multilayer thermal conductivity map [47, 48] that allows for highly accurate heat conduction modeling in the board and package substrate without significantly slowing the flow and heat transfer simulation, as the detailed metal fraction information is mapped onto the CFD mesh just once, before the CFD simulation begins. Similar approaches can be used to model on-chip metal density.

The system-level PCB thermal model may include the detailed metal distribution and the Joule heating on traces and vias calculated through a board-level power calculation tool [49] in a DC IR analysis. The detailed multilayer Joule heating power maps generated in the board-level power calculation tool can be mapped onto the

(a)

(b)

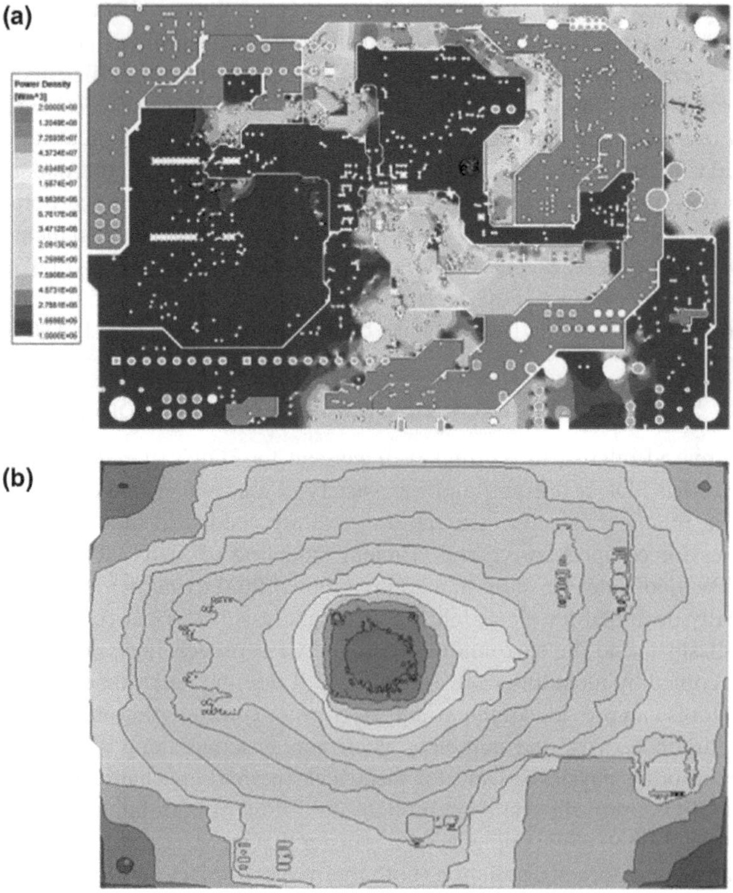

FIGURE 10.6 (a) DC IR losses. (b) Temperature on PCBA.

PCB in the CFD tool (Figure 10.6). Generally, electromagnetic field losses, both on the surfaces and in the volumes, can be mapped onto components in the thermal model, and two-way connections established between the electromagnetic and thermal solvers.

Temperature results from an electrothermal simulation may be used in a structural solver [50] for thermal stress simulations. Thermal stress on package interconnects, for example, can be assessed. Temperature and stress loads can be passed to a PCB-level reliability tool for a reliability physics analysis [51]. PCB reliability can be assessed from components' time to failure predictions.

10.2.2 Simulation Approaches: Chip Thermal Modeling for 3DIC

A detailed chip thermal model (CTM) includes chip size and materials information, a fine-grain tile-based temperature-dependent power map (mainly from temperature-dependent leakage power), e.g., 10 μm resolution, and metal density map per metal

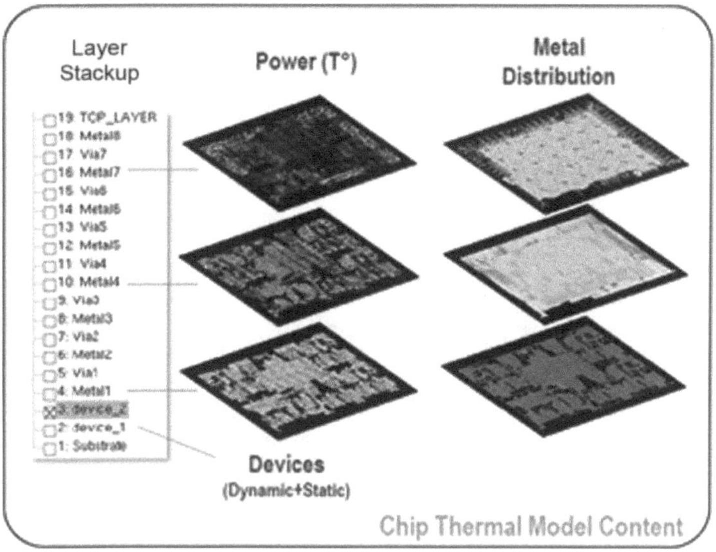

FIGURE 10.7 Chip thermal model content.

layer (Figure 10.7). The transistor or instance power is typically temperature dependent due to the leakage current component, which due to its exponential increase with temperature can risk thermal runaway at hot spots. The total power in a device includes the dynamic switching power, internal power, and the leakage power.

To efficiently capture on-chip hot spots, for a power map like that shown in Figure 10.8, the mesh density and heat source distribution in the system model for the chip can be controlled to match closely to the power pattern, i.e., finer near the hot spots and coarser in the low-power regions to minimize the model's mesh size so that more computational resources can be used in the system-level modeling of heat transfer (i.e., convection, radiation, etc.) [3]. Concentrating the power map detail and meshing resources near hot spots allows efficient hot spot detection.

Significant factors that increase the maximum chip temperature in a system-level CFD model include (a) increasing the number of heat sources in the power map, (b) refining the mesh on the power map, and (c) increasing the fidelity of the die model, e.g., Si block to three-layer die model [10]. For example, in Figure 10.9, a smeared power profile on a silicon die model is compared to a detailed power map on a three-layer die model. A significant increase in hot spot temperature is observed in the model that includes more chip power map and structure detail. The total computational effort required to capture the maximum temperature at hot spots depends on the specific CTM and its power profile. However, for practical purposes, the system-level model does not seek to capture hot spots, but rather has a very coarse representation of the chip power map so that computational resources can be spent on the system-level thermal design's simulation. Guidance for a system-level chip power map is typically resolution on the order of ~1–100 sources.

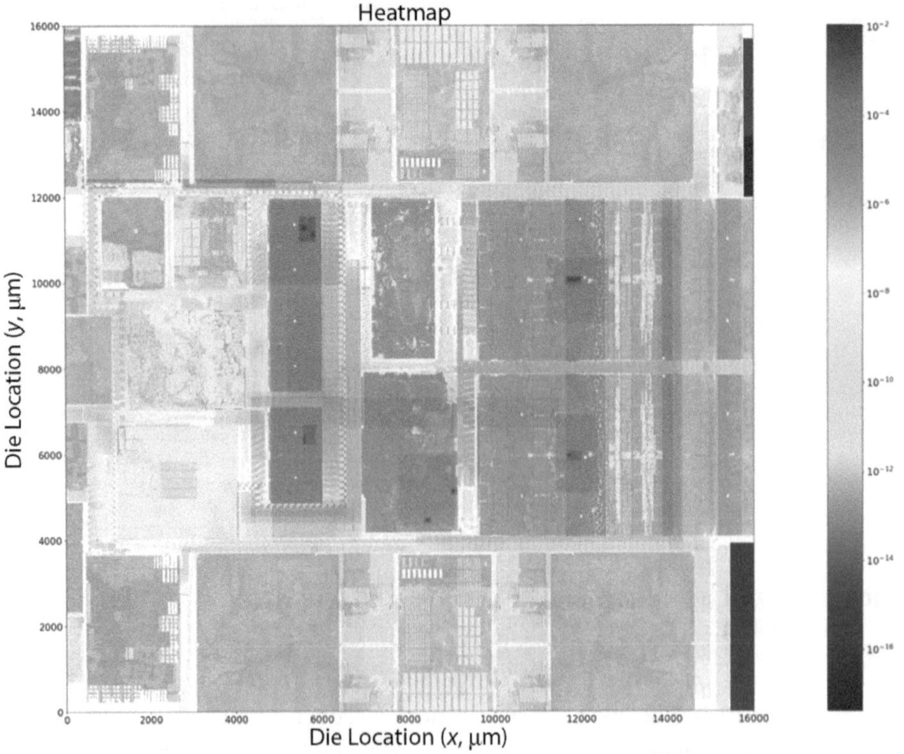

FIGURE 10.8 A CTM power map for thermal modeling.

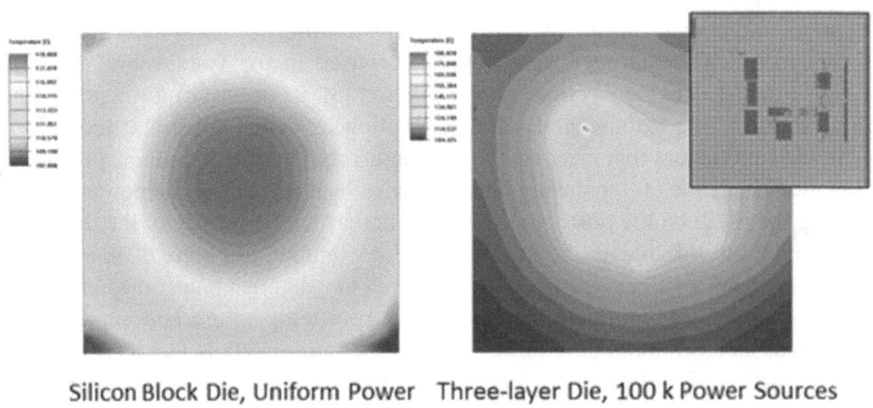

FIGURE 10.9 Comparison of a smeared power profile on a silicon die model (119°C) to a detailed power map on a three-layer die model (186°C). A significant increase in hot spot temperature is observed.

To represent the 3DIC in the system-level thermal model, multiple CTMs are included and arranged, i.e., in a stack, and placed on an interposer. While the interconnects between chips may be modeled in full 3D with Joule heating modeled by solving the electrical field, this is generally not practical. For a system-level thermal model, it is practical to just model the interconnect as a 2D surface, for ease in meshing, with effective thermal resistance and heating (from Joule heating) applied to the surface. (Modeling geometry of modern chip interconnect structures explicitly in 3D requires a thermal submodeling approach.)

10.2.3 Connecting System and CTMs for 3DIC

The CPS thermal model provides (a) chip-aware system-level thermal results for the system designer, and (b) system-aware chip thermal results for the chip designer. The system-level model allows a coarse power map and provides an approximate temperature map for each 3DIC chip in the given system. CTM temperature data files for each chip in the 3DIC can be generated when a design contains CTM components.

However, for detailed chip thermal modeling, a submodeling approach is useful. This allows a system-level model to focus on system-level thermal management effects and not spend excessive computational efforts on chip thermal modeling. The system level thermal information required for detailed chip-level thermal modeling in steady state involves (a) an HTC profile for the 3DIC stack or package, or (b) an HTC profile for each chip's top/bottom surface in the 3DIC, or (c) an R_{ja} for each chip which can be related to the respective die's top/bottom effective heat transfer coefficient(s). These HTC boundary conditions enable system information to be incorporated at a detailed on-chip level, without overconstraining the temperature results with a temperature boundary condition.

For the case of a transient scenario, power maps are imported into the system-level model and thermal information (such as HTC results) are exported out of the system-level model at specified intervals. Full system thermal inertia effects should be included for an accurate on-chip transient response. A chip's thermal response is unrealistically fast without the system mass incorporated. Chip-level thermal submodels may incorporate system thermal inertia effects through, for example, a calibrated capacitive node approach using time to steady-state measurements. A thermal system's step response can be sufficient to characterize it [52]. DTM methods allow the simulation inputs, such as chip power, to be controlled dynamically during the simulation run based on measurements such as on-chip temperature sensors [6].

10.3 STATIC/DYNAMIC THERMAL SIMULATION FOR ADVANCED 3DIC SYSTEM

10.3.1 Static and Dynamic Thermal Analyses

The power consumption in a chip is due to two main sources, the static and the dynamic power of the devices. The static power dissipation in a device is due to subthreshold leakage and gate leakage. The subthreshold leakage is the dominant source of leakage/static power and is temperature dependent, as the subthreshold leakage current increases exponentially with temperature. The dynamic power depends on the switching activity at various nodes in the circuit but is not temperature dependent. These power dissipations act as heat sources which increase the temperature of the chip. The heat generated by these devices can be dissipated through three primary mechanisms: conduction, convection, and radiation. As an example, the steady-state thermal profile of the chip while accounting for thermal conduction can be obtained by solving Equation (10.1):

$$\nabla \cdot [k\nabla T(x,y,z)] = -q(x,y,z) \tag{10.1}$$

Here k is the thermal conductivity, T is the temperature, and q is the power density. For the transient thermal profile, Equation (10.2) needs to be solved:

$$\rho \cdot c_p \cdot \nabla T(t) = k\nabla^2 T\left(x,y,z\right) + q(x,y,z) \tag{10.2}$$

In Equation (10.2), ρ is the density and c_p is the specific heat capacity. The conduction of the heat depends on the thermal conductivity, dimensions of the material, and temperature gradient across the material. In the chip, these materials can be silicon, silicon dioxide, and metal layers of various shapes and geometries. Convection refers to the heat flow through a fluid, either liquid or gas. In context of the chip, the heat can be dissipated to the ambient air either through natural convection or through forced convection. For example, a fan can force cool air over a heat sink to remove the heat. Radiation is the mechanism whereby a body emits thermal energy through electromagnetic radiation and the power radiated by a body depends on the temperature of the body.

The primary mechanisms of the heat transfer in chip are conduction and convection, such as the heat conducting to the heat sink through the chip, and the heat being dissipated from the heat sink to the air through convection. All the above heat flow mechanisms can be mathematically modeled and solved to generate temperature profiles of the chip.

When temporally constant heat sources are applied for a sufficiently long duration, eventually a steady state will be reached where the temperatures at different locations on the chip will not change with time. For such cases, a steady-state or static thermal analysis is sufficient which will provide a final steady-state temperature distribution on the chip. However, if the heat sources are transient, then a transient thermal analysis is required to predict the transient thermal behavior of the chip.

10.3.2 Techniques for Thermal Simulation

To generate the thermal profile of a chip given the various heat sources, material properties, and boundary conditions, the heat transfer equations need to be solved. Given the scale of the problem, numerical simulation techniques have evolved and have become gold standards for computing thermal profiles of a CPS. The three popular methods of generating thermal profiles are finite element methods (FEM), finite difference method (FDM), and thermal RC model.

The FEM is the most accurate method, but it also has capacity and runtime implications. This method involves dividing the domain into smaller elements, with shared boundaries and interconnected nodes [53]. This generates a mesh network and a set of partial differential equations can be generated for the heat transfer modeling. However, these partial differential equations are difficult to solve analytically. Each element is assigned an approximate function to represent the original problem for that region. The governing equations for each element are calculated and combined to form a global set of equations, which can be solved to obtain temperature values for all the elements. The formulation creates a set of equations of the form $AT_h = b$, where A is the stiffness matrix. This equation is solved using numerical techniques for steady-state analysis and numerical integration techniques for transient thermal analysis. The accuracy of the method depends on the number of subregions used for analysis, and the stiffness matrix is formulated based on the material properties of the system. The application of boundary conditions is also an important step in the process. Most of the commercial grade software tools use the FEM methods for thermal analysis.

The finite difference method is a numerical technique used to solve differential equations that are difficult to solve analytically. The domain of interest is divided into a grid of discrete points, and the differential equations are converted into algebraic equations by approximating the derivatives at each point using finite differences. The finite difference approximation expresses the derivative of a function in terms of the function values at adjacent points on the grid. By solving the resulting set of algebraic equations iteratively, the values of the function at each point on the grid can be obtained. The accuracy of the method depends on the size of the grid and the order of the finite difference approximation used. This method is commonly used in areas such as CFD, heat transfer, and structural mechanics.

The thermal RC circuit-based method relies on generating an equivalent electrical RC circuit based on the parameters of the chip and material properties. The current is equivalent to the heat conduction, such that the heat flows through the thermal resistance and thermal capacitance network due to temperature differences. Since there are well-known methods for solving electrical RC circuits, such thermal circuits can be solved to get the temperatures at all the nodes of the circuit.

10.3.3 Recent Advances in Thermal Simulation

The previous sections provided an overview of the classical methods for thermal analyses for CPS. A critical limitation of such methods for thermal analyses is the performance and capacity limits of software tools built on such methods. These limits become more apparent when thermal analyses are performed with higher accuracies and finer granularities. As an example, for chip sizes greater than 1 cm × 1 cm and on-chip thermal analysis granularity set as 10 μm × 10 μm, a transient thermal analysis with several power patterns can take prohibitively large time on the order of weeks using classical methods. Therefore, the traditional FEM-based numerical simulations are unviable to be used for generating fine-grained thermal responses for large 3DIC design. Recent advances in ML and deep learning (DL) have enabled methods which can be used to perform computationally intensive tasks very efficiently. A short overview of such methods is presented in the next section.

10.3.4 A Brief Overview of Published Papers

In Ref. [37], a method is described to perform transient thermal simulations on chips in multi-chip systems and 3DIC packages, while accomplishing the following key objectives: (1) capture on-chip hot spots with detailed, fine-grain power map resolution (i.e., tile-level); (2) account for system-level thermal effects; (3) rely on minimal CPS thermal characterization; (4) provide extremely fast solutions (e.g., in seconds) via a highly distributable and efficient tile-by-tile sweeping procedure; and (5) provide high accuracy compared to full fidelity FEA or CFD models. To achieve these goals, a new multiscale, distributable, system modeling method with black box ROM for transient chip thermal analysis is presented that captures multiple (disparate) length and time scales on chip in system with high accuracy and high speed. Chip thermal coupling among stacked dies in a 3DIC is performed at multiple levels of the multiscale ROM to account for hot spot thermal coupling/effects across chips. The work in Ref. [54] uses a convolutional encoder–decoder generative network for predicting the temperature maps from the input power maps. The paper uses a similar network to predict the IR drop maps additionally. This can be thought of as an image-to-image translation problem, where the input 2D power distribution image is translated to a 2D temperature distribution image. In the examples presented in this paper, the authors use power map images of size 34 × 32 with the tile size of 250 μm × 250 μm on 8.5 mm × 8 mm test cases. The transient temperature is handled by using long short-term memory (LSTM) blocks [55]. The LSTM blocks enable learning long-term data dependencies which in the context of thermal analysis allows keeping track of the impact of power on temperature over time. Figure 10.10 shows the block diagram of the LSTM convolutional encoder–decoder architecture for predicting transient thermal maps. In case the LSTM blocks are not used, then the same encoder–decoder architecture can be used to generate the static thermal analysis results. The results of static thermal analyses show an average error of 0.63°C with a maximum error of 2.93°C. The results of the transient thermal analyses show an average error of 0.53% and maximum error of 6.80%.

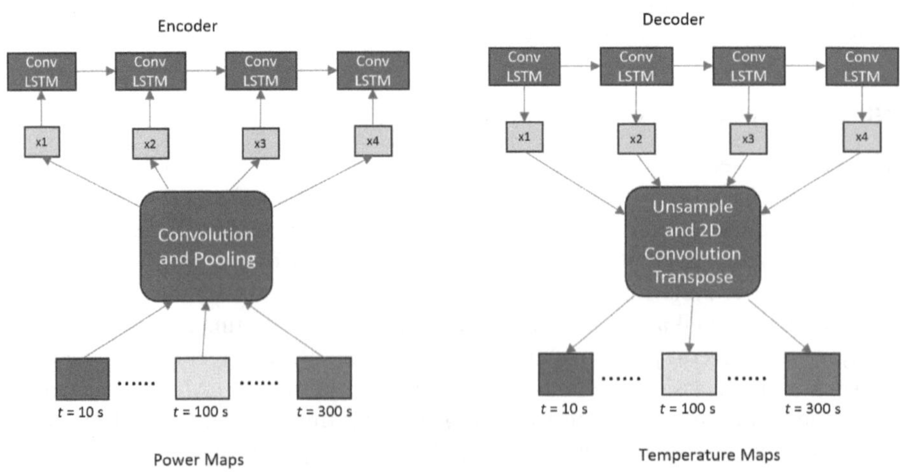

FIGURE 10.10 An LSTM-based encoder–decoder for predicting transient thermal response.

The work in Ref. [56] uses a data-driven generative adversarial learning method to predict the full-chip transient thermal maps for multi-core CPUs. The authors model the thermal map generation problem as image generation problem. The models rely solely on the quantifiable real-time chip utilization and thermal sensor data from commercial chips, and do not presume the need for any extra physical sensors. The method used for estimating thermal maps has the ability to generate precise full-chip transient thermal maps by utilizing performance monitor traces from commercially available multi-core processors. The generator and discriminator used in the process were created using basic convolutional layers, and the Wasserstein distance was used as the loss function. This study uses a significant volume of CPU thermal distribution data (referred to as thermal maps) and corresponding real-time PCM (performance counter monitor) data. The model can acquire knowledge about the transformation process by utilizing this data. A measurement system based on an infrared (IR) camera is used to externally acquire accurate thermal maps of a working CPU.

A generative adversarial network (GAN) is a type of DL network that consists of two neural networks, a generator and a discriminator, that are trained together in a game-theoretic setting. The generator learns to create new data samples that resemble the training data, while the discriminator learns to distinguish between the generated samples and the real training data. Through an iterative process, the generator becomes better at creating realistic samples that can fool the discriminator, while the discriminator becomes better at distinguishing between the generated and real samples. Eventually, the generator can create samples that are indistinguishable from the real training data. The framework used in this work is shown in Figure 10.11. In order to have a better convergence, this work used Wasserstein distance as the loss function to measure the distance between the real and fake image distributions.

Figure 10.12 shows the results of the thermal analysis. This method achieves an average root mean square error (RMSE) of 0.83°C with a standard deviation of 0.52°C and can achieve 240 times speed-up over the FEM method with similar accuracy.

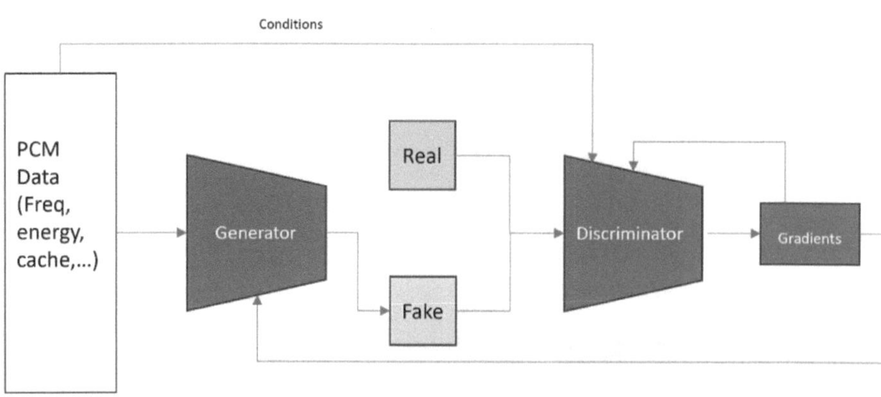

FIGURE 10.11 Modeling framework for thermal analysis using GAN.

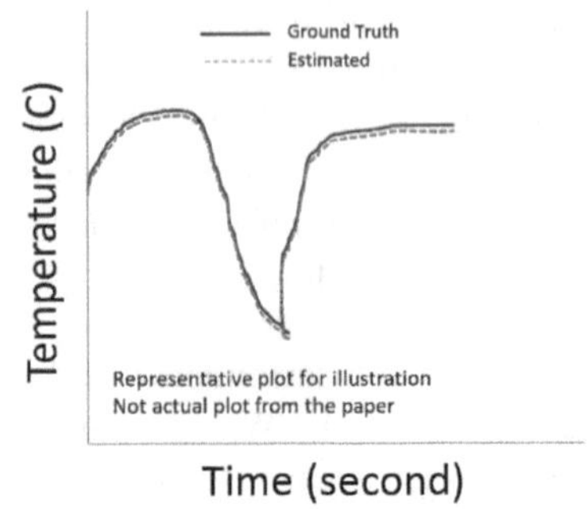

Error	Temperature (°C)
RMSE	0.83
Std Dev	0.52

FIGURE 10.12 Results of the thermal analysis; plot is illustrative and not the actual plot.

10.3.5 DNN-based DeltaT Prediction for Fast Static Thermal Simulation

This section discusses a recently developed deep neural networks (DNN) [55]-based methodology for fast static thermal simulation [8]. This technique uses an innovative data-driven DNN framework for performing static thermal analysis which is 100–1,000× faster than the traditional FEM-based thermal solver while achieving the same level of accuracy and resolution. This enables running static thermal analyses at much higher resolutions for large chip sizes compared to traditional FEM-based methods.

10.3.5.1 Background

For system-level thermal analysis, chips are usually modeled as a pure silicon block with limited number of meshes. Figure 10.13 shows an example of the fine grid power and thermal maps for a 10 mm × 10 mm chip at a resolution close to 10 μm in size. The chip has about 30 million instances and for the fine mesh on the silicon

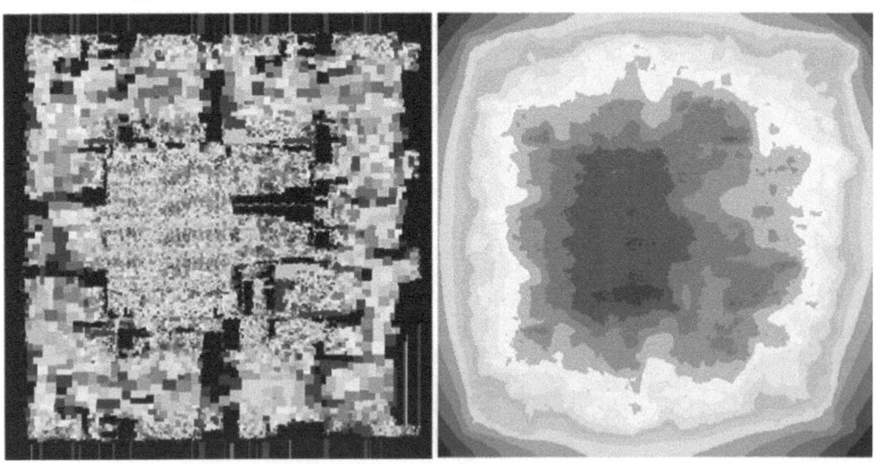

FIGURE 10.13 Tile-based power map (*left*) and thermal profile (*right*) of a 10 mm × 10 mm chip with 30 million instances.

block, it took more than 12 hours to solve for the "chip + package + PCB" model using FEM.

Approximating the chip as a silicon block for thermal analysis ignores the details of the other structures on the chip. Although the main volume of a chip is the silicon substrate, the details in the chip structure such as the thin interconnection layers with embedded wires and insulator and oxide layer, especially in the case of SOI (silicon on insulator), actually have significant impacts on the non-uniform local temperature rise and decay behaviors due to the existence of the dielectric material with low thermal conductivity, e.g., SiO_2, among many other heating objects in the chip. An important observation is that the local temperature rise will decay fast when the dielectric media exists. Hence, the conductive silicon model used in Figure 10.13 enables the thermal effects to reach far from the heat sources, even though this is not happening in reality. The predicted hot spot temperature will also be underestimated, due to fast heat dissipation in silicon materials. From the studies [8], the realistic thermal decay distance for either device or wire segments is about 5 μm in the dielectric layers [17]. Also, the temperature increase on devices is expected to be higher than that in Figure 10.13 as the heat sources are buried in the dielectric materials in the CMOS chip configurations. Note that the thermal responses from a chip model of silicon block in Figure 10.13 are still valuable and can be treated as the thermal profile on the top face of the chip. When the real device junction temperature is of concern, the dielectric media should be included in the analysis model, even in the system-level thermal analysis. Hence, two factors in the analysis affect the thermal results of the chip, i.e., mesh fineness and dielectric media in the chip model. Figure 10.14 shows the temperature decay behavior in a chip model with as much detail in modeling as possible, i.e., via/wire layout, dielectric media, devices, and the silicon substrate. The temperature decay curve describes how the rise in temperature

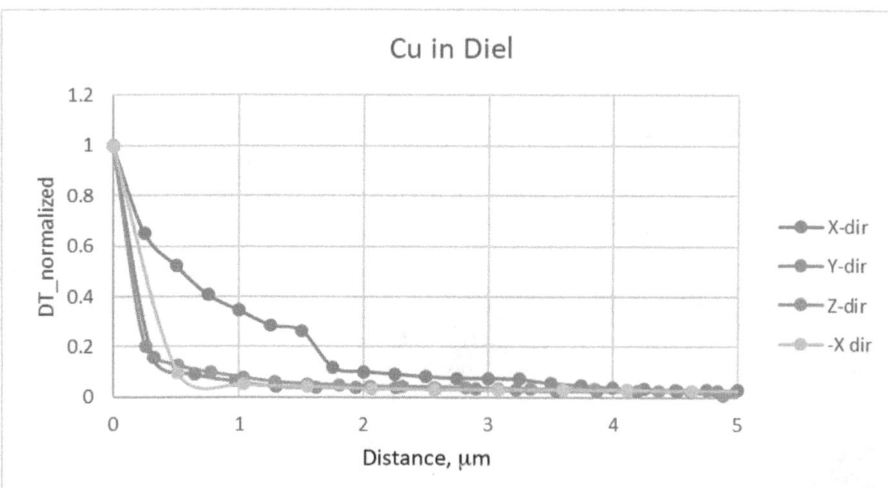

FIGURE 10.14 Temperature decay from a heated object in the device or interconnection layers of an N10 chip [8].

decreases along the distance to a single heat source. The *Y*-axis shows the normalized temperature change against the reference temperature. The decay curves are measured along different directions. For one of the directions along the metal layer (the *X*-direction), its reverse direction has also been measured. The decay behaviors are similar from either wire segment heating or device heating. Layout does have some impact to the thermal decay as shown in Figure 10.6, but the temperature impact will drop to near zero over the distance well before 5 μm.

If there is a local heating object on the chip and included in the finite element (FE) mesh, it is expected that there will be additional temperature rise at this location due to the localized heating. In the real chip design, the effect of localized heating from chip devices/metal buried in dielectric layers is known to be insignificant beyond 5 μm [3], as shown in Figure 10.14. Therefore, it is reasonable to assume that the effect of temperature rises due to localized heat sources across a chip will be limited to their immediate vicinities of 5 μm on all sides. This is referred to as the LHE (localized heating effect, including self-heat and Joule heating)-induced DeltaT. Hence the chip thermal profile resolution can be enhanced by appending the local temperature rise at the cell (or tile in 10 μm × 10 μm) location to the coarse thermal profile from system-level thermal analysis to get the final thermal responses at a high resolution.

10.3.5.2 Thermal Profile Enhancement

This section explains how LHE at multiple locations can be combined with a uniform average power on chip to generate a DeltaT map with a fine resolution. For a multichip package, e.g., 3DIC, the system thermal response is shown in Figure 10.15. The Theta-JA of each die can be extracted by the maximum temperature (T_{max}) on the chip, as shown schematically in Figure 10.16 from the chip thermal profile of die 1, i.e., Theta-JA (1). For each die, T_{max} depends on the heating up of the neighboring chips in the package or from the system, and the thermal profile may not be symmetric to the chip center. The Theta-JA(i) of die (i) is calculated based on the coarse mesh on-chip modeling in the system. This Theta-JA and the coarse chip thermal profile represent the system response when LHE is not considered.

It is assumed that the base coarse thermal profile can be separated from the LHE effect and consider the Theta-JA as the general thermal environment of a chip, regardless of the package details and board/system configurations. Figure 10.17 is the thermal response plotted along a path through the T_{max} location, i.e., center of the chip in

FIGURE 10.15 System thermal response in multi-chip systems.

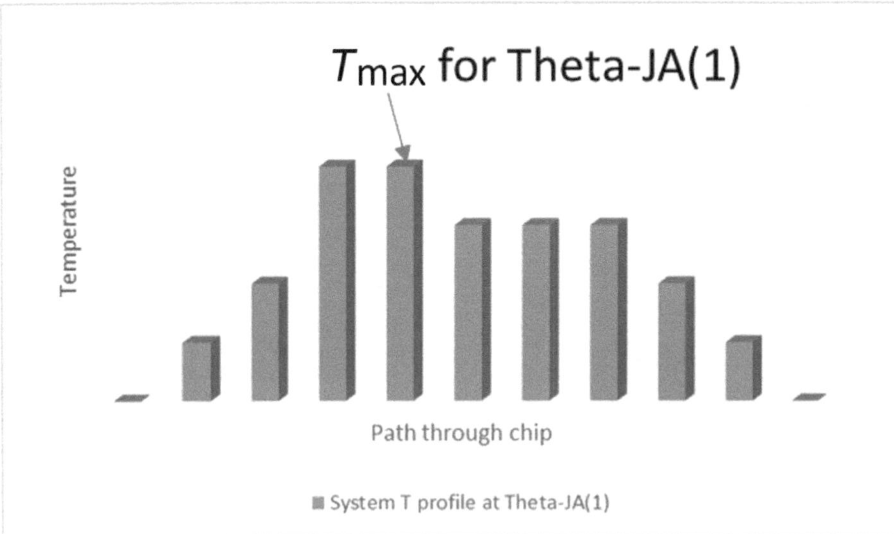

FIGURE 10.16 Thermal profile along a path on the chip.

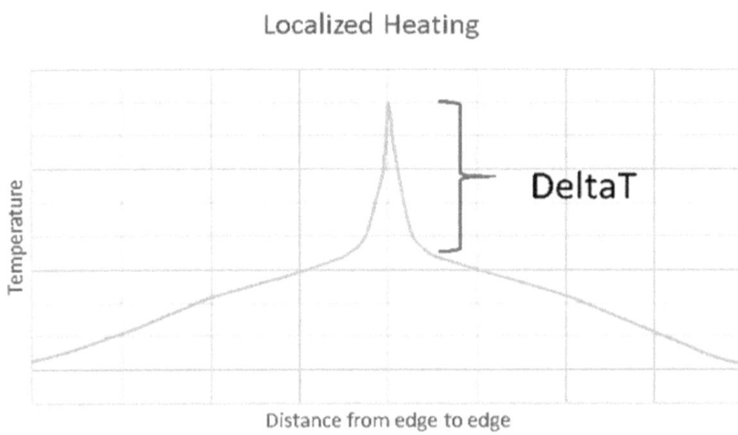

FIGURE 10.17 Temperature increase at a location due to localized heating.

Figure 10.16 after adding a localized heating zone. The zone is 10 μm × 10 μm tile at the T_{max} location with higher power density in addition to the uniform power across the chip, to extract the additional temperature rise (i.e., DeltaT). The DeltaT derived at the die center can be generalized to other chips with similar material properties since this effect is a localized effect and does not depend on the package and system. The on-chip mesh used in Figure 10.15 is with fine grid (such as 10 μm × 10 μm) at the location of the localized heating, i.e., at the die center in this case, to capture the local heating accurately. Then for a general system like the one in Figure 10.15, given the detailed power map on a chip, the DeltaT map can be constructed, by applying

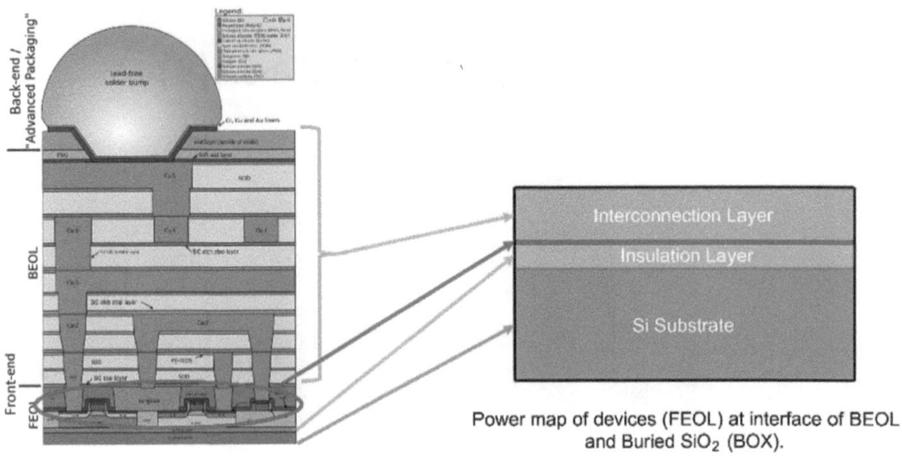

FIGURE 10.18 A three-layer die model.

one localized heating at a time, at fine resolution for any die size with the corresponding Theta-JA assumption. A further enhancement to the silicon block model in Figure 10.15 for localized heating effects is to extend each power on tile to an array of power tiles, e.g., 5 × 5 tiles with localized power pattern with two extra rows or columns to the center tile. This model captures the effect of nearby heat sources on a given victim location. A DNN-based DeltaT Predictor can be trained and used for the high-resolution DeltaT/FinalT map extraction of a new chip in Figure 10.15.

Instead of the single silicon block model, an improved, and a more realistic three-layer chip model with two dielectric layers on top of the thicker silicon substrate can provide the thermal decay behavior similar to that in Figure 10.14, i.e., 5 μm in decay distance. In this three-layer model, as shown in Figure 10.18, the heating sources are buried between the upper dielectric layer and the lower insulation layer, with practical thickness, e.g., 6 μm and 0.1 μm, respectively. The T_{max} predicted is closer to the foundry data than the pure silicon chip model. Studies show that the existence of the insulation layer under the heating object is important to generate practical T_{max} and the thickness of the equivalent interconnection layer has secondary effects only. This three-layer model then becomes the foundation for thermal modeling in this work.

Figure 10.19 shows the overall flow of the thermal profile resolution enhancement using DNN. An inference engine was created using the training cases with variations of power on the 5 × 5 tiles. Given a new chip with detailed power map in 10 μm × 10 μm resolution, DeltaT Predictor can step through each tile of a chip with its surrounding power map on 5 × 5 neighboring patterns.

The linear superposition of the coarse system thermal and the DeltaT map gives the final thermal profile at an enhanced resolution.

10.3.5.3 DNN-based DeltaT Predictor Training

A typical BGA package is used with thermal board configurations (Figure 10.20) at a few Theta-JA's from low to high, e.g., 10–120 C/W, to cover the possible ranges of

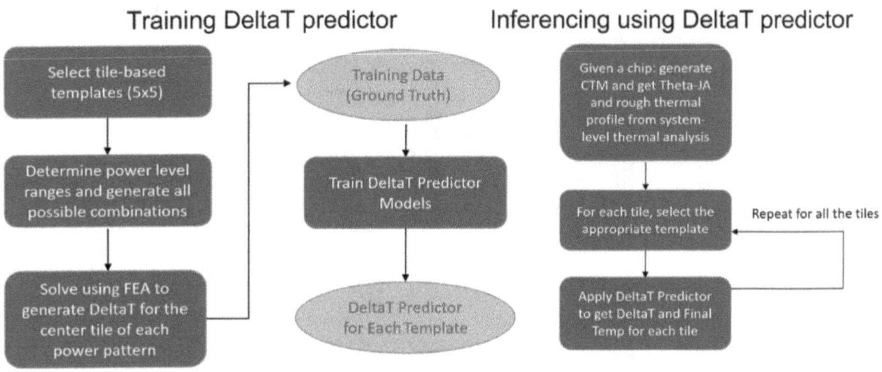

FIGURE 10.19 Training the DNN DeltaT Predictor and inferencing.

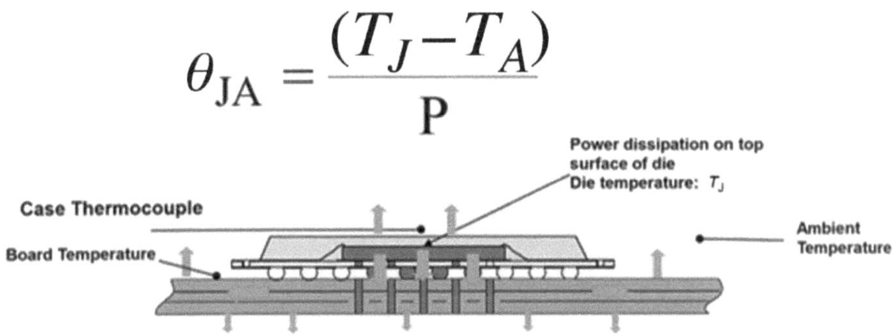

FIGURE 10.20 Package system setup for extracting Theta-JA parameter.

Theta-JA from realistic system-level thermal analysis. These Theta-JA's are used for later interpolation to match the package/system Theta-JA from system-level thermal analysis at the inferencing step on a new chip. For each training case at a selected Theta-JA, one can create a fine grid pattern with location template, e.g., 5 × 5 tiles at 10 μm resolution, by within-the-die, along the border, or at the corner location template to apply the localized power pattern in calculating the DeltaT at the center tile as shown in the flow (Figure 10.21). The specific template is determined by the center tile distance to the nearest chip boundary, and the tile size. The power values on the 5 × 5 tiles are based on the practical tile-base power density at different levels consistent to the technology node of the chip. For the 5 × 5 tile example here, the possible power combinations could be as many as 3^{25}, and proper design schemes could be used to reduce the total number of power patterns to be solved, e.g., using symmetry and DOE (design of experiment) technique (Figure 10.22) to reduce the number of training cases. Parallel processing of running training cases with various inputs of Theta-JA and power patterns on 5 × 5 tiles using FEM thermal analysis to generate the resulting DeltaT in the center tile as ground truth would be helpful to accelerate this step.

After all the training data has been generated, the corresponding DeltaT Predictor based on the appropriate DNN architecture [55] can be trained as in Figure 10.19.

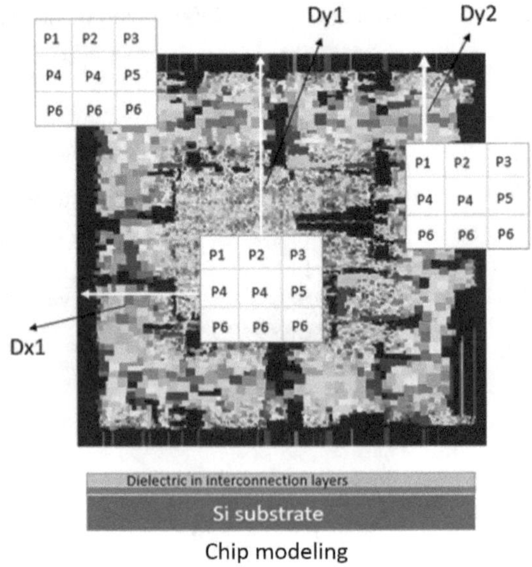

FIGURE 10.21 Examples of template-based 5 × 5 tiles of size 10 μm × 10μm for templates within-the-die, along the border, and at the corner, with LHE assumption such that thermal coupling decays rapidly beyond 50–100 μm depending on the technology node.

FIGURE 10.22 The power-on-tile patterns using DOE (design of experiment) for reduced input training cases for a template, dividing the tiles into center and peripheral groups.

The DNN architecture with six hidden layers, 64–128 neurons per layer, and with ReLU active element is used to create the DeltaT Predictor model in this work.

10.3.5.4 Inferencing Step

Based on the Theta-JA, for example, of the chip from the system-level thermal analysis and the tile-based power map from the chip power analysis tool, the DNN-based DeltaT Predictor in Figure 10.23 efficiently computes the DeltaT at the center tile of the template-based 5 × 5 tiles that has the corresponding power pattern at a tile location in the full-chip power map. The template window is moved across the chip

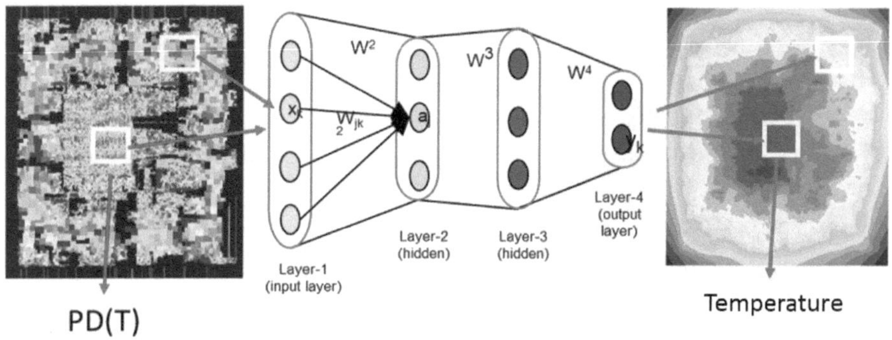

FIGURE 10.23 Inferencing step by applying DeltaT Predictor for DNN-based fast thermal solver.

by centering it at each tile location. When it is centered at a tile location, the effect of all the heat sources inside the window is accounted for at the center tile, thereby calculating the temperature rise, the DeltaT. Adding the DeltaT map from DeltaT Predictor to the coarse Thermal profile from the system-level thermal analysis, the final thermal map with enhanced resolution is generated for use in chip reliability analysis or determination of top-ranked T_{max} locations for the placement of on-chip thermal sensors.

10.3.5.5 Results

About 1.7 million training and 0.2 million test cases with FE models were created and solved in a commercial-grade chip-package thermal analysis tool to generate the training data needed by DeltaT Predictor. Each FE model has 5 × 5 (10 μm × 10 μm) tiles with unique power patterns on each tile embedded in a (4 mm × 4 mm) training chip for LHE (i.e., localized heating) and it takes about 8 seconds to solve in a Linux server with 32 cores using multi-threading.

The number of cases are reduced significantly by considering the factors such as symmetry and levels of power on tiles in power pattern. To generate the large number of training cases, large-scale parallel processing was implemented to run in batches. The results are automatically collected for model training afterward.

For each training case, Theta-JA, power in center 3 × 3 tiles and in ring tiles (the remaining tiles of the 5 × 5 tiles) are the input features. For evaluating the model, tenfold cross-validation is used on the training set for model determination.

The power level on each tile of the power template was assigned to three different levels to cover the practical ranges of power values for a particular node of advanced FinFET technology such as 7 nm.

For different package/system environments, characterized as Theta-JA, the training cases are categorized into multiple Theta-JA values for the implementation of the DeltaT Predictor. The Theta-JA values for the training cases need to cover the realistic ranges of practical packages in a system.

The training cases of DeltaT Predictor with 5 Theta-JA values and about 1.7 million power template patterns for each Theta-JA takes about 11–14 seconds, while

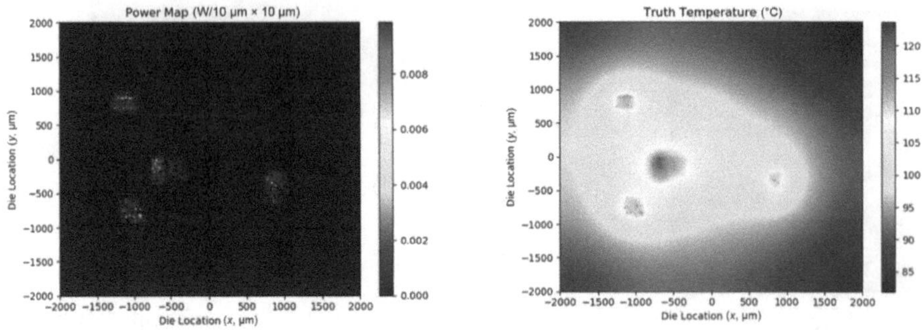

FIGURE 10.24 Power map and ground truth temperature using FEA solver.

modeling and solving all the cases using FEA takes more than 12 hours even with the help of parallel processing.

Apart from the DNN-based DeltaT Predictor, another popular technique was also used to generate the DeltaT Predictor as a part of this experiment. To verify the accuracy in inferencing, the 4 mm × 4 mm chip configuration is reused but with a randomly generated power pattern in a detailed FEA to obtain the golden results of temperature and DeltaT. The maximum differences found in DeltaT on tiles as compared to the "golden results" is about 2%.

Similarly, for the same CPS configuration, a tile-based power map of 4 mm × 4 mm from a test chip is calculated from CTM generated by a commercial dynamic voltage drop analysis tool.

The total power on chip was 3.377 W, as shown in Figure 10.24, and the base Theta-JA (27.86 C/W) of the package/system was calculated from applying uniform 1 W on chip with T_{max} on chip extracted from FEM analysis. The "golden" DeltaT on each tile was calculated by the difference of temperature on each tile and a reference temperature, e.g., T_{max} of the uniform power case. Alternatively, the temperature profile of the uniform power case can also be used as the reference temperature in calculating the "golden" DeltaT map, as this could be more accurate at the border/corner locations of the chip.

The Ground Truth FinalT map was generated using FEA solver and is shown in Figure 10.24. The coarse thermal map from system-level thermal analysis is in Figure 10.25 with 200 μm × 200 μm resolution and can be run efficiently. The DeltaT Predictor generates the FinalT map as in Figure 10.26 with the inputs as the given randomly generated tile-based power map and a Theta-JA values of 27.86 C/W.

The maximum differences found in DeltaT/FinalT on tiles as compared to the "golden" was about <4%, as shown in Figure 10.27, which is a satisfactory result on accuracy.

When the new chip sizes become as large as 10 mm × 10 mm or even 25 mm × 25 mm, the FE model with fine mesh on chip will make the FEA approach less favorable as the runtime could be up to 15 hours or more, while the DeltaT Predictor run can be 1,000× faster. Even though the data is generated on a 3D die, the solutions are retained only on the chip where the power map is applied. Hence, the temperature is predicted only on a surface, but its spatial distribution and magnitude are affected

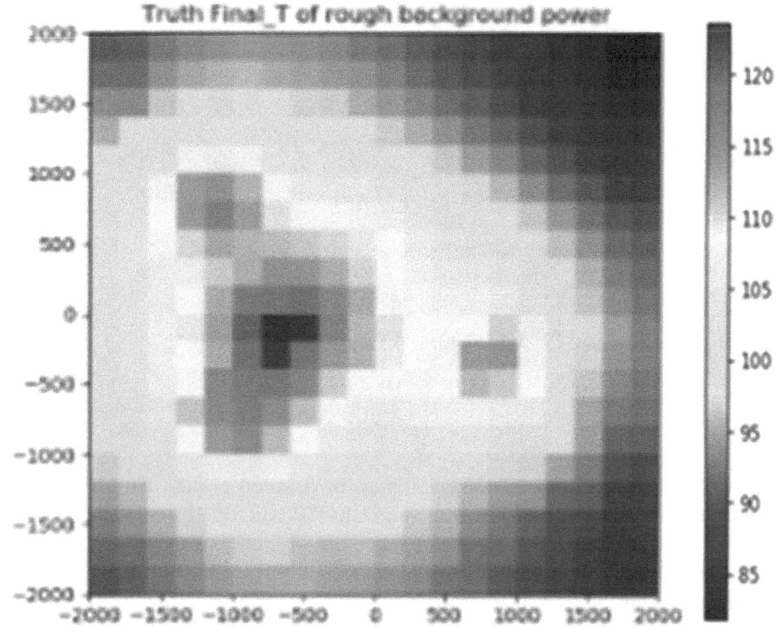

FIGURE 10.25 Coarse T map of the test chip using FEA from system-level thermal analysis and can be run efficiently.

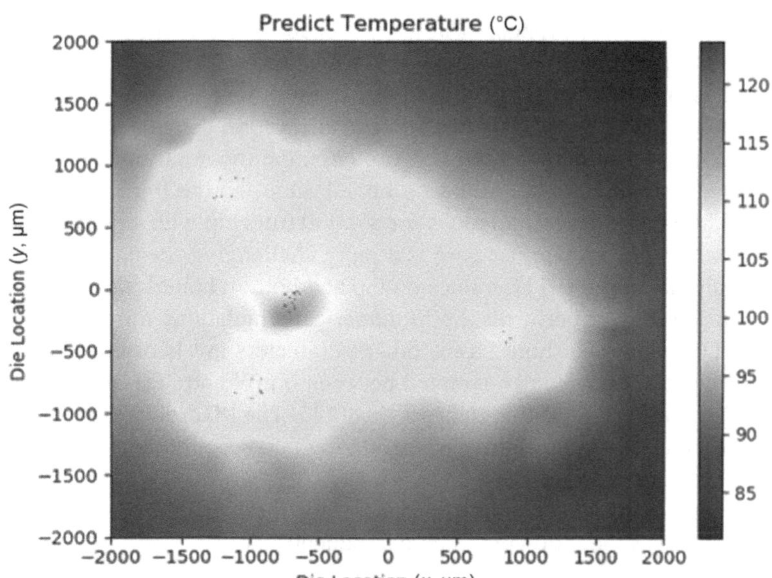

FIGURE 10.26 Final T map from the summation of DeltaT and temperature map of rough thermal profile using coarse grid thermal analysis (FEA).

Chip size	Package size	Theta-JA
4 mm × 4 mm	10 mm × 10 mm	27.86

	Data sets	Runtime	CPU cores	Memory
Training	~1.7 million	~8 hours	256	1.6 G
Rough profile	1	~14 seconds	1	75 million

	Max *T*	DeltaT	Runtime	CPU cores	Memory
Ground truth	123.58°C	12.09°C	42 minutes	32	4 G
Predicted	124.40°C	12.98°C	10 seconds (~600x speed-up)	1	0.6 G

FIGURE 10.27 Runtime and memory statistics of detailed ground truth FEA run, rough thermal profile from coarse grid FEA run, and DeltaT Predictor.

by system parameters such as heat transfer coefficient (HTC) and die thickness. As a result, these parameters are accounted for in the thermal ML-solver along with power map. The division between the training and testing data sets is 80/20.

10.3.6 Autoencoder-Based Fast Static On-Chip Thermal Analysis

10.3.6.1 Background and Architecture

The authors in Ref. [7] propose a novel autoencoder-based framework with constrained equilibrium condition during solution evaluation, similar to partial differential equation (PDE) solvers. This work also uses the simple three-layer die model for thermal analysis. The work describes two use cases for a thermal simulation solver that utilizes ML techniques. In the first use case, the input parameters consist of a power map distribution, HTC values on boundaries, and die thickness. The solver performs 5,000 numerical simulations for a 4,000 μm chip with a grid resolution of 256 × 256 × 32. The second use case is a more challenging scenario with a spatial tile-based distribution for HTC values on top and bottom boundaries, similar to the power map. The solver performs 2,500 numerical simulations for a random power map and HTC values. In both cases, other parameters are kept constant, and the power on each tile is randomly sampled between 0 mW and 300 mW. The division between the training and testing data sets is 80/20. The ML solver predicts the temperature distribution on the chip's surface, taking into account the power map, HTC values, and die thickness.

A schematic of the autoencoder-based fast static thermal solver model is shown in Figure 10.28. It has the following main components.

1. *Decomposition of Computational Domain:* The computational domain corresponding to the 2D chip is decomposed into subdomains of equal physical sizes, where each subdomain consists of $m2$ computational elements.

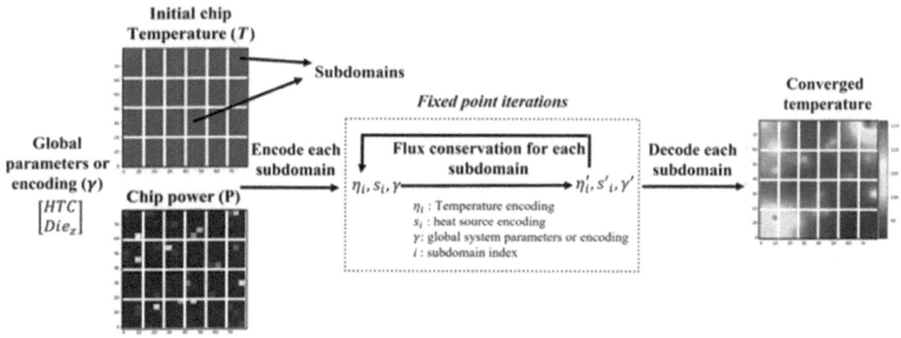

FIGURE 10.28 Schematic of autoencoder-based thermal solver.

The value of m is set to 16, resulting in around 256 subdomains for the 4,000 μm × 4,000 μm chip based on experiments.

2. *Encoding on Subdomains:* At an initial temperature of $T = 300$ K and a given power map, P are encoded into lower dimensional encodings η and s on every subdomain using pretrained encoders. The global system parameters, γ = *HTC*, *Diez*, which are the heat transfer coefficients and die thickness, respectively, are kept constant across all the subdomains in the first use case. The conditioning of all the subdomains with global parameters improves the predictive capability of the solver. In the second use case, HTCs on the top and bottom surfaces are decomposed into subdomains and encoded similarly to the power map, but *Diez* is used as a global parameter. Even though the HTC is applied on the top and bottom surface of the die, they are considered to be applied on the plane of the chip for the sake of modeling a 2D problem.
3. *Constrained Thermal Solver:* A constrained equilibrium problem is solved using fixed point iterations where the encodings (η, s, γ) on each subdomain and its neighbors are evaluated iteratively using a pretrained flux conservation autoencoder. In each iteration, the output encoding corresponding to temperature, η', evaluated by the flux conservation autoencoder, is used to replace the input temperature encoding, η for the next iteration. Other encodings such as s and γ are kept fixed. The fixed encodings serve as hard constraints and steer the temperature encoding toward an equilibrium, which resembles the encoding of converged temperature. The fixed point iteration is stopped when the following condition is met, $L2(\eta - \eta') < 1e^{-8}$.
4. *Decoding on Subdomains:* The converged temperature encodings on subdomains at the end of the fixed point iterations are decoded into temperatures and post-processed on the entire computational domain.

This approach strongly draws ideas and inspiration from traditional PDE solvers. The novel iterative inferencing approach is designed to provide stability and robustness to the solution methodology in comparison with black-box ML methods. The flux conservation autoencoder is useful in exchanging information among neighboring subdomains and from the boundary to the interior of the chip. This is like traditional solvers where flux conservation is employed between computational elements using numerical approximations for the same purpose. Finally, it is important to note that the training portion in this method corresponds to simply training three or four autoencoders depending on the use case, as will be explained later. Autoencoders are an unsupervised ML technique to obtain lower dimensional encodings from higher dimensional fields [55].

10.3.6.2 Autoencoders Used for Thermal Solver

The work uses autoencoders in its approach, which are CNN-based encoder–decoder networks for down-sampling the subdomain field into lower dimensional encoding and up-sampling it back to the original input field. Separate autoencoders are trained for temperature and power map in both use cases, and an autoencoder for distributed HTC is also trained in the second use case, as shown in Figure 10.29. In addition, DNN-based encoder–decoder networks are used for the flux conservation autoencoders. The temperature, power map, and system parameters on a 2D stencil of five subdomains are encoded using the pretrained encoders and stacked together as input. Autoencoders are chosen over other compression techniques because they have powerful generalization and denoising capabilities with reasonable compression ratios. The autoencoders are trained using an optimizer with mean absolute error loss to improve the stability, robustness, and convergence of the constrained equilibrium solver employed during the evaluation of the approach.

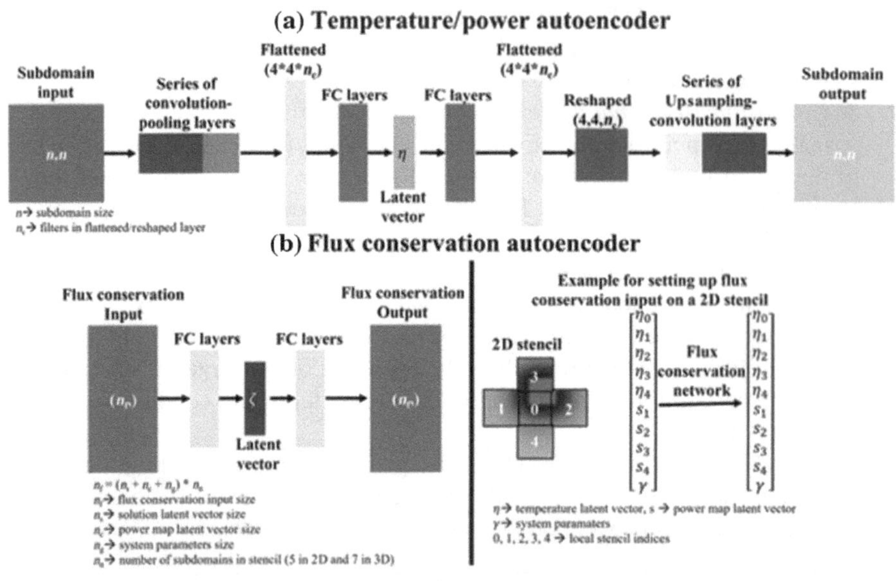

FIGURE 10.29 Autoencoders trained in the thermal solver.

10.3.6.3 Results

This work experiments with the two uses cases described above and compare with the golden results from a commercial industry tool. This work also shows that the results can be generalized for varying chip sizes, tile sizes, and out of range HTC values. The relative L_2 error is defined as $L_2 = \frac{L_2\left(T_{\text{true}} - T_{\text{pred}}\right)}{L_2\left(T_{\text{true}}\right)}$. On the unseen 500 samples, the L_2 error is 0.078. When larger chip sizes are experimented with, the results also have a good agreement with the golden results, as shown in Figure 10.30. The L_2 error for (8,000 μm × 8,000 μm) and (16,000 μm × 16,000 μm) chip sizes are 0.058 and 0.089, respectively. It is also shown that the varying tile sizes also result in good accuracy.

Distributed HTCs can have a greater impact on temperature distribution, making it more challenging to model. Additionally, the smaller tile sizes result in more local heating and shaper peaks in temperature which are also difficult to model. Despite the challenges, this approach performs reasonably well with average relative L_2 errors of 0.045 across all samples. It may also be observed from the contour plots in Figure 10.31 that the temperature predictions between the autoencoder-based model and the commercial industry tool match well. This solver continues to generalize to bigger chip sizes for this use case as well with the same accuracy as the previous case.

The computational time required by the commercial industry tool on a 4,000 μm chip is about 30 minutes on a single CPU. In contrast, the thermal ML-solver converges in less than 10 seconds on a single CPU. Additionally, the approach can be

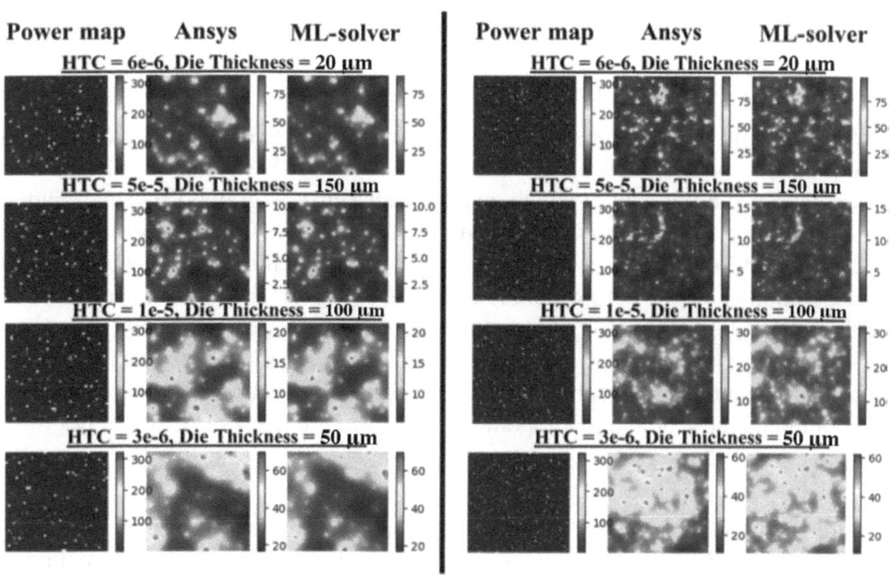

FIGURE 10.30 Results from large die sizes, 8,000 μm × 8,000 μm (*left*) and 16,000 μm × 16,000 μm (*right*).

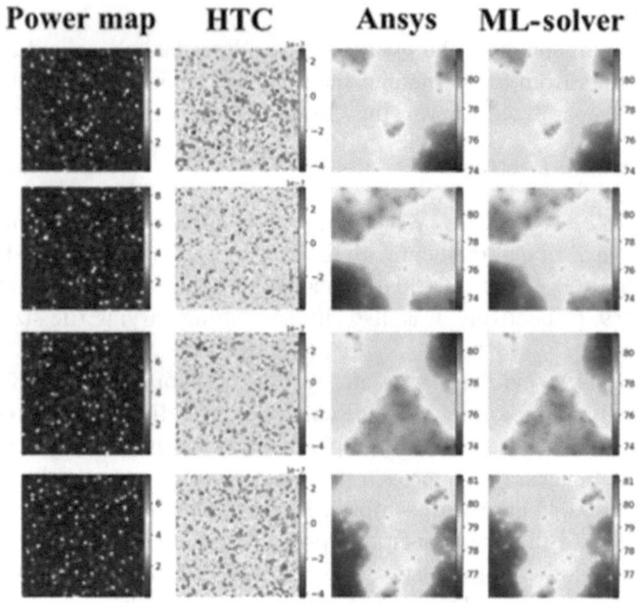

FIGURE 10.31 Distributed HTC on top and bottom with 50 μm tile size.

scaled to multiple CPUs and GPUs for even greater time-saving as the chip size increases. While the approach may be slower than black-box ML models due to iterative evaluations, it offers better accuracy, generalizability, and scalability.

10.3.7 Novel Fast Transient Thermal Simulation

Performing fine-grained transient thermal analysis on large 3DIC designs is virtually impossible using traditional CFD/FEA-based solvers. This section describes a novel fast transient thermal solver which enables transient fine-grained thermal analysis with extremely fast runtimes compared to traditional FEA-based thermal analysis [9]. As shown in Figure 10.3, the power can vary widely and change abruptly during the operation of a chip. For example, in the case of a smartphone, the transient power variation can be because of the boot-up sequence, followed by watching a video, talking on the phone, or browsing the web. Therefore, the static thermal analysis will not be sufficient to predict the actual thermal hot spots, can be too pessimistic, or cannot model the spatial and temporal variations in the hot spots accurately.

10.3.7.1 Overall Flow

Figure 10.32 shows the flow overview of the fast transient thermal solver. The inputs to this flow are the CPS design inputs, boundary conditions, and the transient power profile.

A trained ML model is used to predict the *decay surface models* to characterize the local, intermediate, and global models. The ML predictor and the decay surface

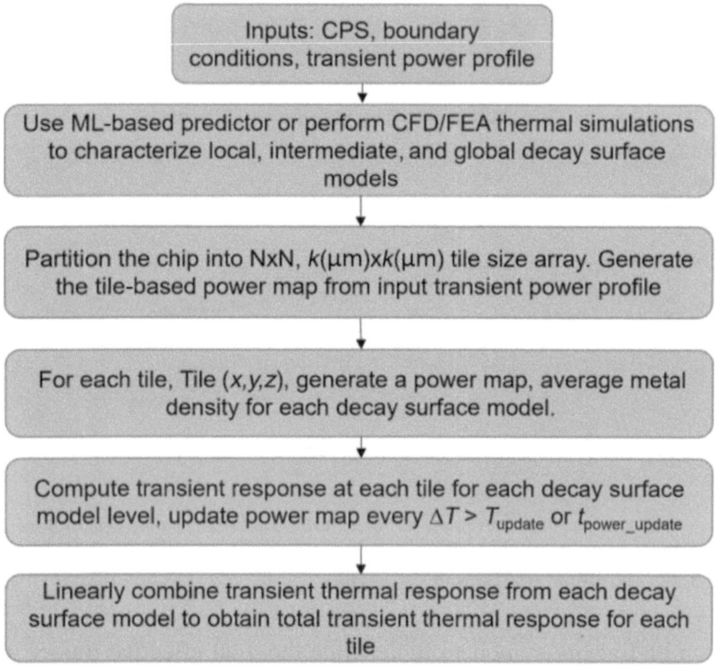

FIGURE 10.32 Fast transient thermal solver flow overview.

models will be described in later sections. The chip is partitioned into N × M tiles of size k (μm) × k (μm) on which the transient power is distributed to generate the fine-grained power map which serves as the heat sources. With this transient power map, the transient temperature curves are generated using a convolution-based approach described later for each tile location. Finally, the transient temperature curves for each of the tiles are linearly superimposed to generate the effective transient response. This work also uses a three-layer die model.

10.3.7.2 Transient Decay Surface

Figure 10.33 shows an example transient decay surface characterized for a 10 μm × 10 μm heat source. This figure shows the temperature change with time at various distances from the heat source. This essentially models the temperature at any location due to a given heat source as a function of time and distance as in Equation (10.3):

$$\Delta T_{ij} = f\left(d_{ij}, t\right) \tag{10.3}$$

$$\Delta T_i = \sum_j T_{ij} \ \forall \ j \in H \tag{10.4}$$

Here H is the set of on-chip heat sources. The term ΔT_{ij} denotes the temperature change at the ith location due to jth on-chip heat source. A linear time invariant (LTI) system is assumed and, therefore, the contribution of each of the on-chip heat

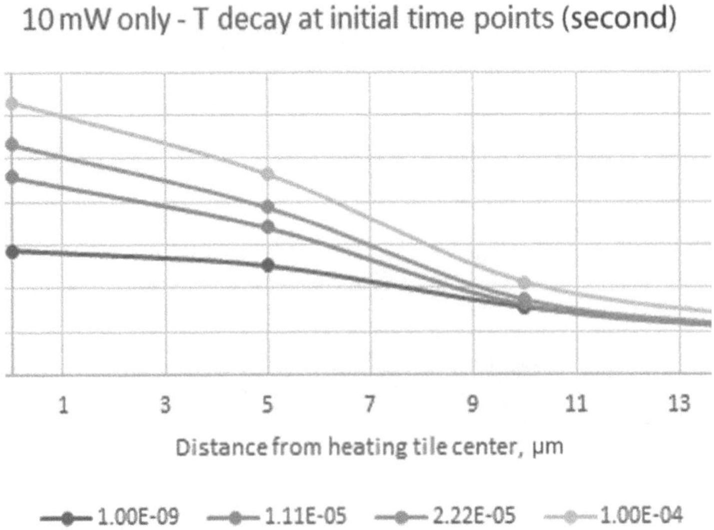

FIGURE 10.33 Example transient decay surface.

sources can be linearly combined to generate the total effective transient response at the *i*th location, as in Equation (10.4). This operation is repeated for all the tiles in the chip to generate the transient response curves for each tile. For a chip of size 1 cm × 1 cm, the number of tiles will be 1 million, and therefore for each tile, the impact of 1 million tiles needs to be computed on a given tile. Repeating this operation for each tile will lead to 1 million × 1 million operations which can be prohibitively expensive computationally. Therefore, in this solution, the number of operations for each tile can be reduced using *multiscale decay surface model.* The thermal response at any given tile location can be computed by considering the impact of heat sources from nearby tiles, heat sources from intermediate distance tiles, and heat sources from far away tiles. These impacts are categorized as local impact, intermediate impact, and global impact corresponding to *local tiles*, *intermediate tiles*, and *global tiles.* These tiles then correspond to *local level*, *intermediate level*, and *global level* decay surface models.

The decay surface at any level can be categorized by considering an M × M array where $M \ll$ no. of tiles on the chip. For example, a local-level decay surface can be characterized with an array of 21 × 21 tiles with a tile size of 10 µm × 10 µm, whereas an intermediate-level decay surface can be categorized with an array of 10 × 10 tiles with a tile size of 210 µm × 210 µm, and the global-level decay surface can simply be the full CPS level response at the center of the chip with the power uniformly distributed on the chip.

Figure 10.34 shows how a decay surface can be characterized for an M × M array. The tile array is placed on the chip, and a heat source of power P_{char} is applied at the center of the array. With this heat source, the transient thermal response is captured at all the M × M tiles. The set $Ds = \{P_{char}, \Delta T_{char}\}$, where ΔT_{char} is the set of all the transient curves for M × M tiles, represents the decay surface model for this level.

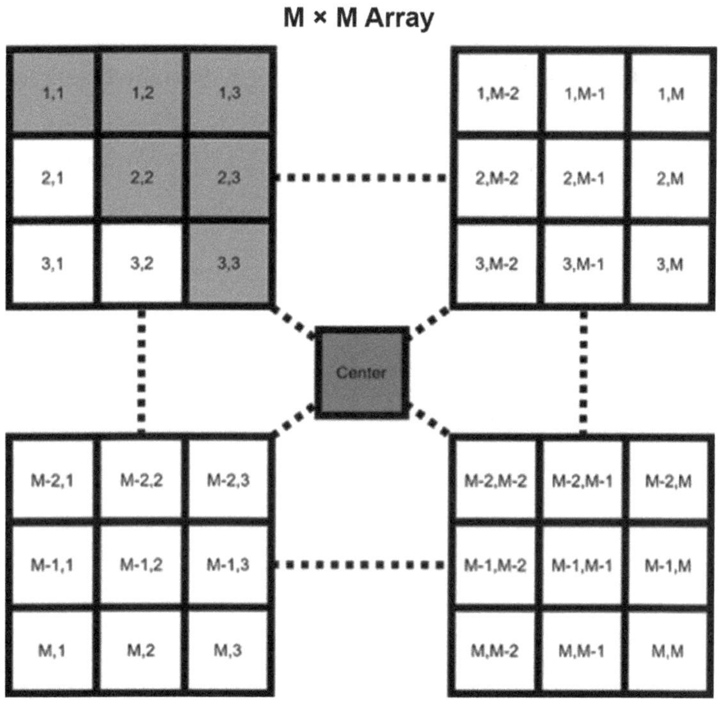

FIGURE 10.34 Decay surface characterization for M × M array.

The guiding principle is that for the local-level decay surface, the smallest tile size is used for high accuracy to capture the impact of local heat sources, and coarser tile sizes are used for intermediate-level decay surfaces since the impact is an aggregated effect. It is observed that the thermal responses in the decay surfaces scale almost linearly with power as follows:

$$\frac{\Delta T'}{\Delta T\text{char}} = \frac{P'}{P\text{char}} \tag{10.5}$$

It should be noted that the decay surface characterizations are performed with the full CPS model.

10.3.7.3 ML-based Decay Surface Predictor

Having described the fundamental idea of transient decay surface in the previous section, it should be noted that for any given CPS, this initial characterization has to be performed before the fast transient thermal analysis solver can be used. To address this challenge, a ML-based predictor was developed which can predict the decay surfaces for a given CPS. This section explains the ML-based transient decay surface characterization. To accurately capture the far-field and near-field transient decay surface components, ML-based models for various levels are constructed and trained. The precision of the models' predictions is evaluated on unseen cases/data. Transient simulations have been carried out utilizing a high-fidelity numerical solver

TABLE 10.1
Parameter for Training Data Generation for Decay Surface Predictor Model

System Parameters	Values/Ranges
Power per tile (mW)	500/1,000/5,000/20,000
Heat transfer coefficient (mW/μm²·K)	$1e^{-9}$–$2e^{-5}$
Die *x*/*y* length (μm)	4,000–30,000
Die *z* height (μm)	[100, 200]
Dielectric/interconnection layer thickness (μm)	1–10
Insulation layer thickness (μm)	0.01–0.05
Tile size (μm)	10/210/3,200/16,000
Si substrate thermal conductivity (mW/μm·K)	0.1305
Insulation layer thermal conductivity (mW/μm·K)	$0.138e^{-2}$
C-scale	0.5–200
Dielectric thermal conductivity interconnection layer (mW/μm·K)	0.00138–0.138

for simulating the transient response of the chip with its center tile heated. By adjusting the input system parameters of the simulation model, different sets of data can be generated. The corresponding system parameter space for the four decay surface levels is provided in Table 10.1.

It is worth noting that the system parameters for the four different levels are almost the same except for the tile size and the power applied on the tile. Besides, one additional parameter, C-scale, for modifying the thermal capacitance of the system, has been introduced as well. Uniform random sampling has been employed to select values from the predefined space for each system parameter. In total, 1,000 distinct cases have been created for each decay surface level. Since the input to the ML model includes different combinations of system parameters along with the time step and the output of the ML model is the corresponding temperature distribution of the chip at the given time step, a DeepONet-based neural network has been built. The network structure is depicted in Figure 10.35, where the branch net will take system parameters as input and the trunk net will take time step as input. The information from the branch net and trunk net are combined through a merging layer and output the temperature distribution on the chip. The trained model is tested on a test

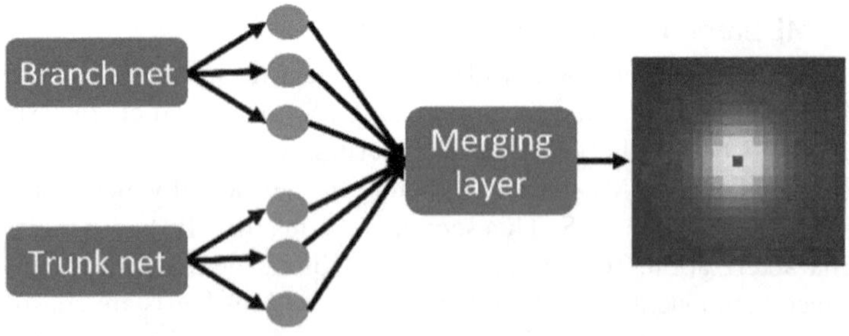

FIGURE 10.35 DeepONet model for decay surface predictor [9].

data set of 100 unseen cases and the overall L_2 relative error is less than 1%, which verifies the accuracy of the trained model.

10.3.7.4 Computing Effective Transient Thermal Responses

Given an on-chip power map with tile-level heat sources, the transient thermal response at any tile i can be constructed by using the decay surface models and convolving it with the power map centered at the tile i with the thermal responses in the decay surface scale based on the actual power distribution as in Equation (10.3). The convolution operation to compute the temperature at any location (x, y) and time t can be written as follows:

$$\Delta T\left(x,y,t\right)=p^{*}\Delta Tds=\sum_{j,k}\Delta Tds\left[j,k\right]p\left[x-j,y-k\right] \tag{10.6}$$

where ΔTds is the convolution kernel from the decay surface model, and p is the scaled power map with respect to the power used for characterizing the decay surface model. This equation will calculate the contribution of the $\Delta T\left(x,y,t\right)$ due to the corresponding decay surface model ΔTds.

Figure 10.36 shows the convolution of the scaled power map decay surface kernel to obtain the temperature response. For a given decay surface model, this operation is repeated for each time point and thus we get the transient thermal response. The convolution with the local decay surface model will capture the effect of the nearby tiles, and the intermediate decay surface model will capture the effect of the heat sources at intermediate distances, whereas the global decay surface model will capture the full-chip heating effects. As each decay surface model *kernels* are limited in size to a few hundred elements, the convolution is fast. While computing the temperature response for a given tile, the power maps for the intermediate decay surfaces are adjusted so that the heat sources which have been accounted for in the convolution with local decay surface are not considered to avoid double counting. Similarly, while performing convolution with the global decay surface, the heat sources which have been accounted for during the convolution with the local and intermediate decay surface are ignored. Figure 10.37 shows how the impact of heat sources are convolved with the decay surface models to generate the effective transient thermal response.

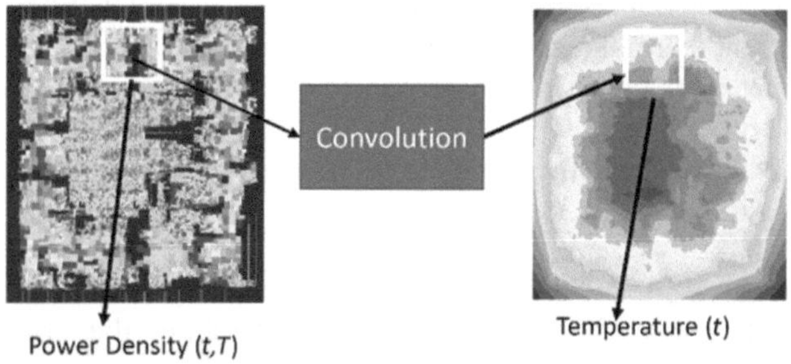

FIGURE 10.36 Convolution on the power map with the decay surface kernel.

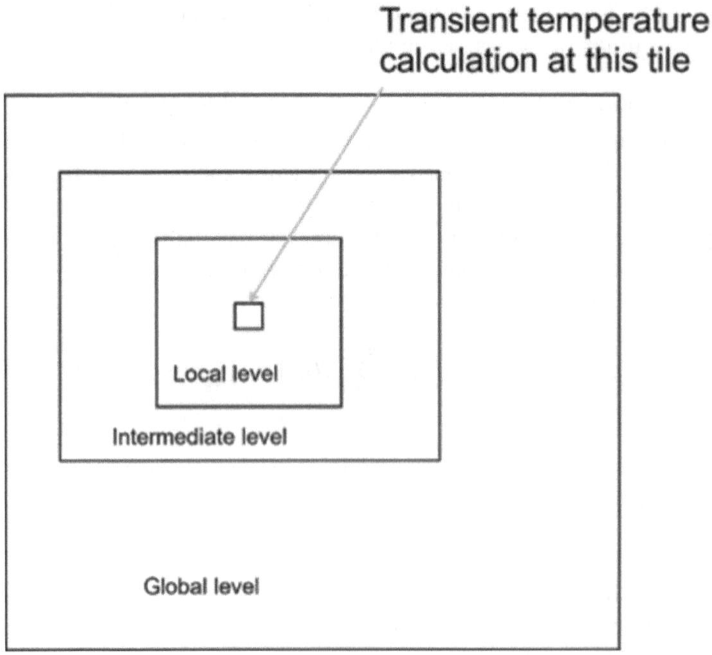

FIGURE 10.37 Combining the thermal responses from decay surfaces.

Therefore, the following steps are used to generate the effective transient thermal response at any given tile location:

1. Generate the power map for the local, intermediate, and global levels at the tile of interest.
2. The local- and intermediate-level power maps are centered at the tile for which transient thermal response is being calculated.
3. For any given level, use the corresponding decay surface to generate the transient thermal responses due to all the heat sources for the level.
4. Linearly combine the transient responses corresponding to each decay surface level to capture the total transient thermal response for the tile.
5. The above steps can be repeated for all the tiles/sensor locations to generate the transient thermal response for all the tiles/sensor locations.
6. Heat sources which have been accounted for at a certain level are excluded from the other levels.
7. With the above steps, the per tile total transient response can be generated as shown in Figure 10.38.

10.3.7.5 Linear Superposition across Time Points

Figure 10.39 describes the methodology for generating the effective transient responses as the power changes with time. As shown in the figure, each of the power values P1 and P2 will have its own transient thermal response curve, TR1 and TR2,

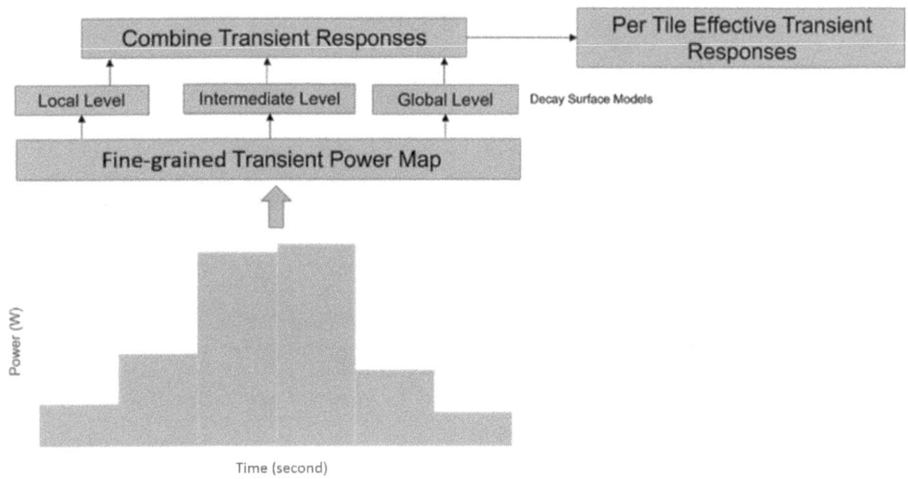

FIGURE 10.38 Generating the total on-chip transient thermal response.

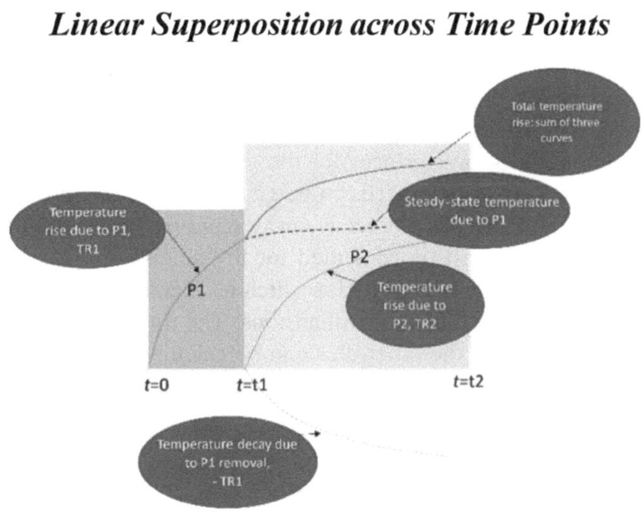

FIGURE 10.39 Linear superposition of transient thermal responses.

respectively. When the power changes from P1 to P2, it can be considered as P1 reducing to zero and a new power P2 being applied. This would mean that the chip will cool down due to P1 reducing to zero but will heat up due to power P2. This is modeled by the negative (mirror image) –TR1, representing cooling effect, starting at time $t = t1$ and the heating curve TR2 starting at time $t = t1$. This can be done for an arbitrary number of power changes such that the transient power profile is represented by the set $\{Pi(ti, tj), i = 1, \ldots, N\}$, where ti, tj are the start and end time, respectively, for power Pi. To compute the effective transient response curve at any tile location, all the curves, including the mirror image curves, are linearly added.

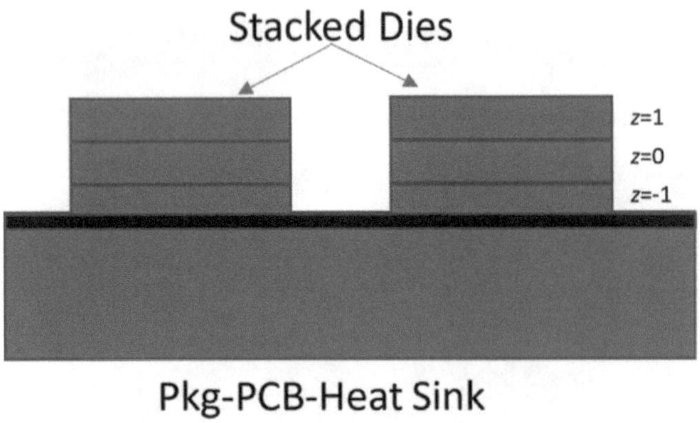

FIGURE 10.40 Stacked die thermal coupling.

10.3.7.5.1 Extension to 3DIC

The methodology described in the previous sections are explained with illustrations on a single die case. However, the methodology can be directly extended to a 2.5D/3DIC case. Figure 10.40 shows an example of 3DIC with stacked dies. The heat sources on a die can affect the neighboring dies through thermal coupling, and hence this effect needs to be modeled to accurately predict the transient thermal response on any die. The decay surface–based transient thermal modeling can be extended by computing decay surfaces for the given die and the neighboring dies. This means that a decay surface needs to be generated in the z dimension as well. Therefore, while characterizing the decay surface models, each such model (local, intermediate, and global) will be generated for the neighboring dies as well. This can also be performed by the ML predictor which has been trained for predicting the decay surface, $dS(x, y, z)$, in the x, y, z dimensions. The transient thermal impact due to a power map on a die on the other dies can be calculated using Equation (10.4). Equation (10.5) then computes the total thermal response on the die of interest.

$$T(dS,\ pMap)_{x,y,z} = \sum_{z=-d}^{d} f\left(dS_z,\ pMap_z\right) \tag{10.7}$$

$$T_{x,y,z} = \sum_{dS,pMap} T_{x,y,z}\left(dS,\ pMap\right) \tag{10.8}$$

In the above equations, dS_z is the decay surface and $pMap_z$ is the power map at die with location at z. Equation (10.8) gives the temperature distribution on any die given the decay surface models and the power maps for each die.

10.3.7.5.2 Results

To demonstrate the dramatic improvement in runtime and capacity, an example chip of size (1.6 cm × 1.6 cm) is selected along with an example package and board. Four different power maps are randomly generated and applied for 1 second duration

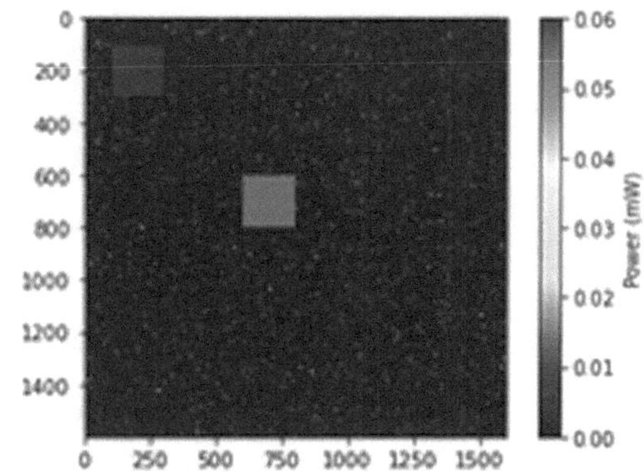

FIGURE 10.41 Example power map for 1.6 cm × 1.6 cm chip.

each. Each of the power map has a resolution of 10 μm × 10 μm with a total of 2.56 million heat sources. An example power map is shown in Figure 10.41. The fast transient thermal analysis is performed for all the 2.56 million locations on the chip for a total simulation time of 4 seconds. The runtime for this analysis is only about 15 minutes given the pre-characterized or ML model predicted decay surface models. Figure 10.42 shows the transient thermal response curves for the two probe locations. Figure 10.43 shows the temperature map at t = 4 seconds for the above transient thermal analysis. Since this is a relatively large chip, the decay surface models at the resolutions of (10 μm × 10 μm, 21 × 21), (210 μm × 210 μm, 21 × 21), and (3,200 μm × 3,200 μm, 9 × 9) and the global decay surface model are used.

To demonstrate the accuracy of the fast transient thermal solver, a small chip of 4 mm × 4 mm is selected and appropriate boundary conditions are applied. An FEA-based solver is used to run transient power-on thermal analysis. This power-on transient thermal analysis simulation is of 5 seconds duration. A randomly generated

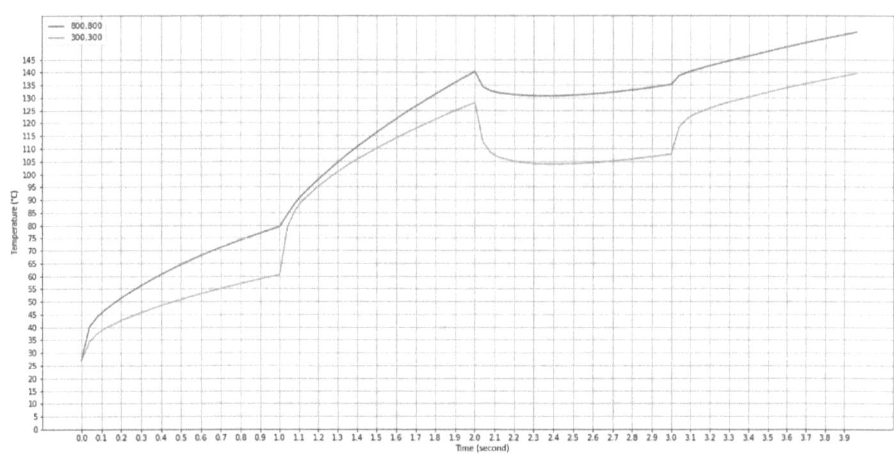

FIGURE 10.42 Transient curves at two different temperature probe locations.

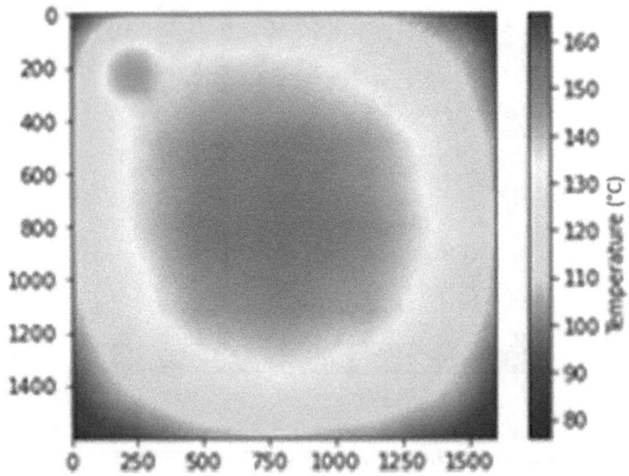

FIGURE 10.43 Temperature map at $t = 4$ seconds.

power map as shown in Figure 10.44 is applied such that the total power is 790 mW. With the above power map, the FEA-based solver and the fast transient thermal solver are used for generating the transient response curves at the center of chip for correlation. The local and global decay surface models are used in the fast transient thermal solver since this is a small chip. As shown in Figure 10.45, the fast transient thermal solver matches the results very closely with the FEA-based solver. For this small case, the fast transient solver is about 120× faster than the FEA-based thermal solver.

Figure 10.46 shows another example with a different power map with the temperatures probed at the center of the chip and an off-center location. The fast transient

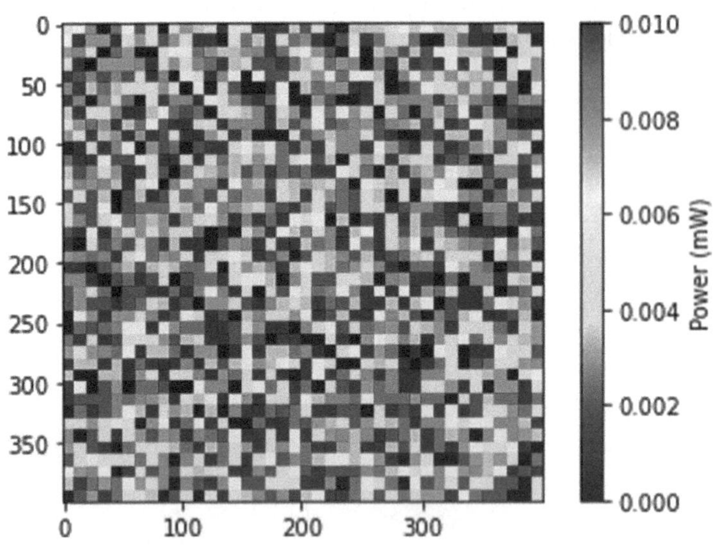

FIGURE 10.44 Power map for a 4 mm × 4 mm chip.

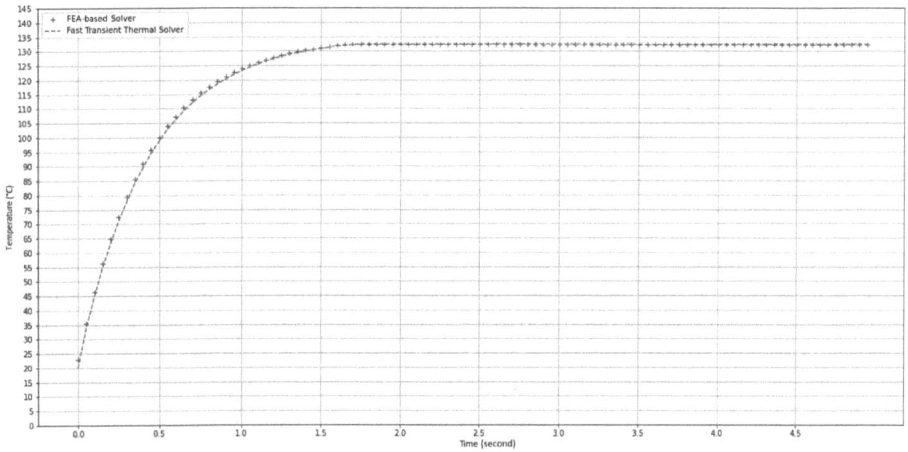

FIGURE 10.45 Comparing the power-on transient between the FEA-based solver and fast transient thermal solver.

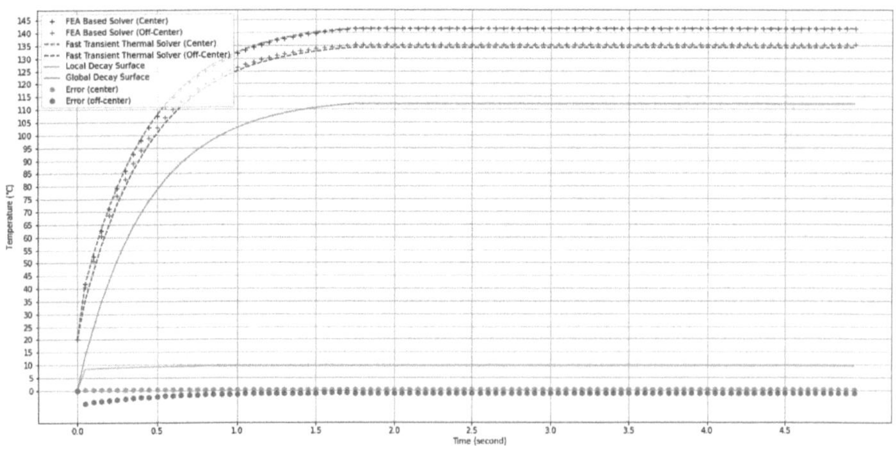

FIGURE 10.46 Temperature components from different decay surface models.

thermal solver matches well with the FEA-based thermal solver for both the probe locations. The figure also shows the contributions of the different components of the decay surface model to the total temperature curve. At the steady state, the local decay surface contributes about 9.5°C, and the global decay surface contributes about 112°C. These two components added with the ambient temperature of 20°C leads to a total temperature of 140.5°C. The contribution of the global decay surface is small since this is a small chip and the local level decay surface covers most part of the chip for the given temperature probe point. The figure also plots the difference between the results from the FEA-based solver (golden) and the fast transient thermal solver.

10.4 Cu–Cu DIRECT BONDING MODELING AND NUMERICAL SIMULATION

While thermal-compression bonding is a key technology to ensure vertical chip/wafer stacking in 3DIC fabrication, Cu–Cu direct bonding offers a preferred alternative when the package size continues to reduce. In addition to thermal management, the mechanical and reliability performance of Cu–Cu-bonded devices must be evaluated. Since the quality of the bonding interface determines the thermal, electrical, and mechanical characteristics of the interconnection, fast and accurate numerical tools are demanded to simulate the thermal mechanical coupling Cu–Cu bonding process as well as to study the mechanical effects of bonding voids after processing.

10.4.1 Overview of Cu–Cu Direct Bonding Modeling

Cu–Cu direct bonding at low temperature [57–60] is a promising way to achieve 3D stacking of integrated circuits [61, 62]. It is a solid-state interconnection technique, also known as a diffusion bonding process [57, 63]. However, the dishing effect and surface roughness from chemical–mechanical polishing (CMP) can significantly deteriorate the mechanical and microstructural properties of bonding [64–67]. Recrystallization in the post-annealing process on the diffusion bonding surface can improve the bonding performance. A proper control of annealing temperature and duration on diffusion bonding can mitigate the surface topographical effects to achieve the desired bonding quality. Nevertheless, migration and growth of voids, induced by improper electromigration and mechanical stress, could cause Cu–Cu bonding to fail. As a result, the void formulation and bond strength have become two main issues influencing bonding's performance, reliability, and yield [68]. To investigate the issues and minimize the need of expensive and time-consuming experiments, complementary computer modeling techniques are required.

In essence, the direct Cu–Cu diffusive bonding process is a typical space–time multiscale problem that describes both fine space scale (atom and void size: nm) and fine timescale (atom diffusion: 10^{-4} second) to coarse space scale (pitch scale: micrometer) and coarse timescale (annealing time: hour). The state-of-the-art computer modeling approach uses a cohesive model to mimic van der Waals forces in bonding process and a non-separation contact condition in simulating the final bonding status [64–67]. However, these numerical models are applied to a FE mesh without considering the voids, which does not address the detail of void formation and how it impacts the bond strength as a multiscale feature in the direct Cu–Cu diffusive bonding process.

In this study, we present a novel computer modeling approach using a three-step workflow that considers the voids formation in predicting the bond strength of the direct Cu–Cu diffusive bonding process. In the first step, an implicit thermal-compression FEM simulation is used to analyze the interfacial contact between two copper pads subjected to the CMP-induced dishing effect and mechanical stress. Given the final geometry of the Cu/Cu contact surfaces in the thermal-compression analysis, a pseudo-bonding model was proposed to replicate the connection of copper pads after the diffusion process. As the name suggests, the pseudo-bonding model which is built based on the Smoothed Particle Galerkin (SPG) [69, 70] method, allows a construction of solution approximation across the bonding surface in pseudo-space and pseudo-time with a specific domain of influence size. In other words, instead of performing a multiscale computational scheme with time and space homogenization, which is computationally infeasible, we assume that the bond status at the final annealing stage is controlled by the domain of influence of bonding zone particles. As a result, particles near the bonding surface have a domain of influence which is a function of annealing temperature and duration of diffusion bonding in SPG computation. This function can be calibrated from experimental data. Subsequently, an implicit transient SPG simulation was conducted to detect the possible formation of voids (unbonded areas) and their locations in the bonded model based on the elastic stress wave analysis. In the third step, an explicit dynamic SPG simulation is conducted to predict the Cu–Cu bonding strength of the connected part considering the influence of voids.

10.4.2 The Theory of Pseudo-bonding Model for Cu–Cu Bonding Simulation

10.4.2.1 Theory of SPG Method

This section introduces the SPG method which is developed based on the Galerkin weak form using the particle discretization methods [69, 70]:

$$\begin{gathered}\int_{\Omega} \delta\hat{u}\cdot\rho\ddot{u}d\Omega=\int_{\Omega}\delta F^{T}:P(F)d\Omega-f^{\text{ext}}(\hat{u})\\ \forall\delta\hat{u}\in H_0^1=\left\{v:v\middle|\Omega\in H^1(\Omega),v=0\text{ on }\Gamma_g\right\}\end{gathered} \tag{10.9}$$

where the domain Ω is discretized by a set of particles $Z_I=\{x_I\}_{I=1}^{NP}\subset\Re^3$; ρ is the material density; $\boldsymbol{F}$ denotes deformation gradient; $\boldsymbol{f}^{\text{ext}}$ represents external force; $\hat{\boldsymbol{u}}$ is the approximating displacement field by the meshfree approximation:

$$\hat{u}(x_I,t)=\sum_{I=1}^{NP}\Psi_I^A(x_I)u(x_I,t)\ \forall x\in\Omega \tag{10.10}$$

where Ψ_I^A is the Lagrangian shape functions of meshfree approximation [71]. SPG employs direct particle integration to discard mesh dependency. The shape function Ψ_I^A of a particle is constructed by its influence zone $A=\{x_J\}_{J=1}^{P}\subset\Re^3$, as shown in Figure 10.47. The size of influence zone is notated as $A(\Delta x,\Delta y,\Delta z)$. The SPG meshless approximation shape function enables handling of large deformation and material failure simulation. In the case of Cu–Cu bonding simulation, the particle-based shape function has the advantage of bonding the two copper pads through the interface, forming one continuous domain, and representing potential voids inside the bonding surface.

The explicit time integration algorithm updates the velocity $\dot{\boldsymbol{u}}(\boldsymbol{x}_I,t)$ through Equation (10.9). The FEM method [72] and other meshfree methods [73] use $\dot{\boldsymbol{u}}(\boldsymbol{x}_I,t)$

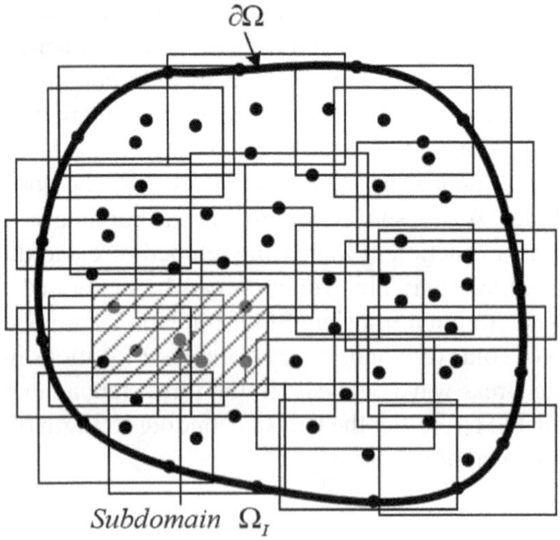

FIGURE 10.47 SPG particle influence zone.

to calculate deformation gradient $\boldsymbol{F}$ directly. The SPG method smooths the particle velocity using a momentum-consistent algorithm. Here, the momentum of particle $\boldsymbol{x}_I$ is evaluated as follows, using a smoothed sense.:

$$\hat{P}(x_I,t)=\sum_{A=1}^{P}\Psi_I^A(x_I)\dot{u}(x_I,t)\rho\Delta V_I \tag{10.11}$$

where ΔV_I is the occupied volume of particle $\boldsymbol{x}_I$ in domain Ω. To keep the momentum consistence and preserve the desired conservation of system momentum, the smoothed velocity $\dot{\hat{u}}(x_I,t)$ is calculated from $\hat{P}(x_I,t)$ with the smoothed mass:

$$\dot{\hat{u}}(x_I,t)=\hat{P}(x_I,t)\Big/\sum_{A=1}^{P}\Psi_I^A(x_I)\rho\Delta V_I \tag{10.12}$$

SPG utilizes the smoothed velocity $\dot{\hat{u}}(x_I,t)$ to calculate deformation gradient. It can be verified that the linear and angular moments are consistent before and after smoothing. The detail of SPG method can be referred to the corresponding literature [69, 70].

10.4.2.2 Thermal–Mechanical Coupling Stress Updating

Due to the significant difference in thermal expansion coefficients between copper and oxide, the geographical vacancy caused by the dishing effect between the top and bottom pads can be considerably reduced when the temperature increases. Hence, thermal–mechanical coupling simulation is necessary for Cu–Cu direct bonding simulation. To update the deformation, the velocity gradient, $\boldsymbol{l}=\dot{\boldsymbol{F}}\boldsymbol{F}^T$, can be split into two parts: a symmetric part and an anti-symmetric part, representing the rate of deformation and the rate of spin, respectively:

$$d = \frac{\left(l + l^T\right)}{2} \tag{10.13}$$

$$w = \frac{\left(l + l^T\right)}{2} \tag{10.14}$$

Taking into account the isotropic thermal expansion property of copper, the rate of deformation can be decomposed to elastic, plastic, and thermal parts:

$$\boldsymbol{d} = \boldsymbol{d}^e + \boldsymbol{d}^p + \boldsymbol{d}^T \tag{10.15}$$

For isotropic hardening, the rate of plastic deformation is defined as follows:

$$\boldsymbol{d}^p = \dot{\boldsymbol{\varepsilon}}^p \hat{\boldsymbol{n}} \tag{10.16}$$

where $\dot{\boldsymbol{\varepsilon}}^p$ is the plastic strain rate which depends on the material constitutive laws; $\hat{\boldsymbol{n}}$ denotes the normal of deviatoric stress.

The rate of thermal deformation is defined as follows:

$$\boldsymbol{d}^T = \alpha \dot{T} \mathbb{I} \tag{10.17}$$

Here $\mathbb{I}$ is the second-order unit tensor and $\dot{T}$ is the temperature increase rate. During the annealing process, the temperature in the entire domain is uniform, meaning that thermal conduction is neglected due to the relatively long thermal loading time.

Coupling with plastic and thermal effect, the stress can be updated by a rate form:

$$\dot{\sigma} = \mathbf{C} : \left(\boldsymbol{d} - \boldsymbol{d}^p - \boldsymbol{d}^T\right) \tag{10.18}$$

where $\mathbf{C}$ is the material elastic matrix.

10.4.2.3 Implicit Dynamics Equations

With SPG approximation, Equation (10.10), or regular FEM approximation [72], the continuous momentum equation, Equation (10.9), can be discretized at time step $n+1$:

$$\boldsymbol{R}\left(\boldsymbol{u}^{n+1}, t^{n+1}\right) = \boldsymbol{M}\ddot{\boldsymbol{u}}^{n+1} - \boldsymbol{f}^{\text{int}}\left(\boldsymbol{u}^{n+1}, t^{n+1}\right) + \boldsymbol{f}^{\text{ext}}\left(\boldsymbol{u}^{n+1}, t^{n+1}\right) = 0 \tag{10.19}$$

Here $\boldsymbol{M}$ is the diagonal lumped mass matrix obtained by row summing assumption:

$$M_{II} = \sum_{A=1}^{P} \Psi_I^A(x_I)\,\rho \Delta V_A \tag{10.20}$$

where ΔV_A denotes the volume of particle A. $\boldsymbol{f}^{\text{int}}$ and $\boldsymbol{f}^{\text{ext}}$ are the internal force vector and external force vector, respectively.

Equation (10.19) is time dependent and can be solved using an incremental time integration algorithm. The most popular time integration algorithm is the Newmark-β integrator [72]. With a known solution at time t^n and time increment, Δt, the updated displacement and velocity at time t^{n+1}can be calculated as follows:

$$\boldsymbol{u}^{n+1} = \boldsymbol{u}^{\beta} + \beta \Delta t^2 \ddot{\boldsymbol{u}}^{n+1} \tag{10.21}$$

$$\dot{\boldsymbol{u}}^{n+1} = \dot{\boldsymbol{u}}^{\beta} + \gamma \Delta t \ddot{\boldsymbol{u}}^{n+1} \tag{10.22}$$

Here $\boldsymbol{u}^{\beta}$ and $\dot{\boldsymbol{u}}^{\beta}$ are the displacement and velocity predictors, respectively, which are related to the known solution at t^n as follows:

$$\boldsymbol{u}^{\beta} = \boldsymbol{u}^n + \Delta t \dot{\boldsymbol{u}}^n + \frac{\Delta t^2}{2}(1-2\beta)\ddot{\boldsymbol{u}}^n \tag{10.23}$$

$$\dot{\boldsymbol{u}}^{\beta} = \dot{\boldsymbol{u}}^n + (1-\gamma)\Delta t \ddot{\boldsymbol{u}}^n \tag{10.24}$$

The Newmark-β integrator is a conditional stabilized algorithm depending on the free parameters of β and γ.

Equation (10.23) leads to an updated acceleration at t^{n+1}:

$$\ddot{\boldsymbol{u}}^{n+1} = \frac{\boldsymbol{u}^{n+1} - \boldsymbol{u}^{\beta}}{\beta \Delta t^2} \tag{10.25}$$

Substituting Equation (10.25) to Equation (10.19), the discretized momentum equation can be casted as a function of displacement $\boldsymbol{u}^{n+1}$

$$\boldsymbol{R}\left(\boldsymbol{u}^{n+1}, t^{n+1}\right) = \frac{\boldsymbol{M}}{\beta \Delta t^2}\boldsymbol{M}\left(\boldsymbol{u}^{n+1} - \boldsymbol{u}^{\beta}\right) - \boldsymbol{f}^{\text{int}}\left(\boldsymbol{u}^{n+1}, t^{n+1}\right) + \boldsymbol{f}^{\text{ext}}\left(\boldsymbol{u}^{n+1}, t^{n+1}\right) = 0 \tag{10.26}$$

Equation (10.26) is called the residual function in which the force terms, $\boldsymbol{f}^{\text{int}}$ and $\boldsymbol{f}^{\text{ext}}$, could be non-linear functions due to the non-linearity of deformation, material properties, and loading conditions. The Newton–Raphson method is widely applied to linearize Equation (10.26). The Taylor expansion of residual function is as follows:

$$\boldsymbol{R}\left(\boldsymbol{u}^{n+1}, t^{n+1}\right) = \boldsymbol{R}\left(\boldsymbol{u}, t^{n+1}\right) + \frac{\partial \boldsymbol{R}\left(\boldsymbol{u}, t^{n+1}\right)}{\partial \boldsymbol{u}}\left(\boldsymbol{u}^{n+1} - \boldsymbol{u}\right) + \mathrm{O}\left((\boldsymbol{u}^{n+1} - \boldsymbol{u})^2\right) \tag{10.27}$$

Dropping the higher order term, the linearized Equation (10.27) can be solved by an iterative procedure. With a known solution at t^n, the iteration starts from $\boldsymbol{u}_0 = \boldsymbol{u}^{\beta}$. Then the iterative step $\upsilon + 1$ can be constructed to fulfill the residual function:

$$\frac{\partial \boldsymbol{R}\left(\boldsymbol{u}_{\nu}, t^{n+1}\right)}{\partial \boldsymbol{u}}\Delta \boldsymbol{u} = -\boldsymbol{R}\left(\boldsymbol{u}_{\upsilon}, t^{n+1}\right) \tag{10.28}$$

where $\Delta \boldsymbol{u} = \boldsymbol{u}_{\upsilon+1} - \boldsymbol{u}_{\upsilon}$. Equation (10.28) is a linear function of $\Delta \boldsymbol{u}$ to update the displacement from $\boldsymbol{u}^n$ to $\boldsymbol{u}^{n+1}$. The linearization for $\frac{\partial \boldsymbol{R}\left(\boldsymbol{u}_v, t^{n+1}\right)}{\partial \boldsymbol{u}}$ is related to the deformation and material constitutive model.

10.4.2.4 Pseudo-bonding Model for Cu–Cu Bonding Model

As mentioned earlier in this section, we assume that the particle-based shape functions in SPG have their domain of influence to be a function of the annealing temperature and duration of diffusion bonding. This assumption is made due to the fact that the contamination-free surface bonding status in the direct Cu–Cu diffusive bonding is mainly influenced by the annealing temperature and time [64, 67]. The annealing temperature determines the geometric shape of the deformed interface due to thermal expansion and plays a crucial role in copper diffusion. The annealing time controls the recrystallization and grain evolution of copper. The thermal–mechanical coupling FEM simulation can capture the deformed interface with the original dish effect. However, copper diffusion, recrystallization, and grain evolution are strong multiscale phenomena. Direct simulation of these effects is computationally expensive. The proposed pseudo-bonding model in Figure 10.48 utilizes a specific influence zone size, denoted as $r(T,t)$, to search neighbor particles at the interface for SPG computation. The influence zone size is determined from experimental data at different temperatures and annealing times. A neural network can be employed to determine $r(T,t)$ from test data, but it is not within the scope of this study. Given $r(T,t)$, the SPG shape function is constructed using the same SPG approximation, Equation (10.10), for particles at the interface. The regular SPG influence zone, $A(\Delta x, \Delta y, \Delta z)$, is a numerical parameter, whereas the pseudo-bonding influence zone, $r(T,t)$, is a process-dependent material parameter that represents the multiscale bonding physics. Using the pseudo-bonding model at the contact surface, the upper and lower pads are interconnected and form a continuous computational domain. In Figure 10.48, blue and black circles represent the pseudo-bonding influence zone and the regular SPG influence zone, respectively. The solid dark arrow indicates that the adjacent surface particles are bonded together, and a dashed dark arrow implies that these two points cannot be bonded because the diffusion and

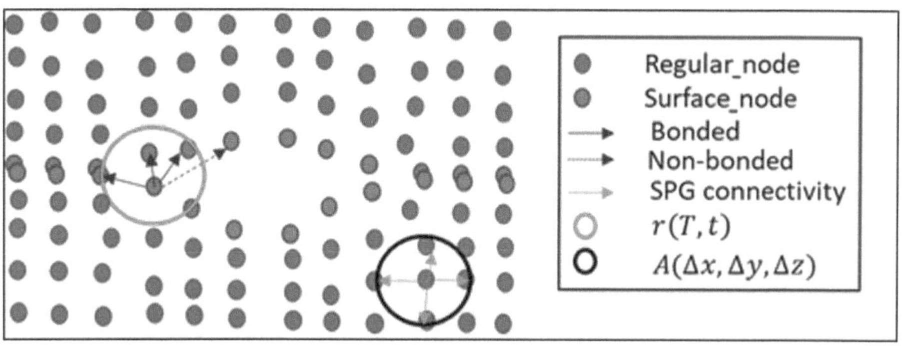

FIGURE 10.48 Illustration of pseudo-bonding model.

grain growth process cannot fill the vacancy between them. Even if two surface particles are at the bonded status, the shape function, Equation (10.10), is determined by their distance due to the kernel function [73] used in the SPG shape function calculation. A commonly used Gaussian kernel function is also used:

$$\varnothing(l_i) = \frac{1}{\left(\pi h_i^2\right)^{n/2}} \exp\left(-l_i^2 / h_i^2\right) \tag{10.29}$$

where l_i and h_i, $i = 1, 2, 3$, are the distance between two particles and radius of influence zone along *X*-, *Y*-, *Z*-directions, respectively. *n* is the dimension of space. The distribution of the Gaussian kernel function is illustrated in Figure 10.49. As the bonding status of two interfacial particles is defined by $r(T, t)$, the interconnection between two pads is constructed autonomously through the construction of SPG shape functions in computation. Correspondingly, the resultant bond strength of the interface zone reflects the influence from the formation of voids.

The formation of voids is the main issue that influences the performance, reliability, and yield of bonding. In theory, bonding occurs due to a combination effect of diffusion and power law creep that shrinks the voids formed by the differences between the mating surfaces affected by dishing effects from CMP. Nevertheless, it is difficult to directly visualize the voids at the bonding surface from the SPG simulation result. Therefore, the numerical capability to capture the voids in terms of locations and sizes becomes important in simulation. With the pseudo-bonding model, voids and defects can be shown to have a strong impact on the bond strength of the Cu–Cu joints. Motivated by non-destructive sound wave scanning, the voids

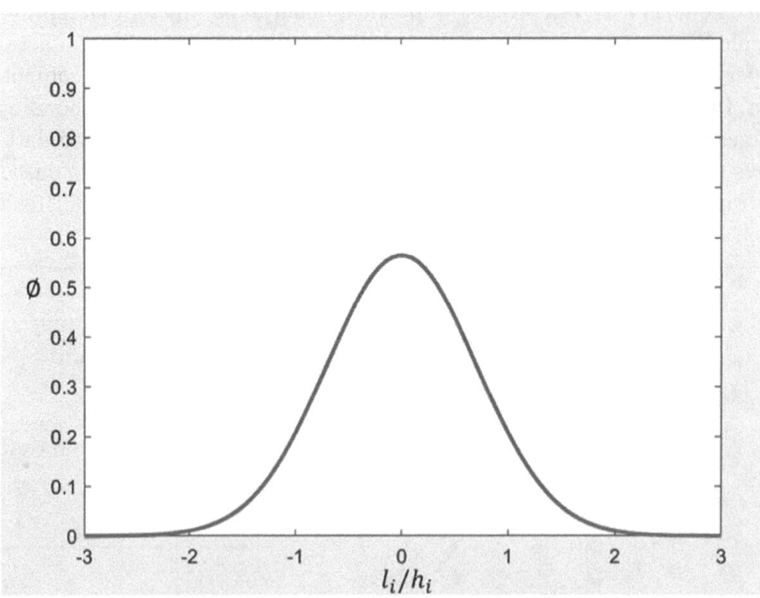

FIGURE 10.49 The distribution of Gaussian kernel function.

are detected by utilizing the elastic stress wave analysis in this study. It is crucial to numerically capture the correct stress state across the bonding interface. The explicit time integration apparently results in inferior wave interaction in the computational domain. Therefore, the implicit time integration scheme, Equation (10.21), is adopted in this study to reach the equilibrium status at each time iteration for accurate simulation results.

10.4.3 The Three-step Workflow of Void Detection and Failure Analysis of Hybrid Bonding at Feature Level

A three-step computer modeling workflow is proposed to evaluate the performance of a direct Cu–Cu diffusive bonding process. First, the implicit FEM simulation is used to analyze the interfacial contact between two copper pads subjected to the CMP-induced dishing effect and the thermal-compression loading conditions. Based on the final geometry of the Cu/Cu contact surfaces obtained from Step 1, the pseudo-bonding model is applied at the interface zone to connect the two copper pads into one piece of material mimicking the result of post-bonding annealing. Subsequently, an implicit transient SPG simulation is conducted in Step 2 to detect the possible formation and locations of voids in the bonded model using the elastic stress wave analysis. In Step 3, an explicit dynamic SPG simulation is used to predict the bond strength of the Cu–Cu interconnection considering the effect of interface voids. The three-step workflow presented in this study is summarized in Figure 10.50.

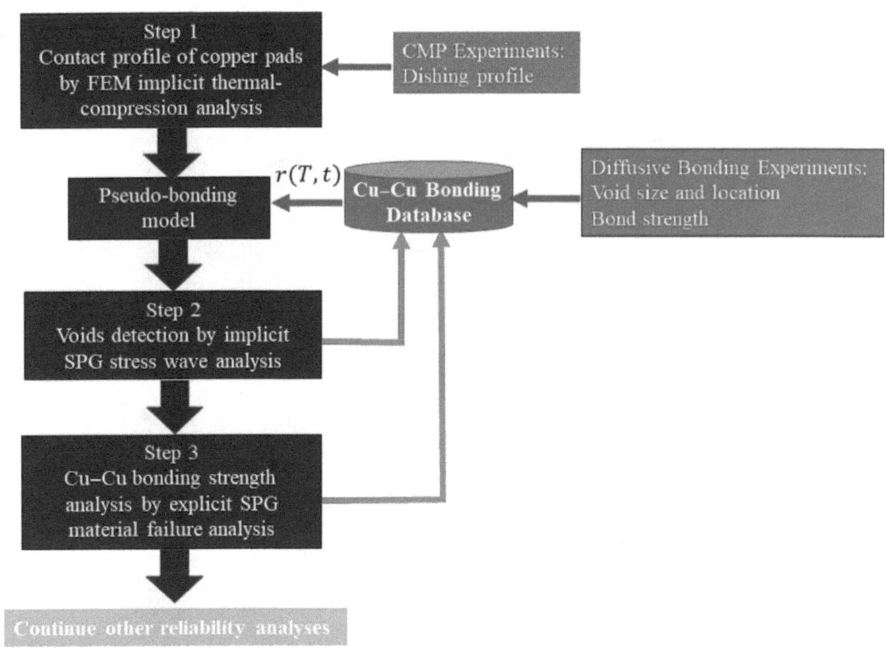

FIGURE 10.50 The three-step workflow for the Cu–Cu bonding analysis.

10.4.3.1 Step 1: Implicit Thermal Compression Analysis of Copper Pads by FEM

The dishing effect induced by the CMP process plays an important role in determining the geometric shape of the contacted Cu–Cu interface under thermal-compression loading. Comparing with pad size (~μm), the depth of dishing shape is at a fine scale (~ nm). The dishing shape profile can be found from experimental or numerical approaches [67]. With the dishing shape profile, A CAE pre-process software can generate a constructive pad mesh with a small indentation, as shown in Figure 10.51.

Two copper pads are assembled with dishing indentations facing each other, as shown in Figure 10.52. The purpose of this simulation is to demonstrate how the dishing and thermal–mechanical coupling effects impact the final geometry of copper pads including the shape of contact surfaces in a thermal-compression process. Therefore, the length unit used in this study is AU (Artificial Unit). The copper pads have dimensions of 50 AU × 50 AU × 30 AU, with a maximum vacancy height of 3.2 AU due to dishing effects. The thermal-compression analysis of copper pads is performed using FEM with a kinematic hardening plasticity model. The material constants are given in Table 10.2. The thermal expansion coefficient is 1.85E-5. The temperature starts from 300 K and reaches 600 K after 1.0 second, and then remains constant. Since the barrier and oxide have a small thermal expansion, a fixed boundary at the *X*- and *Y*-axes is applied to the copper pads. A small compression of 0.25 AU in *Y*-displacement is imposed on the copper pads model, as shown in Figure 10.52. A regular FEM surface–surface contact is defined between top and bottom pads. Previous studies have found that the creep of copper has less influence on the results compared to the temperature [58, 66]. Therefore, the creep model is not included in this study because the effect of creep deformation is reflected in the pseudo-bonding model via the parameter $r(T,t)$.

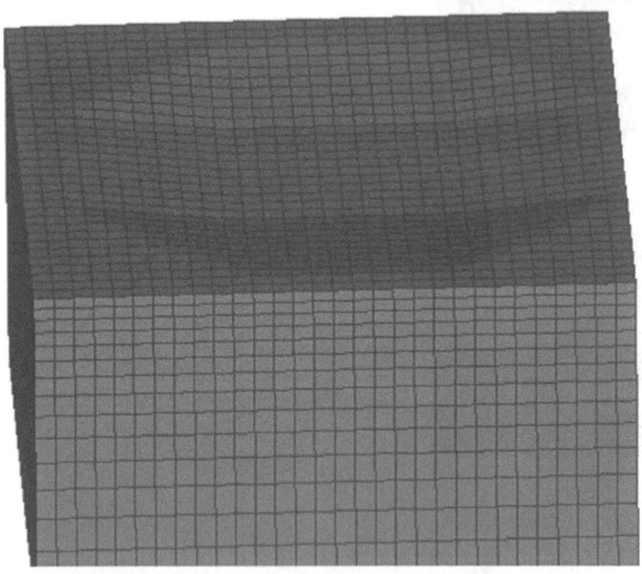

FIGURE 10.51 Copper pad with dishing effect.

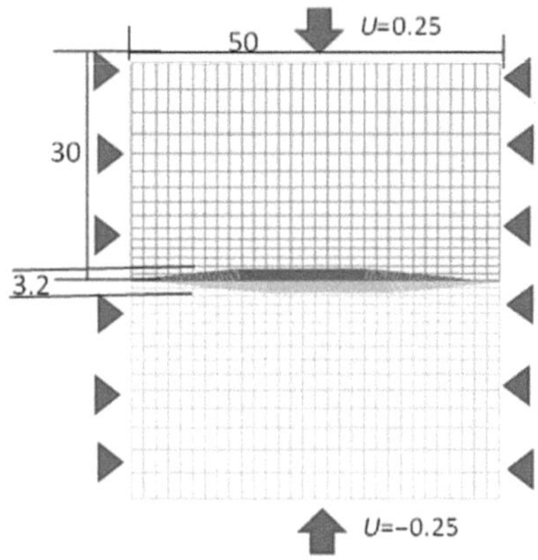

FIGURE 10.52 Computational model in thermal-compression analysis.

TABLE 10.2
Material Constants for Copper (ton/mm/s)

Density	Modulus	Poisson's Ratio	Yield Stress	Tangent Modulus	Hardening Parameter
8.96E-9	117,000	0.3	85	11,700	1

To obtain the correct relative geometric position of the copper pads before annealing, this model is conducted using an FEM implicit solver. The geometry of the two pads is shown in Figure 10.53a. The maximum vacancy height drops from 3.2 AU to 1.14 AU. Most of the dishing vacancies remain uncontacted. The contact-based bonding model will consider this area as non-bonded. Physically, diffusion and grain growth phenomena can fill small vacancies depending on the annealing temperature and time. Therefore, the proposed pseudo-bonding model, $r(T,t)$, is introduced to

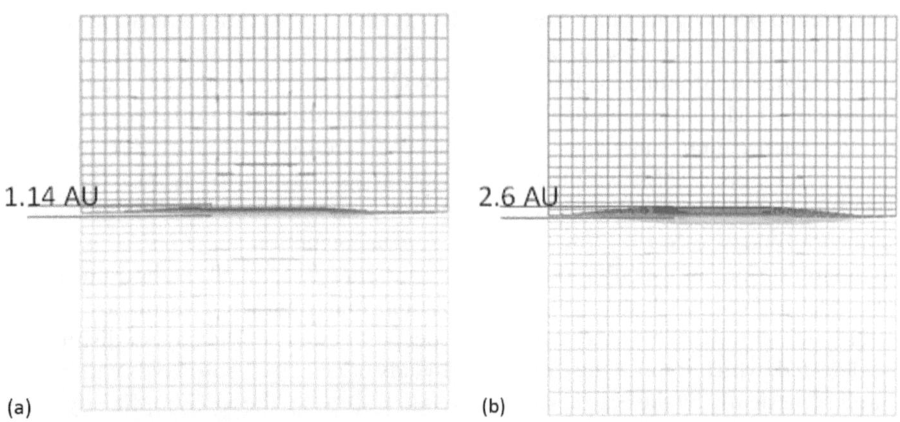

FIGURE 10.53 Geometry of pads after (a) thermal-compression and (b) compression only.

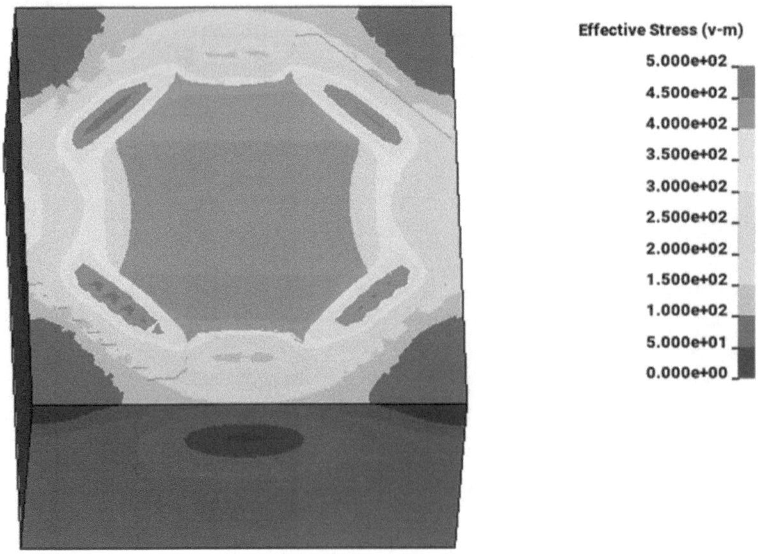

FIGURE 10.54 Effective stress distribution at interface in the implicit thermal-compression analysis.

consider if a vacancy has been filled or not. To compare with the thermal-compression simulation, a compression-only case is conducted, as shown in Figure 10.53b. Here, the maximum vacancy height reduced from 3.2 AU to 2.6 AU. This result implies the importance of thermal expansion, which supports the conclusion of existing investigations [65, 66]. The effective stress distribution of the bottom pad is shown in Figure 10.54. The high-stress zones at the interface imply the geometrically contacted zone due to thermal expansion.

10.4.3.2 Step 2: Voids Detection by Stress Wave

The pseudo-bonding model is developed to detect the potential voids and defects at the bonding surface by analyzing the reflected elastic stress waves using an SPG implicit dynamic simulation. To obtain the SPG analysis model, the final nodal coordinates of FEM analysis model in Step 1 are exported as the initial particle coordinates for the Step 2 analysis. The copper pads are discretized by SPG particles (shown as Figure 10.55a) with the same material constants in Table 10.2. The regular SPG influence zone is set as $\Delta x = 1.8, \Delta y = 1.8, \Delta z = 1.5$, and the pseudo-bonding influence zone is set as $r = 1.7$ for this case. A pausing motion of 2.4d-8 AU is applied to the top surface of the SPG model shown in Figure 10.55a and the bottom surface of model is fixed. With this SPG discretization, four significant vacancies (the red circles in Figure 10.55b) located at the four deepest corners of the copper dishing can be detected, which implies the existence of voids or defects. Those voids can be captured numerically by the elastic stress wave analysis using the SPG implicit dynamic simulation. Figure 10.56 shows the time sequence of stress waves passing through the bonded interface. It illustrates four stress concentration zones at the positions of

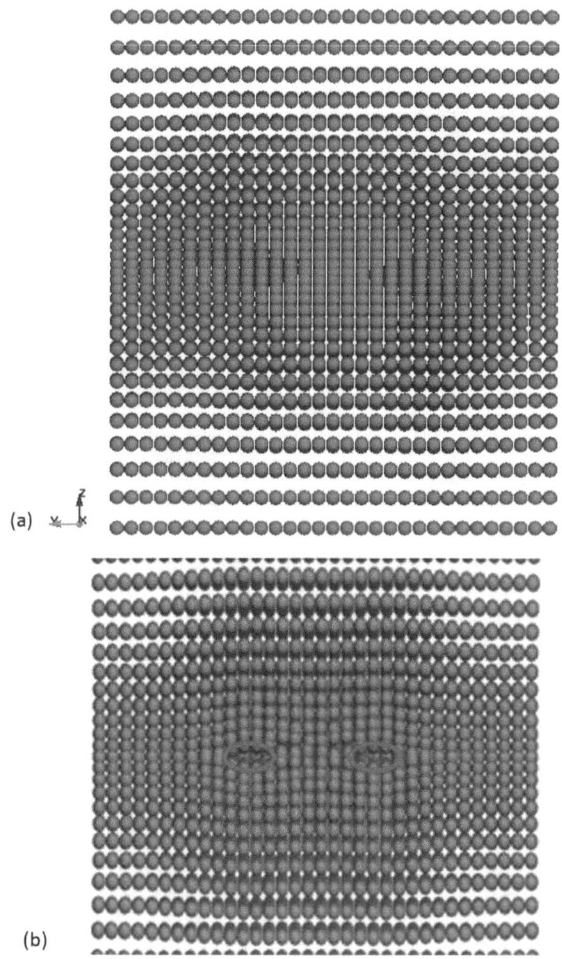

FIGURE 10.55 SPG discretization: (a) whole model; (b) middle surface, red circle: vacancy.

vacancies, as shown in Figure 10.56b–d. The location and size of these high-stress islands at the bonding interface indicate the undesired voids or defects.

10.4.3.3 Step 3: Bonding Strength Test of Bonded Pad with SPG

Given the information of interface voids in Step 2, we would like to predict their influence on the Cu–Cu bond strength using a material dynamic failure analysis. Using the explicit SPG solver [69, 70], a tension-to-fail simulation is conducted to test the strength of the bonded copper pads considering the existence of voids. The computational mesh, regular SPG influence zone, pseudo-influence zone, and material constants are the same as those in the Step 2 simulation. Two opposite displacement boundary conditions ($U_z = \pm 2$ AU) are imposed on the top and bottom surfaces. A dynamic failure process is illustrated in Figure 10.57. Because the strength of the bonding surface is weaker than that of normal copper pads, effective plastic strain

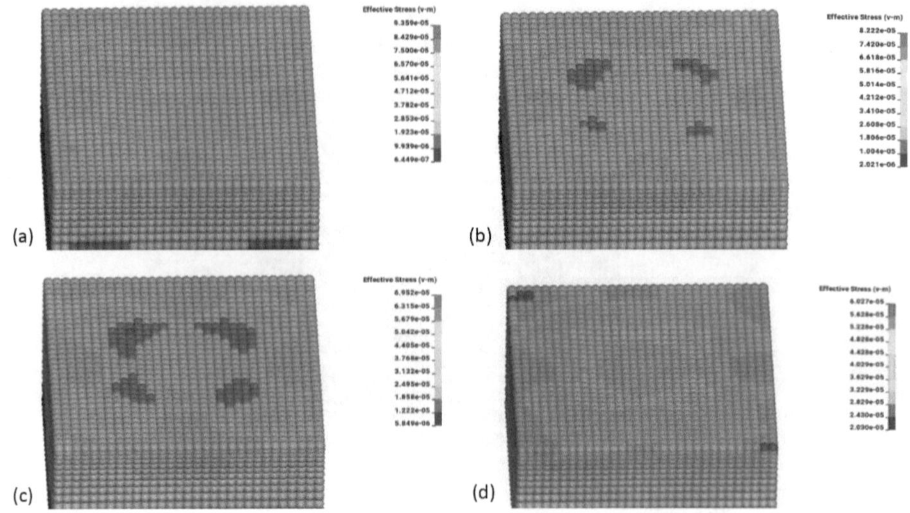

FIGURE 10.56 Effective stress distribution at bonding interface. Hight-stress zones indicate the presence of voids detected in the implicit transient stress wave analysis.

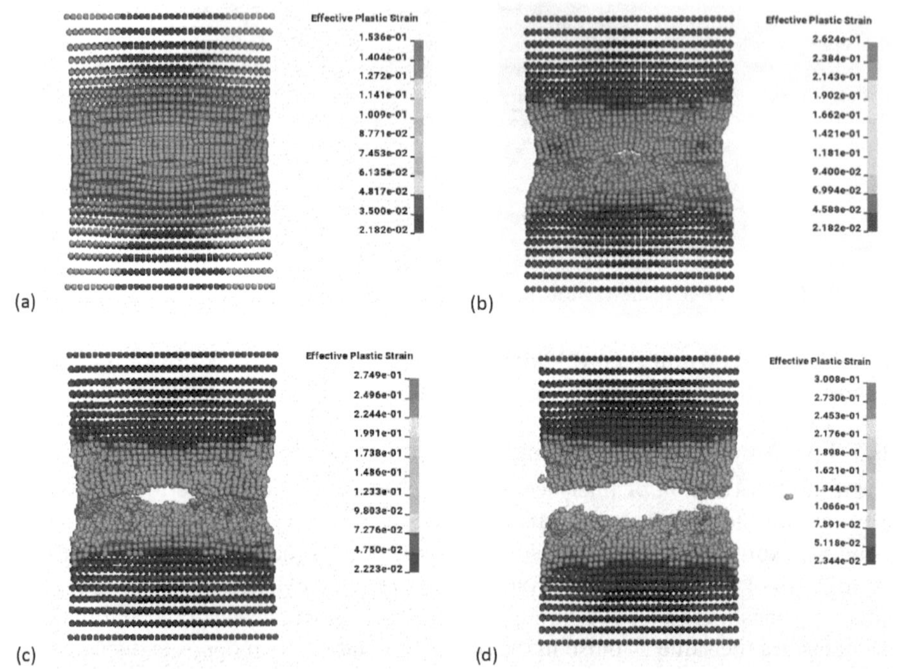

FIGURE 10.57 The dynamic failure process of the bonded copper pads. Background contour: effective plastic strain.

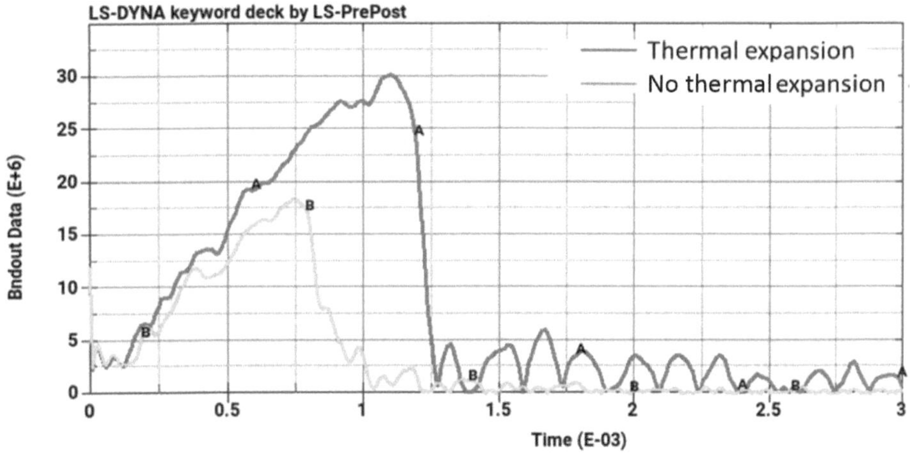

FIGURE 10.58 Reaction force for the bonding strength tests.

accumulates during the tensile deformation nearby the bonding surface. The material separation is initiated at the position of the dishing vacancy because this is the weakest bonding position. Finally, the two pads are torn apart along the bonded surface. It is not surprising that thermal expansion in Step 1 has a big influence on bonding strength. Figure 10.58 shows the comparison of tension reaction force curves in the strength test simulation. The red and green lines are the cases with and without thermal expansion, respectively, described in Step 1. The higher peak value of the red line implies a stronger bond is formed.

10.5 SUMMARY

While 3DIC technology brings a number of opportunities, its thermal management presents a number of challenges. As 3DICs can have higher power densities due to the stacking of multiple chips as in 3DIC with hybrid bonding, heat dissipation for the vertically stacked dies becomes a challenge. Some of the biggest challenges include minimizing on-chip temperature gradients and localized hot spots, thermal coupling between the stacked dies, thermal effects due to interconnect, thermal stress and reliability issues due to CTE mismatch between integrated heterogeneous components, and solution cost given small form factor constraints.

A variety of techniques have been investigated for thermal management of 3DICs. In particular, increased attention has turned toward thermal management techniques for 3DICs integrated directly into the 3D die stack or package. Such embedded thermal management techniques can provide more efficient cooling for 3DICs than traditional cooling methods. However, they may also require more complex manufacturing processes and design considerations.

Given the thermal management challenges, it is important to assess the state-of-the-art thermal modeling methods for 3DIC CPS. CFD tools provide an integrated electronics cooling solution at the system level. Electrical CAD designs (e.g., 3DIC, PCB) and mechanical CAD designs (e.g., enclosures, fans, etc.) are leveraged within CFD system-level tools. Advancements in ROM allow a complex (e.g., CFD) thermal model to be reduced to an equivalent simple fast running model.

It is necessary to include multi-physics effects for high fidelity and accuracy. Electrothermal co-simulation can be performed by the CFD tool and electromagnetic solvers. Temperature results from an electro-thermal simulation may then be used in a structural solver for thermal–stress simulations. Temperature and stress loads can be passed to a PCB-level reliability tool for a reliability physics analysis, such that components' time to failure predictions can be assessed.

To represent the 3DIC in the system-level thermal model, multiple CTMs are included and arranged, i.e., in a stack, and placed on an interposer. In the system-level model, the CTM is coarsened and lacking detail, to allow system level effects to be modeled with high fidelity.

The system-level thermal information required for detailed chip-level thermal modeling is extracted and applied to a detailed chip-level model. DTM methods allow the simulation inputs, such as chip power, to be controlled dynamically during the simulation run based on measurements such as chip temperature sensors. The whole approach offers a comprehensive CPS thermal modeling methodology.

CPS thermal analyses, both static and transient, are computationally expensive, and for modern 3DIC designs, traditional FEM techniques cannot be scaled. Fine-grained thermal analysis is virtually impossible for large designs with traditional CFD/FEA methods in a reasonable time. Such techniques will run into severe run-time and capacity limitations for large 3DIC designs. Recently, there have been efforts to take advantage of developments in ML and DL to make thermal analyses for CPS computationally efficient. ML-based techniques have shown great promise for thermal analyses and will be part of next generation commercial thermal analyses tools. This chapter presents an overview of some of the recent cutting-edge research on ML-based thermal analyses from both academia and industry. Two such techniques for fast static thermal analysis and one technique for fast transient thermal analysis are described in detail. The DNN-based technique uses a trained DNN model for predicting localized heating. It takes a power pattern from a fine-grained set of tiles and predicts the DeltaT at the center tile. The tiles are selected based on a set of predefined templates. This localized heating is then combined with the global coarse static thermal analysis from a commercial thermal analysis tool which uses the accurate FEM-based method. This can be thought of as providing corrections to the coarser analysis from the FEM-based tool and this method shows very good accuracies. Since the DNN model needs to be trained only once, the method is very scalable and can provide speed-up of 100–1,000×. The second method for the fast static thermal analysis employs autoencoders and decoders to convert input power maps into temperature maps. This method divides the design into subdomains and down-samples the input power initial temperature and system parameters using trained autoencoders. These autoencoders use convolutional neural network (CNN) architectures. Then it performs the fixed-point iterations for flux conservation on each subdomain. Finally, the encodings from each subdomain are decoded to generate the final converged temperature values. The speed-up can be greater than 100–1,000× compared to classical thermal analysis methods. The results demonstrate a very good accuracy, scalability, and generalizability of the ML-based solver.

The fast transient thermal analysis method utilizes the combination of global, intermediate, and local transient response curves generated by an ML predictor or pre-characterized as an initial step using an accurate FEM-based thermal solver. These response curves at any location on the die are obtained by scaling the local, intermediate, and global curves using the far-field and near-field transient decay surface components. The final step involves linearly superimposing the response curves for each power value in the transient power profile to generate the effective transient response curve. This approach enables efficient and accurate analysis of transient thermal behavior and the runtime for generating thermal results for a large chip is on the order of minutes, compared to several hours/days while using CFD/FEA-based tools with good accuracy correlation.

A novel computer modeling approach is also proposed for Cu–Cu bonding which includes a three-step workflow considering the dishing effect and thermo-mechanical coupling response to predict the voids formation and subsequently to test the bonding strength of the direct Cu–Cu diffusive bond. A pseudo-bonding model is proposed to replicate the connection of copper pads after the diffusion process. The pseudo-bonding model homogenizes the multiscale process of diffusion and recrystallization in the Cu–Cu interface by a function which can be calibrated from experimental data.

While the voids are formed because of the copper dishing in this study, some voids can be formed due to the wafer surface contaminations such as the presence of slurry particles or residual abrasive particles. The consideration of the voids in the latter case requires the development of a different pseudo-bonding model. It is also possible that one can integrate the Cu–Cu bonding results with the insulating (dielectric or adhesive) materials and upper-scale structures using a two-scale co-simulation technique [52] for future 3DIC reliability analyses. Therefore, other key areas of research such as the multiscale computation and modeling of the misalignment and thermal-induced wafer warpage should also be considered.

ACKNOWLEDGMENTS

The authors wish to thank Dr. Yuxi Xie for his help with the implementation of implicit SPG solver. We would also like to thank J. Wen, R. Ranade, H. He, M. Abarham, W. Xia, C. Ortiz, S. Asgari, P. Vennam, J. Yang, L. Lin, P. Gupta, J. Toublanc, Y. Li, J. Pathak, P. Banerjee, Z. Liang, J. Li, and R. Jang for discussions.

REFERENCES

1. Y. H. Chen, C. A. Yang, C. C. Kuo, M. F. Chen, C. H. Tung, W. C. Chiou, and D. Yu, "Ultra High Density SoIC with Sub-micron Bond Pitch," *2020 IEEE 70th Electronic Components and Technology Conference (ECTC)*, 2020.
2. R. Mahajan, and S. Sane, "Advanced Packaging Technologies for Heterogeneous Integration (HI)," Intel HotChips, 2021.
3. N. Chang, S. Pan, K. Srinivasan, Z. Feng, W. Xia, T. Pawlak, and D. Geb, "Emerging ADAS Thermal Reliability Needs and Solutions," *IEEE Micro*, vol. 38, pp. 66–81, 2018.

4. S. Krishnaswamy, P. Jain, M. Saeidi, A. Kulkarni, A. Adhiya, and J. Harvest, "Fast and Accurate Thermal Analysis of Smartphone with Dynamic Power Management Using Reduced Order Modeling," *2017 16th IEEE Intersociety Conference on Thermal and Thermomechanical Phenomena in Electronic Systems (ITherm)*, 2017.
5. M. Dogruoz, M. Abarhama, and G. Shankaran, "Transient Thermal Behavior of SOIC Packages: An Optimization Study," *15th IEEE Intersociety Conference on Thermal and Thermomechanical Phenomena in Electronic Systems (ITherm)*, 2016.
6. Y. Im, W. Kim, T. An, H. Lee, Y.-S. Cho, J. Yoo, H.-J. Lee, Y. Shin, M. Lee, and V. K Yaddanapudi, "Thermal Sensor Placement Based on Meta-model Enhancing Observability and Controllability," *19th IEEE Intersociety Conference on Thermal and Thermomechanical Phenomena in Electronic Systems (ITherm)*, 2020.
7. R. Ranade, H. He, J. Pathak, N. Chang, A. Kumar, and J. Wen, "A Thermal Machine Learning Solver For Chip Simulation," *Proceedings of the ACM/IEEE Workshop on Machine Learning for CAD*, New York, NY, 2022.
8. J. Wen, S. Pan, N. Chang, W.-T. Chuang, W. Xia, D. Zhu, A. Kumar, E.-C. Yang, K. Srinivasan, and Y.-S. Li, "DNN-based Fast Static On-chip Thermal Solver," *36th Semiconductor Thermal Measurement, Modeling & Management Symposium (SEMI-THERM)*, 2020.
9. A. Kumar, N. Chang, D. Geb, H. He, S. Pan, J. Wen, S. Asgari, M. Abarham, and C. Ortiz, "ML-based Fast On-chip Transient Thermal Simulation for Heterogeneous 2.5D/3D IC Designs," *International Symposium on VLSI Design, Automation and Test (VLSI-DAT)*, 2022.
10. D. Geb, S. Asgari, A. Kumar, J. Wen, N. Chang, S. Pan, M. Abarham, H. He, and V. Gandhi, "On-chip Transient Hot Spot Detection with a Multiscale ROM in 3DIC Designs," *IEEE 72nd Electronic Components and Technology Conference (ECTC)*, 2022.
11. S. G. Kandlikar, "Review and Projections of Integrated Cooling Systems for Three-dimensional Integrated Circuits," *Journal of Electronic Packaging*, vol. 136, no. 2, p. 4, 2014.
12. T. Brunschwiler, A. Sridhar, C. Lee Ong, and G. Schlottig, "Benchmarking Study on the Thermal Management Landscape for Three-dimensional Integrated Circuits: From Back-side to Volumetric Heat Removal," *Journal of Electronic Packaging*, vol. 138, no. 1, p. 3, 2016.
13. F. Tavakkoli, S. Ebrahimi, S. Wang, and K. Vafai, "Analysis of Critical Thermal Issues in 3D Integrated Circuits," *International Journal of Heat and Mass Transfer*, vol. 97, pp. 337–352, 2016.
14. S. V. Garimella, T. Persoons, J. A. Weibel, and V. Gektin, "Electronics Thermal Management in Information and Communications Technologies: Challenges and Future Directions," *IEEE Transactions on Components, Packaging and Manufacturing Technology*, vol. 7, no. 8, pp. 1191–1205, 2017.
15. A. Bar-Cohen, M. Asheghi, T. Chainer, S. V. Garimella, K. Goodson, C. Gorle, R. Mandel, J. J. Maurer, M. Ohadi, J. W. Palko, P. Parida, Y. Peles, J. Plawsky, M. Schultz, J. A. Weibel, and Y. Joshi, "The ICECool Fundamentals Effort on Evaporative Cooling of Microelectronics," *IEEE Transactions on Components, Packaging and Manufacturing Technology*, vol. 11, no. 10, pp. 1546–1564, 2021.
16. DARPA Microsystems Technology Office BAA, "Miniature Integrated Thermal Management Systems for 3D Heterogeneous Integration (Minitherms3D)," DARPA BAA, January 23, 2023. Available at: https://sam.gov/opp/bb25df0332e14c62a528aeb496cb19ac/view.
17. M. Schultz, F. Yang, E. Colgan, R. Polastre, B. Dang, C. Tsang, M. Gaynes, P. Parida, J. Knickerbocker, and T. Chainer, "Embedded Two-phase Cooling of Large Three-dimensional Compatible Chips with Radial Channels," *Journal of Electronic Packaging*, vol. 138, no. 2, p. 4, 2016.

18. A. Bar-Cohen, "Thermal Management of On-chip Hot Spots and 3D Chip Stacks," *2009 IEEE International Conference on Microwaves, Communications, Antennas and Electronics Systems (COMCAS 2009)*, 2009.
19. Y. J. Kim, Y. K. Joshi, A. G. Fedorov, Y.-J. Lee, and S.-K. Lim, "Thermal Characterization of Interlayer Microfluidic Cooling of Three-dimensional Integrated Circuits with Nonuniform Heat Flux," *Journal of Heat Transfer*, vol. 132, no. 4, p. 2, 2010.
20. T. J. Chainer, M. D. Schultz, P. R. Parida, and M. A. Gaynes, "Improving Data Center Energy Efficiency with Advanced Thermal Management," *IEEE Transactions on Components, Packaging and Manufacturing Technology*, vol. 7, no. 8, pp. 1228–1239, 2017.
21. P. Birbarah, T. Gebrael, T. Foulkes, A. Stillwell, A. Moore, R. Pilawa-Podgurski, and N. Miljkovic, "Water Immersion Cooling of High Power Density Electronics," *International Journal of Heat and Mass Transfer*, vol. 147, p. 118918, 2020.
22. B. Ramakrishnan, H. Alissa, I. Manousakis, R. Lankston, R. Bianchini, W. Kim, R. Baca, P. A. Misra, I. Goiri, M. Jalili, A. Raniwala, B. Warrier, M. Monroe, C. Belady, M. Shaw, and M. Fontoura, "CPU Overclocking: A Performance Assessment of Air, Cold Plates, and Two-phase Immersion Cooling," *IEEE Transactions on Components, Packaging and Manufacturing Technology*, vol. 11, no. 10, pp. 1703–1715, 2021.
23. C. Green, P. Kottke, X. Han, C. Woodrum, T. Sarvey, P. Asrar, X. Zhang, Y. Joshi, A. Fedorov, S. Sitaraman, and M. Bakir, "A Review of Two-phase Forced Cooling in Three-dimensional Stacked Electronics: Technology Integration," *Journal of Electronic Packaging*, vol. 137, no. 4, p. 9, 2015.
24. Y. Madhour, B. P. d'Entremont, J. Braz Marcinichen, B. Michel, and J. Richard Thome, "Modeling of Two-phase Evaporative Heat Transfer in Three-dimensional Multicavity High Performance Microprocessor Chip Stacks," *Journal of Electronic Packaging*, vol. 136, no. 2, p. 4, 2014.
25. B. Kelly, G. Moreno, S. Myers, S. Narumanchi, Y. Joshi, and S. Graham, "Dielectric Fluids for the Direct Forced Convection Cooling of Power Electronics," *InterSociety Conference on Thermal and Thermomechanical Phenomena in Electronic Systems, ITHERM*, pp. 308–313, 2021.
26. S. Lu and K. Vafai, "Optimization of the Thermal Performance of Three-dimensional Integrated Circuits Utilizing Rectangular-shaped and Disk-shaped Heat Pipes," *Journal of Heat Transfer*, vol. 144, no. 6, p. 3, 2022.
27. M. Manno, B. Yang, and A. Bar-Cohen, "Near-junction "Hot Spot" Suppression with Integral SiC Microcontact TEC," *International Journal of Heat and Mass Transfer*, vol. 115, pp. 530–536, 2017.
28. C. E. Green, L. Prinzi, and B. A. Cola, "Design and Evaluation of Polymer-Carbon Nanotube Composites for Reliable, Low Resistance, Static and Dynamic Thermal Interface Materials," *2016 15th IEEE Intersociety Conference on Thermal and Thermomechanical Phenomena in Electronic Systems (ITherm)*, pp. 657–661, 2016.
29. S. Kochupurackal Rajan, B. Ramakrishnan, H. Alissa, W. Kim, C. Belady, and M. S. Bakir, "Integrated Silicon Microfluidic Cooling of a High-power Overclocked CPU for Efficient Thermal Management," *IEEE Access*, vol. 10, pp. 59259–59269, 2022.
30. R. van Erp, R. Soleimanzadeh, L. Nela, G. Kampitsis, and E. Matioli, "Co-Designing Electronics with Microfluidics for More Sustainable Cooling," *Nature*, vol. 585, no. 7824, pp. 211–216, 2020.
31. K. P. Drummond, D. Back, M. D. Sinanis, D. B. Janes, D. Peroulis, J. A. Weibel, and S. V. Garimella, "Characterization of Hierarchical Manifold Microchannel Heat Sink Arrays under Simultaneous Background and Hotspot Heating Conditions," *International Journal of Heat and Mass Transfer*, vol. 126, pp. 1289–1301, 2018.
32. C. Koroglu, and E. Pop, "High Thermal Conductivity Insulators for Thermal Management in 3D Integrated Circuits," *IEEE Electron Device Letters*, vol. 44, no. 3, pp. 496–499, 2023.

33. J. Pomeroy, M. Bernardoni, A. Sarua, A. Manoi, D. C. Dumka, D. M. Fanning, and M. Kuball, "Achieving the Best Thermal Performance for GaN-on-Diamond," *2013 IEEE Compound Semiconductor Integrated Circuit Symposium (CSICS)*, pp. 1–4, 2013.
34. M. Tyhach, D. Altman, S. Bernstein, R. Korenstein, J. Cho, K. E. Goodson, D. Francis, F. Faili, F. Ejeckam, S. Kim, and S. Graham, "Next Generation Gallium Nitride HEMTs Enabled by Diamond Substrates," *2014 Lester Eastman Conference on High Performance Devices (LEC)*, pp. 1–4, 8 2014.
35. J. W. Palko, H. Lee, C. Zhang, T. J. Dusseault, T. Maitra, Y. Won, D. D. Agonafer, J. Moss, F. Houshmand, G. Rong, J. D. Wilbur, D. Rockosi, I. Mykyta, D. Resler, D. Altman, M. Asheghi, J. G. Santiago, and K. E. Goodson, "Extreme Two-phase Cooling from Laser-etched Diamond and Conformal, Template-fabricated Microporous Copper," *Advanced Functional Materials*, vol. 27, no. 45, 12 2017.
36. A. Agrawal, S. Huang, G. Gao, L. Wang, J. Delacruz, and L. Mirkarimi, "Thermal and Electrical Performance of Direct Bond Interconnect Technology for 2.5D and 3D Integrated Circuits," *Proceedings: Electronic Components and Technology Conference*, pp. 989–998, 8 2017.
37. M. F. Chen, C. S. Lin, E. B. Liao, W. C. Chiou, C. C. Kuo, C. C. Hu, C. H. Tsai, C. T. Wang, and D. Yu, "SoIC for Low-temperature, Multi-layer 3D Memory Integration," *Proceedings: Electronic Components and Technology Conference*, pp. 855–860, 2020.
38. R. Chen, M. Lofrano, G. Mirabelli, G. Sisto, S. Yang, A. Jourdain, F. Schleicher, A. Veloso, O. Zografos, P. Weckx, G. Hiblot, G. V. d. Plas, G. Hellings, J. Ryckaert, and E. Beyne, "Power, Performance, Area and Thermal Analysis of 2D and 3D ICs at A14 Node Designed with Back-side Power Delivery Network," *2022 International Electron Devices Meeting (IEDM)*, pp. 23.4.1–23.4.4, 2022.
39. K. Matin, A. Bar-Cohen, and J. J. Maurer, "Modeling and Simulation Challenges in Embedded Two Phase Cooling: DARPA's ICECool Program," *Volume 3: Advanced Fabrication and Manufacturing; Emerging Technology Frontiers; Energy, Health and Water: Applications of Nano-, Micro- and Mini-Scale Devices; MEMS and NEMS; Technology Update Talks; Thermal Management Using Micro Channels, Jets, Sprays*, 7 2015.
40. P. R. Parida, M. D. Schultz, and T. Chainer, "Sim2Cool: A Two-phase Cooling System Simulator and Design Tool," *17th IEEE Intersociety Conference on Thermal and Thermomechanical Phenomena in Electronic Systems (ITherm)*, pp. 271–280, 2018.
41. Ansys, Inc., *Ansys Icepak User Manual 2023 R1*, 2023.
42. C. Gorlé, H. Lee, F. Houshmand, M. Asheghi, K. Goodson, and P. R. Parida, "Validation Study for VOF Simulations of Boiling in a Microchannel," *Volume 3: Advanced Fabrication and Manufacturing; Emerging Technology Frontiers; Energy, Health and Water: Applications of Nano-, Micro- and Mini-Scale Devices; MEMS and NEMS; Technology Update Talks; Thermal Management Using Micro Channels, Jets, Sprays*, 7 2015.
43. Z. Pan, J. A. Weibel, and S. V. Garimella, "A Saturated-Interface-Volume Phase Change Model for Simulating Flow Boiling," *International Journal of Heat and Mass Transfer*, vol. 93, pp. 945–956, 2016.
44. V. K. Yaddanapudi, A. A. Kulkarni, A. Adhiya, M. Nagulapally, and D. Tan, "DELPHI Style Extraction and Validation of Compact Model for a Multichip Module Package," *IEEE 17th Electronics Packaging and Technology Conference (EPTC)*, pp. 1–6, 2015.
45. G. V. Shankaran, M. B. Dogruoz, and M. Abarham, "Thermal Analysis and Design of Electronics Systems across Scales Using State-Space Modeling Technique," *IEEE Transactions on Components, Packaging and Manufacturing Technology*, vol. 11, no. 8, pp. 1223–1234, 2021.
46. Y. Im, H. Jo, C. Oh, Y.-S. Cho, J. Yoo, H. Lee, M. Lee, and V. K. Yaddanapudi, "Thermal Model Simplification of Mobile Device with Adaptive Metamodel of Optimal Prognosis (AMOP)," *21st IEEE Intersociety Conference on Thermal and Thermomechanical Phenomena in Electronic Systems (iTherm)*, pp. 1–6, 2022.

47. M. B. Dogruoz, and G. Shankaran, "Spatial Variation of Temperature on Printed Circuit Boards: Effects of Anisotropic Thermal Conductivity and Joule Heating," *IEEE Transactions on Components, Packaging and Manufacturing Technology*, vol. 2, no. 10, pp. 1649–1658, 2012.
48. V. K. Yaddanapudi, S. Krishnaswamy, R. Rath, and R. Gandhi, "Validation of New Approach of Modelling Traces by Mapping Mechanical Properties For A Printed Circuit Board Mechanical Analysis," *2015 IEEE 17th Electronics Packaging and Technology Conference (EPTC)*, pp. 1–6, 2015.
49. E. J. Bracken, S. V. Polstyanko, S. G. Pytel, and I. Waldron, "Coupled Thermal-Fluid-Electrical Simulation for Printed Circuit Board Design," *2011 International Conference on Electromagnetics in Advanced Applications*, pp. 1005–1008, 2011.
50. G. Refai-Ahmed, H. Shi, Y. Bhartiya, T. Pawlak, M. Keshavamurthy, B. Boots, S. Shah, D. Ostergaard, and S. G. Pytel, "Electronic PCB and PKG Thermal Stress Analysis," *2016 IEEE 66th Electronic Components and Technology Conference (ECTC)*, pp. 1402–1408, 2016.
51. M. Serebreni, G. Sharon, N. Blattau, and C. Hillman, "Reliability Physics Approach for High-Density Ball Grid Arrays in Autonomous Vehicle Applications," *SAE International Journal of Advances and Current Practices in Mobility*, vol. 1, no. 4, pp. 1640–1647, 2019.
52. W. Hu, X. Pan, D. Lyu, A. Srivastava, S. Shah, and C.-T. Wu, "A Two-scale Approach for the Drop Shock Simulation of a Printed Circuit Board Package Considering Reflowed Solder Ball Geometries," *International Journal for Multiscale Computational Engineering*, vol. 18, no. 4, pp. 455–476, 2020.
53. H. Sultan, A. Chauhan, and S. R. Sarangi, "A Survey of Chip-level Thermal Simulators," *ACM Computing Surveys*, vol. 52, no. 2, pp. 1–35, 2019.
54. V. A. Chhabria, V. Ahuja, A. Prabhu, N. Patil, P. Jain, and S. S. Sapatnekar, "Thermal and IR Drop Analysis Using Convolutional Encoder–Decoder Networks," in *Proceedings of the 26th Asia and South Pacific Design Automation Conference*, New York, NY, USA, 2021.
55. Ian Goodfellow and A. Courville, *Deep Learning*, MIT Press, 2016.
56. W. Jin, S. Sadiqbatcha, J. Zhang, and S. X. D. Tan, "Full-chip Thermal Map Estimation for Commercial Multi-core CPUs with Generative Adversarial Learning," in *Proceedings of the 39th International Conference on Computer-Aided Design*, New York, NY, USA, 2020.
57. H. Han, C. Lee, Y. Kim, J. Lee, R. Kim, J. Kim, and B. Yoo, "Cu to Cu Direct Bonding at Low Temperature with High Density Defect in Electrodeposited Cu," *Applied Surface Science*, vol. 550, p. 149337, 2021.
58. Y. W. Tsau, J. D. Messemaeker, A. Salahouelhadj, M. Gonzalez, L. Witters, B. Zhang, M. Seefeldt, E. Beyne, and I. D. Wolf, "Simulation of Cu Pad Expansion in Wafer-to-Wafer Cu/SiCN Hybrid Bonding," *Microelectronics Reliability*, vol. 138, p. 114716, 2022.
59. Y.-H. Kuo, D.-P. Tran, J.-J. Ong, K. Tu, and C. Chen, "Hybrid Cu-to-Cu Bonding with Nano-twinned Cu and Non-conductive Paste," *Journal of Materials Research and Technology*, vol. 18, pp. 859–871, 2022.
60. S. Lee, "Fundamentals of Thermal Compression Bonding Technology and Process Materials for 2.5/3D Packages," in *3D Microelectronic Packaging*, Springer Cham, 2017, p. 157:203.
61. Y. Kagawa, N. Fujii, K. Aoyagi, Y. Kobayashi, S. Nishi, N. Todaka, S. Takeshita, J. Taura, H. Takahashi, Y. Nishimura, K. Tatani, M. Kawamura, H. Nakayama, T. Nagano, K. Ohno, H. Iwamoto, S. Kadomura, and T. Hirayama, "Novel Stacked CMOS Image Sensor with Advanced Cu2Cu Hybrid Bonding," in *2016 IEEE International Electron Devices Meeting (IEDM)*, San Francisco, CA, 2016.
62. E. Beyne, D. Milojevic, G. V. d. Plas, and G. Beyer, "3D SoC Integration, beyond 2.5D Chiplets," in *2021 IEEE International Electron Devices Meeting (IEDM)*, San Francisco, CA, 2021.

63. C. S. Tan, D. F. Lim, S. G. Singh, S. K. Goulet, and M. Bergkvist, "Cu–Cu Diffusion Bonding Enhancement at Low Temperature by Surface Passivation Using Self-assembled Monolayer of Alkane-thiol," *Applied Physics Letters*, vol. 95, no. 19, p. 92108, 2009.
64. Y. Beilliard, R. Estevez, G. Parry, P. McGarry, L. D. Cioccio, and P. Coudrain, "Thermomechanical Finite Element Modeling of Cu-SiO_2 Direct Hybrid Bonding with a Dishing Effect on Cu Surfaces," *International Journal of Solids and Structures*, vol. 117, pp. 208–220, 2017.
65. T. Wlanis, R. Hammera, W. Eckera, S. Lhostisb, C. Sartb, S. Gallois-Garreignotb, B. Rebhanc, and G. A. Maiera, "Cu-SiO_2 Hybrid Bonding Simulation Including Surface Roughness and Viscoplastic Material Modeling: A Critical Comparison of 2D and 3D Modeling Approach," *Microelectronics Reliability*, vol. 86, pp. 1–9, 2018.
66. R. I. Made, C. L. Gan, L. Yan, K. H. B. Kor, H. L. Chia, K. L. Pey, and C. V. Thompson, "Experimental Characterization and Modeling of the Mechanical Properties of Cu–Cu Thermocompression Bonds for Three-dimensional Integrated Circuits," *Acta Materialia*, vol. 60, pp. 578–587, 2012.
67. C. Sart, R. Estevez, V. Fiori, S. Lhostis, E. Deloffre, G. Parry, and R. Gonella, "Cu/SiO_2 Hybrid Bonding: Finite Element Modeling and Experimental Characterization," *2016 6th Electronic System-Integration Technology Conference (ESTC)*, Grenoble, 2016.
68. Y.-S. Kim, T. H. Nguyen, and S.-H. Choa, "Enhancement of the Bond Strength and Reduction of Wafer Edge Voids in Hybrid Bonding," *Micromachines (Basel)*, vol. 13, no. 4, p. 537, 2022.
69. X. Pan, C. T. Wu, W. Hu, and Y. Wu, "A Momentum-consistent Stabilization Algorithm for Lagrangian Particle Methods in the Thermo-mechanical Friction Drilling Analysis," *Computational Mechanics Volume*, vol. 64, pp. 625–644, 2019.
70. C. T. Wu, Y. Wu, D. Lyu, X. Pan, and W. Hu, "The Momentum-consistent Smoothed Particle Galerkin (MC-SPG) Method for Simulating the Extreme Thread Forming in the Flow Drill Screw-driving Process," *Computational Particle Mechanics*, vol. 7, pp. 177–191, 2020.
71. B. Ren, C. Wu, and D. Lyu, "An h-Adaptive Meshfree-enriched Finite Element Method Based on Convex Approximations for the Three-dimensional Ductile Crack Propagation Simulation," *Computer Aided Geometric Design*, vol. 76, p. 101795, 2020.
72. T. Belytschko, W. K. Liu, B. Moran, and K. Elkhodary, Nonlinear Finite Elements for Continua and Structures, 2nd Edition, Wiley, 2014.
73. S. Li and W. K. Liu, Meshfree Particle Methods, Berlin, Springer, 2004.

11 Characterization, Modeling, and Reliability for Direct Copper Interconnection

Stéphane Moreau and Dongkai Shangguan

11.1 INTRODUCTION

Newer forms of interconnects have emerged in advanced semiconductor packaging to meet the demand for high density and high performance, and interconnect reliability is becoming not only more complex but also more critical. Hybrid bonding (HB) and chiplets present new topics for the long-term reliability of the interconnect and the system. This chapter will outline and discuss the key reliability issues in advanced packaging and direct copper interconnection, including failure modes and mechanisms, reliability testing, and performance. The most recent results of reliability testing and studies will be reviewed.

While reliability and quality are often discussed in the same context and both are critical aspects for direct copper interconnection, reliability is distinguished from quality in that quality usually refers to time-zero compliance or conformance issues for the material/device, and reliability refers to the time dependence of material/device degradation as all devices are known to degrade with time. For that reason, "reliability" is defined as the "ability/probability that a product (system) will perform its intended function for a specific time period when operating under normal (or stated) environmental conditions" [1, 2]. Measuring and modeling the degradation rate, the time to failure, and the failure rate are the subjects of reliability engineering.

According to the JEDEC[1] JEP148B standard [3], "qualification" is the "process of demonstrating that an entity is capable of meeting or exceeding the specified requirements." Regarding the "robustness," it is defined as the "capability of functioning correctly or not failing under varying application and production conditions." The preceding terms often imply different tests: on the one hand, we have "pass" or "fail" tests, (qualification/robustness tests) and on the other hand, lifetime tests (reliability tests). The associated vocabulary includes failure criterion, lifetime, mission profile, failure mode, failure mechanism, etc. Since some of the expressions are sometimes ambiguous, their definitions are given below:

- *Mission Profile:* "The simplified representation of all of the relevant conditions to which a device will be exposed in its intended application throughout the full life cycle" [3]. Example: For power converters used in airborne applications and positioned in the close vicinity of the jet engine,

DOI: 10.1201/9781003408673-11

the power devices are required to survive, at the same time, static environment temperatures of about 200°C during the flight and extreme thermal cycles between –55°C and 200°C with a rate of 10°C/minute during landing and take-off. The required lifetime of the equipment is 50,000 hours corresponding to about 500 landing/take-off cycles [4].

- *Failure Mode:* Recognizable (electrical) symptom by which failure is observed [2]. Examples: short circuit, open circuit, etc.
- *Failure Mechanism:* The specific microscopic physical, chemical, metallurgical, environmental phenomena or processes that cause device degradation or malfunction [2]. Examples: corrosion, dielectric breakdown, electromigration (EM), etc.

It is clear that both quality and reliability depend on capable and optimized manufacturing processes. As discussed in previous chapters, HB demands strict process control and methodical process optimization, to ensure quality and reliability of the semiconductor device. Simulation and modeling have been used extensively for characterization and optimization of processes and performance (including thermal, electrical, and reliability) for advanced semiconductor packaging. In this chapter, progress in the development of modeling methodology and significant simulation results will be reviewed as related to the direct copper interconnection process and the performance of the HB interconnects.

A thorough understanding of failure modes and mechanisms, as well as reliability testing, in combination with computer modeling and simulation, are critical elements to ensure quality and reliability of advanced packaging for semiconductor devices using HB interconnects.

11.2 3D INTEGRATION: INTERCONNECTS, FAILURE MECHANISMS, AND RELATED TESTS

11.2.1 3D Interconnects

11.2.1.1 Solder Bump to Copper Pillar

Even though the through-silicon via (TSV) has brought a technological breakthrough and is currently seen as the interconnection method emblematic of 3D integration, wire bonding [5,6] and microbumps or their equivalents [7,8] are the first interconnections allowing vertical stacking of chips.

The advent of flip chip technology was an important step in this miniaturization. This technology, of collective soldering (Figure 11.1), allowed the chip designers to increase the number of inputs/outputs (I/O) by distributing them over the entire surface of the chip compared to the wire bonding technology, which only used the periphery. The use of the entire surface of the die has made it possible to reduce the size of the die and at the same time increase its functionality.

The evolution of the flip chip technology over the past 60 years can be divided into two generations. The first generation generally has an interconnect pitch greater than 40 μm and corresponds, more or less, to the emblematic Controlled Collapse Chip Connection or C4 from IBM (Figure 11.2). The second generation,

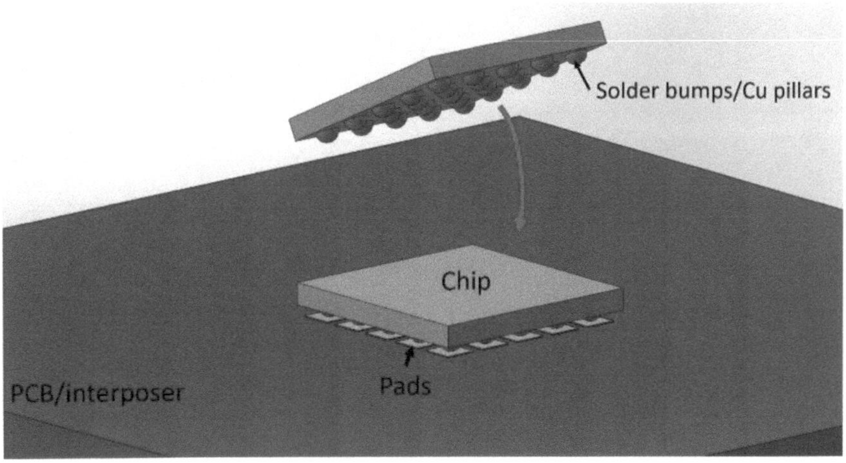

FIGURE 11.1 Schematic illustration of the flip chip technology.

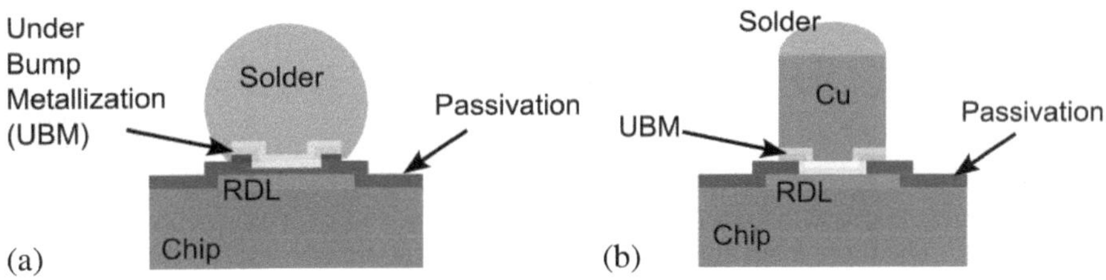

FIGURE 11.2 (a) Electroplated solder bump structure. (b) Cu pillar bump structure.

solder-capped microbump, usually called "copper pillar," was developed to achieve interconnect pitches of less than 40 μm (Figure 11.2). The shape of the pillar allows a high ratio between the height and the pillar diameter, thus allowing a very tight interconnect pitch even when the cylinder heights are large, and a solder cap is formed on the top of the pillar to aid connectivity. The ability to dissipate heat is another advantage of this technology, making them good candidates for high-performance applications. The overall height can range from 5 μm to 100 μm for a diameter of 10–100 μm or even more. Most pillars currently in production have a height greater than 20 μm [9, 10].

11.2.1.2 TSV

TSV is not a recent idea since it is at the heart of the U.S. patent #3,044,909, dated 1958, of the 1956 Nobel Prize in Physics, William Shockley [11], even though, at that time, it was not necessarily about stacking chips.

Generally, a TSV is an interconnect that electrically connects the backend-of-line (BEOL) on the front-side to the redistribution layer (RDL) on the backside, and its height (from a few to a few hundred micrometers) is equal to the silicon thickness of the chip that is itself defined according to the final application.

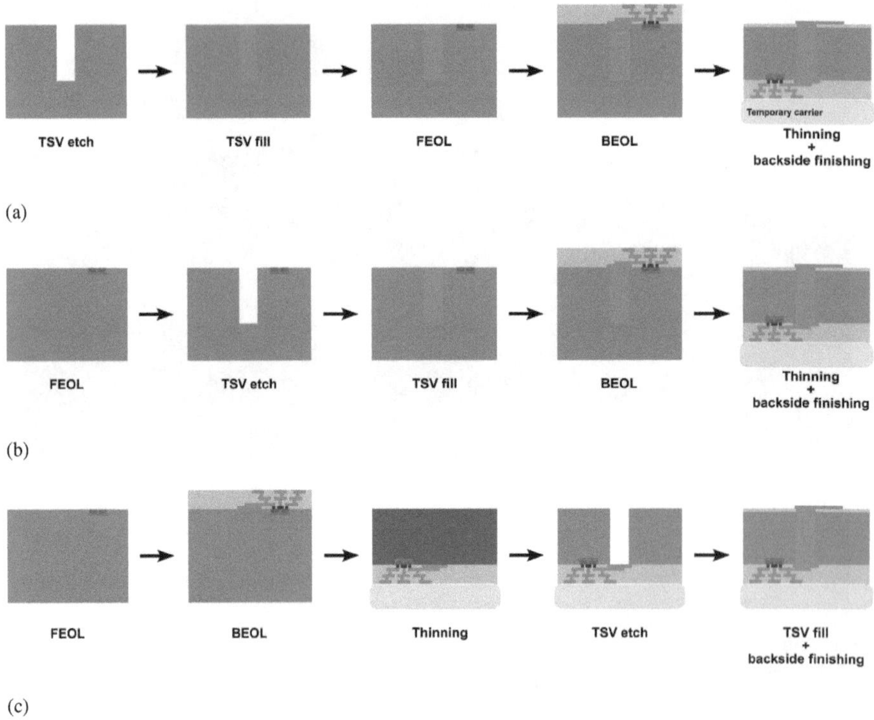

FIGURE 11.3 The three main approaches for TSV fabrication: (a) via first, (b) via middle, and (c) via last, depending on when the TSV is fabricated.

Various TSV configurations have been developed in the industry, and three integration paths for TSV are generally defined [12] and illustrated in Figure 11.3.

11.2.1.3 RDL

The corollary of this densification method was the development of a new metallic layer, on the surface of the chip, in order to ensure a redistribution of the signals between the buried layers and the TSVs and/or the microbumps/copper pillars (Figure 11.4). The RDL plays this role.

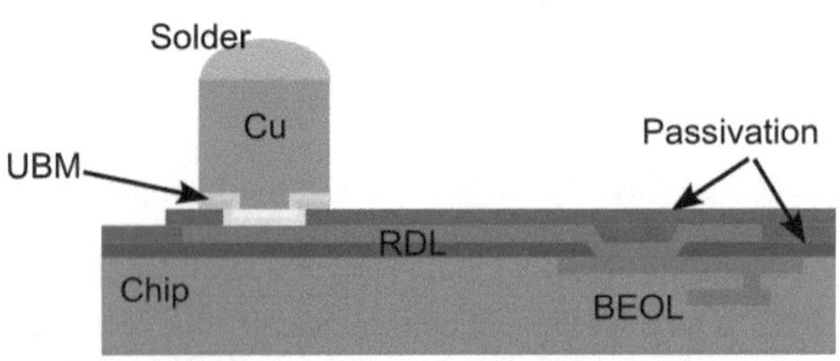

FIGURE 11.4 Schematic diagram of an RDL, seen in the cross section, connecting the BEOL interconnects and Cu pillars.

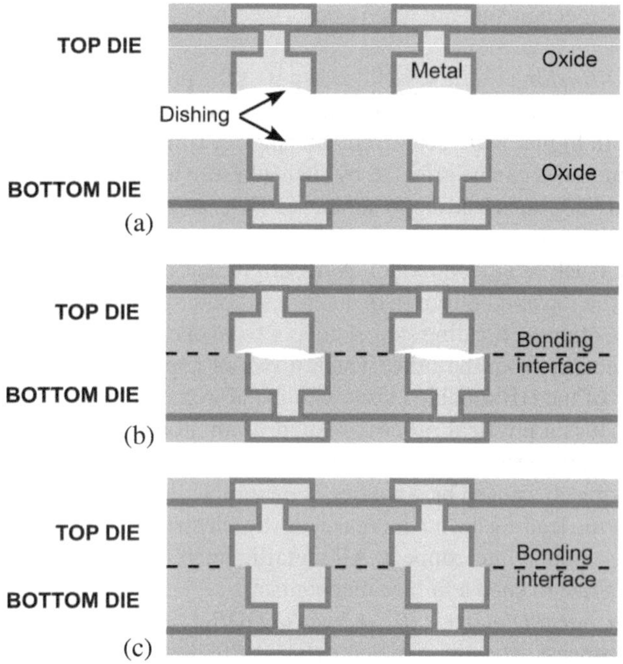

FIGURE 11.5 HB between two dies. (a) Before bonding. (b) Oxide-to-oxide bonding at room temperature. (c) Annealing step to close dishing gaps (coefficient of thermal expansion of metal is larger than that of oxide).

11.2.1.4 HB-based Interconnect

To further reduce the pitch and increase the interconnect density, a new way of stacking wafers and chips is the HB technology which combines local oxide-to-oxide bonding with local direct metal-to-metal bonding, in order to obtain an electrical connection between a top tier (wafer or die) and a bottom tier (wafer mainly) (Figure 11.5).

11.2.2 3D Integration-related Failure Mechanisms

The failure mechanism is the physical, chemical, metallurgical, and environmental phenomena that cause device degradation over time. Compared with traditional 2D integrated circuit (IC) packaging, 3DIC integration involves several new elements in design, manufacturing, and supply chain processes. This results in new reliability challenges as 3D heterogeneous integration potentially combines micro-electro-mechanical systems (MEMS), optoelectronic devices, digital circuits, etc. that are built on different manufacturing process technologies. In addition to the classical failure mechanisms such as Cu diffusion (TSV, HB), EM (Cu pillar, TSV, RDL, HB), dielectric breakdown (TSV, RDL, HB), and moisture ingress (RDL, HB), phenomena such as thermomechanical stress (e.g., TSV-induced mechanical stress on BEOL and/or front-end-of-line [FEOL]), Joule heating, and chip-package interaction (CPI) are also of great importance.

These failure mechanisms are briefly described here:

- *Copper Diffusion* [13]: It is well known that the presence of metal contamination inside dielectrics can dramatically degrade dielectric reliability, resulting in higher leakage current and shorter time to breakdown [14, 15].

  TSV modules can be affected by such a failure mechanism in case of diffusion barrier issues leading to adjacent FEOL and BEOL reliability issues.

  The surfaces in contact at the HB interface contain both copper and a dielectric (such as silicon oxide). A misalignment or overlay error can exist between the top and bottom tiers due to the bonding tool accuracy, and consequently, after the bonding step, if copper from one wafer is in direct contact with dielectrics from the other wafer, a risk of copper diffusion within the dielectric of the HB module exists with the aforementioned consequences.
- *EM* [16]: It is a physical process related to an atomic flow in a conductor (mainly metals) driven by the electrons and enhanced by temperature. Any flux divergence, caused by a barrier, for example, leads to a void nucleation and growth, leading to an increase in the electrical resistance, and ultimately to an open interconnect. All metallic interconnects are thus intrinsically sensitive to such a failure mechanism.
- *Time-dependent Dielectric Breakdown (TDDB)* [13, 17]: TDDB refers to the physical process whereby a dielectric stored under a constant electric field, less than the materials breakdown strength, will break down with time.

  As mentioned above, the presence of metal contamination inside dielectrics dramatically degrades dielectric reliability leading ultimately to shorter time to breakdown. Thus, TDDB degradation can be a signature of Cu diffusion. Excluding Cu pillar, all 3D interconnects can potentially suffer from such degradation. However, due to certain particularities (barrier issues caused by the TSV aspect ratio, Cu facing SiO_2 at the bonding interface, for example), TSV and HB-related interconnects are more prone to TDDB degradation.
- *Moisture Ingress* [18]: Moisture can come from different sources, either internal (e.g., materials or substances used in manufacturing) or external (e.g., water from the dicing step, atmospheric contamination). Moisture ingress could result in a decrease in the isolation performances (current leakage increase, lower TDDB lifetimes) and/or metallic corrosion, leading to high-resistance interconnects.

  Indeed, organic passivation materials, which are typically silicone-based, are often used at the RDL level. All organic materials are non-hermetic and, therefore, cannot completely prevent the moisture ingression.

  As silicon dioxide can be hydrophilic, a potential risk of moisture sensitivity exists for HB-based interconnects, if Cu is directly in contact with silicon dioxide at the bonding interface due to intrinsic misalignment between the top and bottom chips.
- *Thermomechanical Stresses* [19, 20]: These stresses originate from the thermal expansion or contraction of materials when they are mechanically constrained. During their lifetime, interconnects experience many thermal cycles, due to current- and power-cycling (Joule heating–related aspects) and

environmental changes. Differences in the thermal expansion coefficients of the different materials comprising the interconnect levels can lead to stress changes in interconnects. Such a repetitive stresses can result in fatigue damage such as adhesive or cohesive ruptures leading to larger leakage current and/or high-resistance interconnects, and ultimately open circuits.

- *Joule Heating:* Due to the stacking of thin chips with dense features, Joule heating is extremely critical and the generated heat must be carried away efficiently, as poor thermal dissipation will have dramatic consequences. An increase of the IC temperature will lead to an aggravation of the negative effects of thermo-activated phenomena (species diffusion, EM, thermomigration, and stress-induced voiding). This temperature increase will potentially cause an increase of the thermal stresses due to different coefficients of thermal expansion (CTE) at play, leading potentially to adhesive or cohesive failures and chip warpage.
- *CPI* [21, 22]: This is related to complex electro-thermomechanical interactions leading to local to global mechanical stresses. Three major CPI failure modes are often cited: (1) chip failures (chip corner/edge cracking, white bump [23]), (2) package failures (e.g., underfill or solder mask cracking/delamination), and (3) transistor performance shift (caused by stresses due to a thin die, Cu pillars, or TSV).

Table 11.1 gives an overview of the potential reliability threats for HB interconnects and the associated typical degradations.

11.2.3 Reliability Testing and Associated Standards

The aforementioned failure mechanisms are addressed by means of advanced electrical characterization and standardized accelerated tests, as listed in Table 11.2. The corresponding JEDEC standards are given for reference, but other reliability standards (MIL-STD, IEC, EIAJ, etc.) can be used as well.

TABLE 11.1
Potential Reliability Threats and Associated Failure Modes for HB Interconnects

Potential Reliability Threat		Degradation
Moisture ingress	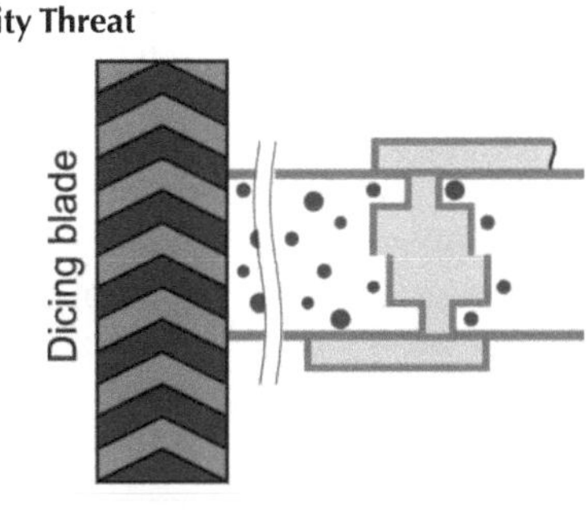	Cu corrosion → high resistance interconnect → open circuit Large leakage current → lower TDDB performance → dielectric breakdown

(Continued)

TABLE 11.1 ***(Continued)***
Potential Reliability Threats and Associated Failure Modes for HB Interconnects

Potential Reliability Threat		Degradation
Thermomechanical stress	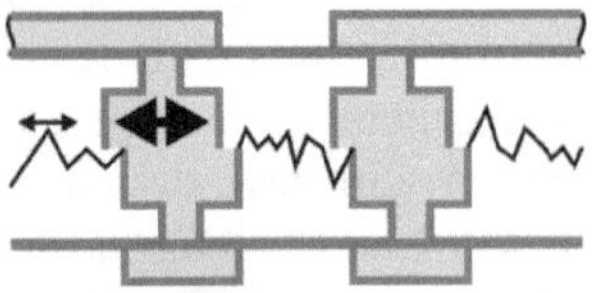	Delamination (Cu/Cu, Cu/dielectric) Leakage current High resistance interconnect → open
EM	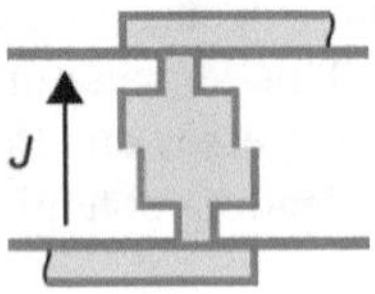	High resistance interconnect Open
Copper diffusion	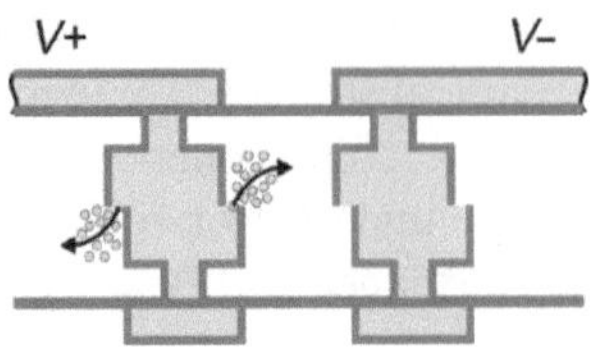	Large leakage current → lower TDDB performance → dielectric breakdown
Dielectric breakdown	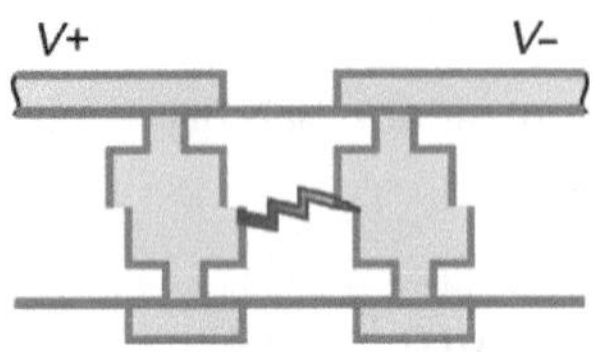	Large leakage current TDDB

TABLE 11.2
Failure Mechanism, Corresponding Accelerated Tests, and JEDEC Standards

Failure Mechanism	Advanced Characterization/ Accelerated Test	JEDEC Standard
Cu diffusion	(1) Triangular voltage sweep / bias temperature stress (TVS/BTS) [24] (2) TDDB test	JEP159
EM	EM test	JESD202
Dielectric breakdown	TDDB	JEP159
Moisture ingress	Unbiased temperature/humidity stress test (highly accelerated stress test [HAST], autoclave)	JESD22-A102 JESD22-A110 JESD22-A118
Thermomechanical stress	Thermal cycling test (TC)	JESD22-A104
Thermally activated phenomena[2]	High-temperature storage (HTS) with or without bias	JESD22-A103 JESD214-01
	Stress-induced voiding (SiV)	JEP139, JESD87, JESD214.01
Chip-package interaction		JEP156A

11.3 OVERVIEW OF ROBUSTNESS/RELIABILITY STUDIES ON HB

This overview presents the studies published over the past decade on HB qualification/reliability testing, by various authors, listed in an alphabetic order. The context of each study is presented followed directly by the robustness/reliability testing data, and the results are presented roughly in a chronological order from a historical perspective.

11.3.1 ADEIA

In 2018, for a die-to-wafer (D2W) option, Gao et al. [25] demonstrated the feasibility of pitch scaling from 40 μm to 10 μm for a pad width, respectively, ranging from 15 μm to 5 μm. Copper was used as the conductive material and a "pad-only" integration was used (i.e., bonding pads are directly connected to BEOL metal lines without any vias). An electrical yield greater than 90% was reached in an ISO 6 (class 1000) clean room prototype laboratory with manual operation for most of the equipment, and a better yield is expected with an industrial operation. After resistance measurements at the wafer level, dicing was done and qualification tests performed for four different lots. For the temperature cycling test (–40/+150°C, 2,000 cycles), most of the dies underwent a resistance shift of –1.5% to –1.7% with one exception of a 3.4% drop. Two sample sets with 22 units each were for high-temperature storage test at 225°C and 275°C, respectively, tested for up to 2 × 1,000 hours. All 44 parts showed a slight resistance drop of at least 1.2–1.4%. Another set of 22 dies went through 168 hours of unbiased autoclave test (121°C, 100% relative humidity [RH], water vapor pressure 205 kPa, 168 hours). All parts showed no resistance increase and no sign of Cu corrosion.

In 2019, in the framework of a wafer-to-wafer (W2W) process, Theil et al. [26] demonstrated an electrical yield between 87.5% and 100%, with post-bond anneal with a thermal budget as low as 300°C for 30 minutes, for 3.8 μm-pitch daisy chains. Copper was used for the bonding pads as well as for the BEOL metal layers. A "pad-only" integration was used there. All the test structures under investigation passed both thermal cycling test (–55/+125°C, 10–14°C/minute, 10 minute dwell, 1,000 cycles) and unbiased autoclave test (121°C, 100% relative humidity, water vapor pressure 205 kPa, 168 hours).

In 2021, Gao [27] presented a five-die stack hybrid-bonded module with TSV. This paper deals with fabrication, assembly process, electrical testing, and reliability performance. The qualification/reliability tests include autoclave, high-temperature storage, and temperature cycling with preconditioning at moisture sensitivity level 3 (MSL3) per JEDEC specifications. The targeted application is a DRAM (Dynamic Random Access Memory) HBM (High Bandwidth Memory) die. The die size is 8 mm × 12 mm × 50 μm, the TSV is 5 μm wide at a 35 μm pitch, and the HB pads are 15 μm wide. The anneal temperature was held at 200–250°C for 1 hour to ensure compatibility with temperature-sensitive applications such as DRAM.

Twenty-two samples were subjected to the autoclave test (121°C, 100% relative humidity, water vapor pressure 103.4 kPa, 168 hours). All parts passed the test with a failure criterion of 10% increase in resistance (with actual increase <1%).

Twenty-two samples were subjected to high-temperature storage testing (150°C). The requirement for DRAM packages is 1,000 hours with a resistance increase of less than 10%. Electrical testing data at 168, 500, and 1,000 hours resulted in no resistance increase. Instead, a slight drop of ~1% was observed. This slight drop in resistance is attributed to improved metallurgical bonding at the joint interface with additional Cu annealing during the environmental test.

Twenty-two stacks were also subjected to moisture sensitivity test according to JEDEC J-STD-020D (MSL3) standard. The samples were first baked at 125°C for 24 hours, then soaked at 30°C, 60% RH for 192 hours, followed by three times reflow at 260°C peak temperature. The samples showed no change in resistivity and no change in voiding patterns or void size.

Forty-five additional samples were preconditioned according to JEDEC JESD22-A113D (MSL3) requirements. Following the precondition, the samples were subjected to a temperature cycling test (–40°C/+125°C, 1,000 cycles). At the time of publication, the test was in progress, but at the first reading point (250 cycles), no change in resistivity was detected.

11.3.2 IMEC

De Messemaeker et al. [28] published a paper in 2018 dedicated to EM tests of Cu/SiCN-SiCN/Cu HB pads. A "pad only" configuration was used with an asymmetry (540 nm-wide top pads bonded to 1,620 nm wide bottom pads) with a pitch of 3.24 μm. Cu was used as the conductive material at both the bonding level and the BEOL layers. To mitigate the challenge of Cu diffusion and migration at fine pitches, a SiCN–SiCN bonding process was used [29]. After bonding, the wafer stack was annealed, first at 250°C to strengthen the SiCN–SiCN bonding, then at 350°C to obtain stable Cu–Cu bonds. Dedicated test structures were designed for upstream EM tests and for downstream EM tests, respectively. The failure criterion was defined as a 20% resistance increase.

1. *Upstream Electron Flow:* Two temperatures (200–240°C) and two current densities (8.2–13.7 MA/cm^2) and less than ten samples per test were used to extract the times-to-failure (TTF) and consequently the Black's parameters [30]. An activation energy of 0.93 eV was extracted. This value is comparable to activation energies found for surface diffusion for Cu damascene lines (0.7–0.9 eV) but also compatible with those for grain boundaries (0.88–0.95 eV) [31]. According to the killer void shape, the capping/Cu interface could be the diffusion path with the highest diffusivity, thus confirming the effective activation energy value extracted. A current exponent of 1.9 was also obtained, which is quite high compared to conventional values obtained for Cu-based interconnects ($n \approx 1$). Values in the range between 1 and 2 are interpreted as a failure based on a mixture of the well-known EM models: void nucleation and void growth [32]. Another hypothesis is related to a non-negligible Joule effect. This hypothesis is partially confirmed by temperature increases (10–20°C) caused by the imposed current densities as characterized by the authors. Physical failure analysis

identified the killer void in the top BEOL layer as other teams did for smaller or larger HB pitches and different types of integration [33, 34]. As the failure does not occur in the HB pads, De Messemaeker et al. [28] can only calculate a lower bound to the EM performance of the HB pads. For a 0.01% failure rate and ten years at an operating temperature of 100°C, the lower bound maximum allowable current density, in the 540 nm top HB pads, is 8.9 MA/cm^2.

2. *Downstream Electron Flow:* The lifetime of the samples is significantly increased (roughly ×5 at 50% failures) compared to the upstream one with the same stress conditions (240°C, 13.7 MA/cm^2). This can be easily explained by the wider bottom pad. Nevertheless, the lifetime gain is larger than expected from the geometry point of view, as the size ratio between the top and the bottom pads is only 1:3. Physical failure analyses showed voiding in the bottom BEOL line, analogous to the voiding in the top BEOL line for the upstream tests, and a large void at the top of the top HB pad. Focused ion beam and scanning electron microscope (FIB–SEM) slice-and-view technique of a tested sample, which had not yet failed, revealed how voids originally at the bonding interface (i.e., bonding voids due to the reconstruction at the pad-to-pad interface [35], fang-like voids [36]) are collected into a large void, moving up along the back sidewall of the top HB pad. This suggests that in EM, the empty volume behaves as a preexisting void moving up along the barrier–Cu interface at the sidewall(s) of the top HB pad, and becomes a dominant contributor to EM failures.

This study clearly highlights a potential new EM-related degradation scheme for sub-micrometric HB interconnects.

The same year, Cavaco et al. [37] published a paper on the morphological, electrical, and qualification characterization of a 200 mm W2W HB integration using SiN as the dielectric material. SiN was investigated for its feasibility to replace SiO_x or SiCN in the HB. Up to five BEOL layers are used on the top and bottom tiers. The conductive material is Cu, and the HB integration is of "pad + via" type. In this work, 2 μm-wide copper HB pads with a pitch of 10 μm were used. The test vehicle includes different types of test structures (resistance structures, Kelvin structures, daisy chains with different geometries) but the presented results are for 2,500-HB pad chains. Two accelerated stress tests were used: HTS (+125°C, 1,000 hours) and TC (−40/+125°C, 1,000 cycles). By using SiO_x or SiCN as the dielectric, no yield reduction or a change of the resistance distribution for the daisy chain of 2,500 HB pads was observed. However, by using SiN as the dielectric layer, a yield reduction down to 95% (−5%) was observed. This can be explained by a comparison of the bonding strengths for the three dielectric materials (SiCN, SiO_x, SiN) used at the bonding level. SiCN and SiO_x exhibited the highest bonding strength, which was on the order of 2 J/m^2. In contrast, SiN exhibited the lowest bond strength, which was below 0.8 J/m^2. While increasing the post-bonding anneal temperature, the bonding strength for SiCN and SiO_x remained approximately constant, whereas the SiN bonding strength dropped to almost half of its initial value. This can lead to unbonded test structures as mentioned by Cavaco et al. [37].

11.3.3 On Semi

In 2017, Gambino et al. [38] presented the development and characterization on HB interconnects between two wafers built from two different technologies: 180 nm Al BEOL and 110 nm Cu BEOL. A two-step bonding annealing was performed: oxide–oxide bonding at 100°C followed by Cu–Cu bonding at ~300°C. A "pad-only" configuration was used.

The test structures consist of daisy chains and leakage structures. The HB pad size is 3 μm and the pitch is 8 μm, and the top HB pad is thicker than the one below. The largest daisy chain has a total of 102,656 links. The leakage structures consist of parallel daisy chains, with 3,200 links in each chain.

Three qualification tests were included: unbiased HAST (+130°C, 85% RH, 96 hours), TC (−40/+85°C, 1,000 cycles), and HTS (200°C, 500 hours). The qualification evaluation showed no failures for the unbiased HAST and for the HTS tests. The TC test showed one failure, which was noted as related to a packaging issue. The resistance showed no measurable change during the HTS test.

11.3.4 Sony

In 2016, mass production of stacked back-illuminated CMOS image sensors (BI-CIS) using a Cu/SiO_2 HB technique was announced [39]. The stacked BI-CIS includes a 22.5 Mpx CIS featuring a 1.0 μm unit pixel size and an image signal processor (ISP). For the HB module, a "via + pad" integration was used. The teardown reveals a five metal (Cu) CIS die and a seven metal (6 Cu + 1 Al) ISP die [40]. The Cu vias are 3.0 μm wide and have a 14 μm pitch in the peripheral regions. In the active pixel array, they are also 3.0 μm wide, but have a pitch of 6.0 μm. Kagawa et al. [39] presented the electrical, qualification, and reliability testing results from this study. A so-called SiV test, which is a HTS test, at 175°C for 1,000 hours on 4 μm pitch daisy chains with 3 million links, was done. No resistance shift was identified. TDDB tests were carried out and the extrapolated performance under operating conditions led to a lifetime of more than ten years.

In 2019, Kagawa et al. [41] published a new paper discussing the improvements of their HB module, with a HB pitch of 3 μm. Electrical characterizations, qualification, and reliability studies were done on 3 μm pitch daisy chains with 3 million links. A new SiV test (175°C, 1,000 hours) did not evidence any more resistance shift.

In 2023, Kagawa et al. [42] presented a novel structure, "Cu–Cu wiring," formed through HB of Cu lines. A new chemical–mechanical polishing (CMP) process was developed to form various wiring patterns at the bonding interface: square shape, line shape, and mixtures thereof. Two failure mechanisms were investigated: stress migration and EM. For the stress migration investigation, a daisy chain was used with 2,000 vias. The width and length of the Cu–Cu wiring are 1.5 μm and 20 μm, respectively. Such a test structure with a 0.3 μm misalignment exhibited no resistance shift after a storage of 1,000 hours at 175°C. For the EM test, a daisy chain was also used with a 0.7 μm wide and 100 μm long Cu-to-Cu wiring with different misalignment values (0, 0.3, and 0.45 μm). The authors claim that the misalignment degrades the EM TTF as the extrapolated lifetimes under Sony's device specification decreased from 40 years to 27 and 23 years, respectively, for a misalignment value of 0, 0.3, and 0.45 μm.

11.3.5 ST/CEA-Leti

In 2016, Lhostis et al. [43] presented qualification results based on TC (–55/+150°C, 15°C/minute, 1,000 cycles) and HTS (+175°C, 2,000 hours) tests on daisy chains with up to 30,000 links and 3.6 μm-wide HB pads. The failure criterion is a relative increase of 10% in the electrical resistance. The actual maximum increase is +0.6% of the electrical resistance after 1,000 cycles. The evolution of leakage current and capacitance of the combs were also investigated. No significant evolution was observed with a leakage current lower than 10^{-10} A. After 2,000 hours, a very small decrease (–2%) of the electrical resistance of the daisy chains was observed after HTS. This evolution is attributed to a microstructure stabilization. No significant evolution of the leakage current and capacitance of the combs were observed after HTS.

Moreau et al. [34] give detailed EM results for different configurations: HB pad size (3.6, 4.4, and 5.4 μm wide), HB via redundancy (4 and 16), HB via positioning versus HB pad (centered and not centered), and test structures (a 100-link daisy chain and a via line test structure designed for HB). The stress temperature was set to 350°C and the stress current was fixed in order to keep the Joule heating lower than 5°C (5–30 mA). The failure criterion was set at 10% resistance increase. Physical failure analyses (lock-in thermography + FIB–SEM) demonstrate that the HB module has no impact on the EM resistance of the present integration, and the weakest link is always at the BEOL level (top or bottom depending on the electron flow). There is no HB-related failure, and the bonding interface is immune regarding the electron flow direction. In other words, bonding defects do not move with the electron flow. The reservoir effect [44] is an explanation of the results with regard to the HB pad size and HB via redundancy. Due to the HB pad size, the positioning of the HB via has no influence from the EM point of view.

Jourdon et al. [45] in 2017 studied the impact of the passivation-annealing step on the EM performance. This annealing, under different atmospheres such as forming gas (H_2/N_2) or deuterium [46], was used in addition to the bonding annealing to improve the dark current of CMOS image sensors. Three variants were investigated: no passivation step, with forming gas, and with deuterium. For the first variant, the bonding annealing temperature was modulated according to three flavors: 300°C/1 hour, 300°C/2 hours, and 400°C/2 hours. This study suggests that the bonding annealing temperature does not affect the EM lifetime, as EM tests performed on a HB stack before the final passivation annealing leads to typical values of Black's parameters for a Cu BEOL ($E_a = 0.99$ eV $\pm$ 0.034 eV, $n = 1.36 \pm 0.11$). However, significant influence of the final passivation annealing conditions was observed on the EM lifetime. In the case of a high-pressure deuterium annealing, the degradation (–31%) is attributed to the presence of deuterium accumulated at the barriers and capping layers. Indeed, deviation of the characteristic lifetime is usually caused by variations of metal/capping interface properties. Numerous studies deal with how these properties are affected by interface treatment, or the nature of the cap [47, 48]. This hypothesis was evidenced by time-of-flight secondary ion mass spectroscopy (ToF-SIMS) characterizations to avoid time-consuming EM tests to extract the Black's parameters. In the case of forming gas annealing, the degradation (–24%) is lower but still significant.

In 2018, Jourdon et al. [49] presented results of TC and EM tests. Since the HB module is made of dissimilar materials, with different CTEs, thermomechanical stresses occur. By means of 2D FEM (finite element modeling), Jourdon identified a potential risk associated with thermomechanical stresses with pitch reduction (from 7.2 to 1.44 μm). Indeed, as pitch goes smaller and smaller, oxide spaces are not wide enough to release the stress, so delamination is more inclined to occur at the Cu–SiO_2 bonding interface for 1.44 μm pitch and below. In order to evaluate this risk, TC tests at the wafer level were conducted at −65/+150°C and −55/+150°C, for 500 cycles, respectively, for HB pad pitches of 7.2 and 1.44 μm. All test structures remained fully functional after the test, and the resistance variation was lower than 1%.

Ayoub et al. [50] in 2020 presented results from HTS (+175°C, 2,000 hours) and TC (−65/+150°C, 1,000 cycles) tests for 1.44 μm pitch test structures. The variation is less than 0.5% for both tests. These minor variations in resistance confirm that the bonding interface is stable and that none of the process variations affects the structure's resistance. For capacitance, the variation after 1,000 cycles was less than 1% for the comb–comb test structures of 13,000 and 57,000 links. The negligible leakage current variation after both tests demonstrates that there is no evidence of copper diffusion.

In 2021, Moreau et al. [24] published a paper dedicated to Cu diffusion at 7.84 μm pitch. Indeed, at the bonding interface, due to the bonding tool accuracy, an inherent misalignment exists between the top and bottom tiers leading to a direct contact between copper and the neighboring dielectrics. The presence of metal contamination inside the dielectrics can dramatically degrade the dielectric reliability, resulting in higher leakage current and shorter time to breakdown as discussed previously. Consequently, Cu diffusion arises as a potential reliability concern for HB integration. From their HB integration scheme [43], a dedicated 300 mm test vehicle was manufactured in order to fabricate metal-insulator-semiconductor (MIS)-like test structures. Bias temperature stress test (BTS) in conjunction with triangular voltage sweep (TVS) electrical characterization were used to evidence mobile ion diffusion. The temperature for both tests was fixed at 250°C, and the sweep rate was fixed at 221 mV/s. In the end, whatever the BTS/TVS conditions (up to +20 V, up to 20 minutes), no particular peak or hump can be identified. In other words, no diffusion of ionic species (Cu+, Na+, K+, …) can be identified.

Moreau et al. [24] give different hypotheses to explain this observation:

- The advanced process used for the HB surface preparation (CMP and cleaning) suppresses all (or nearly all) of the contamination sources.
- If there is any diffusion, it only leads to mobile ion concentration below 10^9–10^{10} 1/cm^2 (detection sensitivity of TVS/BTS techniques).
- A diffusion barrier is present in the region of interest. As proposed by Aboelfotoh from IBM in 1990, cuprous oxide (Cu_2O) could be an effective diffusion barrier for copper [51].

To explore this last hypothesis, a monochromated electron energy loss spectroscopy (EELS) in a transmission electron microscope (TEM) was performed to identify any Cu_xO compound material at the Cu HB pads/SiO_2 layer level. The STEM–EELS spectra showed unequivocally a thin layer (~3 nm) of cuprous oxide (Cu_2O) at the copper/dielectric interface. The Cu_2O formation could be explained in different manners, for example:

- After the CMP step and before bonding, HB Cu pads are exposed to the ambient atmosphere (and in particular oxygen).
- After bonding, the dielectric material, SiO_2, can be a source of corrosive elements (and in particular oxygen). This situation associated with appropriate environmental conditions can lead to the formation of a Cu_xO compound, as described by Choudhary et al. [52].

In 2022, Ayoub et al. [53] presented a new method to perform TDDB tests in the context of HB interconnects in order to take into account the misalignment variability and pad shape. This method was applied to a 1.44 μm pitch 3D stacked test vehicle to demonstrate its accuracy. A comb–serpentine–comb test structure was used for this study. This new method consists of performing I-V measurements on targeted structures for the TDDB analysis with a stop condition at a specified leakage current point. Estimation of V_{BD} and therefore the distance between the comb and the serpentine can be done with an error as small as ±5%. By ensuring the same electric field value per test structure at the wafer scale despite an intrinsic manufacturing process variability, a robust TDDB study (three electric fields and three temperature values) can be ensured. Times-to-failure thus obtained follows a $1/E$ model at electric fields ranging from 1.6 MV/cm to 4 MV/cm. Unlike the $\sqrt{E}$ model usually applicable to classical BEOL interconnects, a $1/E$ model suggests that the role of Cu in breakdown is negligible. The cuprous (Cu_2O) oxide layer, previously identified by Moreau et al. [24], might still be playing a role in decelerating Cu diffusion, in the misaligned Cu/SiO_2 areas at the HB interface. The acceleration parameter is temperature dependent and increases by 35% when the temperature decreases from 195°C to 155°C. The activation energy was found to linearly depend on $1/E$ for the investigated electric field values. Unlike values reported for other BEOL dielectrics at identical electric fields, higher activation energy values were extracted (1.29 eV at 2.1 MV/cm and 0.95 eV at 4 MV/cm versus 1.1 eV and 0.7 eV, respectively, for a fluorosilicate glass (FSG) dielectric with a SiN capping). An additional EELS acquisition confirms the hypothesis that the cuprous oxide is specific to this integration structure.

In 2022, Ayoub et al. [54] presented a methodology to achieve submicrometer HB pitch by analyzing single Cu pad thermomechanical behaviors and by quantifying the Cu–Cu contact resistivity, and a 100% electrical yield was achieved for interconnect pitches down to 0.81 μm. To address the RC delay increase with pitch reduction, a methodology is proposed based on (1) reducing the HB level thickness, and (2) lowering the Cu pad width for the same interconnect pitch to increase the spacing between adjacent pads. In the robustness tests (TC and HTS) on a daisy chain structure with 300 nm-wide Cu pads and a HB thickness of 500 nm, whatever the test or measured value (electrical resistance or capacitance), the relative variation was negligible (±1%) and well below the proposed 10% failure criterion, using daisy chains with 5,000 links and comb–comb structures with 57,000 links.

In 2023, Ayoub et al. [55] pushed further the investigation on Cu diffusion in the framework of the Cu–SiO_2 interface. The total reflection X-ray fluorescence spectrometry (TXRF) and the liquid-phase decomposition–inductively coupled plasma mass spectrometry (LPD-ICPMS) techniques were used to investigate the Cu atomic diffusion in the context of Cu–SiO_2 integration. The authors demonstrated that the Cu oxide layer has similar barrier characteristics in decelerating Cu atomic species diffusion as compared to a silicon nitride inserted in a BEOL stack. Cu concentration is similar to the one for a classical BEOL stack. This result confirms the negligible impact observed on the reliability of Cu–SiO_2 HB integration. The study of Cu ionic diffusion, after long-term storage (4,000 hours at 175°C, a 4× increase in test duration as defined by the corresponding JEDEC standard), by BTS/TVS characterization, highlighted the lack of Cu diffusion up to 200°C and for different voltage ramp rate (125 and 250 mV/s). TDDB results were in line with this result. Indeed, the times-to-failure followed a $1/E$ model for a temperature of 175°C. This result suggests that the role of Cu in assisting breakdown is negligible. The extracted acceleration parameter (14.2) is quite similar to the one (14.6) obtained on non-aged samples, suggesting that the integration, including the cuprous oxide, is thermally stable. The EELS technique was again used in the misaligned region to chemically analyze this region. Once again, a homogeneous Cu_2O layer was identified there confirming its thermal stability. The authors concluded that the Cu diffusion reliability concerns for the Cu–SiO_2 HB configuration are overcome by the self-formed stable Cu oxide layer. The authors warned that the present conclusion corresponds to the specific SiO_2 used in their experiments and may vary depending on the SiO_2 properties.

In 2023, Moreau et al. [56] presented the latest advances in EM for the Cu–SiO_2 integration. The paper is mainly divided into three parts:

1. *Effect of Design on EM Lifetime:* This part attempts to highlight any optimum design configuration regarding the reservoir effect associated with redundant vias (2 × 2 matrix), and via/feed line positioning (relative angle: 45°, 90°, 180°), for the test structure based on 6.84 μm pitch. In both cases, among the investigated configurations, no particular configurations exhibited any advantage.
2. *Relevance of Routing Signals at the Bonding Interface:* This study was done in the context of architecture optimization, such as dimensional shrinkage (interconnect sizes 3.42 um pitch and 1.44 μm pitch, and routing length equivalent to one or two pitches). For the first time for this integration, physical failure analyses highlighted a change in the failure mode. Indeed, voids were identified in HB Via/HB Metal levels (HBV/HBM) in addition to their classical location at the BEOL level. Furthermore, the weakest link is clearly the bonding interface as small voids are identified in the misalignment area, even though this does not imply that these defects will be the killer ones.
3. *Effect of Pitch Reduction on EM Performance:* The goal is to investigate a possible impact of pitch reduction on EM failure modes in addition to the EM performance for HB-pitch down to 1.44 μm. An asymmetric via line test structure designed for the HB technology was used with 1 via per bonding pad whatever the pitch. This particular configuration (1 link versus

4 links) was used to ease the failure analysis step by analyzing only the single link side. The authors confirm a change in the failure modes. Indeed, the failure mode moves from a killer void at the BEOL level for pitches above ~3.5 μm (conservative value) to a killer void at the HB module level below ~3.5 μm. Nevertheless, according to the authors, the projected performance under normal use conditions (10 years, 0.01%, 125°C) are not affected by this change in the failure mode. The authors noted that since the killer void is located at the same location for the routing configuration, the projections made should also be identical. Consequently, signals can be reliably routed at the bonding interface.

11.3.6 TSMC

In 2019, Chen et al. [57] presented the System on Integrated Chips platform using in particular the HB technology, with the bonding pad pitch down to sub-10 μm. Numerous qualification and reliability tests were done at the chip or package level. At the wafer level, tests performed included Pressure Cooker Test (PCT, 168 hours), HTS (1,000×), TC (1,000×), uHAST (168 hours), EM, SiV, and breakdown voltage (V_{bd}) tests, and SoIC bonds (i.e., HB pads) demonstrated robust reliability performances. At the package level, all samples passed the standard JEDEC test conditions as well as tests with extended test cycles and time.

In 2020, Chen et al. [58] discussed an evolution of the SoIC technology to address ultrahigh density (UHD) 3D technology by shrinking the bonding pitch from 9 μm to 0.9 μm. Wafer level reliability was assessed using SiV, dielectric breakdown (V_{bd}), HTS and TC tests. 46 samples from one wafer were used per test. Test durations of SiV, TC, and HTS are respectively 500 hours, 1,000 cycles, and 168 hours. A 1k daisy chain was used for SiV, TC, HTS tests, and a 1k-to-1k comb structure was used for the V_{bd} measurement. The UHD SoIC technology exhibited stable and high yield, robust reliability, and process maturity throughout the different pitch densities and daisy chain lengths.

In 2022, Liang et al. [59] presented a 3 μm bond pitch, face-to-face, chip-on-wafer SoIC integration study with a low thermal budget. Daisy chains and Kelvin structures were built in for chip-level leakage test and resistance measurements. For qualification and reliability of this advanced SoIC technology, the tests included EM, SiV, breakdown voltage (V_{bd}), uHAST, and TC tests. The technology passed uHAST (130°C, 85% RH, water vapor pressure 0.229 MPa, 168 hours) and TC (−65/+150°C, 300×) tests.

In 2023, Liang et al. [60] gave further insights related to the SoIC technology qualification and reliability. Tests included EM, SiV, and breakdown voltage (V_{bd}), in addition to qualification tests such as uHAST and TC tests. For the uHAST test (130°C, 85% RH, water vapor pressure 0.229 MPa, 168 hours, $\Delta R/R_0$ <10%), no significant increase in resistance was observed. For the TC test (−65/+150°C, 300 cycles, $\Delta R/R_0$ <10%), no obvious resistance change of a daisy chain occurred.

EM test was done at 200°C for a constant current density of 1.5 MA/cm^2 and the failure criterion is a relative resistance increase of 10%. No resistance change was observed after 1,500 hours.

For the SiV test, the temperature was kept at 200°C for 500 hours, and 30 daisy chains were used. Less than 5% increase in normalized resistance was observed after the test.

V_{bd} tests were done up to 90 V and 200 V with 0.1 V/step at 125°C for more than 30 samples. No leakage current above 10^{-4} A was measured. The effect of temperature (100°C, 150°C, and 200°C) on V_{bd} results was also investigated for only one voltage condition (90 V). The leakage current under 90 V with temperatures up to 150°C was undetectable and certainly significantly lower than 10^{-4} A.

11.3.7 WDC/Imec

In 2022, Hou et al. [61] presented a paper dealing with dielectric breakdown and the associated leakage mechanism. The integration of interest is W2W Cu/SiCN HB. The design is asymmetric according to the bonding interface to improve the bonding alignment tolerance: the bottom Cu pad is designed to be larger than the top one, respectively 600 nm and 200 nm, with a pitch of 1 μm resulting in a critical pad-to-pad distance of 400 nm. Physically, the bottom pad width along the bonding interface is measured to be 680 nm resulting in a corrected critical pad-to-pad spacing of 320 nm. A thin layer of SiCN is capped on top of the SiO_2 layer. The thickness of the SiCN cap layer after CMP is measured to be 98 nm. A comb–comb test structure with 22,500 pad pairs was used for evaluating the TDDB performance. Two configurations were evaluated in this work: full pad structure and bottom half structure (where the top metal and top pad are excluded). The breakdown voltage (V_{BD}) along the bonding interface was evaluated for temperatures ranging from 100°C to 200°C using a controlled I–V methodology. Three different ramp rates (1.15, 0.26, and 0.06 V/s) were applied to the two different pad configurations to record the current change versus voltage. Assuming an E-model ($TTF \propto e^{-\gamma E}$), the measured acceleration factors (γ) are found to be in the range of 11–16 for temperatures ranging from 100°C to 200°C, and the extrapolated lifetime under use conditions exceeds ten years. A decrease of V_{BD} values and Weibull slope could be observed when the temperature reached 200°C. The conduction mechanism along the Cu/SiCN HB interface was characterized and found to follow a Poole–Frenkel emission model. According to the extracted dielectric constants, the leakage current is mainly located in the SiCN bonding interface. For an electric field of 2.5 MV/cm and a temperature of 200°C, a transition point of current density versus temperature is noticeable. When the temperature increases above 200°C, the observed increased leakage current density suggests that there is a different dominant mechanism besides the Poole–Frenkel emission. Since the increase of leakage current density could only be observed at "high" temperatures (>200°C) and "high" electrical field (>1.5 MV/cm), one potential hypothesis is that copper starts to drift under these stress conditions. This hypothesis is further verified by applying a constant current stress at 200°C. As noted, the observed electrical behavior is the typical signature of a soft breakdown resulting from the formation of a copper filament. Since the degradation of reliability and increase of leakage current only happen under conditions more severe than the normal use conditions, copper drift is not currently considered as a concern for the HB scheme studied.

11.3.8 YMTC

In 2021, Ouyang et al. [33] published a paper dealing with aging tests on the technology used for 3D NAND flash memories, in which memory cells and peripheral circuits are bonded by millions of pairs of HB pads. TEM pictures confirm that Cu is used as the conductive material at both the bonding level and the BEOL layers. In addition, a "via + pad" configuration is identified with an asymmetry: HB pads from the top wafer are narrower (~300 nm) than those from the bottom wafer (~450 nm), to mitigate the overlay error, at 1 μm pitch.

Qualification and reliability tests were conducted at both the test structure level and the product level:

- *Test Structure Level:* Five test structures were used: a Kelvin-like test structure, a 40,000-link daisy chain with redundancy, a 3D stacked comb-to-comb test structure, an upstream test structure, and a downstream one for EM testing. Qualification tests included temperature cycling (−65/+150°C, 2 cycles per hour, 15,000 cycles), high-temperature storage (200°C, 7,000 hours), and thermal humidity storage (85°C, 85% RH, 7,000 hours). EM testing was performed at the package level. The failure criteria are a relative increase above 10%, and a leakage current above 0.1 μA.
- *Product Level:* Samples were packaged in a ball grid array package (BGA) with four different die stacking configurations: 2, 4, 8, and 16 die per package. Qualification tests included preconditioning (125°C for 24 hours, 30°C and 60% RH for 192 hours, and 3 reflows), TC (−65/+150°C, 2 cycles per hour, 500 cycles), HAST (110°C, 85% RH, 264 hours + bias), and HTS (150°C, 1,000 hours). The failure criterion is defined as no functional failure and no scanning acoustic microscope (SAM) abnormality.

The samples exhibited no (HAST) or a slight (<3% for TC and HTS tests) negative resistance shift, which was attributed to an annealing effect. Cross sections performed on aged samples did not reveal any cracks or voids. In addition, no breakdown voltage reduction was observed after 15,000 thermal cycles.

Summary of Reliability Tests on HB

These reliability tests on HB, which cover different configurations and various test methods and conditions, demonstrate the robustness of the HB technology, offer insights into the relevant failure mechanisms, and provide certain quantitative parameters which are important for reliability characterization and simulation.

11.4 MODELING AND SIMULATION

Computer modeling and simulation capabilities for the HB process can be a very useful tool for understanding and optimizing the process and assessing performance and reliability. Fundamentally, numerical modeling can help reveal the mechanisms underlying the bonding processes. Practically, modeling and simulation can help identify the relative roles of various parameters in materials, design, and processes in the bonding process and in the performance of the interconnect, and parametric studies can be performed for optimization once the numerical model has been developed, calibrated, and validated. By modeling the critical process steps (such as CMP, pre-bonding of the dielectric layers, and Cu–Cu bonding), design attributes and process parameters can be optimized for individual process steps and for the whole process. Furthermore, electrical, thermal, and reliability performance can be modeled, assessed, and optimized.

11.4.1 Modeling Framework

Sitaraman et al. [62] presented a holistic "modeling framework" for HB (Figure 11.6), a platform which starts from materials modeling and optimization, through systems modeling and optimization, in four "co-optimization loops." It starts with the "interconnect loop" which focusses on materials, process, and interconnect co-optimization, followed by the "die/channel loop" which targets an aggregate of interconnects to capture effects (such as crosstalk, signal integrity, die warpage, etc.) The next loop is the design loop, which encompasses the interposer, package, and PCB, to understand the electrical performance and thermomechanical reliability of the subsystem, including assembly process flow optimization. The last loop is the system loop which includes the impact of algorithms and software and the impact of the previous optimizations on their performance [62].

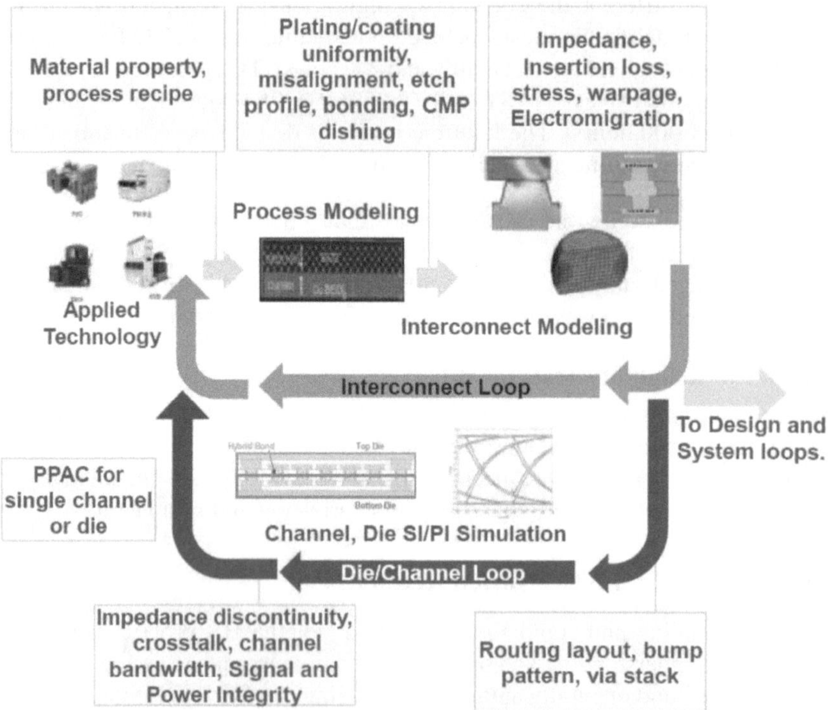

FIGURE 11.6 A modeling framework for HB proposed by Sitaraman et al. (© 2023, IEEE. Reprinted with permission from Ref. [62].)

Indeed, it is going to take considerable efforts to develop capabilities for every element of the four loops. Each of the elements in the loops can be useful on its own, and some examples will be discussed elsewhere in this chapter. This platform offers an interesting framework for developing and constructing holistic modeling capabilities for HB applications in particular.

11.4.2 Process Modeling

Process modeling addresses the critical process steps (including CMP, dielectric bonding, and Cu–Cu bonding) to help achieve fundamental understanding of the bonding mechanism and develop tools for process optimization in practical applications. Dishing depth, bonding area evolution, peeling stress, etc. are critical process attributes that need to be addressed.

11.4.2.1 Process Modeling: CMP

Sitaraman et al. [62] modified the empirical Preston's model and calibrated against experimental data, for CMP processes at the feature scale. The material removal rate (MRR) equation is as follows:

$$\text{MRR}(x,y) = k \times P(x,y) \times V(x,y) + \text{MRR}_0 \tag{11.1}$$

where MRR at a location (x, y) on the wafer surface is defined as a function of the local pressure $P(x, y)$ and the local pad velocity $V(x, y)$, and k is a constant that includes all of the other factors. MRR_0 accounts for the chemical reaction caused by the chemicals in the slurry and is pressure independent. Using experimental data measured from both a blanket wafer and a patterned wafer, k and MRR_0 values were calibrated for each material.

Using this model, the dishing dependence on the Cu pad pattern was studied (Figure 11.7), and CMP process simplification and optimization were carried out using the validated model.

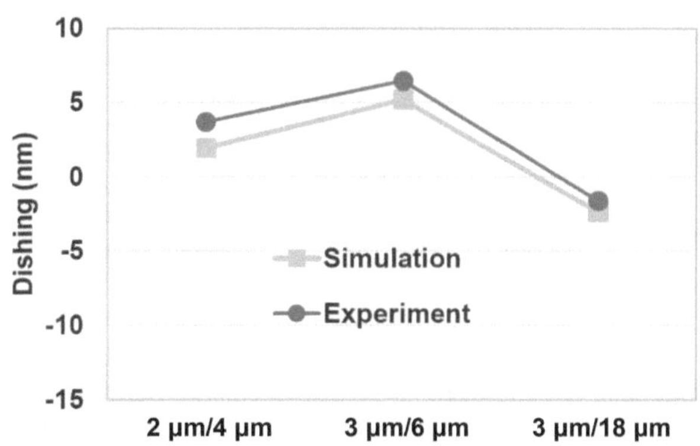

FIGURE 11.7 Post-CMP dishing depth for different pad patterns. (© 2023, IEEE. Reprinted with permission from Ref. [62].)

11.4.2.2 Process Modeling: Bonding Mechanism

Sitaraman et al. [62] studied the interactions at the dielectric interface during the HB formation, using computational ab initio simulations-based atomistic modeling, to understand, at the atomic level, the mechanisms underlying dielectric surface activation and bonding processes. Ab initio simulations-based atomistic modeling can qualitatively describe the exact nature of the van der Waals dispersive interactions between hybrid bonded surfaces, which is helpful for understanding the effect of the chemical nature of dielectric surfaces on the cohesive interaction during the HB formation. In the ab initio simulations-based modeling [62], the interaction energy is calculated using interatomic potentials based on the density functional theory (DFT) [63], while the kinetic energy is calculated from the Kohn–Sham kinetic energy using the linear combination of atomic orbitals (LCAO) basis set for wave function definition [64], and the Grimme DFT-D2 dispersion correction of the energy and forces was used to qualitatively add the effect of the van der Waals interaction [65].

The computational ab initio simulations in this work have shown that bonding between dielectric interfaces can be improved by changing the surface terminations [62]. For example, surface wetting followed by the possible process conditions can produce OOH on the surface, and OOH formation will be beneficial for increasing the adhesive interactions between slabs (Figure 11.8). Simulations also

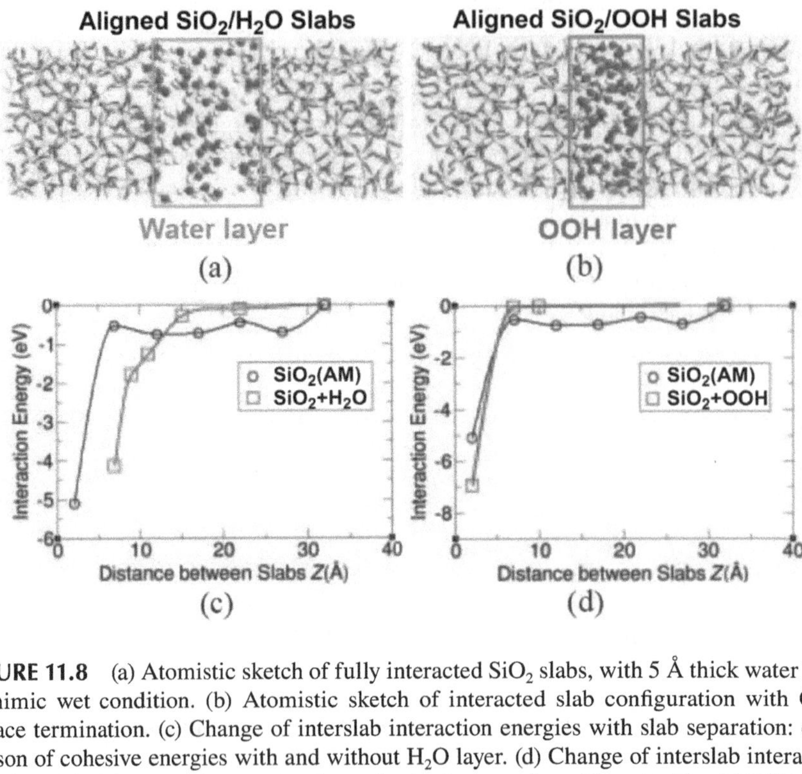

FIGURE 11.8 (a) Atomistic sketch of fully interacted SiO_2 slabs, with 5 Å thick water layer to mimic wet condition. (b) Atomistic sketch of interacted slab configuration with OOH surface termination. (c) Change of interslab interaction energies with slab separation: comparison of cohesive energies with and without H_2O layer. (d) Change of interslab interaction energies with slab separation: comparison of cohesive energies with and without OOH termination. (© 2023, IEEE. Reprinted with permission from Ref. [62].)

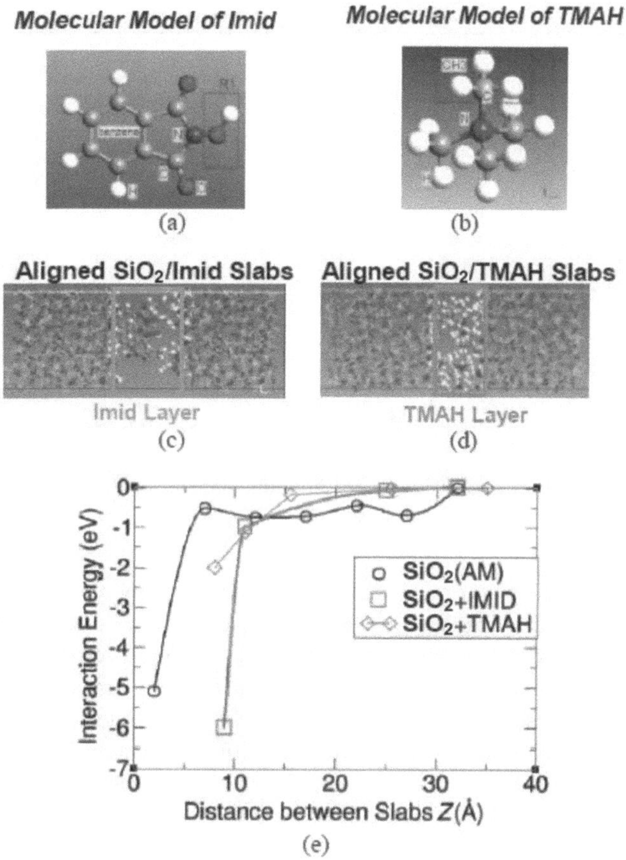

FIGURE 11.9 Atomistic sketch of (a) IMID, (b) TMAH molecule, (c) fully interacted SiO_2 slabs with monolayer of IMID, (d) interacted slab configuration with a monolayer of TMAH, and (e) change of interslab interaction energies with slab separation: comparison of cohesive energies with and without polymeric monolayers. (© 2023, IEEE. Reprinted with permission from Ref. [62].)

show that polymeric molecules can be a strong candidate to enhance dielectric–dielectric bonding [62]. For example, polymers with carbon rings can strengthen interfacial bonds between slabs and increase the cohesive bonding (Figure 11.9).

Kim et al. [66] used atomistic molecular dynamics (MD) simulation (Figure 11.10) to study the chemical and structural composition of SiO_2 surfaces during surface activation with N_2 or O_2 plasma treatments and hydration processes, and to elucidate how bonding energy between SiO_2 wafers can be enhanced with surface activation. The work reveals that Si–O–H formation on the SiO_2 surface during the O_2 plasma treatment and the hydration process, and then the reaction of two Si–O–H bonds, could spontaneously convert to Si–O–Si bonds during the following bonding process. In the N_2 surface activation, formed NO-groups on the SiO_2 film are hydrolyzed to form additional OH functional groups on the surface

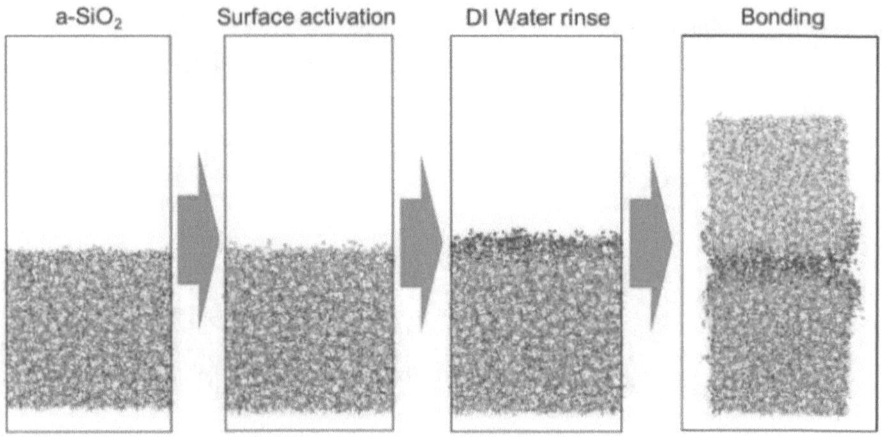

FIGURE 11.10 A four-step fusion wafer bonding process at the atomic scale. (© 2023, IEEE. Reprinted with permission from Ref. [66].)

during the deionized water rinsing process, leading to increased Si–O–Si formation during the bonding process and eventually enhanced bonding strength [66].

Beilliard et al. [67] presented thermomechanical finite element simulations of Cu–SiO_2 HB (Figure 11.11), with a cohesive interaction model (CIM) mimicking non-linear interactions between Cu–Cu and SiO_2–SiO_2 interfaces and the related bonding mechanism implemented within a non-linear contact mechanics framework. The bonding process along with the influence of the annealing conditions and the Cu plastic behavior on the closure of the Cu–Cu interface are investigated.

The work demonstrates that cohesive interactions combined with the thermal expansion of Cu during annealing are the main mechanics of the bonding process of

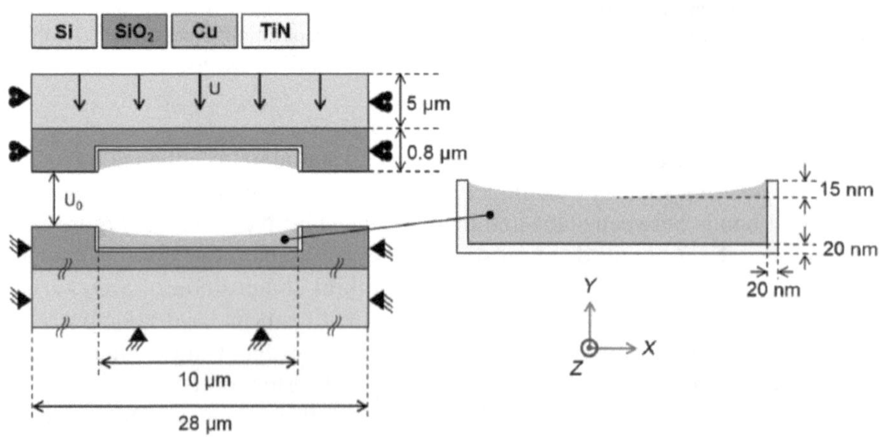

FIGURE 11.11 Schematic description of the copper lines with their dimensions and boundary conditions. (© 2023, Elsevier. Reprinted with permission from Ref. [67].)

Cu lines with a dishing effect [67]. As cohesive forces tend to induce plastic strains in the vicinity of the bonding wave front, the plastic response at the bonding interface of Cu with a low yield stress could act as a mechanical fuse that reduces the effect of the cohesive forces, and plastic deformation has therefore a detrimental effect on the bonding process, as shown in Figure 11.12.

Sart et al. [68] developed a 3D finite element model for the thermal annealing of Cu–SiO_2 HB pads (Figure 11.13). To investigate the effect of thermomechanical deformation, cohesive interactions are not accounted for, and instead non-separation contact is enforced to mimic perfect bonding (i.e., surfaces that come into contact remain bonded).

The simulation results (Figure 11.14), for a case study of a representative unit cell centered about a bonding pad of 4 μm in width and 500 nm in thickness with an interconnect pitch of 8 μm, suggest that with 10 nm dishing, thermomechanical deformation alone is not sufficient to entirely close the interface. For 200°C annealing, after

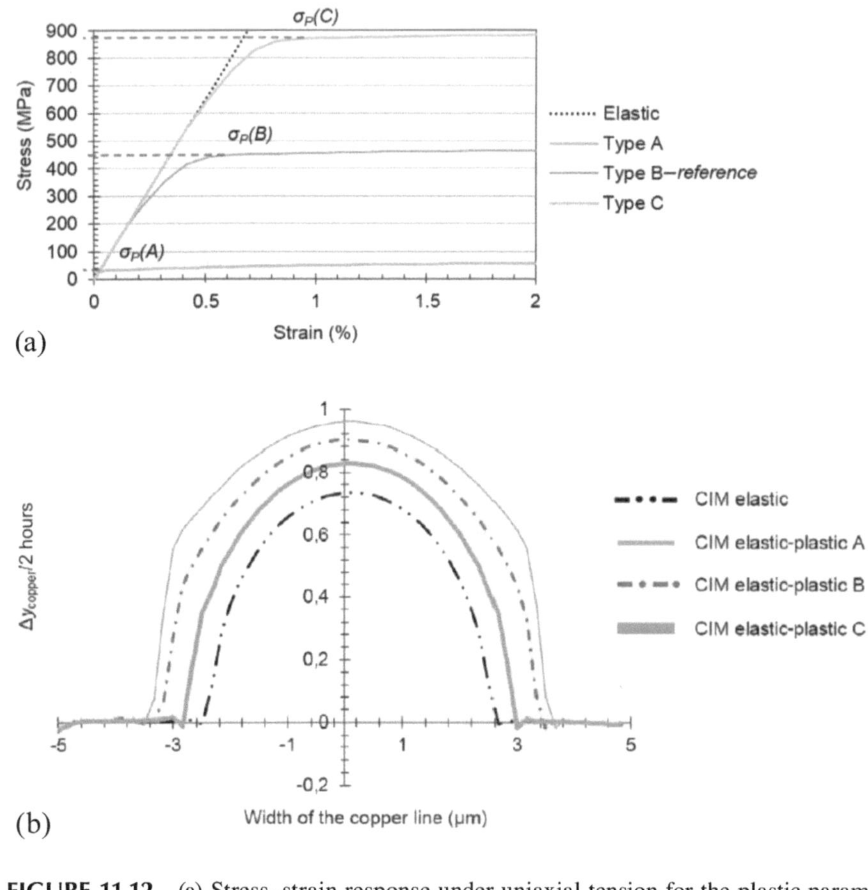

FIGURE 11.12 (a) Stress–strain response under uniaxial tension for the plastic parameter of the Cu types. (b) The distance separating the top and bottom Cu surfaces as a function of the position on the Cu line for different Cu properties. (© 2023, Elsevier. Reprinted with permission from Ref. [67].)

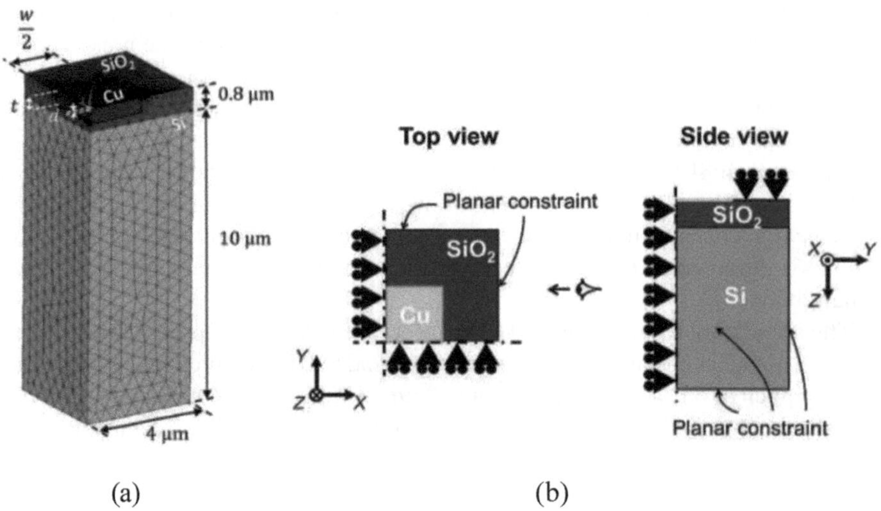

FIGURE 11.13 (a) Dimensions and typical mesh of the quarter model. (b) Boundary conditions. (© 2023, IEEE. Reprinted with permission from Ref. [68].)

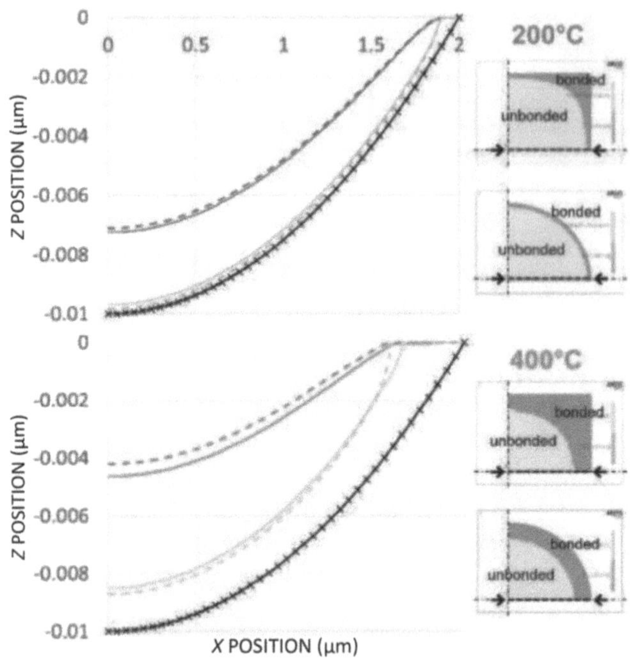

FIGURE 11.14 *Left:* Evolution of dishing profile for square (plain) and circular pads (dashed) during the annealing step at 200°C (blue) and 400°C (red) (black: reference; dark: during annealing; light: cooling to room temperature). *Right:* Variation of the bonded area (orange). (© 2023, IEEE. Reprinted with permission from Ref. [68].)

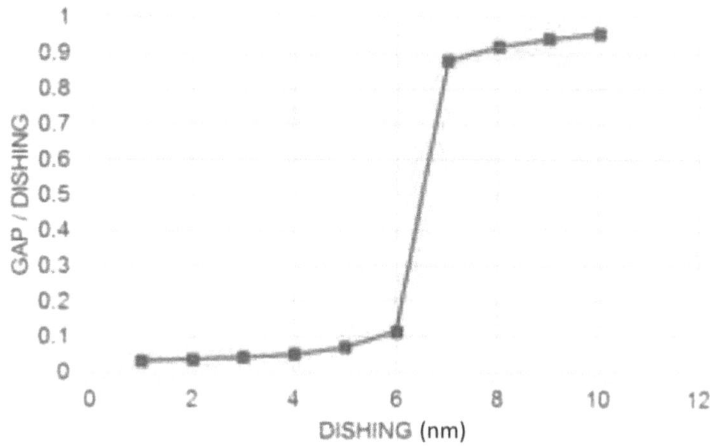

FIGURE 11.15 Effect of initial dishing depth on the bonding at room temperature. (© 2023, IEEE. Reprinted with permission from Ref. [68].)

returning to room temperature, the dishing profile remains almost unchanged, while after 400°C annealing, the gap separating the top and bottom pads at the center is reduced by about 15%. Regarding the pad shape ("circular" versus "square"), the amplitude of the deformation is higher with the circular pads, i.e., the gap in between the circular pads is lower than that of the square pads during annealing but higher after returning to room temperature, and the bonded area is much higher with the square pads than with the circular pads, especially for lower annealing temperatures. These differences are due to the difference in the distribution of the elastic strains due to the CTE mismatch between copper and the surrounding silicon dioxide, as well as the difference in the shape of the dishing surface, for the two types of pads. Through a parametric study, it has also been found that during annealing, reducing the pad width at a constant dishing depth hinders contact between the pads, and increasing the pad thickness helps bridge the gap between the two pads [68].

Then, to study the cohesive interaction during the room temperature pre-bonding, the attractive part of the interaction is described using a non-linear contact mechanics strategy with a traction–separation law derived from the interaction potential [69], and the repulsive part is defined using the standard penalty contact method [69]. It is found that for the case with 10 nm dishing, the effect of cohesive interaction at room temperature is modest; however, below a certain threshold value of initial dishing, cohesive tractions are high enough to bring the two pads into contact even before the annealing step (Figure 11.15). Generally, while both thermomechanical deformation and cohesive tractions contribute to successful bonding of the Cu interconnects, the former is key to bringing the two surfaces sufficiently close for cohesive interaction to occur [69].

11.4.2.3 Process Modeling: W2W Bonding

Ip et al. [70, 71] developed a multi-physics model for the W2W bonding dynamics which includes the air viscosity damping and solid mechanics of silicon wafer

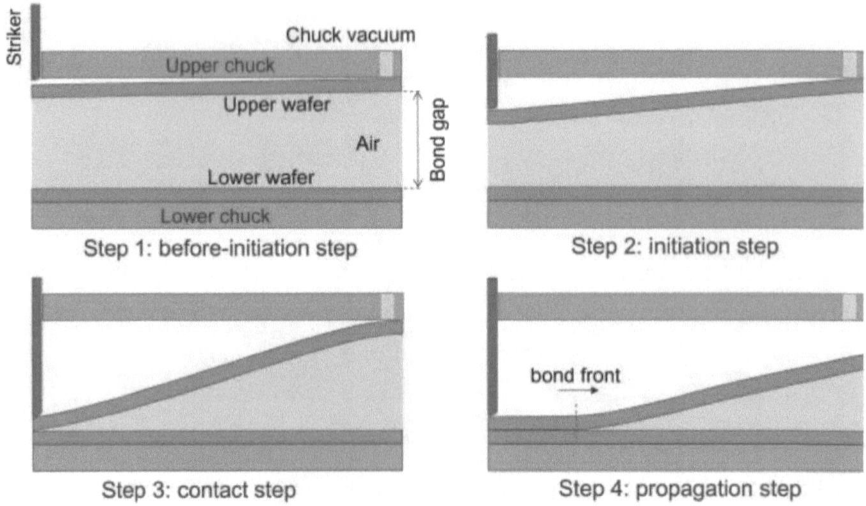

FIGURE 11.16 Schematic of the wafer bonding process. The lower wafer is considered flat and clamped to the lower chuck, and only the upper wafer deformation is considered. (© 2023, IEEE. Reprinted with permission from Ref. [70].)

deformation, for direct investigation of post-bond wafer distortion under different process conditions consisting of four process steps (Figure 11.16). The direct wafer bonding process is mathematically modeled as well as using finite element analysis (FEA) as a dynamic coupled system of governing differential equations describing the underlying mechanical mechanisms, namely, the mechanical deformation of the upper wafer, fluid effect, and contact and adhesion forces, and the wafer height during the bonding process is obtained by solving the moving boundary problem numerically using the finite difference discretization in space and time, and by applying appropriate boundary conditions and initial conditions [70, 71]. A height value, which was tuned based on experimental results, was applied to switch from the initiation step to the contact step.

The strong agreement between the simulation and experiment results (Figure 11.17) suggests that the solid–fluid interaction is responsible for the time-dependent bond propagation. The results also show that the bond time does not vary significantly, and that the deformed wafer shape at the bond initiation does not significantly impact the bond propagation rate, for the design configuration in this study [70, 71]. On the other hand, the time to initiate the bonding decreases as the bond gap increases, suggesting that the viscous air damping effect decreases as a larger bond gap allows the air to escape more readily. The wafer is released after bond initiation, and the wafer release time increases as the upper chuck vacuum pressure increases. These parameters are important for optimizing the bonding process [70, 71].

The work also indicates (Figure 11.18) that the incoming wafer shapes play an important role in wafer bonding process and must be accounted for in order to achieve optimal process performance.

Hirano et al. [72] developed a multiscale simulation of bonding wave propagation incorporating a microscale Cu pad structure in a macroscale wafer-level structure.

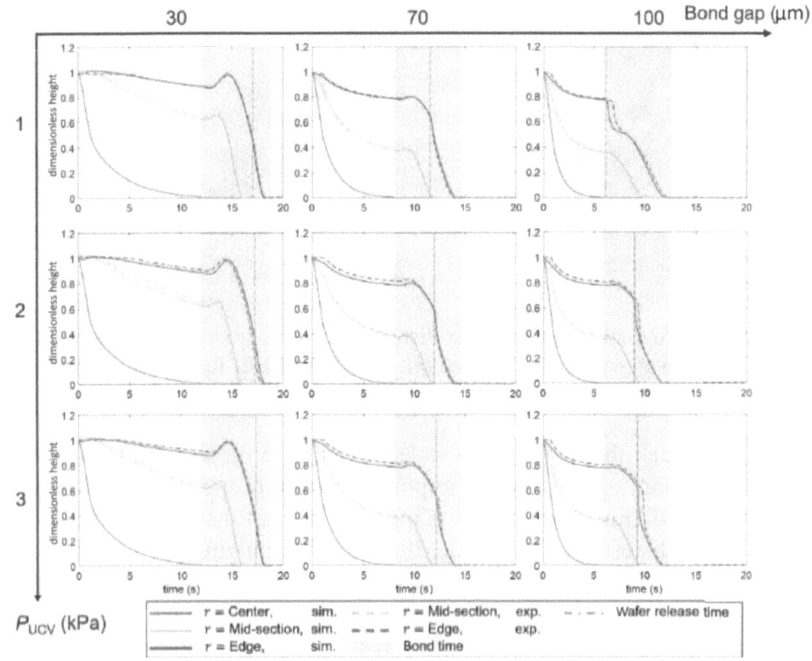

FIGURE 11.17 Comparison of simulation results and experimental data of wafer height over time during the bonding process, for different bond gaps and upper chuck vacuum pressures. (© 2023, IEEE. Reprinted with permission from Ref. [70].)

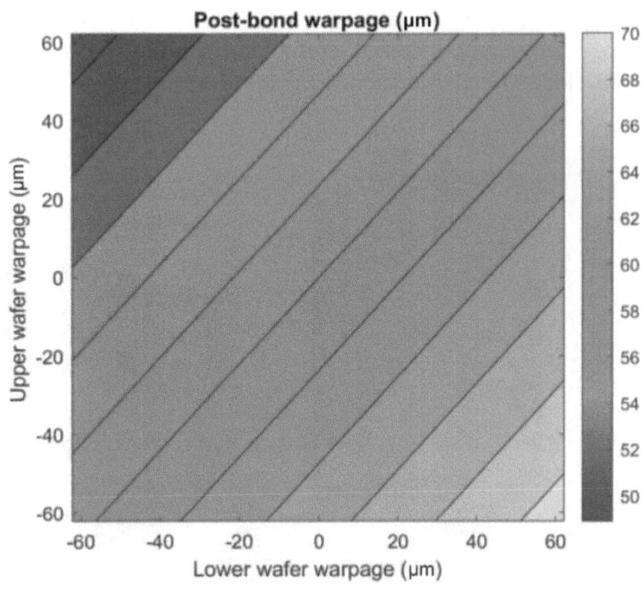

FIGURE 11.18 Simulated post-bond warpage as a function of incoming upper and lower wafer warpage. (© 2023, IEEE. Reprinted with permission from Ref. [71].)

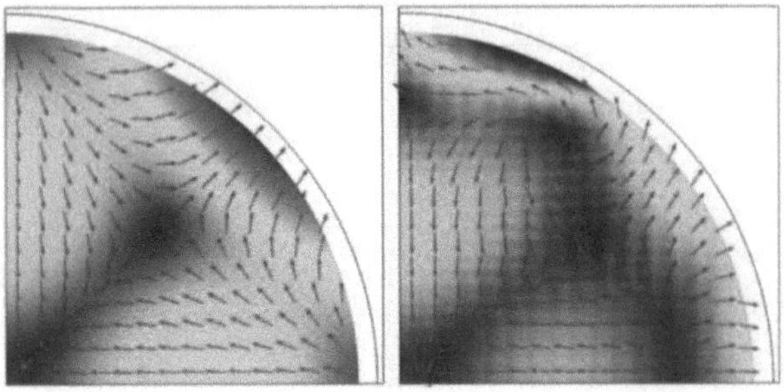

FIGURE 11.19 Simulated post-bonding distortion of a quarter of a Si wafer: a bare Si wafer (*left*), and a wafer with Cu pads (*right*). (© 2023, IEEE. Reprinted with permission from Ref. [72].)

The airflow between the wafers is modeled to analyze the precise behavior of the bonding propagation, which plays an important role in post-bonding distortion (and consequently misalignment of the Cu–Cu pads). The results suggest (Figure 11.19) that the bonding distortion changes significantly by changing the layout of Cu–Cu pads, due to the difference in the bonding wave front propagation shape. For example, for the wafer that has Cu–Cu pads in the active region and no Cu–Cu pads in the scribe region, the bonding front propagates in a rectangular shape and the velocity of the propagation varies in each direction, in contrast to a wafer with no Cu–Cu pads, for which a circular bonding front propagation shape exists [72].

Ji et al. [73] investigated the impact of various design and process parameters on two-layer W2W HB using a FEA tool (Figure 11.20), in order to achieve a better understanding of the design and process parameters which could be used to establish process guidelines. Cu deformation during the annealing process is captured by using an elastic–plastic-creep material model for Cu, and Cu-to-Cu protrusion and bonding area development are modeled.

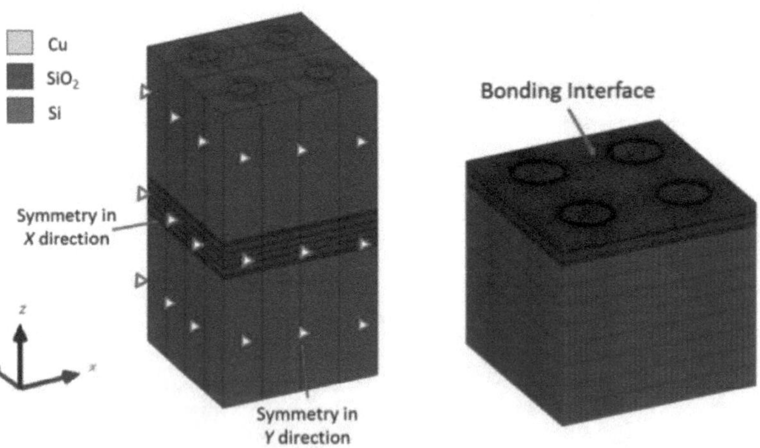

FIGURE 11.20 Finite element model for 2 × 2 Cu pads. *Left:* Full geometry. *Right:* Bottom structure showing bonding interface. (© 2023, IEEE. Reprinted with permission from Ref. [73].)

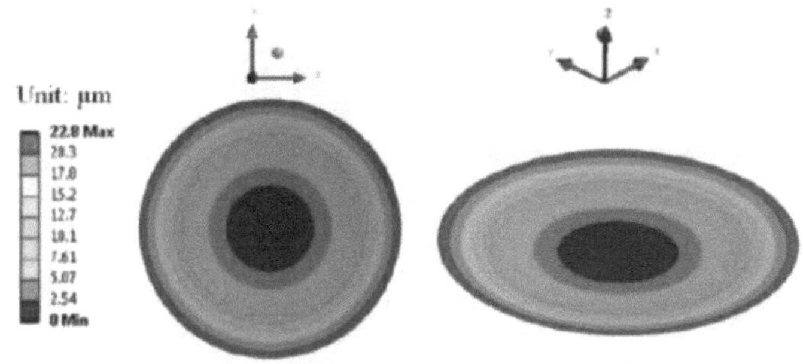

(a) Case A: Cu Ø=3 μm, Pitch=6 μm, T_{ref}=350°C

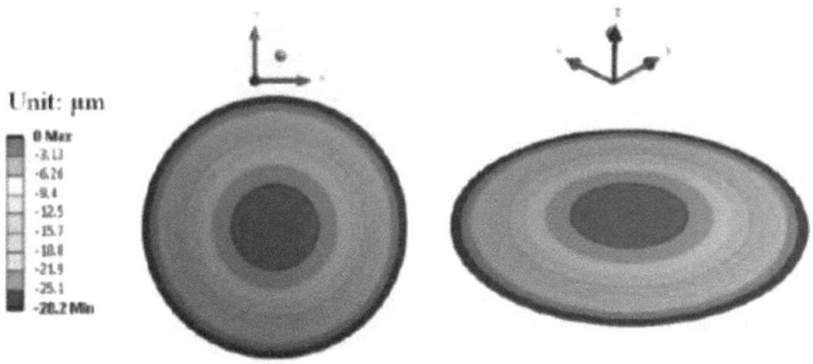

(b) Case B: Cu Ø=3 μm, Pitch=12 μm, T_{ref}=350°C

FIGURE 11.21 Example of wafer warpage contours. *Left:* Top view. *Right:* Isometric view. (© 2023, IEEE. Reprinted with permission from Ref. [73].)

The wafer warpage was modeled under various conditions; examples are shown in Figure 11.21, with a "smile" for case A and a "frown/cry" for case B. A parametric study shows that the warpage direction and magnitude are associated with the Cu area density (calculated from Cu pad diameter and pitch, and Cu pad depth), as shown in Figure 11.22. Zero warpage is observed when the equivalent CTE of the Cu pad structure matches the CTE of the Si wafer. The more the Cu area density deviates from this zero-warpage Cu area density value, the more the warpage observed. The zero-warpage Cu area density value decreases slightly with the deeper Cu pad, and the zero-warpage Cu area density values remains the same for a given Cu pad depth, even with different reference temperatures. Regardless of the warpage direction, a deeper Cu pad leads to larger warpage, and the warpage increases with the increasing reference process temperature [73].

In the same work, the formation of Cu-to-Cu bonding area is simulated, and the analysis is based on only mechanical considerations and does not consider metal diffusion. Figure 11.23 shows the contour plots for the case with 3 μm Cu pad diameter, 6 μm pitch, 10 nm dishing for a 2 hours/350°C annealing process (0.5 hours ramp-up, 1 hour dwell, and 0.5 hour cool-down). After the annealing process, the unbounded area is found to be located at the peripheral area of the Cu pad (as highlighted in yellow). This study shows that the Cu bonding area increases rapidly with both decreasing dishing values and increasing annealing temperatures, as expected, while the Cu pitch plays a less significant role for forming the Cu bonding area [73].

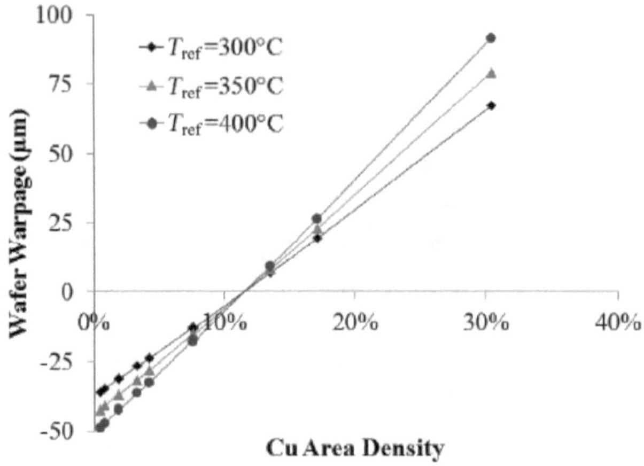

FIGURE 11.22 Example of wafer warpage versus Cu area density (for 1 μm deep Cu pad). (© 2023, IEEE. Reprinted with permission from Ref. [73].)

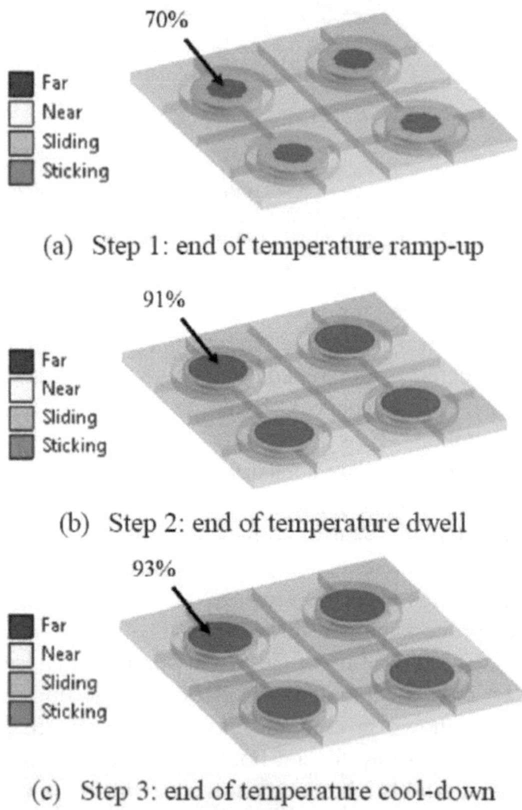

FIGURE 11.23 Cu bonding areas at various time instants (highlighted by red color). (© 2023, IEEE. Reprinted with permission from Ref. [73].)

Correspondingly, the stresses along the *Z* direction (normal to the bonding interface) on both SiO_2-to-SiO_2 and Cu-to-Cu bonding interfaces at the same three time instants for this case are shown in Figures 11.24 and 11.25, respectively. Parametric study shows that the peeling stress on the SiO_2–SiO_2 interface increases with smaller dishing values and increasing annealing temperatures, whereas the peeling stress on the Cu–Cu interface, once the Cu-to-Cu bonding area is formed, increases with larger dishing values [73].

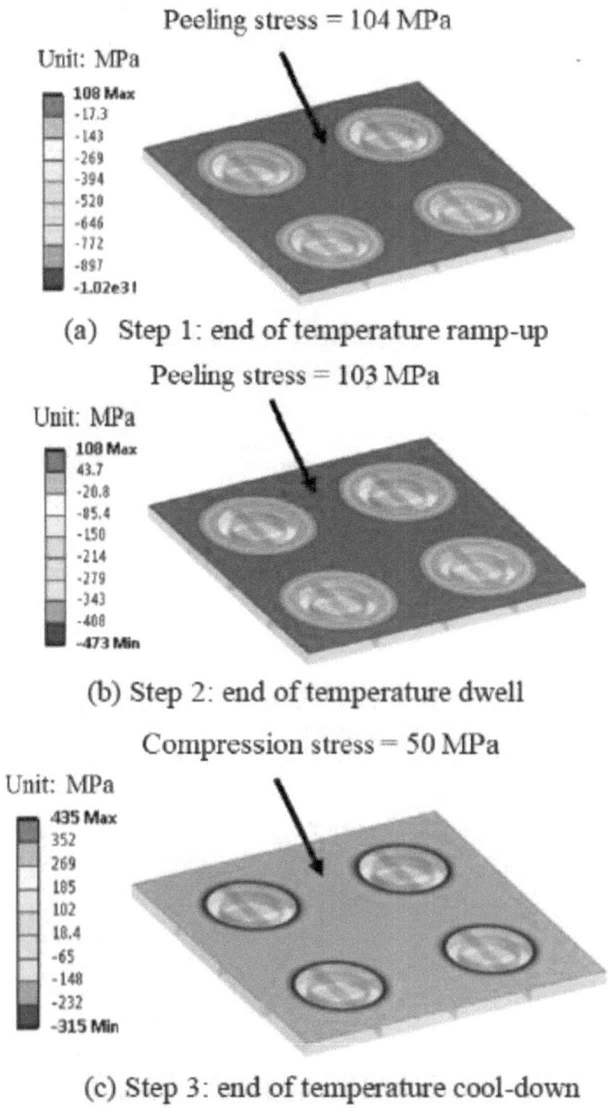

FIGURE 11.24 Stresses on SiO_2-to-SiO_2 bonding interface. (© 2023, IEEE. Reprinted with permission from Ref. [73].)

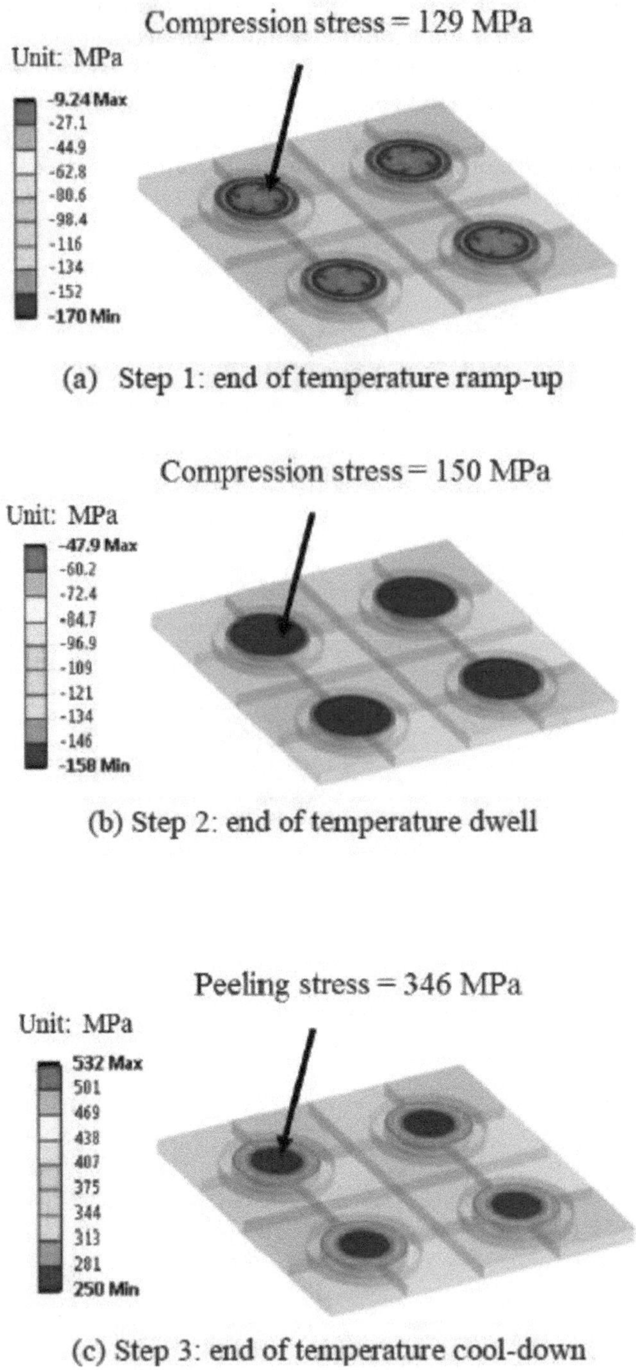

FIGURE 11.25 Stresses on Cu-to-Cu bonding interfaces. (© 2023, IEEE. Reprinted with permission from Ref. [73].)

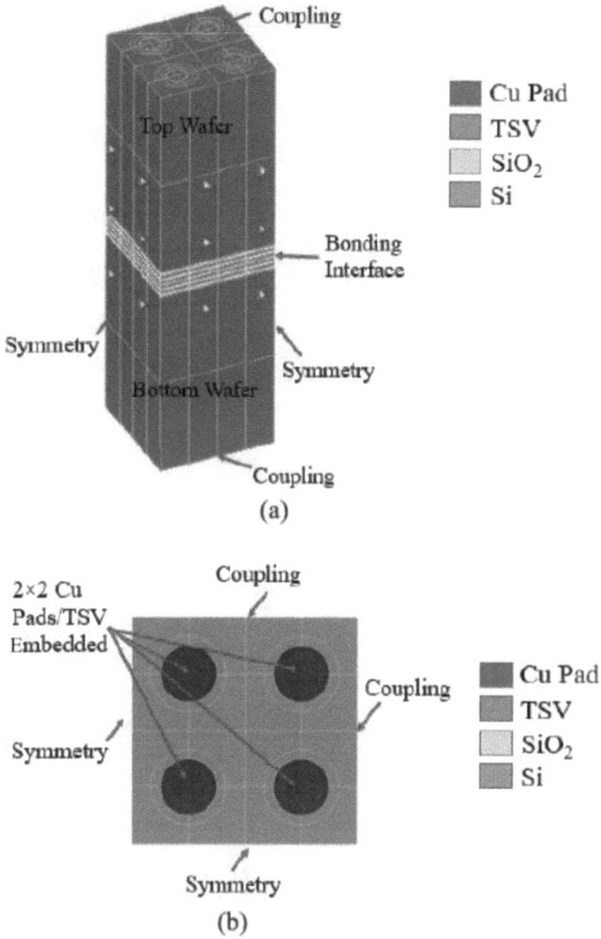

FIGURE 11.26 Finite element model representing a localized unit containing 2 × 2 Cu pads/TSV pairs facing each other. (a) Isometric view. (b) Top view at the bonding interface. (© 2023, IEEE. Reprinted with permission from Ref. [74].)

Ji et al. [74, 75] further developed a 3D non-linear finite element model (Figure 11.26) for the W2W HB annealing process, and performed simulation for various design and process conditions: (a) design A: two wafers with Cu pads only; (b) design B: top wafer (TSVs and Cu pads) and bottom wafer (Cu pads only); and (c) design C: two wafers with TSVs and Cu pads.

The simulation results highlight the annealing temperature and the dishing value as the most critical parameters among all the factors considered [74, 75]. For 350°C annealing, the simulation shows that the Cu-to-Cu bonding area increases with reduced dishing values (Figure 11.27), and if the dishing values are 10 nm or less, the difference in the Cu bonding area is small (<5%) between wafers with TSV (designs B and C) and wafers without TSV (design A). However, with a Cu dishing value of

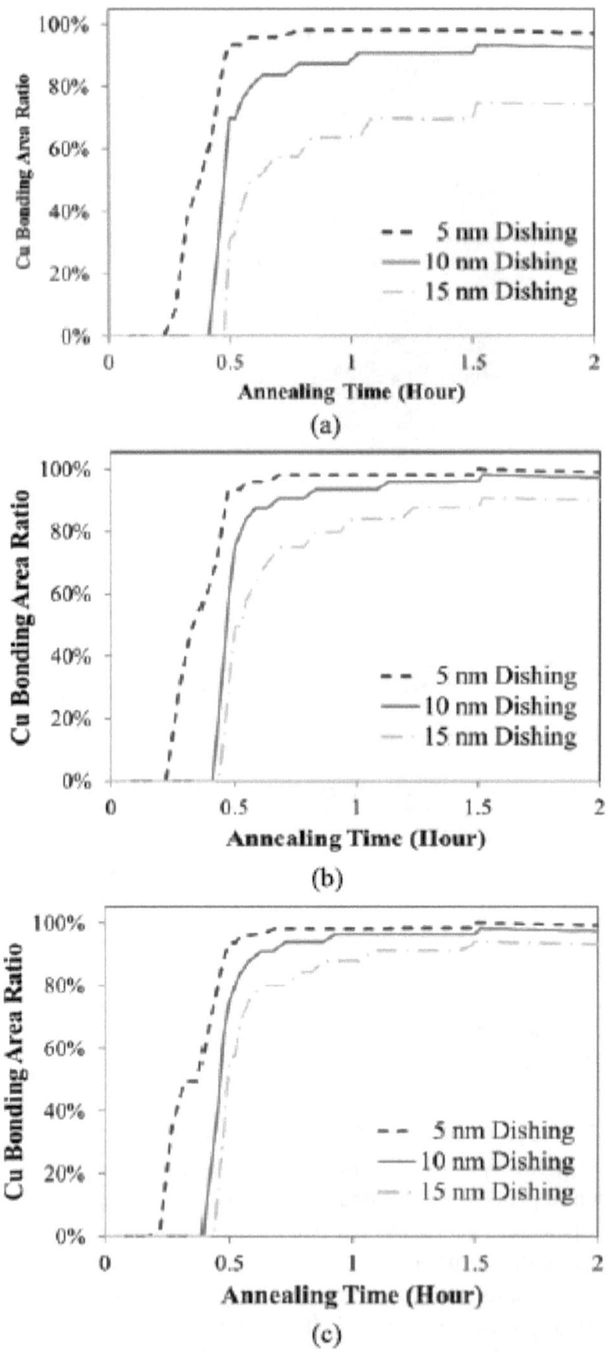

FIGURE 11.27 Simulation results of the Cu-to-Cu bonding area during the annealing process at 350°C. (a) Design A. (b) Design B. (c) Design C. (© 2023, IEEE. Reprinted with permission from Ref. [74].)

15 nm, the TSV under the Cu pad will help increase the Cu bonding area by approximately 15% due to more Cu protrusion generated by TSV. Furthermore, regardless of the dishing value, design B has a similar Cu bonding area when compared with design C despite the fact that design B has TSV in the top wafer only while design C has TSV in both the top and bottom wafers. Overall, a good Cu bonding area (>90%) could be obtained by designs with Cu pads only at low dishing values. If, on the other hand, the CMP process could not achieve the dishing value of 10 nm or less, the design with TSV under the Cu pad could increase the Cu bonding area significantly. For fixed process conditions of 350°C annealing and 10 nm dishing, the annealing dwell duration and the TSV pitch and depth have negligible influences on the Cu-to-Cu bonding area for designs B and C, whereas for design A, only a small variation (<3%) in the Cu-to-Cu bonding area is observed. The results also show that the difference in the Cu bonding area among the three designs becomes less obvious when the annealing temperature increases. However, for annealing temperatures at or below 350°C, dishing plays a more critical role, and it is beneficial to have the additional Cu protrusion contributed by TSV expansion for a larger Cu bonding area, especially for the cases with a dishing value of 10 nm and above. For the worst scenario of design A with 15 nm dishing and 300°C annealing, no Cu-to-Cu bonding area is formed [74, 75].

In terms of the peeling stress, among the three designs for 350°C annealing and 10 nm dishing (Figure 11.28), design A has the largest peak peeling stress of 346 MPa on the Cu bonding interface. With the additional TSV structures inside the wafers for designs B and C, the peak Cu peeling stresses are slightly lower (323 MPa for design B and 327 MPa for design C), as designs B and C have slightly larger Cu bonding areas, as shown in Figure 11.27. The results also show (Figure 11.29) that the peak peeling stress on the Cu bonding interface generally increases with lower annealing temperatures. For example, for the case with a large dishing value of 15 nm and a low annealing temperature of 300°C, due to the sharp decline in the Cu bonding area, the peak Cu peeling stress increases significantly as compared with the cases with smaller dishing values and higher annealing temperatures. While designs B and C have a smaller peak peeling stress on the Cu bonding interface when compared with design A, there is no significant difference in the peak Cu peeling stress between designs B and C, as long as both designs achieve a similar Cu bonding area. If there is a substantial difference in the Cu bonding area formed, such as the case with 300°C annealing and 15 nm dishing, design B will have a much higher peak Cu interfacial peeling stress due to its much smaller Cu bonding area formed. This suggests that with TSV underneath the Cu pad, that is, more Cu volume involved to form Cu-to-Cu bonding, the risk of Cu debonding is lower [74, 75].

As for the SiO_2-to-SiO_2 bonding interface, design C is found to have the largest peak peeling stress (89 MPa), while design A has the smallest (56 MPa), as more Cu protrusion due to the presence of TSV increases the peeling stress on the surrounding dielectric bonding interface, and the risk of the dielectric layer debonding is hence increased. On the other hand, the peak peeling stress on the SiO_2 bonding interface rises with smaller dishing and higher annealing temperatures, due to the more severe Cu protrusion and the resulting larger pushing forces between the opposite Cu pads under these conditions. Therefore, a trade-off on the dishing and

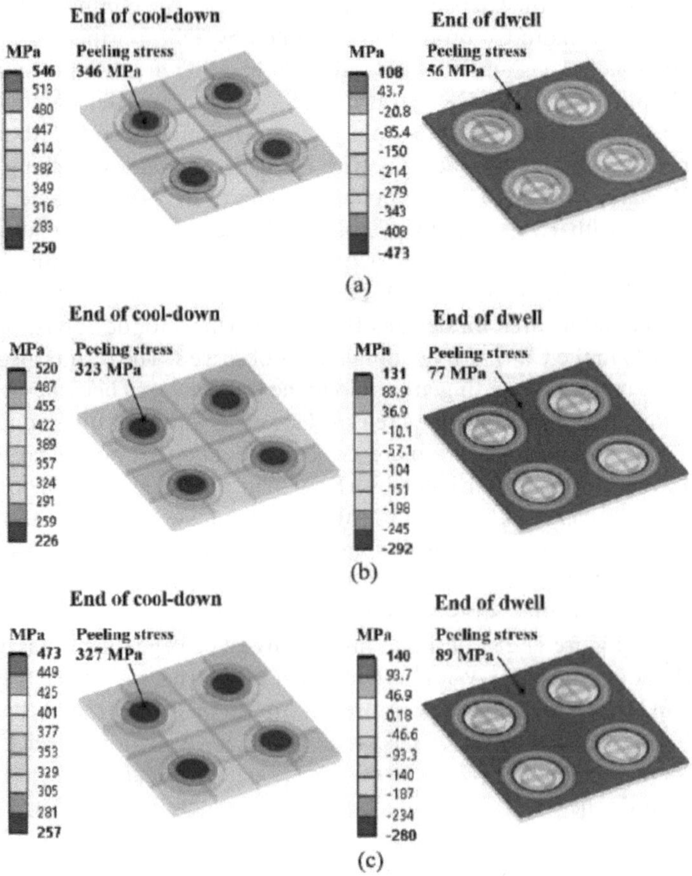

FIGURE 11.28 Contours of the simulated peak peeling stress on the Cu-to-Cu bonding interface (*left:* at the end of the cool-down stage) and SiO_2-to-SiO_2 bonding interface (*right:* at the end of the dwell stage) for 350°C annealing and 10 nm dishing. (a) Design A. (b) Design B. (c) Design c. (© 2023, IEEE. Reprinted with permission from Ref. [74].)

the annealing temperature has to be made between achieving good Cu bonding area and low Cu debonding risk, and avoiding high dielectric peeling stress to lower the dielectric debonding risk [74, 75].

No significant influences on the Cu and dielectric bonding interfaces are observed for other process variables such as the annealing dwell duration, the TSV/Cu pad pitch, and the TSV depth, for all the designs, and the only exception is the impact induced by the TSV/Cu pad pitch on the SiO_2 interfacial peeling stress, as a small pitch design has a higher peak SiO_2 peeling stress [74, 75].

11.4.2.4 Process Modeling: Die-to-Wafer Bonding

Ren et al. [76] developed a finite element model to gain insights into the process window for void-free and reliable interconnection for die-to-wafer HB, for both

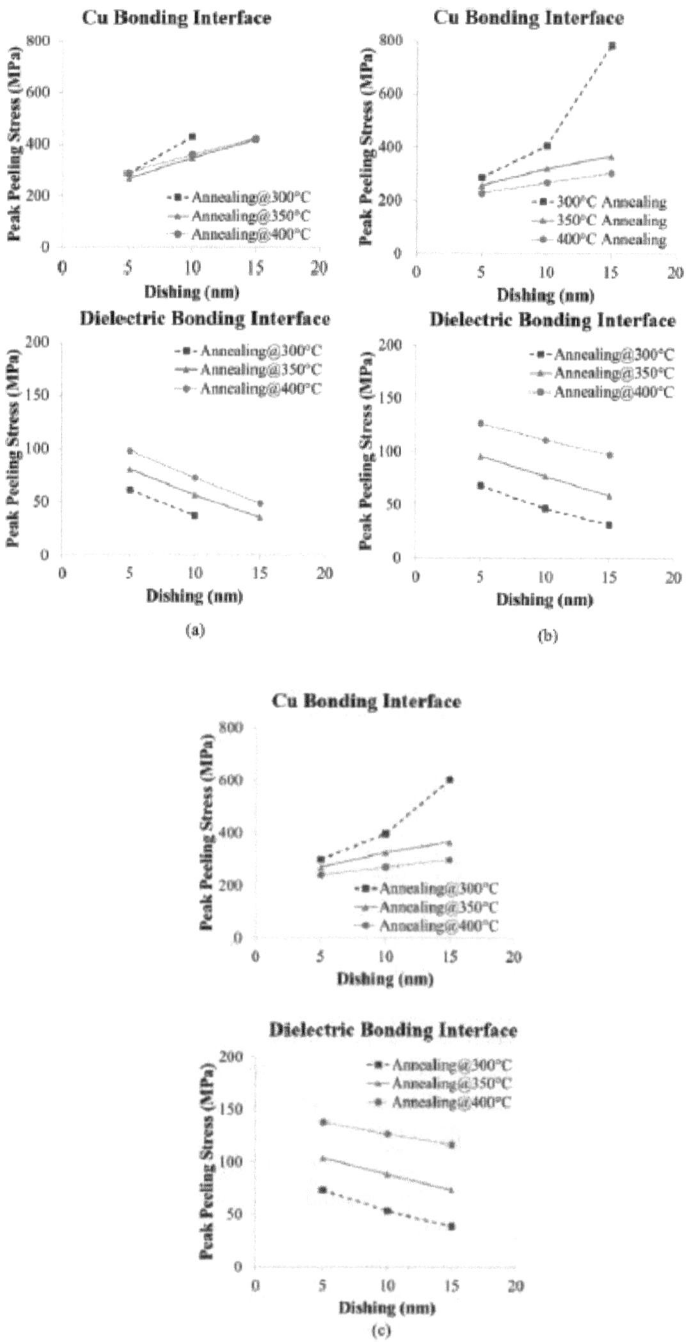

FIGURE 11.29 Simulation results showing the effects of dishing values and annealing temperatures on the peak peeling stresses. (a) Design A. (b) Design B. (c) Design C. (© 2023, IEEE. Reprinted with permission from Ref. [74].)

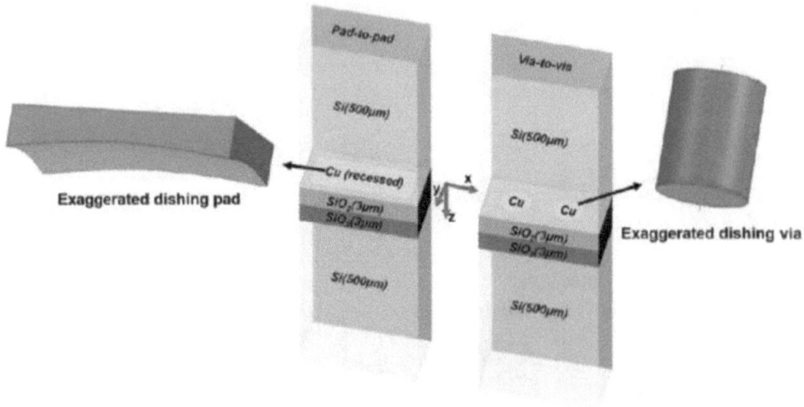

FIGURE 11.30 Schematic of Cu pad/via models [76]. (© The Electrochemical Society. Reproduced with permission from IOP Publishing Ltd.)

via-to-via bonding and pad-to-pad bonding (Figure 11.30). The accumulated stress at bonding interfaces during the heating step for the frictionless metal–dielectric interfaces is calculated. The temperature window for the annealing process is constrained by the critical stress on the dielectric and the contact pressure of the two mating metal pads to ensure good electrical connection.

In the experimental work, the critical stress on the dielectric is determined to be 78.1 MPa; the risk of dielectric debonding arises when the tensile stress on the dielectric material bonding interface is higher than this value. The average peeling stress on dielectric interfaces is calculated, and the critical stress is utilized as the upper boundary of the heating temperature window [76].

For via-to-via bonding, during annealing, the metal expansion causes the two Cu surfaces to make contact. As the metal is constrained in the X and Y directions by the dielectric materials, it expands only in the Z direction and eventually meets the Cu that is expanding from the other surface. As the edges of metal (under compressive stress) touch, the Cu deforms plastically and bonds to each other. As such, the void is most likely to occur at the center of the Cu via when the temperature is low enough. Therefore, the expansion and stress on the Cu via interface should be adequately high to guarantee the void-free metal bonding across the contact interface, and the temperature when the stress at the center of metal is larger than 0 is employed as the lower boundary of the heating temperature window. As can be seen from the process window shown in Figure 11.31, the annealing temperature should be higher for larger dishing values [76].

For metal pad–metal pad bonding, the results (Figure 11.32) show that the short edges, but not the pad center, are at the highest risk for void formation. The lower boundary of the process window is set to ensure this area on Cu with the minimum compression stress (or maximum tensile stress) to touch while the upper boundary of the process window is limited by the critical stress on the dielectric, and then the process windows for different dishing values can be predicted (Figure 11.33) [76].

Ji and Tippabhotla [77] developed a finite element model for the annealing process for SiO_2/Cu chip-to-wafer HB (Figure 11.34). As is shown in Figure 11.35,

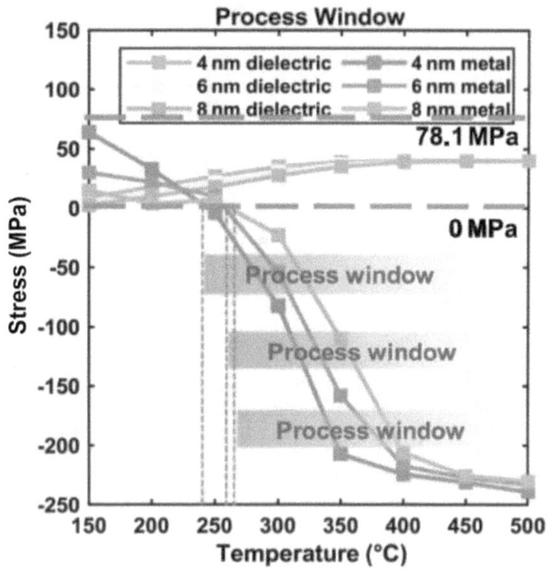

FIGURE 11.31 Process window for $\Phi = 5$ μm vias with different dishing depths [76]. (© The Electrochemical Society. Reproduced with permission from IOP Publishing Ltd.)

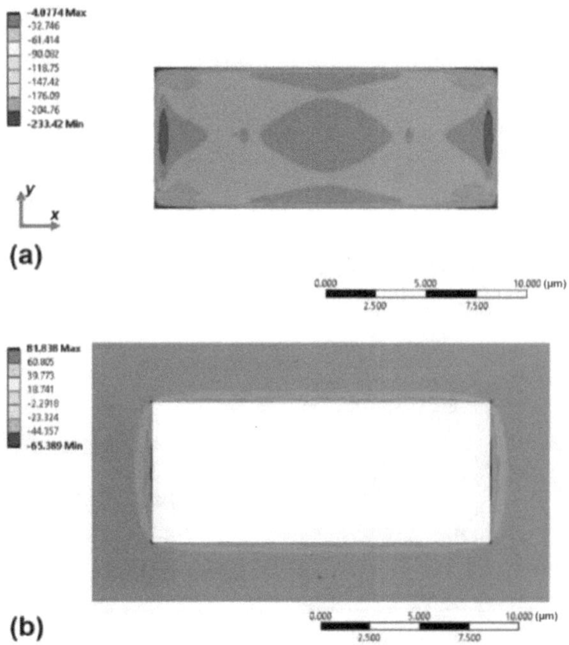

FIGURE 11.32 Recessed Cu pads of 4 nm are heated to 200°C. (a) Stress distribution at the mated Cu surface. (b) Stress distribution at the bonded dielectric interface [76]. (© The Electrochemical Society. Reproduced with permission from IOP Publishing Ltd.)

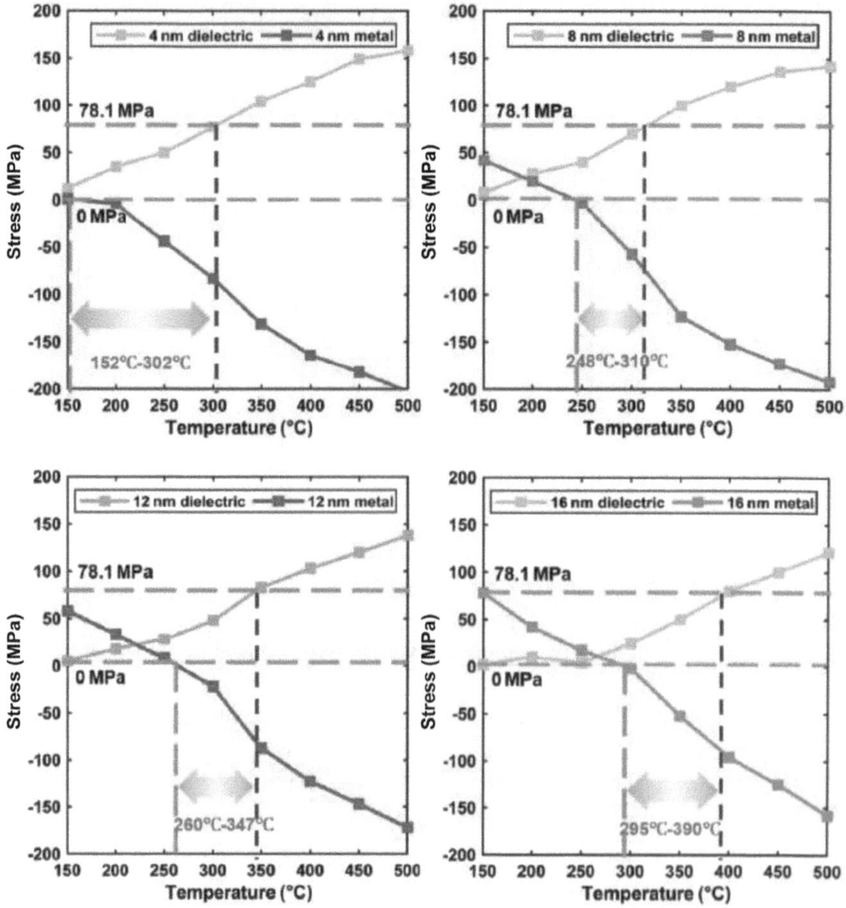

FIGURE 11.33 Process window for 17 μm × 7 μm pads with different dishing depths [76]. (© The Electrochemical Society. Reproduced with permission from IOP Publishing Ltd.)

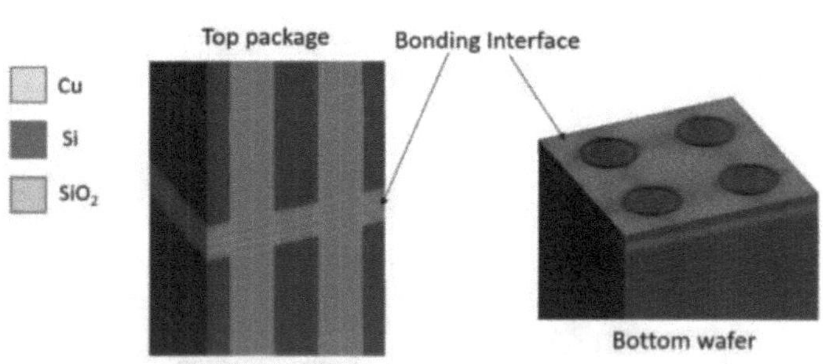

FIGURE 11.34 The finite element model developed in Ref. [77]. *Left:* The configuration and the bonding interface. *Right:* Geometry and associated mesh of the bottom wafer. (© 2023, IEEE. Reprinted with permission from Ref. [77].)

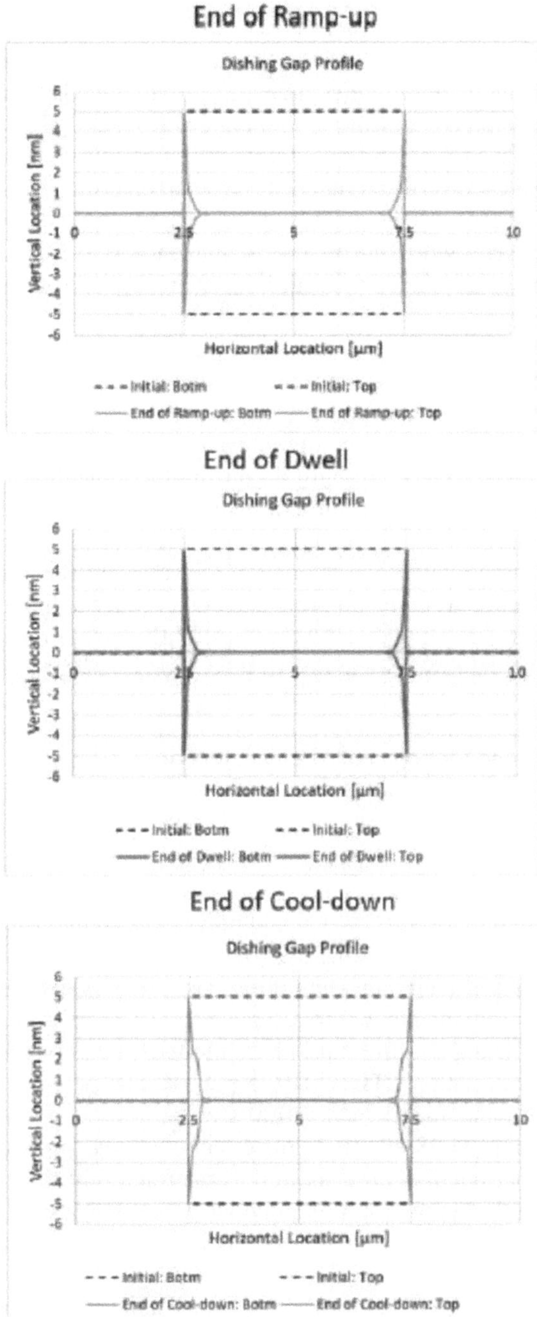

FIGURE 11.35 Dishing gap closure at various time instants (1 μm thick and 5 μm diameter Cu pad, 5 nm dishing, and 250°C annealing temperature for 1 hour dwell) [77]. (© 2023, IEEE. Reprinted with permission from Ref. [77].)

the dishing gap closure occurs mostly during the annealing ramp-up stage. At the end of ramp-up, most of the Cu pad surfaces are bonded together except for a thin periphery zone at the Cu pad surfaces, and no further bonding occurs after the dwell. Less Cu dishing helps achieve more Cu bonding areas, and Cu bonding areas can be improved by increasing the annealing temperature from 250°C to 300°C.

The predicted Cu-to-Cu bonding contact forms mostly during the temperature ramp-up stage, and Cu protrusion induced by the creep effect during the dwell stage is very limited. The model does not include Cu–Cu metal diffusion under Cu creep during the dwell stage, which forms the permanent Cu-to-Cu bonding and determines the bonding performance to ensure no separation or delamination during the cool-down stage.

The dielectric peeling stress, which can cause delamination and cracking, occurs mainly in the annealing ramp-up and dwell stages (Figure 11.36). As less dishing leads to earlier Cu–Cu contact formation during annealing, it induces more peeling stress on the dielectric bonding interface as more compression stresses on the Cu pad bonding interface are built up during the annealing ramp-up stage. In order to avoid dielectric peeling, a proper dishing value should be applied to allow for the Cu pad expansion during the annealing process. The peak peeling stress on the dielectric bonding interface at the end of the ramp-up stage increases with the annealing temperature, as more Cu protrusion takes place at higher annealing temperatures. A longer annealing dwell duration slightly relaxes the dielectric peeling stress at the end of the dwell stage due to the Cu creep effect, but the influence is very small [77].

11.4.2.5 Process Modeling: Polymer Dielectric

Tippabhotla et al. [78] developed a localized 3D finite element model (Figure 11.37) for a single bonding interface with a bottom wafer and a top die to simulate the bonding condition and stress during thermocompression bonding (TCB) and post-bond annealing at ambient pressure, for Cu/polymer-dielectric chip-to-wafer HB. Due to the mismatch in the CTE of Cu and polymer, a slight protrusion of Cu pads is needed, instead of dishing as in the case of Cu–SiO_2 HB. As this process uses elevated temperatures and long annealing durations, plastic yielding and creep deformation of Cu may occur, and therefore, Cu plasticity and creep behavior are included in this simulation, using a bilinear plasticity model. Two polymers are included in the study: polymer A with a CTE of 119.45 ppm/K, and polymer B with a CTE of 11.5 ppm/K at 25°C and 46.0 ppm/K at 400°C.

The results (Figure 11.38) highlight the effect of Cu pad protrusion and polymer CTE on the polymer bonding area, as well as the role of the Cu pad/polymer layer thickness particularly in the case of a polymer with a lower CTE and a higher Cu pad protrusion. For polymer B, thickness reduction from 3 μm to 2 μm results in reduction of the bonding area from 84% to zero for the 15 nm Cu protrusion case (whereas the effect is negligible for the 5 nm protrusion case) while the lower bound of the bond head temperature (200°C) is used. While the effect of the bond head temperature on the polymer interface bonding area appears to be minimal at lower values of protrusion, the effect could be significant at higher values of protrusion. The bonding load appears to have no effect on the polymer bonding area, as the

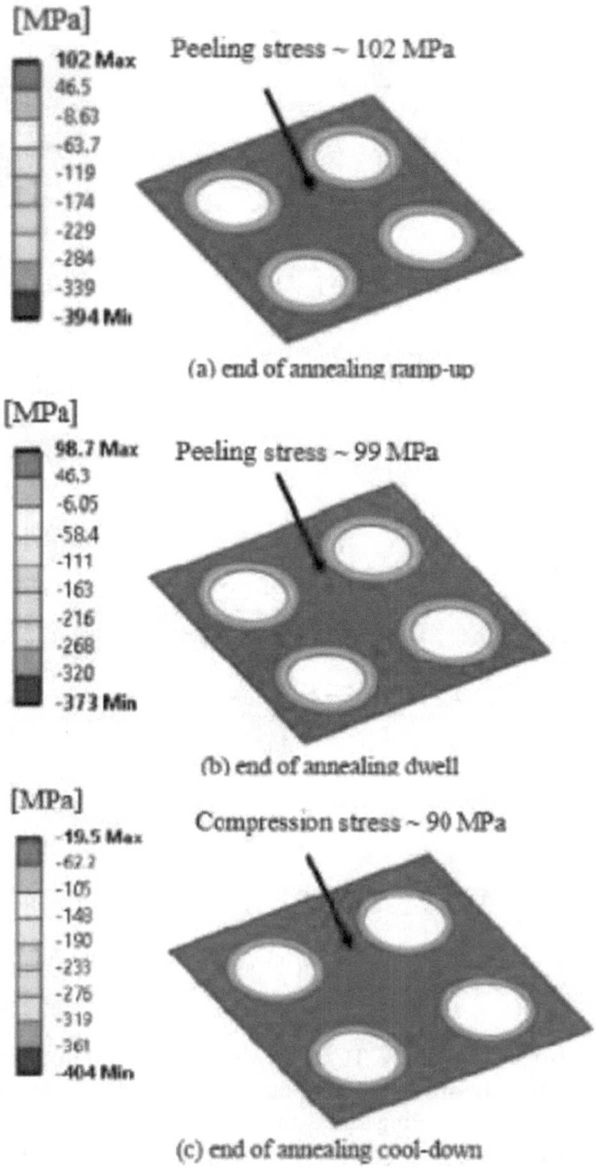

FIGURE 11.36 Dielectric interface stresses during annealing for the case in Figure 11.35. (© 2023, IEEE. Reprinted with permission from Ref. [77].)

magnitude of forces induced by the expansion of the polymer is much more significant; however, in practice, the bonding load is still significant in the TCB process for initial tacking of the Cu pads and uniform thermal conduction across the bonding interfaces. The effect of the bottom chuck temperature on the bonding area appears to be insignificant [78].

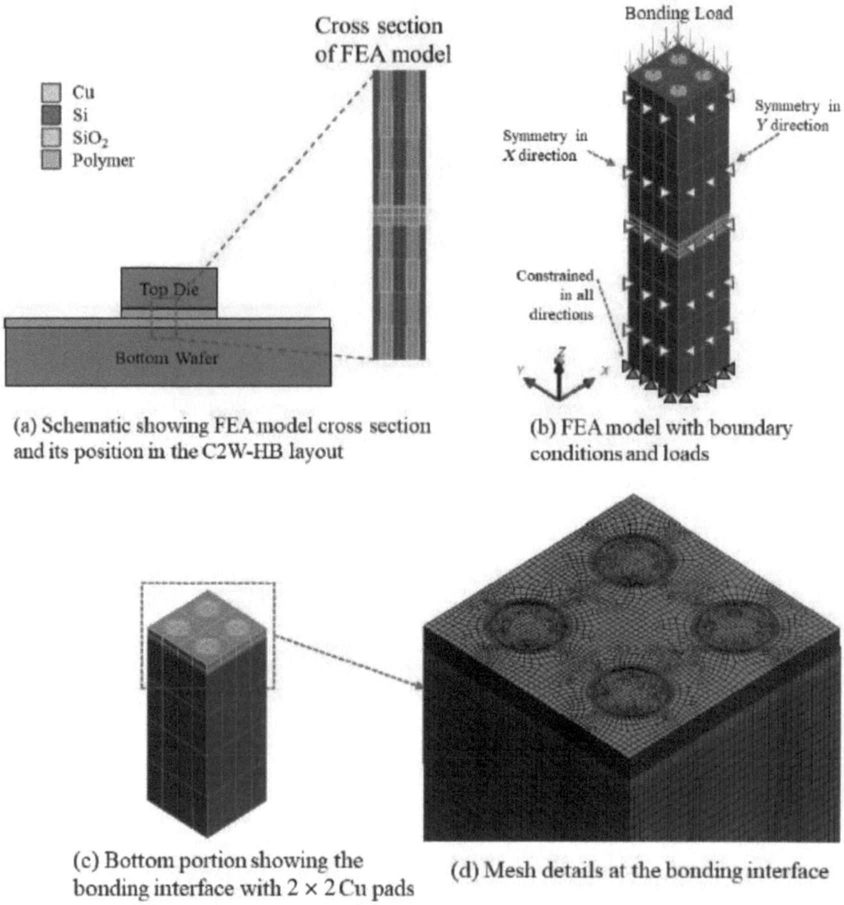

FIGURE 11.37 Details of the FEA model developed in Ref. [78]. (© 2023, IEEE. Reprinted with permission from Ref. [78].)

The results also show that the TCB bond head temperature and the annealing temperature significantly affect the peeling stresses in the Cu pads. Figure 11.39 shows that the maximum peeling stress in the Cu pad during the TCB process is quite high (130–375 MPa) for polymer A but negligible for polymer B (<50 MPa), and the maximum Cu pad peeling stress increases with the increase in the bond head temperature and decreases with the increase in the Cu pad protrusion. The maximum peeling stress in the polymer interface after the TCB process for polymers A and B is negligible (<20 MPa), except for polymer A at the bond head temperature 350°C, as the polymer interface peeling stress increases with the increase in the bond head temperature [78].

Figure 11.40 shows that the maximum peeling stress in the Cu pads during the annealing process for polymer A is quite high (230–450 MPa) but low for polymer B (<100 MPa), and the maximum Cu pad peeling stress increases with the increasing

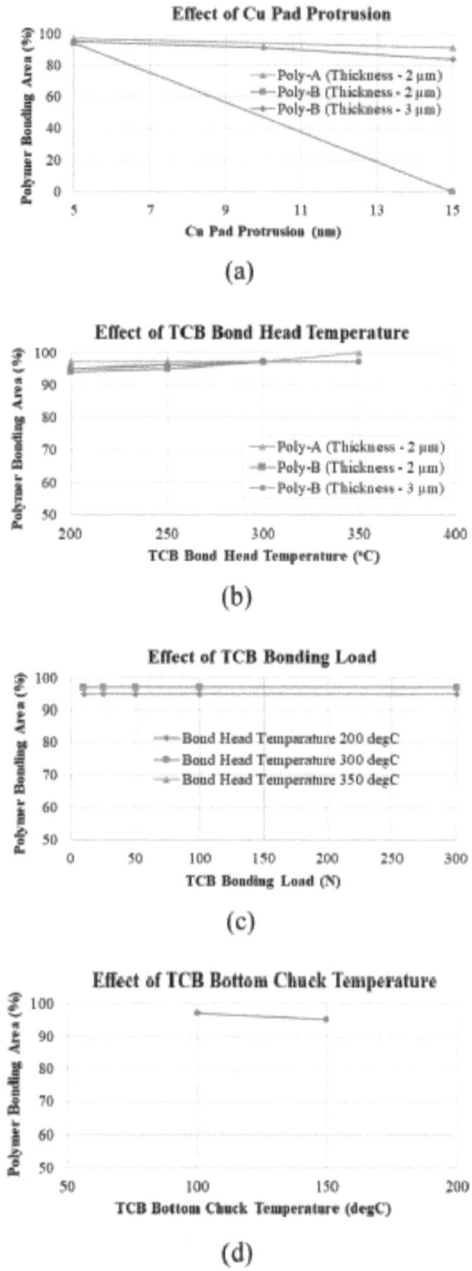

FIGURE 11.38 Effect of different design and process parameters on the polymer interface bonding area (unless otherwise specified, the results correspond to polymer B, Cu pad/polymer thickness 3 µm, Cu pad protrusion 5 nm, bond head temperature 200°C, bottom chuck temperature 100°C, and bonding load 100 N). (© 2023, IEEE. Reprinted with permission from Ref. [78].)

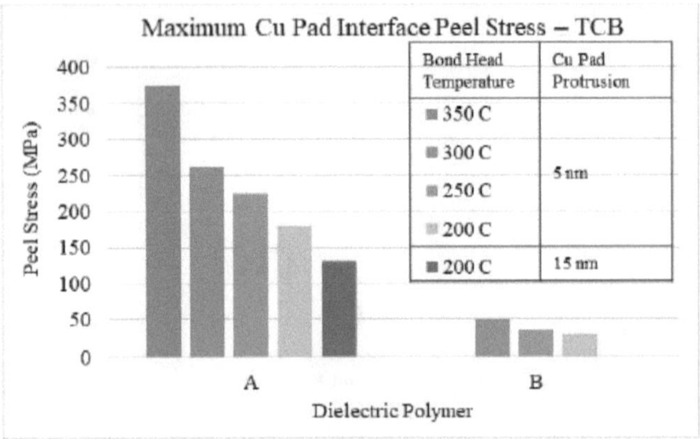

(a)

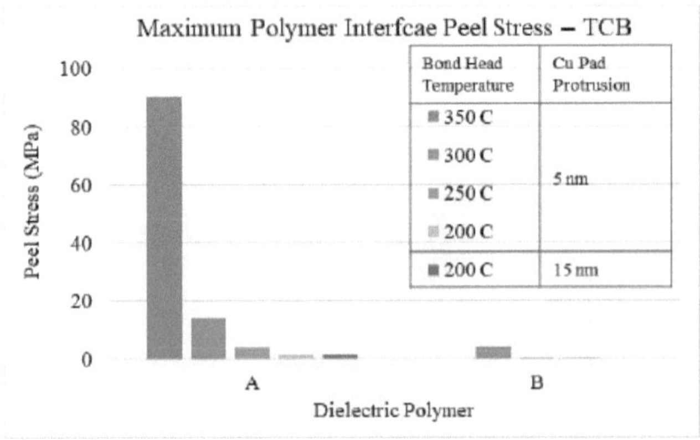

(b)

FIGURE 11.39 Maximum peeling stress in (a) Cu pad and (b) polymer interfaces, due to TCB process (bottom chuck temperature 100°C, bonding load 100 N, Cu pad/polymer thickness 2 μm). (© 2023, IEEE. Reprinted with permission from Ref. [78].)

annealing temperature. In the case of annealing at 350°C, the Cu pad peeling stress for polymer A is lower than that of the 300°C case, as the glass transition temperature of polymer A is 290°C. The maximum peeling stress in the polymer interface after the annealing process is negligible (<20 MPa), except for polymer A at the bond head temperatures 300–350°C, and the polymer interface peeling stress also increases with the increase in the annealing temperature till 300°C. For annealing at 350°C, it decreases due to the glass transition of polymer A.

Tippabhotla et al. [79] performed numerical simulation on the bonding progression for a special case of chip-to-wafer HB where SiO_2 dielectric is used for the bottom wafer and polymer dielectric is used for the top die. As these two materials are quite different in their thermomechanical behavior, bonding of the Cu–Cu interfaces depends on the relative expansion/contraction of the surrounding SiO_2 (bottom wafer) and the dielectric polymer (top die).

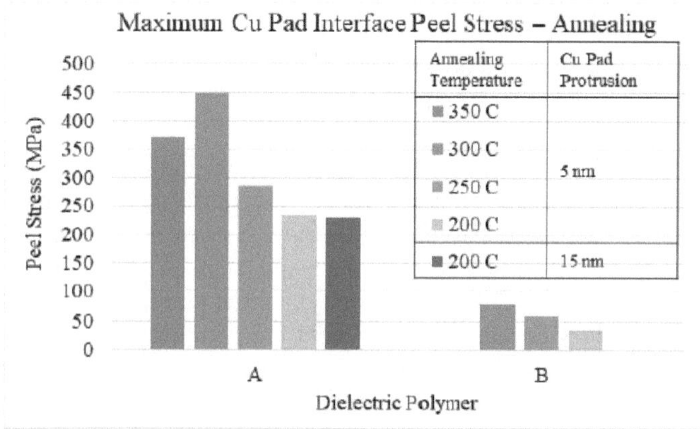

(a)

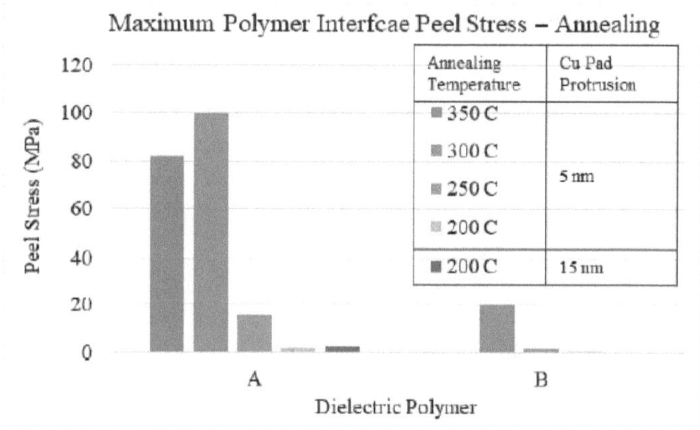

(b)

FIGURE 11.40 Maximum peel stress in (a) Cu pad and (b) polymer interfaces, during the annealing process (TCB bond head temperature is same as the annealing temperature, bottom chuck temperature 100°C, bonding load 100 N, Cu pad/polymer thickness 2 μm). (© 2023, IEEE. Reprinted with permission from Ref. [78].)

The simulation results show [79] that protrusion of the Cu pads is required for a successful Cu–Cu direct bonding in this configuration, similar to the case of die-to-wafer HB with a polymer–polymer interface. The simulation results also show that the glass transition temperature of the dielectric polymer needs to be higher than the TCB interface temperature and the annealing temperature, and the CTE mismatch between the dielectric polymer and Cu needs to be reduced such that a small protrusion of Cu pads can be designed to compensate for the differential thermal expansion and keep the Cu pads in contact during the TCB and annealing processes. Even if pre-bonding of Cu pads is achieved by protrusion of Cu pads at the interface, there could be a high chance of debonding during the annealing process due to the differential thermal expansion between the dielectric polymer and the Cu pads.

11.4.3 Thermal Modeling

Thermal modeling and simulation for advanced 3D integrated circuit (3DIC) systems based on HB is the topic for another chapter in this book, and is only addressed briefly in this chapter.

There have been several studies on the thermal simulation of 3DIC systems based on HB. For example, Agrawal et al. [80] performed thermal simulations (Figure 11.41) for HBM applications, with solder capped microbumps in TCB and compared to direct bond Cu–Cu interconnects in HB, in a JEDEC standard environment for forced convection (3 m/s air velocity), with no radiation effect included.

Under the boundary conditions of ambient temperature of 45°C and an operating power of 2 W on each die, the predicted temperature distributions are shown in Figure 11.42. The junction temperature difference between (4 + 1) HBMs with a microbump and HBMs with a Cu–Cu interconnect was predicted to be 8°C. For an eight-die-stack HBM configuration, the direct bond Cu–Cu stack has a 25% reduction of the junction temperature compared to the microbump stack. More importantly, the difference between the junction temperature of the hottest and coolest dies in the stack is reduced from 33°C with microbumps to 5°C for the Cu–Cu stack [80]. The enhanced thermal performance is due to the all inorganic materials in the stack for the Cu–Cu direct interconnect connections, as compared with the

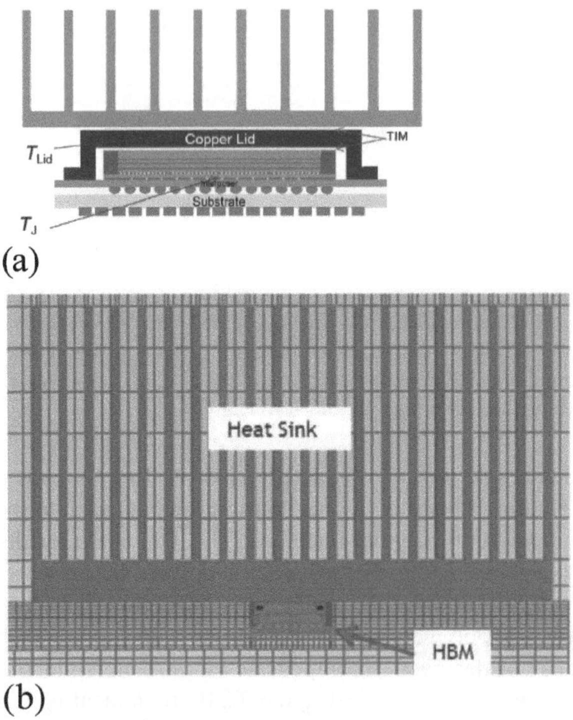

FIGURE 11.41 (a) HBM structure. (b) Finite element model. (© 2023, IEEE. Reprinted with permission from Ref. [80].)

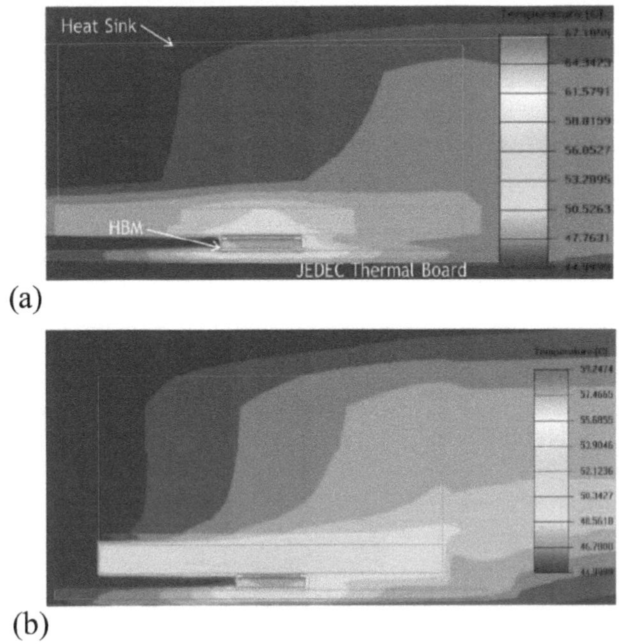

FIGURE 11.42 Predicted temperature distribution for four-die-stack HBM, with forced convection (3 m/s air velocity). (a) Microbump stack (with underfill). (b) Cu–Cu direct bonding. (© 2023, IEEE. Reprinted with permission from Ref. [80].)

microbump stack (with underfill), and it is expected to significantly improve the performance of the die, such as the timing margin requirements and refresh rate for the memory cells.

11.4.4 Electrical Modeling

As the interconnects have a significant impact on the electrical performance of electronic devices, such as signal integrity (SI) and power integrity (PI), especially for high-speed signaling, various studies have been performed for modeling and simulation of high-speed interconnects.

Agrawal et al. [80] compared the electrical performance between a conventional microbump TCB joint and a direct Cu–Cu interconnect through numerical simulation (Figures 11.43 and 11.44), using a 3D full-wave simulation tool to simulate the S parameters and TDR (time domain reflectometry) impedances, up to 6 GHz.

The results indicate that the parasitic capacitance of the Cu–Cu joint is only 30% of a TCB joint. In terms of signal integrity, the Cu–Cu interconnect has smaller loss for the signal interconnect (net 1) at higher frequency ranges (>3 GHz) as compared with TCB. For the TDR response, the discontinuity of net 1 is about 0.5 Ω for direct Cu–Cu interconnect, and >1 Ω for TCB. Similarly, the power grid net of the Cu–Cu interconnect is smaller than that of TCB, because the structure of the Cu–Cu interconnect is smaller than that of the TCB interconnect. Overall, the simulation

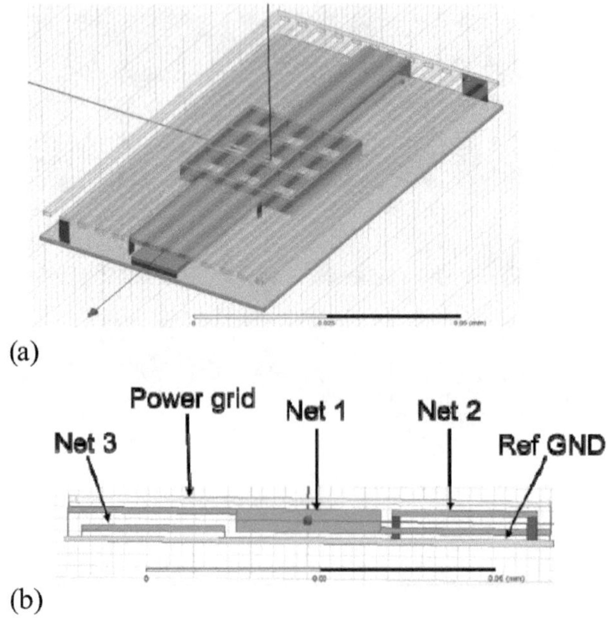

FIGURE 11.43 Three-dimensional models for direct Cu–Cu interconnect. (a) 3D view. (b) Side view. (© 2023, IEEE. Reprinted with permission from Ref. [80].)

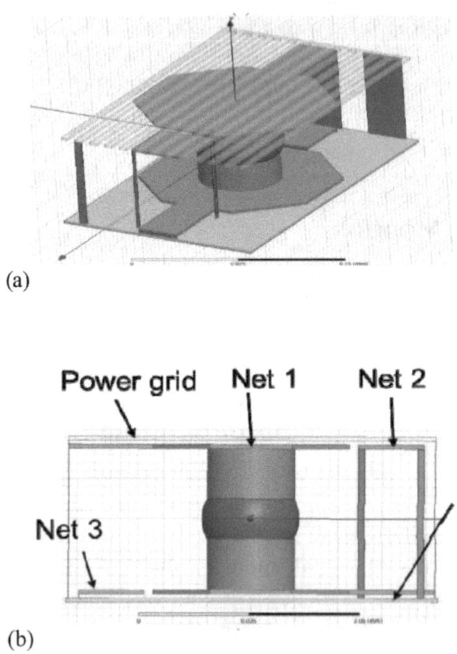

FIGURE 11.44 Three-dimensional models for TCB. (a) 3D view. (b) Side view. (© 2023, IEEE. Reprinted with permission from Ref. [80].)

shows that for high-speed applications (>5 GHz), direct Cu–Cu interconnects offer a smaller insertion loss and less impedance discontinuities; however, the crosstalk is somewhat higher, yet well below typical requirements. Misalignment (close to 45% by area) has little impact on the inductance and impedance discontinuity of Cu–Cu interconnects [80].

Sitaraman et al. [62] performed electrical modeling of HB interconnects using 3D full-wave electromagnetic simulations to study the insertion loss and impedance discontinuities. Of particular interest is the surface contact resistance, which has been experimentally verified to be 140 mΩ·μm² for a Cu–Cu HB interconnect when the bond is annealed at 200°C. This DC contact resistance is included in the finite element model and extended to high frequencies by introducing a thin metal layer (50 nm) with modified conductivity (Figure 11.45), and by altering the permeability of this layer, the skin depth of the new layer can be equated to that of copper. This technique enables capturing the effect of the skin depth along with the contact resistance, at high frequencies (up to 20 GHz). HB interconnects with a 3 μm interconnect bump diameter and 5 μm pads are used in the case study [62].

The simulation results (Figure 11.46) suggest that once the contact resistance is considered, the resistance of the HB interconnect becomes more significant, leading to perceptible change in the insertion loss. The impact of area reduction due to CMP and misalignment on the insertion loss can also been seen. The S-parameter data then is exported to a SPICE (Simulation Program with Integrated Circuit Emphasis, a general-purpose, open-source analog electronic circuit simulator) program to perform TDR analysis on the interconnect. Because of its small dimensions, the HB interconnect offers the least change from the nominal 50 Ω impedance, as compared with solder microbump interconnects (Figure 11.47).

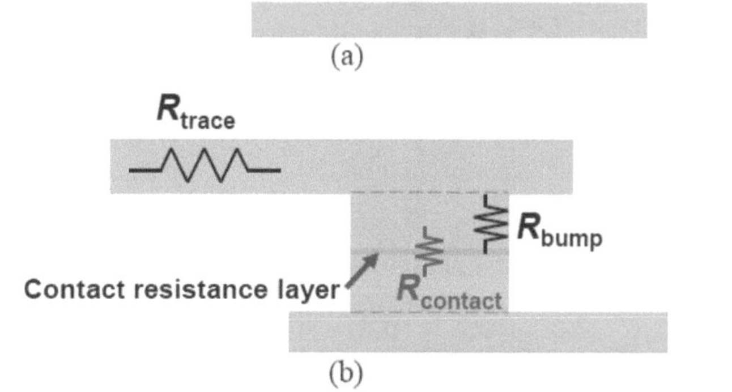

FIGURE 11.45 Hybrid bond interconnect model. (a) Cross section view. (b) Contact resistance. (© 2023, IEEE. Reprinted with permission from Ref. [62].)

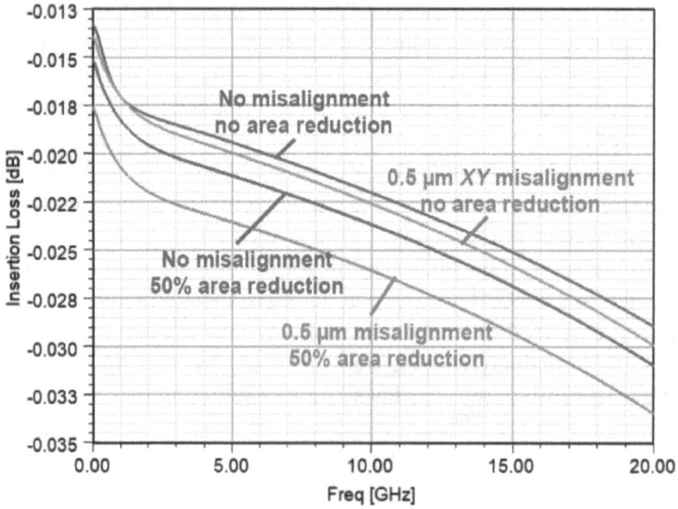

FIGURE 11.46 HB interconnect simulation results for insertion loss. (© 2023, IEEE. Reprinted with permission from Ref. [62].)

Channel simulations were carried out to evaluate the impact of crosstalk between multiple adjacent interconnects and how signals are affected while propagating through the interconnects, as well as the impact of process-induced variations on the bandwidth performance of the interconnects [62]. Studying a set of five adjacent interconnects provides a good idea of the performance of the entire channel, as crosstalk effects fade away beyond two adjacent interconnects on either side. In this study, a HBM simulation model was used (Figure 11.48), and the channel is terminated using capacitive terminations and simulated for a speed of 2 Gbps for different bump areas and misalignments. The results (Figure 11.49) show that across all

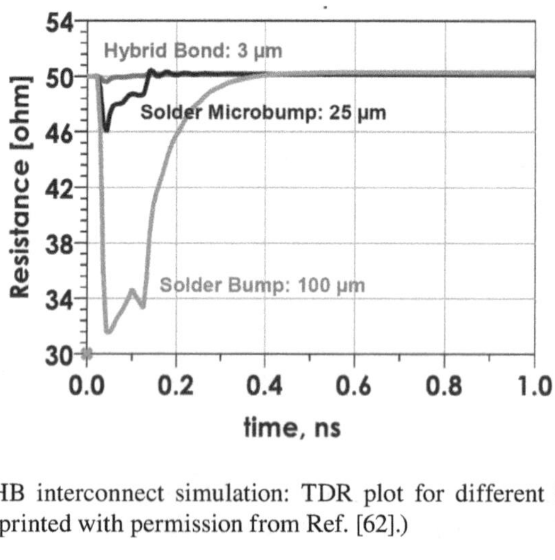

FIGURE 11.47 HB interconnect simulation: TDR plot for different interconnect types. (© 2023, IEEE. Reprinted with permission from Ref. [62].)

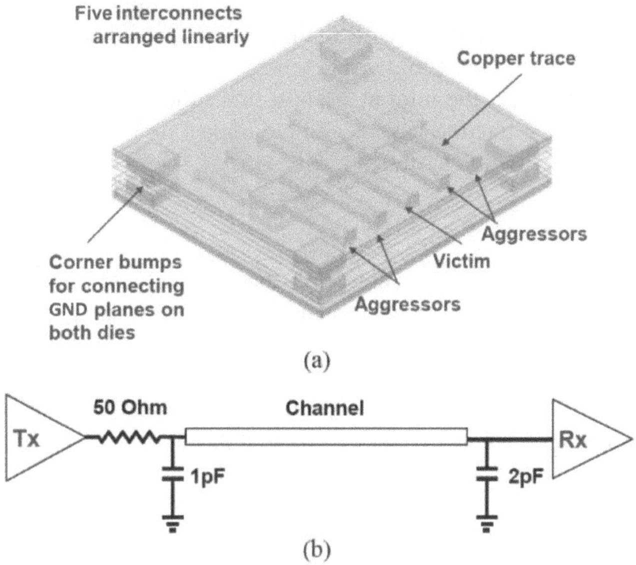

FIGURE 11.48 Channel simulations for HB interconnects. (a) 3D finite element model. (b) Circuit schematic. (© 2023, IEEE. Reprinted with permission from Ref. [62].)

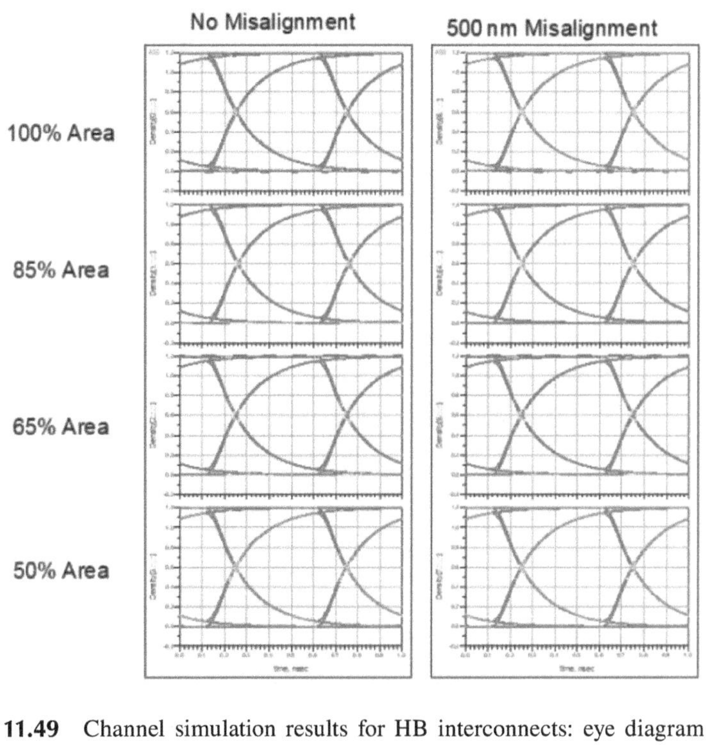

FIGURE 11.49 Channel simulation results for HB interconnects: eye diagram plots for 2Gbps. (© 2023, IEEE. Reprinted with permission from Ref. [62].)

the variations, there was negligible impact on the eye parameters (height, jitter, and width), because a variation of 0.002 dB in the insertion loss is not significant enough to impact the eye height substantially, and the crosstalk between the interconnects is low enough to not impact the jitter at 2 Gbps, indicating that 0.5 μm misalignment and 50% bump area reduction have negligible impact on the performance of these interconnects in a real-world system at 2 Gbps [62].

11.4.5 EM Modeling

In an EM simulation by Sitaraman et al. [62], the current density in the interconnects and traces is calculated using the electrical resistivity of the assigned materials and the electrical boundary conditions. Based on the electrical current flow, heat generation due to Joule heating is calculated, and the temperature increase in the structure is then calculated. Considering the structural boundary condition, the stress and displacement distribution are calculated in different parts of the HB structure. Finally, the atomic concentration is predicted for EM simulation.

In a transient thermo-electromechanical analysis for a HB interconnect assembly (Figure 11.50), the simulation results (Figures 11.51–11.53) show that the current

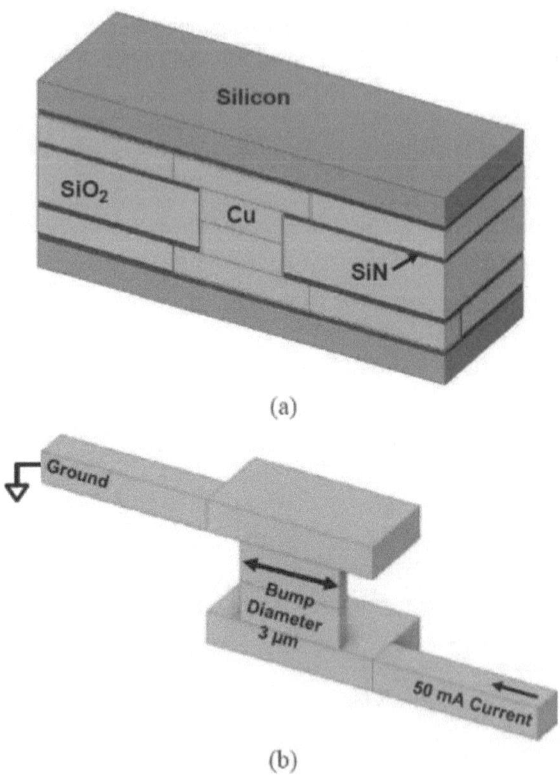

FIGURE 11.50 (a) 3D half-model of the HB assembly. (b) Electrical boundary conditions. (© 2023, IEEE. Reprinted with permission from Ref. [62].)

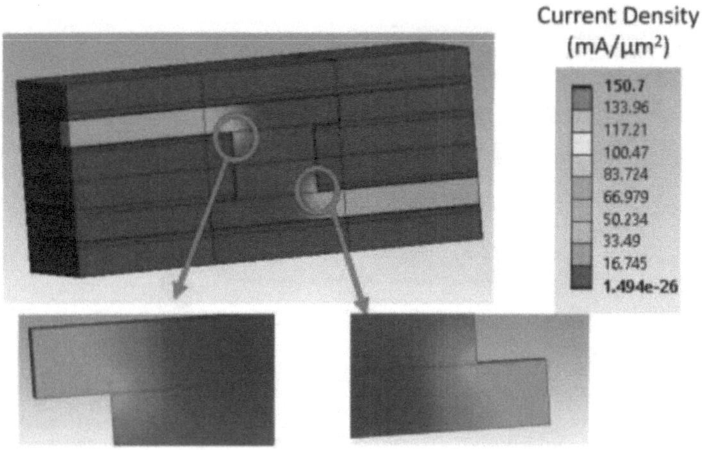

FIGURE 11.51 Electrical current density distribution in the HB model. (© 2023, IEEE. Reprinted with permission from Ref. [62].)

density in the traces is higher than in the interconnect. Within the interconnect, the current density is considerably higher at the edge where the interconnect connects to the trace. Energy is mainly generated in the top and bottom traces, due to the higher current density, and the heat energy gets dissipated to the top and bottom silicon layers. The atomic concentration (Figure 11.53) is below "1" (which will result in "voids") at the top left where the current exits the interconnect, and it is above "1" (which will result in "hillocks") at the bottom right where it enters the interconnect. When a 10 nm Ti barrier layer is used, the atomic concentration at the bottom right decreased from 1.2 to 1.1 and increased at the top left from 0.3 to 0.9, which will result in reduced voids at the top left and reduced hillocks at the bottom right [62].

Ceric et al. [81] developed a modeling approach to study the emergence, growth, and migration of voids through Cu-to-Cu hybrid bonds due to EM, in an effort to understand the complex dynamics of voids with the influencing factors (such as the current density, temperature, and mechanical stress). The EM model is used in

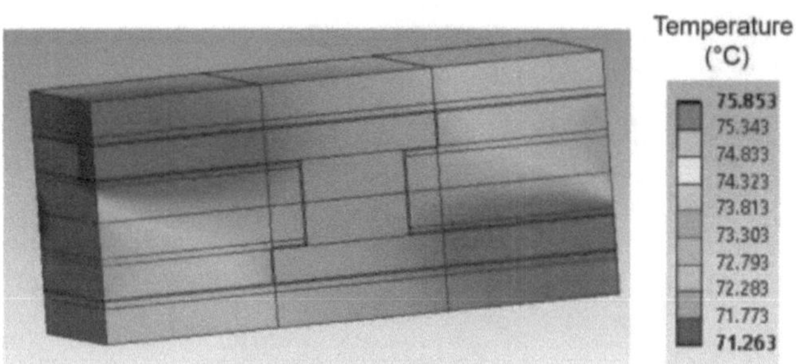

FIGURE 11.52 Temperature distribution in the HB model. (© 2023, IEEE. Reprinted with permission from Ref. [62].)

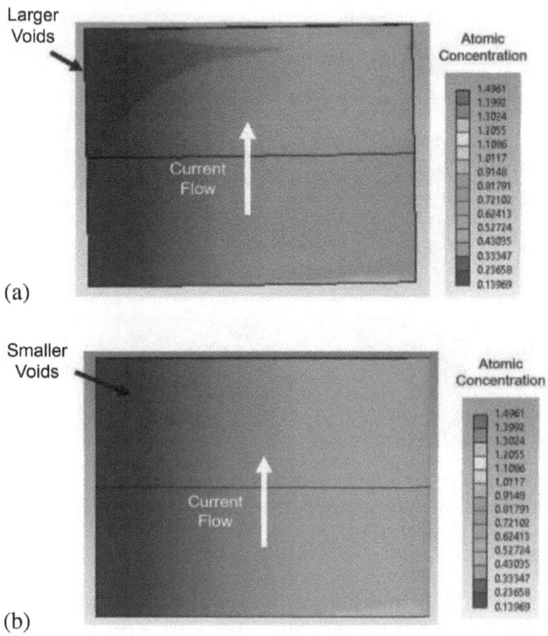

FIGURE 11.53 Atomic concentration in the interconnect model. (a) Without the Ti layer. (b) With 10 nm Ti layer. (© 2023, IEEE. Reprinted with permission from Ref. [62].)

combination with an electrothermal model to account for Joule heating and with a particular diffusivity model to include the effects of the metal microstructure (Figure 11.54).

For hybrid Cu/SiCN bonding pads (Figure 11.55), with electrons flowing from the top pad to the bottom pad, the total vacancy flux, which redistributes vacancies throughout the bulk of the interconnect metal, was calculated to study the vacancy dynamics without voids. It can be seen from Figure 11.56 that the intensities of EM and

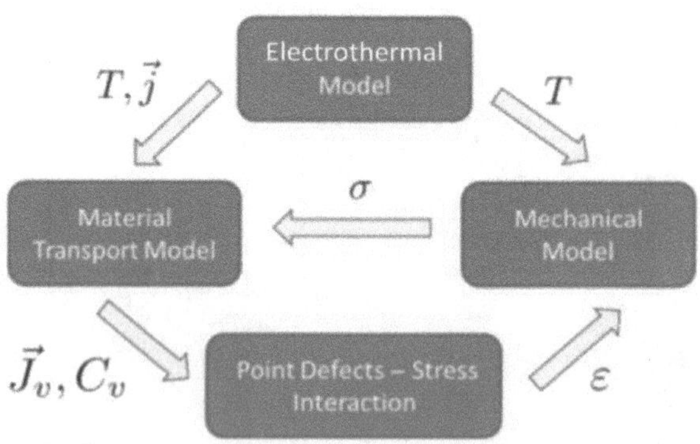

FIGURE 11.54 Schematic representation of the model used for simulation. (Reprinted from Ref. [81], with the permission of AIP Publishing.)

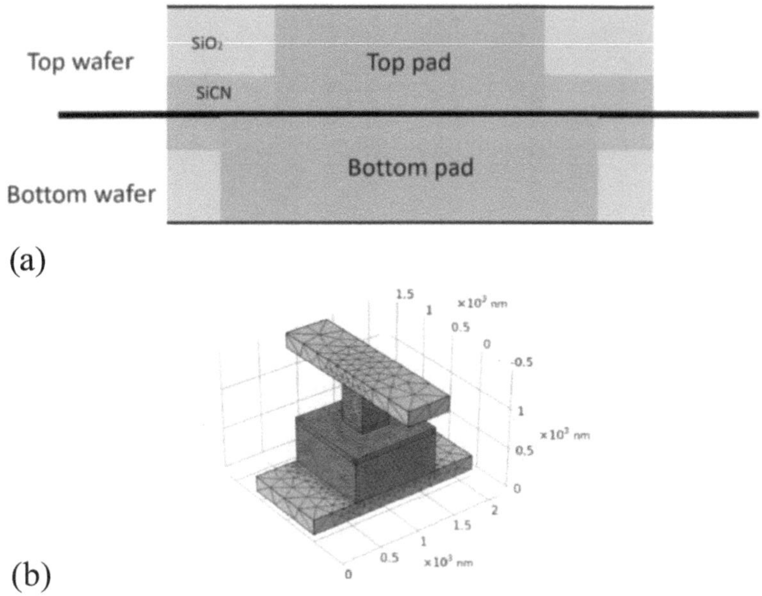

(a)

(b)

FIGURE 11.55 (a) Schematic of the modeled layers and the stack. (b) Copper metallization geometry meshed for FEM. (Reprinted from Ref. [81], with the permission of AIP Publishing.)

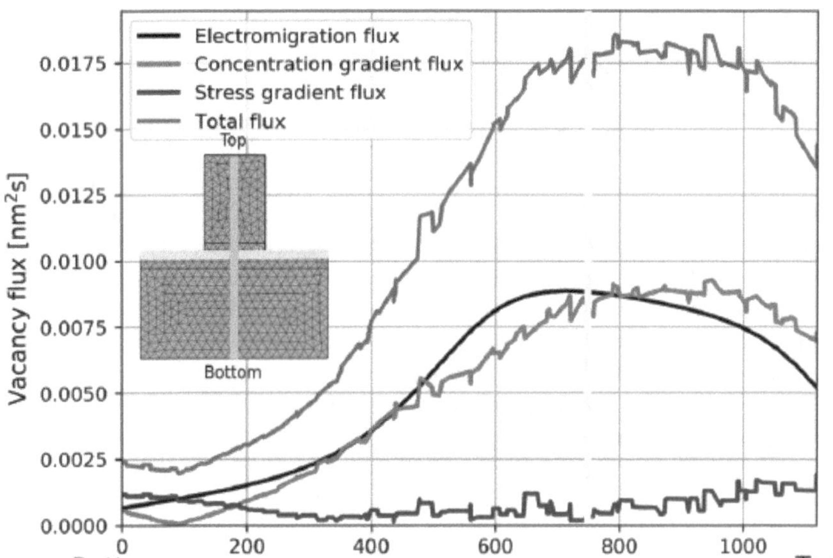

FIGURE 11.56 Vacancy flux profiles along the bottom–top geometry axis (green line) after 1 hour EM stressing. The yellow line marks the top of the bottom pad. (Reprinted from Ref. [81], with the permission of AIP Publishing.)

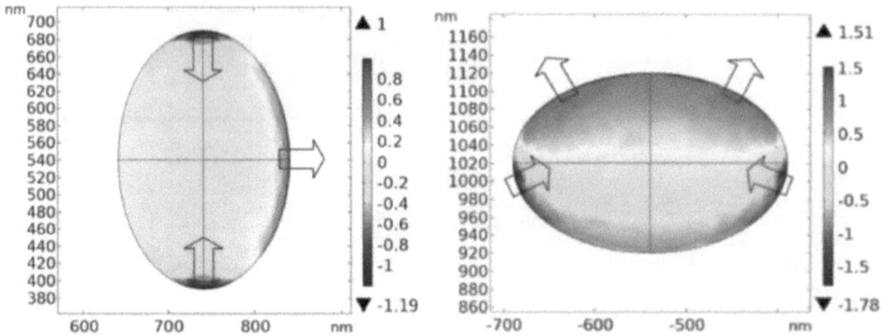

FIGURE 11.57 Normalized total velocity at the void surface. The main trends of the void surface development are indicated by arrows. *Left:* The void is in the bottom interconnect. *Right:* The void is in the top interconnect. (Reprinted from Ref. [81], with the permission of AIP Publishing.)

the concentration gradient–related flux components are about five times larger than the stress gradient–related flux component, and the intensities of the EM and the concentration flux components are approximately 8 times and 15 times higher, respectively, in the top pad than in the bottom pad, as can be expected due to a smaller cross section of the top pad and correspondingly higher current density, in this configuration [81].

Further, to study how the presence of a void changes the stress and the vacancy distribution in the interconnect and subsequently how the stress and vacancy gradients together with EM impact the normal velocity at the void surface, the normalized total velocity at the void surface (assuming an ellipsoidal shape) was calculated (Figure 11.57). The simulation results indicate that in the bottom pad, at the side of the void pointing toward the contact with the top pad (the right side), the normal velocity strives to expand the void. In the top pad, the normal velocities, from both the front and the back of the void, push the void upward. The migration velocity of a void in the upper pad is 20% higher than the migration velocity of a void in the bottom pad for the configuration under study, and the calculated velocities of the evolving void surface fully correspond to the experimentally observed behavior of voids migrating from the lower pad to the upper diffusion barrier of the upper pad, where they can cause an electrical failure [81].

EM simulations can provide further insights into the impact of layout geometry, materials properties, and operating conditions on void dynamics, and can enable targeted adjustments of the influencing factors to inhibit void migration and growth, in order to mitigate Cu-to-Cu HB failures due to EM [81].

11.5 SUMMARY

As newer forms of interconnects emerge to meet the demand for high density and high performance, interconnect reliability becomes more complex and more critical. Certainly, HB and chiplets present many new topics for the long-term reliability of the interconnect and the system. This chapter reviews the most significant qualification and reliability achievements for HB-based interconnects, over the last ten years by the industry and the scientific community.

Critical topics such as interconnects, failure mechanisms, and related tests and standards for the 3D packaging technology are reviewed in this chapter. The potential reliability threats and their consequences are presented: moisture ingress, thermomechanical stresses, EM, Cu diffusion, dielectric breakdown, Joule heating, and chip-package interaction (CPI).

Further, this chapter summarizes and discusses qualification and reliability testing results of the HB technology by the industry and the scientific community. Overall, the electrical resistance shift measured after the qualification tests is limited to a few percent under a variety of test conditions included. EM and TDDB are of particular importance for reliability of HB interconnects. The EM performances (Black's parameters) are similar to classical BEOL-like interconnects. Nevertheless, with pitch reduction, the killer defect is no longer at the BEOL level but at the HB level, especially for submicrometer HB pitches. For Cu diffusion/TDDB, a clear trend emerges that there is no significant Cu diffusion under normal use conditions. Overall, the behaviors of HB interconnects against common reliability threats are very similar to that of the BEOL in the semiconductor device.

Simulation and modeling provide useful tools for characterization and optimization of direct copper interconnection processes and performance. In this chapter, progress in the development of modeling methodology is reviewed as related to the direct copper interconnection process and the performance of the HB interconnects. Efforts in the modeling and simulation of the critical process steps (such as CMP, pre-bonding of the dielectric layers, and Cu-to-Cu bonding) are reviewed, for the purpose of optimizing packaging design and processes. The most recent work on the modeling and simulation of the interconnect performance (including thermal, electrical, and reliability), particularly with direct copper interconnection, is also summarized in this chapter.

As HB becomes more widely practiced in the industry, more data will become available to help validate and refine the computer models. Materials requirements, design guidelines, and process windows can be established based on the modeling and simulation studies to accelerate process development and optimization and ensure performance and reliability. Concurrent co-optimization of design, materials, processes, and performance, in a holistic and integrated framework of design technology co-optimization (DTCO) and system technology co-optimization (STCO) enabled by digital twins and artificial intelligence (AI),, is critical to the successful deployment of the HB technology in a high-volume environment.

ACKNOWLEDGMENTS

Stéphane MOREAU would like to thank the French National Research Agency (ANR) for partially funding this work under the "Investissements d'avenir" programs: ANR 10-AIRT-0005 (IRT NANO-ELEC).

NOTES

1 JEDEC: Joint Electron Device Engineering Council.

2 Thermally activated phenomena are typically diffusion processes, stress-induced voiding (SiV), electromigration, thermomechanical stresses, etc.

REFERENCES

1. W. R. Blischke, D. N. P. Murthy, "An Overview," in Reliability: Modeling, Prediction, and Optimization, Wiley, 2000, pp. 1–29.
2. M. Ohring, L. Kasprzak, "An Overview of Electronic Devices and Their Reliability," in *Reliability and Failure of Electronic Materials and Devices* (Second Edition), M. Ohring, L. Kasprzak (eds.), Academic Press, Boston, 2015, pp. 1–38.
3. JEDEC, "Reliability Qualification of Semiconductor Devices Based on Physics of Failure Risk and Opportunity Assessment: JEP148B," JEDEC Solid State Technology Association, Arlington, VA, 2019, p. 30.
4. M. Ciappa, "Lifetime Prediction on the Base of Mission Profiles," in *Microelectronics Reliability*, vol. 45, pp. 1293–1298, 2005.
5. J. Handy, "Toshiba 128 GB eMMC: How Do They Do That?," April 18, 2023. https://www.edn.com/toshiba-128gb-emmc-how-do-they-do-that/
6. I. Qin, O. Yauw, G. Schulze, A. Shah, B. Chylak, N. Wong, "Advances in Wire Bonding Technology for 3D Die Stacking and Fan Out Wafer Level Package," in *2017 IEEE 67th Electronic Components and Technology Conference (ECTC)*, 2017, pp. 1309–1315.
7. S. W. Yoon, J. H. Ku, N. Suthiwongsunthorn, P. C. Marimuthu, F. Carson, "Fabrication and Packaging of Microbump Interconnections for 3D TSV," in *2009 IEEE International Conference on 3D System Integration*, 2009, pp. 1–5. doi: 10.1109/3DIC.2009.5306554.
8. J. H. Lau, "Recent Advances and New Trends in Flip Chip Technology," *Journal of Electronic Packaging*, vol. 138, 2016. doi: 10.1115/1.4034037.
9. Amkor, "Copper Pillar," April 19, 2023. https://c44f5d406df450f4a66b-1b94a87d576253d9446df0a9ca62e142.ssl.cf2.rackcdn.com/2018/02/Copper_Pillar_Flip_Chip_TS106.pdf
10. Shinko, "IC Assembly: Copper Pillar Bumping," April 19, 2023. https://www.shinko.co.jp/english/product/package/assembly/cu-pillar.php
11. J. H. Lau, "Evolution, Challenge, and Outlook of TSV, 3D IC Integration and 3D Silicon Integration," *2011 International Symposium on Advanced Packaging Materials (APM)*, 2011, pp. 462–488. doi: 10.1109/ISAPM.2011.6105753.
12. D. Malta, "TSV Formation Overview," Handbook of 3D Integration, Wiley, 2014, pp. 65–78.
13. M. He, T.-M. Lu, *Metal-Dielectric Interfaces in Gigascale Electronics: Thermal and Electrical Stability*, Springer, New York, NY, 2012.
14. H. Miyazaki, K. Hinode, Y. Homma, N. Kobayashi, "Time-dependent Dielectric Breakdown of Interlevel Dielectrics for Copper Metallization," in *Japanese Journal of Applied Physics*, vol. 35, pp. 1685–1689, 1996. doi: 10.1143/jjap.35.1685.
15. R. Gonella, P. Motte; J. Torres, "Time-dependent-dielectric breakdown used to assess copper contamination impact on inter-level dielectric reliability," *2000 IEEE International Integrated Reliability Workshop Final Report (Cat. No.00TH8515)*, 2000, pp. 189–190. doi: 10.1109/IRWS.2000.911938.
16. J. Lienig, M. Thiele, "Fundamentals of Electromigration," in *Fundamentals of Electromigration-aware Integrated Circuit Design*, Springer International Publishing, Cham, 2018, pp. 13–60.
17. J. W. McPherson, "Time-dependent Dielectric Breakdown Physics: Models Revisited," in *Microelectronics Reliability*, vol. 52, pp. 1753–1760, 2012. doi: 10.1016/j.microrel.2012.06.007.
18. M. Ohring, L. Kasprzak, " Environmental Damage to Electronic Products," in *Reliability and Failure of Electronic Materials and Devices* (Second Edition), M. Ohring, L. Kasprzak (eds.), Academic Press, Boston, 2015, pp. 387–441.
19. M. Ohring, L. Kasprzak, " Defects, Contaminants, and Yield," in *Reliability and Failure of Electronic Materials and Devices* (Second Edition), M. Ohring and L. Kasprzak (eds.), Academic Press, Boston, 2015, pp. 111–179.

20. J. W. McPherson, "Time-to-Failure Models for Selected Failure Mechanisms in Mechanical Engineering," in *Reliability Physics and Engineering: Time-to-Failure Modeling*, J. W. McPherson (ed.), Springer International Publishing, Cham, 2019, pp. 227–304.
21. HIR, "Single Chip and Multi Chip Integration," *Heterogenous Integration Roadmap*, 2021.
22. HIR, "Reliability," *Heterogenous Integration Roadmap*, 2021.
23. C. Feger, N. LaBianca, M. Gaynes, S. Steen, Z. Liu, R. Peddi, M. Francis, "The Over-Bump Applied Resin Wafer-level Underfill Process: Process, Material and Reliability," in *59th Electronic Components and Technology Conference*, 2009, pp. 1502–1505. doi: 10.1109/ECTC.2009.5074211.
24. S. Moreau, H. Manzanarez, N. Bernier, J. Jourdon, S. Lhostis, H. Frémont, "From Electrical to Physical-Chemical Characterization of the Cu/SiO_2 Hybrid-Bonding Interface: A Cu_2O-layer as a Cu Diffusion Barrier?" in *IEEE Electron Device Letters*, vol. 42, pp. 731–734, 2021. doi: 10.1109/LED.2021.3067069.
25. G. Gao, L. Mirkarimi, T. Workman, G. Guevara, J. Theil, C. Uzoh, G. Fountain, B. Lee, P. Mrozek, M. Huynh, R. Katkar, "Development of Low Temperature Direct Bond Interconnect Technology for Die-to-Wafer and Die-to-Die Applications: Stacking, Yield Improvement, Reliability Assessment," *2018 International Wafer Level Packaging Conference (IWLPC)*, 2018, pp. 1–7. doi: 10.23919/IWLPC.2018.8573278.
26. J. A. Theil, L. Mirkarimi, G. Fountain, G. Gao, R. Katkar, "Recent Developments in Fine Pitch Wafer-to-Wafer Hybrid Bonding with Copper Interconnect," *2019 International Wafer Level Packaging Conference (IWLPC)*, 2019, pp. 1–6. doi: 10.23919/IWLPC.2019.8913862.
27. G. Gao, L. Mirkarimi, G. Fountain, D. Suwito, J. Theil, T. Workman, C. Uzoh, G. Guevara, B. Lee, M. Huyhn, P. Mrozek, "Low Temperature Hybrid Bonding for Die to Wafer Stacking Applications," *IEEE 71st Electronic Components and Technology Conference (ECTC)*, 2021, pp. 383–389. doi: 10.1109/ECTC32696.2021.00071.
28. J. De Messemaeker, S. W. Kim, M. Stucchi, G. Beyer, E. Beyne, K. Croes, "Electromigration Behavior of Cu/SiCN to Cu/SiCN Hybrid Bonds for 3D Integrated Circuits," *2018 International Conference on Solid State Devices and Materials (SSDM)*, Tokyo, Japan, 2018. doi: 10.7567/SSDM.2018.G-1-04.
29. E. Beyne, S. Kim, L. Peng, N. Heylen, J. D. Messemaeker, O. O. Okudur, A. Phommahaxay, T. Kim, M. Stucchi, D. Velenis, A. Miller, G. Beyer, "Scalable, Sub 2μm Pitch, Cu/SiCN to Cu/SiCN Hybrid Wafer-to-Wafer Bonding Technology," *IEEE International Electron Devices Meeting (IEDM)*, 2017, pp. 32.4.1–32.4.4. doi: 10.1109/IEDM.2017.8268486.
30. J. R. Black, "Mass Transport of Aluminum by Momentum Exchange with Conducting Electrons," *6th Annual Reliability Physics Symposium (IEEE)*, 1967, pp. 148–159. doi: 10.1109/IRPS.1967.362408.
31. C. S. Hau-Riege, "An Introduction to Cu Electromigration," in *Microelectronics Reliability*, vol. 44, pp. 195–205, 2004.
32. J. R. Lloyd, "Black's Law Revisited: Nucleation and Growth in Electromigration Failure," in Microelectronics Reliability, vol. 47, pp. 1468–1472, 2007. doi: 10.1016/j.microrel.2007.07.094.
33. Y. Ouyang, S. Yang, D. Yin, X. Huang, Z. Wang, S. Yang, K. Han, Z. Xia, "Excellent Reliability of Xtacking™ Bonding Interface," *IEEE International Reliability Physics Symposium (IRPS)*, 2021, pp. 1–6. doi: 10.1109/IRPS46558.2021.9405115.
34. S. Moreau, D. Bouchu, V. Balan, A. L. L. Berrigo, A. Jouve, Y. Henrion, C. Besset, D. Scevola, S. Lhostis, F. Guyader, E. Deloffre, S. Mermoz, J. Pruvost, "Mass Transport-Induced Failure of Hybrid Bonding-based Integration for Advanced Image Sensor Applications," *IEEE 66th Electronic Components and Technology Conference (ECTC)*, 2016, pp. 1940–1945. doi: 10.1109/ECTC.2016.27.

35. L. Di Cioccio, P. Gueguen, R. Taibi, D. Landru, G. Gaudin, C. Chappaz, F. Rieutord, F. de Crecy, I. Radu, L. L. Chapelon, L. Clavelier, "An Overview of Patterned Metal/ Dielectric Surface Bonding: Mechanism, Alignment and Characterization," in *Journal of the Electrochemical Society*, vol. 158, pp. P81–P86, 2011. doi: 10.1149/ 1.3577596.
36. G. Banerjee, R. L. Rhoades, "Chemical Mechanical Planarization Historical Review and Future Direction," in *ECS Transactions*, vol. 13, pp. 1–19, 2008. doi: 10.1149/1.2912973.
37. C. Cavaco, K. Chatzinis, B. v. Lijnschoten, S. Guerrieri, "Hybrid Cu–SiN and Cu–SiOx Direct Bonding of 200 MM CMOS Wafers with Five Metal Levels: Morphological, Electrical and Reliability Characterization," *IEEE 20th Electronics Packaging Technology Conference (EPTC)*, 2018, pp. 608–611. doi: 10.1109/EPTC.2018.8654300.
38. J. P. Gambino, R. Winzenread, K. Thomas, R. Muller, H. Truong, D. Defibaugh, D. Price, K. Goshima, T. Hirano, Y. Watanabe, M. Breen, N. Oldham, "Reliability of Hybrid Bond Interconnects," *IEEE International Interconnect Technology Conference (IITC)*, 2017, pp. 1–3.
39. Y. Kagawa, N. Fujii, K. Aoyagi, Y. Kobayashi, S. Nishi, N. Todaka, S. Takeshita, J. Taura, H. Takahashi, Y. Nishimura, K. Tatani, M. Kawamura, H. Nakayama, T. Nagano, K. Ohno, H. Iwamoto, S. Kadomura, T. Hirayama, "Novel Stacked CMOS Image Sensor with Advanced Cu2Cu Hybrid Bonding," *IEEE International Electron Devices Meeting (IEDM)*, 2016, pp. 8.4.1–8.4.4. doi: 10.1109/IEDM.2016. 7838375.
40. J. Morrison, R. Fontaine, D. Jame, D. Yang, "Samsung Galaxy S7 Edge Teardown," June 25, 2021. https://www.techinsights.com/blog/samsung-galaxy-s7-edge-teardown
41. Y. Kagawa, S. Hida, Y. Kobayashi, K. Takahashi, S. Miyanomae, M. Kawamura, H. Kawashima, H. Yamagishi, T. Hirano, K. Tatani, H. Nakayama, K. Ohno, H. Iwamoto, S. Kadomura, "The Scaling of Cu–Cu Hybrid Bonding for High Density 3D Chip Stacking," *Electron Devices Technology and Manufacturing Conference (EDTM)*, 2019, pp. 297–299.
42. Y. Kagawa, T. Kamibayashi, N. Fujii, S. Furuse, T. Yamada, T. Hirano, H. Iwamoto, "Cu–Cu Wiring: The Novel Structure of Cu–Cu Hybrid Bonding," *IEEE 73th Electronic Components and Technology Conference (ECTC)*, 2023.
43. S. Lhostis, A. Farcy, E. Deloffre, F. Lorut, S. Mermoz, Y. Henrion, L. Berthier, F. Bailly, D. Scevola, F. Guyader, F. Gigon, C. Besset, S. Pellissier, L. Gay, N. Hotellier, A. L. L. Berrigo, S. Moreau, V. Balan, F. Fournel, A. Jouve, S. Chéramy, M. Arnoux, B. Rebhan, G. A. Maier, L. Chitu, "Reliable 300 mm Wafer Level Hybrid Bonding for 3D Stacked CMOS Image Sensors," *IEEE 66th Electronic Components and Technology Conference (ECTC)*, 2016, pp. 869–876.
44. W. Shao, A. V. Vairagar, C.-H. Tung, Z.-L. Xie, A. Krishnamoorthy, S. G. Mhaisalkar, "Electromigration in Copper Damascene Interconnects: Reservoir Effects and Failure Analysis," in *Surface and Coatings Technology*, vol. 198, pp. 257–261, 2005. doi: /08/01/2005..
45. J. Jourdon, S. Moreau, D. Bouchu, S. Lhostis, N. Bresson, D. Guiheux, R. Beneyton, S. Renard, H. Frémont, "Effect of Passivation Annealing on the Electromigration Properties of Hybrid Bonding Stack," *IEEE International Reliability Physics Symposium (IRPS)*, 2017, pp. MR-3.1–MR-3.6.
46. K. Hyuck In, O. J. Kwon, S. Hyungcheol, P. Byung-Gook, P. Sang Sik, L. Jong Duk, "The Effects of Deuterium Annealing on the Reduction of Dark Currents in the CMOS APS," in *IEEE Transactions on Electron Devices*, vol. 51, pp. 1346–1349, 2004.
47. M. H. Lin, Y. L. Lin, J. M. Chen, M. Yeh, K. P. Chang, K. C. Su, W. Tahui, "Electromigration Lifetime Improvement of Copper Interconnect by Cap/Dielectric Interface Treatment and Geometrical Design," in *IEEE Transactions on Electron Devices*, vol. 52, pp. 2602–2608, 2005.

48. C.-K. Hu, L. G. Gignac, J. Ohm, C. M. Breslin, E. Huang, G. Bonilla, E. Liniger, R. Rosenberg, S. Choi, A. H. Simon, "Microstructure, Impurity and Metal Cap Effects on Cu Electromigration," *AIP Conference Proceedings*, vol. 1601, pp. 67–78, 2014. doi: 10.1063/1.4881341.
49. J. Jourdon, S. Lhostis, S. Moreau, J. Chossat, M. Arnoux, C. Sart, Y. Henrion, P. Lamontagne, L. Arnaud, N. Bresson, V. Balan, C. Euvrard, Y. Exbrayat, D. Scevola, E. Deloffre, S. Mermoz, A. Martin, H. Bilgen, F. Andre, C. Charles, D. Bouchu, A. Farcy, S. Guillaumet, A. Jouve, H. Fremont, S. Cheramy, "Hybrid Bonding for 3D Stacked Image Sensors: Impact of Pitch Shrinkage on Interconnect Robustness," in *IEEE International Electron Devices Meeting (IEDM)*, 2018, pp. 7.3.1–7.3.4. doi: 10.1109/IEDM.2018.8614570.
50. B. Ayoub, S. Lhostis, S. Moreau, E. L. Perez, J. Jourdon, P. Lamontagne, E. Deloffre, S. Mermoz, C. d. Buttet, V. Balan, C. Euvard, Y. Exbrayat, H. Frémont, "Impact of Process Variations on the Capacitance and Electrical Resistance Down to 1.44 μm Hybrid Bonding Interconnects," *IEEE 22nd Electronics Packaging Technology Conference (EPTC)*, 2020, pp. 453–458. doi: 10.1109/EPTC50525.2020.9315028.
51. M. O. Aboelfotoh, "Cuprous Oxide (Cu_2O) as an Effective Diffusion Barrier for Copper," IBM Technical Disclosure Bulletin, October 1990.
52. S. Choudhary, J. V. N. Sarma, S. Pande, S. Ababou-Girard, P. Turban, B. Lepine, S. Gangopadhyay, "Oxidation Mechanism of Thin Cu Films: A Gateway towards the Formation of Single Oxide Phase," in *AIP Advances*, vol. 8, p. 055114, 2018.
53. B. Ayoub, S. Moreau, S. Lhostis, P. Lamontagne, H. Combeau, J. G. Mattei, H. Frémont, "New Method to Perform TDDB Tests for Hybrid Bonding Interconnects," *IEEE International Reliability Physics Symposium (IRPS)*, 2022, pp. 4C.3-1–4C.3-6. doi: 10.1109/IRPS48227.2022.9764446.
54. B Ayoub, S. Lhostis, S. Moreau, E. Souchier, E. Deloffre, S. Mermoz, M. G. G. Cacho, N. Szekely, C. Rey, E. Aybeke, V. Gredy, P. Lamontagne, O. Thomas, H. Frémont, "Sub 1 μm Pitch Achievement for Cu/SiO_2 Hybrid Bonding," in *IEEE 24th Electronics Packaging Technology Conference (EPTC)*, 2022, pp. 418–424. doi: 10.1109/EPTC56328.2022.10013180.
55. B. Ayoub, S. Lhostis, S. Moreau, J.-G. Mattei, A. Mukhtarov, H. Frémont, "Investigation into Cu Diffusion at the Cu/SiO_2 Hybrid Bonding Interface of 3D Stacked Integrated Circuits," in *Microelectronics Reliability*, vol. 143, p. 114934, 2023.
56. S. Moreau, D. Bouchu, J. Jourdon, B. Ayoub, S. Lhostis, H. Frémont, P. Lamontagne, "Recent Advances on Electromigration in Cu/SiO_2 to Cu/SiO_2 Hybrid Bonds for 3D Integrated Circuits," *IEEE International Reliability Physics Symposium (IRPS)*, 2023, pp. 1–7. doi: 10.1109/IRPS48203.2023.10118173.
57. M. F. Chen, F. C. Chen, W. C. Chiou, D. C. H. Yu, "System on Integrated Chips SoIC™ for 3D Heterogeneous Integration," *IEEE 69th Electronic Components and Technology Conference (ECTC)*, 2019, pp. 594–599.
58. Y. H. Chen, C. A. Yang, C. C. Kuo, M. F. Chen, C. H. Tung, W. C. Chiou, D. Yu, "Ultra High Density SoIC with Sub-micron Bond Pitch," *IEEE 70th Electronic Components and Technology Conference (ECTC)*, 2020, pp. 576–581.
59. S. W. Liang, G. C. Y. Wu, K. C. Yee, C. T. Wang, J. J. Cui, D. C. H. Yu, "High Performance and Energy Efficient Computing with Advanced SoIC™ Scaling," *IEEE 72nd Electronic Components and Technology Conference (ECTC)*, 2022, pp. 1090–1094.
60. S. W. Liang, Y.R. Liang, Gene C. Y. Wu, K. C. Yee, C. T. Wang, D. C. H. Yu, "Reliability Performance on Fine-pitch SoIC™ Bond," in *IEEE 73rd Electronic Components and Technology Conference (ECTC)*, 2023.
61. L. Hou, E. Chery, K. Croes, D. Tierno, S. A. Chew, Y. Chen, P. Rakbin, E. Beyne, "Reliability Investigation of W2W Hybrid Bonding Interface: Breakdown Voltage and Leakage Mechanism," *IEEE International Reliability Physics Symposium (IRPS)*, 2022, pp. 4C.2-1–4C.2-6.

62. S. Sitaraman, L. Jiang, S. Dag, M. Masoomi, Y. Wang, P. Lianto, J. An, R. Wang, G. See, A. Sundarrajan, E. M. Bazizi, B. Ayyagari-Sangamalli, "A Holistic Development Framework for Hybrid Bonding," in *IEEE 72nd Electronic Components and Technology Conference (ECTC)*, 2022, pp. 691–700.
63. P. Hohenberg, W. Kohn, "Inhomogeneous Electron Gas," in *Physical Review*, vol. 136, pp. B864–B871, 1964.
64. X. Ren, P. Rinke, V. Blum, J. Wieferink, A. Tkatchenko, A. Sanfilippo, K. Reuter, M. Scheffler, "Resolution-of-Identity Approach to Hartree–Fock, Hybrid Density Functionals, RPA, MP2 and GW with Numeric Atom-centered Orbital Basis Functions," in *New Journal of Physics*, vol. 14, p. 053020, 2012. doi: 10.1088/1367-2630/14/5/053020.
65. S. Grimme, "Semiempirical GGA-type Density Functional Constructed with a Long-range Dispersion Correction," in *Journal of Computational Chemistry*, vol. 27, pp. 1787–1799, 2006. doi: 10.1002/jcc.20495.
66. H. Kim, Y.-H. Tsai, S. Hoshino, I. Son, M. Kaoru, P. Biolsi, S. Arkalgud, "Atomistic Simulation Analysis of Plasma Surface Activation in Wafer-to-Wafer Oxide Fusion Bonding," *IEEE 73rd Electronic Components and Technology Conference (ECTC)*, 2023, pp. 2034–2039. doi: 10.1109/ECTC51909.2023.00349.
67. Y. Beilliard, R. Estevez, G. Parry, P. McGarry, L. Di Cioccio, P. Coudrain, "Thermomechanical Finite Element Modeling of Cu–SiO_2 Direct Hybrid Bonding with a Dishing Effect on Cu Surfaces," in *International Journal of Solids and Structures*, vol. 117, pp. 208–220, 2017. doi: 10.1016/j.ijsolstr.2016.02.041.
68. C. Sart, R. Estevez, V. Fiori, S. Lhostis, E. Deloffre, G. Parry, R. Gonella, "Cu/SiO_2 Hybrid Bonding: Finite Element Modeling and Experimental Characterization," *6th Electronic System-Integration Technology Conference (ESTC)*, 2016, pp. 1–7. doi: 10.1109/ESTC.2016.7764484.
69. X. P. Xu, A. Needleman, "Numerical Simulations of Fast Crack Growth in Brittle Solids," in *Journal of the Mechanics and Physics of Solids*, vol. 42, pp. 1397–1434, 1994.
70. N. Ip, N. Nejadsadeghi, C. Fonseca, N. Kohama, K. Motoda, "Multi-physics Simulation of Wafer-to-Wafer Bonding Dynamics," *IEEE 72nd Electronic Components and Technology Conference (ECTC)*, 2022, pp. 502–506. doi: 10.1109/ECTC51906.2022.00085.
71. N. Ip, N. Nejadsadeghi, N. Kohama, H. Tanoue, K. Motoda, "Distortion Simulation for Direct Wafer-to-Wafer Bonding Process," *IEEE 73rd Electronic Components and Technology Conference (ECTC)*, 2023, pp. 694–698. doi: 10.1109/ECTC51909.2023.00121.
72. T. Hirano, T. Yamada, S. Kobayashi, Y. Hagimoto, H. Iwamoto, "Simulation of Device Structure Impacts on Bonding Wave and Strain in Wafer-to-Wafer Cu–Cu Hybrid Bonding," *IEEE 73rd Electronic Components and Technology Conference (ECTC)*, 2023, pp. 1314–1318.
73. L. Ji, F. X. Che, H. M. Ji, H. Y. Li, M. Kawano, "Modelling and Characterization on Wafer to Wafer Hybrid Bonding Technology for 3D IC Packaging," *IEEE 21st Electronics Packaging Technology Conference (EPTC)*, 2019, pp. 87–94. doi: 10.1109/EPTC47984.2019.9026578.
74. L. Ji, F. X. Che, H. M. Ji, H. Y. Li, M. Kawano, "Wafer-to-Wafer Hybrid Bonding Development by Advanced Finite Element Modeling for 3-D IC Packages," in *IEEE Transactions on Components, Packaging and Manufacturing Technology*, vol. 10, pp. 2106–2117, 2020.
75. L. Ji, F. X. Che, H. M. Ji, M. Kawano, "Bonding Integrity Enhancement in Wafer to Wafer Fine Pitch Hybrid Bonding by Advanced Numerical Modelling," *IEEE 70th Electronic Components and Technology Conference (ECTC)*, 2020, pp. 568–575.

76. H. Ren, Y.-T. Yang, G. Ouyang, S. S. Iyer, "Mechanism and Process Window Study for Die-to-Wafer (D2W) Hybrid Bonding," in *ECS Journal of Solid State Science and Technology*, vol. 10, p. 064008, 2021.
77. L Ji, S. K. Tippabhotla, "Numerical Evaluation on SiO_2 Based Chip to Wafer Hybrid Bonding Performance by Finite Element Analysis," *IEEE 72nd Electronic Components and Technology Conference (ECTC)*, 2022, pp. 524–530. doi: 10.1109/ECTC51906.2022.00089.
78. S. K. Tippabhotla, L. Ji, Y. Han, "Numerical Simulation of Cu/Polymer-Dielectric Hybrid Bonding Process Using Finite Element Analysis," *IEEE 72nd Electronic Components and Technology Conference (ECTC)*, 2022, pp. 1695–1703. doi: 10.1109/ECTC51906.2022.00267.
79. S. K. Tippabhotla, L. Ji, C. S. Choong, "Numerical Simulations to Assist Chip-to-Wafer Hybrid Bonding Process Development," *IEEE 24th Electronics Packaging Technology Conference (EPTC)*, 2022, pp. 650–654. doi: 10.1109/EPTC56328.2022.10013143.
80. A. Agrawal, S. Huang, G. Gao, L. Wang, J. DeLaCruz, L. Mirkarimi, "Thermal and Electrical Performance of Direct Bond Interconnect Technology for 2.5D and 3D Integrated Circuits," *IEEE 67th Electronic Components and Technology Conference (ECTC)*, 2017, pp. 989–998. doi: 10.1109/ECTC.2017.341.
81. H. Ceric, H. Zahedmanesh, K. Croes, R. Lacerda de Orio, S. Selberherr, "Electromigration-induced Void Evolution and Failure of Cu/SiCN Hybrid Bonds," *Journal of Applied Physics*, vol. 133, 2023. doi: 10.1063/5.0134692.

12 Applications of Hybrid Bonding and Chiplets for Heterogeneous Integration

Chandrasekhar Mandalapu, John Wuu, Raja Swaminathan, and Liwei Wang

12.1 INTRODUCTION

The explosion of connected devices over the last 40 years in semiconductor industry has driven an explosion of semiconductor content riding the back of Moore's law. Starting with the personal computer era and continuing into the smartphones and Internet of Things (IoT) waves, silicon has permeated every aspect of human lives. The insatiable demand for semiconductors that spans from the cloud to the edge, and from artificial intelligence (AI) to 5G communications, is driving the microelectronics industry to design and technology innovations beyond what the slowing Moore's law can support with raw silicon scaling.

12.1.1 Demand for Computation Is Outpacing Moore's Law

As is widely known, silicon technology node introductions have been slowing down and simultaneously delivering less benefit, while at the same time the costs per yielded mm^2 of silicon are going up. This is particularly challenging because the semiconductor industry has thrived on delivering more performance and features in each generation by adding transistors. With these trends, the cost per transistor will stop scaling in the next few years, which creates notable economic headwinds to meeting the demand. These costs are not just a result of inflationary pressure but based on the underlying physics and complexity of these new nodes.

The next aspect of the slowdown in node introductions is that scaling factors are diverging between different intellectual property (IP) types, with static random-access memory (SRAM) and especially analog circuits lagging well behind the scale factors of logic. This leads to the chip-level view of area scaling where, with a mix of logic, SRAM, and analog content, it is not feasible to shrink chip designs appreciably toward the end of this decade. This illustrates that the irresistible force of compute demand is colliding with the immovable object of device physics, creating an environment where *new architecture approaches and non-device innovations* are critical for the semiconductor ecosystem.

DOI: 10.1201/9781003408673-12

So, the semiconductor industry has the bright future of exploding compute demand and simultaneously the dark cloud of technology headwinds. The trillion-dollar question is how to architect, design, and build future systems that solve these challenges. The answer is increasingly clear that modular, multi-chip design/chiplets is a fundamental enabler. The following sections will have a closer look at what modular design can do, and what the enabling technology requirements are.

12.1.2 Introduction to Chiplets

An initial motivation for chiplets was economics. Back in the day, Moore's law enabled a doubling of transistors and capability in each generation, and all was good. Lately, this has not worked out as shrink factors slow down while compute demand has not, and die sizes have been growing at an unsustainable rate.

With chiplets, a formerly monolithic system-on-chip (SoC) can be split into multiple components/dies to improve performance. However, this results in a non-trivial overhead associated with "chipletizing" the design. Each die needs test capability, power management, and an interface so that it can talk to the other chiplets. These interfaces will not be as small, low latency, or power efficient as on-die wires; therefore, the architecture needs to accommodate new boundaries and complexity.

The chiplet approach provides various benefits like product flexibility, yield improvement due to smaller die size, etc. For example, a single large die with a fixed number of defects on a wafer will yield a small set of functional SoCs for a wafer's worth of chips (Figure 12.1). As soon as that big SoC split up into, say, four chiplets, the same number of defects now just take out a small chiplet, and by using wafer sort capabilities to select the good dies, more functional SoCs from the same silicon can be built.

As mentioned earlier, the chiplets approach also provides the flexibility of building chiplet SoCs with varying numbers of chiplets to address different markets. Perhaps a less obvious benefit is that faster (speed/performance) chiplets can be picked from the wafer and assembled into higher performance and higher priced SoCs for customers who want, and will pay for, the greatest performance possible.

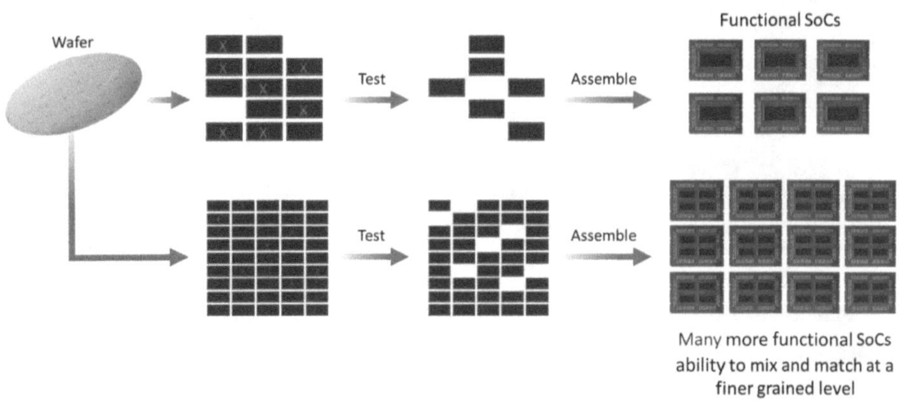

FIGURE 12.1 High-level approach to chiplets.

The benefits described above are substantial, though modular design is bigger than just decomposing an SoC into chiplets. By mixing and matching chiplet types, tailored products can be built for specific markets. Some chiplets can be general-purpose CPUs, while others can be more specialized for specific domains and can be included more or less in a given product depending on the requirements. However, the success of this approach is heavily dependent on the package technologies used to assemble these dies and enable them to communicate with each other.

12.2 TECHNOLOGY OPTIONS

Various product segments (e.g., mobile, PC, server, and desktops) require different die-to-die interconnect technologies depending on bandwidth and interconnect density requirements. As shown in Figure 12.2, various package architectures exist in the industry.

Examples include:

- Multi-chip module (MCM) architectures from various industry players
- 2D architectures based on redistribution layer (RDL)-like interconnects or (2D organic)-like integrated fan-out with redistribution layer (INFO-R), and fan-out chip-on-substrate (FoCoS)
- 2.5D silicon-based architectures like embedded multi-die interconnect bridge (EMIB), elevated fan-out bridge (EFB), integrated fan-out with integration of a local silicon interconnect (LSI) like INFO-L, and silicon interposer where the die-to-die interconnect is achieved using passive silicon
- 3D architectures—defined as active-on-active silicon stacking, such as the AMD 3D V-Cache™, Intel's Foveros/omni-directional interconnect (ODI), wafer-on-wafer (WoW) architecture found in CMOS image sensors, and the memory markets

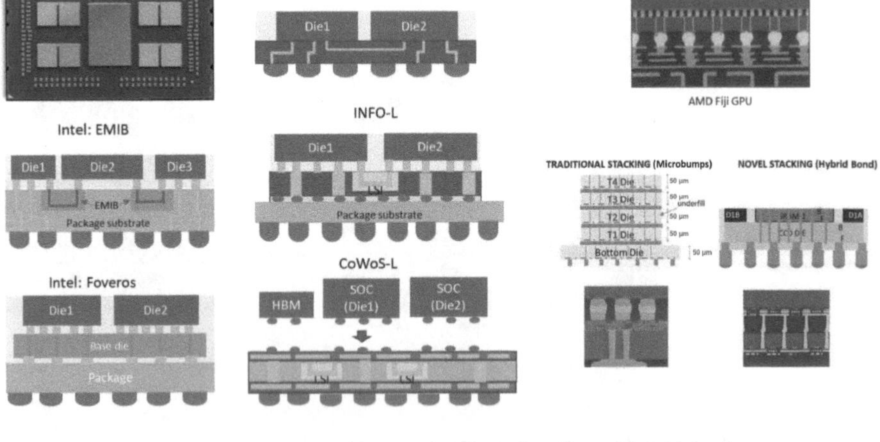

FIGURE 12.2 Sample package architecture options for die-to-die chiplet interconnects.

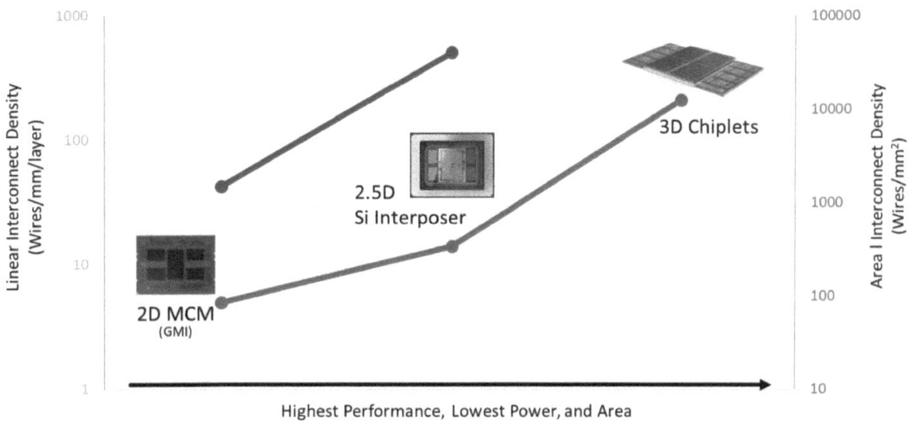

FIGURE 12.3 Linear (*left Y-axis*) and areal (*right Y-axis*) interconnect densities of various package architectures.

Product-specific requirements such as power, performance, area, and cost (PPAC) determine the choice of chiplet package architecture, rather than a one-size-fits-all approach. The key driver to make all this work is decreasing interconnect pitches and increasing interconnect density. Different packaging approaches, liner interconnect density per layer, and areal interconnect density are tabulated in Figure 12.3.

MCMs are great, low-complexity designs, but only suitable for specific chiplet application of lower to medium bandwidth requirements due to lower interconnect density, as shown in Figure 12.3 (bottom left). For example, AMD used MCM architecture in EPYC™ and Ryzen™ lines, with CPU cores on one chiplet and the IO and memory interfaces on another chiplet. MCM architectures works very well in this application, as CPU bandwidth requirements are relatively modest and can be supplied across high-speed SERDES routes.

Medium to higher bandwidth applications require more exotic SoC chiplet configurations/architectures like silicon interposer, EFB, etc., where interconnect density is >10× compared to MCM. For instance, AMD Instinct™ designs require high-bandwidth memory (HBM) to feed compute engines. For the higher density interconnects required to support over a terabyte per second (TB/s) of memory bandwidth, AMD used passive silicon interposer for the first instance and EFB approach on MI-200 Instinct™ product.

12.2.1 Chiplet Architecture Using 3D Stacking

The interconnect density of 2D and 2.5D technologies is limited by the lengths of the neighboring die edges, which can be overcome by 3D stacking the dies vertically. In addition to interconnect density improvements, 3D stacking can also offer smaller silicon footprint in a package, along with reductions in interconnect latency and power. Traditional 3D stacking uses solder-based microbumps with tall through-silicon vias

(TSVs), which can scale from 50 μm to ~36 μm, and is acceptable for the low end of higher bandwidth applications. Taller TSVs suffer from large parasitics and also have larger pitch than the requirements of high-performance applications.

The holy grail of chiplet architecture is of course 3D stacking using hybrid bonding (HB). This technology uses silicon fab-like manufacturing methods with back-end design rule-based TSVs with copper-only interconnects without the presence of solder. Using silicon fab-based techniques to enable interconnect technology for extreme bandwidth applications is a turning point in the industry's advanced packaging journey. Figure 12.4 shows a 3D architecture with hybrid bond in scale relative to best-in-class solder-based microbump architecture. Due to this extreme scaling, >3× higher interconnect energy efficiency, >15× higher interconnect density, as well as better signal and power performance compared to microbump 3D architectures can be achieved [1].

In the AMD 3D V-Cache™, a 64 MB SRAM die is 3D stacked on top of a CPU core die of 32 MB L3 cache [1]. The packaging choice of HB is very specific to this architecture, as it requires thousands of signals and power connections, which provides dramatically higher bandwidth density. Higher density technologies are more expensive due to precise patterning requirements and more processing steps; however, it comes with an interface area reduction and lower energy for data transfer.

In addition to the above-mentioned benefits, chipletizing also comes with overheads, including IO area, additional design effort and complexity, and testing steps. Getting to the right architecture requires that the value of the proposed newly modular solution, with its configuration flexibility and yield benefits, more than outweighs the costs. Getting this right is a highly multidisciplinary endeavor, requiring engineers from different domains to rapidly iterate and provide solutions in new ways.

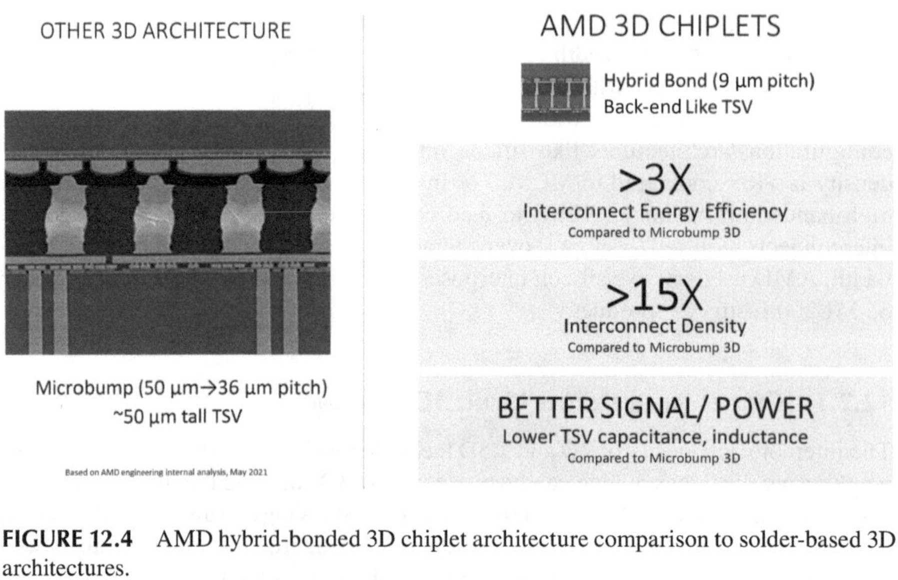

FIGURE 12.4 AMD hybrid-bonded 3D chiplet architecture comparison to solder-based 3D architectures.

12.3 DESIGN IMPLICATIONS

When 3D stacking dies with high power density, such as microprocessors, thermal dissipation and power delivery are usually two of the top design considerations. Three-dimensional stacking worsens thermal dissipation as the power densities of the overlapping dies are combined, and the presence of top die(s) along with the bonding process introduce thermal barriers between the dies and the heat sink. Similarly, due to the combined power densities, the substrates for 3D designs need to deliver higher current per unit die area than in their 2D counterparts. In addition, the bottom die needs to provision for frequent power TSVs in its floorplan, while the top die must account for the degradations to its supply from being delivered through the bottom die.

The first step in tackling thermal dissipation and power delivery trade-offs is determining which die should be the top die closer to the heat sink, and which die should be the bottom die closer to the substrate. The bottom die, with its direct interface to the substrate assuming a face-down configuration, can enjoy solid power delivery; however, it must absorb the top die's power TSV area overhead and the associated disruptions to its IPs. In addition, with the top die serving as a thermal barrier to the heat sink, the bottom die's operating temperature will be higher than its 2D counterpart's. On the flip side, the top die can enjoy the thermal dissipation advantage from its proximity to the heat sink, and it does not require TSVs in its design. However, its power delivery will have additional impedance compared to 2D designs, which can limit its performance and/or power efficiency. As a result, power delivery and thermal dissipation trade-offs must be carefully considered when determining in what order to place dies in a 3D stacked design.

After the die order is determined, floor planning also plays an indispensable role in mitigating power delivery and thermal dissipation penalties. The stacked dies' floorplans should be co-optimized to avoid overlapping power density and thermal hot spots over each other.

Using the AMD 3D V-Cache™ as an example, the high-power-density CPU die was allocated as the bottom die to ensure its robust power delivery, and the low power density cache die (L3D) as the top die can limit the number of power TSVs the bottom die must support to a low level. Then, to mitigate the bottom CPU die's thermal penalty, the top L3D die was designed to only overlap the CPU die's low power density L3 cache region. Then, for the high-power-density CPU cores, structural silicon dies were placed over them to allow for efficient heat dissipation, as shown in Figure 12.5.

Timing closure is another key consideration in 3D stacking high-performance dies, which is related to die matching. When two dies are randomly paired together, the speed can be limited by the slower of the two dies. In a chip-on-wafer (CoW) technology where known-good-die (KGD) testing can allow for dies with similar performance parameters to be selectively paired, there exists more flexibility to trade-off between product performance, binning complexity, and yield. However, in a WoW technology, while wafers can be selectively paired based on parametric reading, individual dies on the two wafers are always paired with their counterparts, providing little opportunities for die matching.

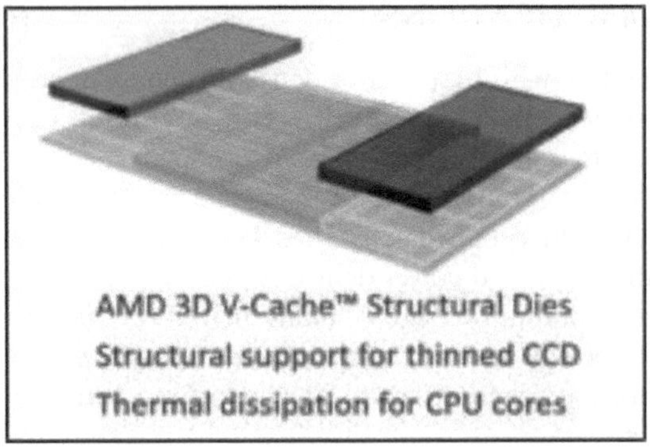

FIGURE 12.5 AMD 3D V-Cache™ design for thermal and power optimization.

12.4 APPLICATIONS USING HB

As shown in Figure 12.6, there are multiple ways to enable 3D stacking and die orientations depending on design and product architectures. Various factors need to be considered when making the choice of stacking technologies and/or stacking orientation. Some of the considerations include die sizes, wafer yield/need for KGD, interconnect pitch requirements, die matching need, power delivery need, die–die signal density requirements, high-speed IO on top/bottom tier, thermal requirements, etc.

HB applications can be segregated into WoW or CoW based on the above-mentioned factors such as wafer yield, hybrid bond pitch, and die matching requirements. WoW bonding was used in image sensors and memory applications, where yields are higher and tight HB pitch is required. CoW bonding was used in more challenging HPC applications, where KGDs and bin matching are used for bonding.

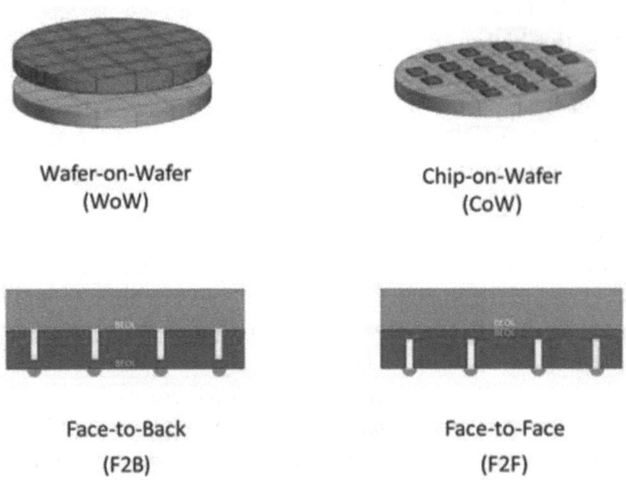

FIGURE 12.6 Schematics of various 3D integration schemes.

12.4.1 CMOS Image Sensors

The CMOS image sensors (CIS) market is the early adopter of the direct bonding technology to create the backside illumination (BI) process. In this process, the device wafer is oxide bonded to the supporting (handle) wafer. Then, the device wafer is thinned down, followed by the color filter array and the micro-lens formations on the backside of device wafer [4]. The BI process has been widely adopted as the promising technology to improve the sensitivity and image quality of CMOS image sensors from a pixel size of 1.75 and smaller [5]. Due to the strong growth in the smartphone market, cell phone cameras need to continuously incorporate new functionalities to support features beyond the traditional image quality, speed, and pixel count improvements [2]. To support this high demand of new functionalities, the supporting wafer was replaced with a logic wafer in a conventional BI image sensor process to create a stacked BI-CIS process. Figure 12.7 shows the structure and cross section of both conventional and stacked BI-CIS technologies.

Pixel circuitry requires a lesser number of metal layers when compared to highly integrated analog to digital converters (ADC) metal layers. The photodiode requires contacts with low leakage to detect smaller signal changes, whereas the logic of ADC requires low resistance contacts to support low-noise and high-speed operation. The stacking process provides great advantages to choose different process/ metal schemes for different functionalities and eliminates process trade-offs in image sensors. TSVs are used to connect between the top CIS and the bottom logic

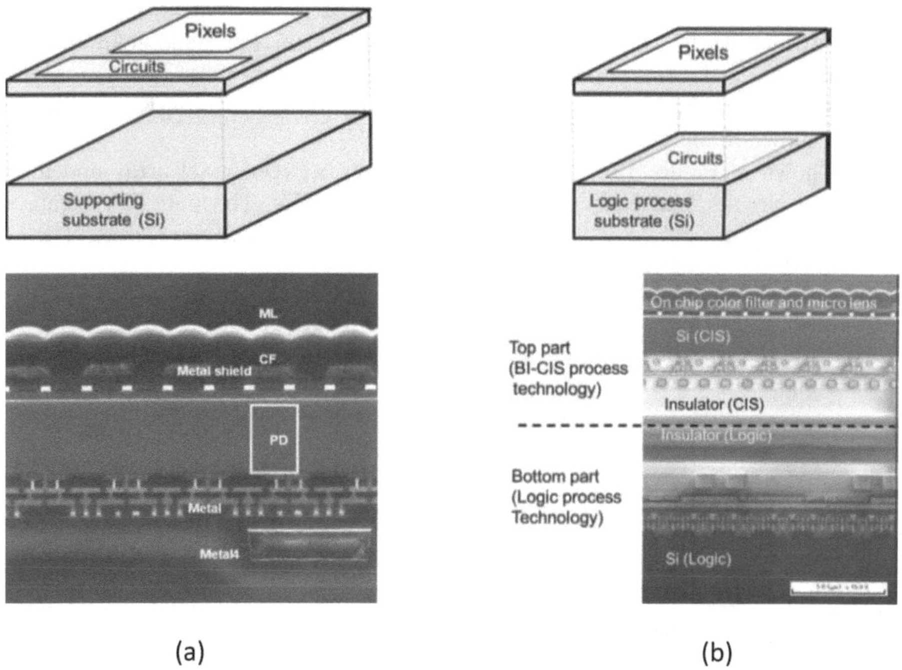

FIGURE 12.7 Structures and cross sections of (a) conventional and (b) stacked BI-CIS [2, 3].

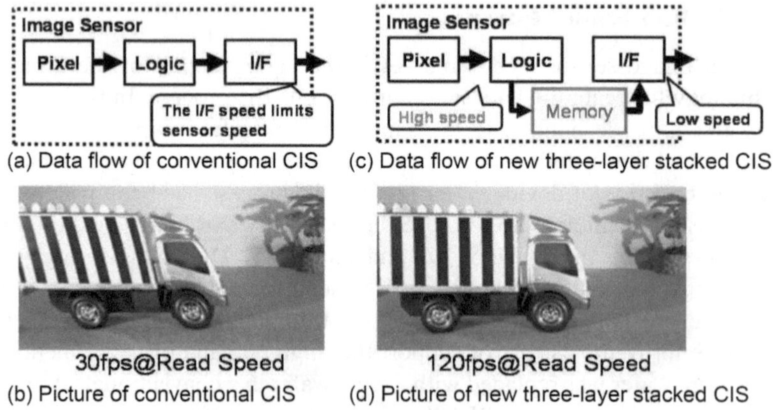

FIGURE 12.8 Data flow between conventional CIS and DRAM-stacked CIS chip [6].

circuitry, and the number of TSVs is approximately equal to the number of rows and column signals [2].

The speed of the CIS is limited by the output speed of the interface (I/F), as signal data from pixel flows through the logic circuits and output through the I/F serially. In one example of conventional CIS implementation [6], the reading speed could only be increased up to 30 frames per second (fps) and rolling shutter distortion occurs due to reading time difference between the first and last pixels, as shown in Figure 12.8a and b. DRAM was added as a further improvement to the stacked BI-CIS, which works as frame memory with high transmission bandwidth and as a temporary buffer for image data. As a result, I/F speed is not limiting anymore and reading speed can be increased up to 120 fps and eliminates rolling shutter distortion, as shown in Figure 12.8c and d [6, 7].

As shown in Figure 12.9, DRAM is stacked in between pixel array and logic periphery circuits, and each layer is connected by TSVs. Through TSVs, data is

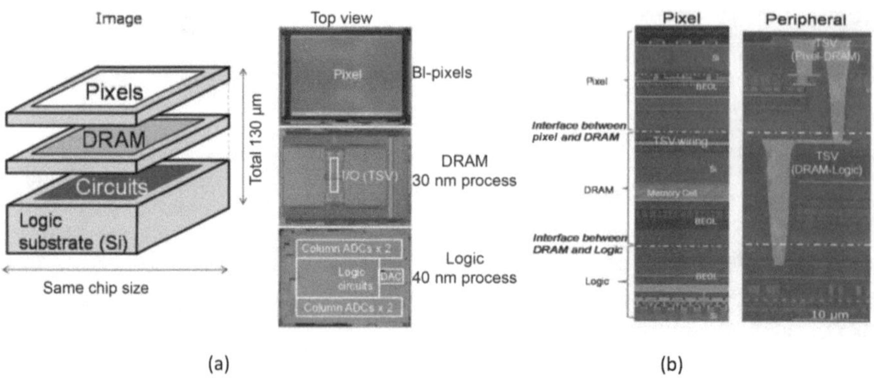

FIGURE 12.9 (a) Three-layer stacked CIS chip schematic. (b) Cross section of a pixel and peripheral area [6].

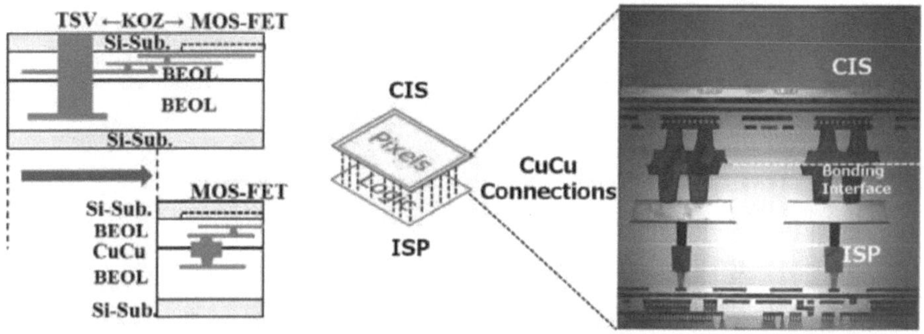

(a) (b)

FIGURE 12.10 (a) Schematic diagrams of stacked BI-CIS with TSV (*upper*) and with HB (*lower*). (b) Cross-sectional view of stacked BI-CIS with Cu–Cu HB [8].

transmitted from pixel to ADC converters, and the digitally converted image is transmitted to DRAM for storage. A low-power and high-bandwidth link was achieved in between DRAM and image processor by stacking them on top of each other.

TSV's require keep out zones (KOZ) to prevent them from incurring performance penalties onto the neighboring devices. In addition, the number of TSVs per SOC continues to increase with the decrease in pixel sizes. Both of these factors will affect the circuit design and the die size/silicon area. Cu–Cu HB was introduced in CMOS image sensors to avoid the above-mentioned issues. As shown in Figure 12.10, copper pad was added on top of back-end-of-line (BEOL) layer using single or dual damascene process, followed by face-to-face (F2F) WoW bonding and annealing to form rigid copper connections. Stacked BI-CIS using Cu–Cu HB helped to achieve fine-pitch and large-scale connections, which helped to improve circuit design flexibility and to reduce chip size [8].

Cu–Cu HB was further extended to three-layer stacked structures. As shown in Figure 12.11a–c, the top and middle wafers are F2F wafer bonded, the middle wafer backside is thinned down, and TSVs and copper connection pads are fabricated. Finally, the middle wafer and bottom wafer are electrically connected by face-to-back (F2B) Cu–Cu HB, as shown in Figure 12.11d. A wide range of 3D stacked

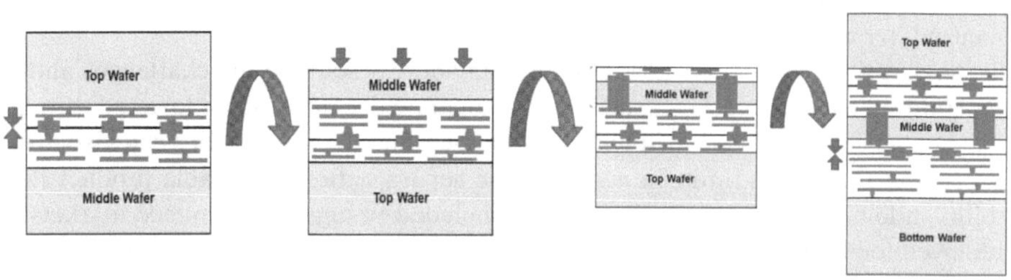

(a) (b) (c) (d)

FIGURE 12.11 Process flow and schematic representation of three-wafer-stacked module [9].

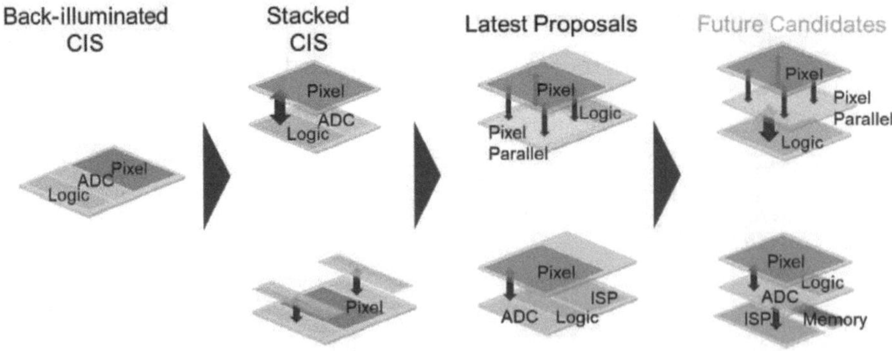

FIGURE 12.12 Evolution and future prospects of stacked BI-CIS technology [9, 10].

structures to enhance the device functionalities can be enabled due to the flexibility to use F2F and F2B HB [9].

Image sensor performance has improved, particularly at high frame rates and high pixel resolutions through stacking technology. Pixel parallel stacked architectures have been proposed and demonstrated in the several latest proposals. As shown in Figure 12.12, multi-hi-stacking HB can be a key enabler to expand the functionalities of device stacking technology for future CIS architectures [10].

12.4.2 Compute Applications

Cache size remains an important knob for CPU performance, especially given the increase in applications' dataset sizes. As a result, on-die cache capacities have steadily grown over time for high-performance CPUs. However, this traditional approach is facing multiple headwinds. First, the slowdown in Moore's law scaling requires larger die sizes to accommodate the growing cache capacities, which in some cases are pushing against the reticle limit. The slowdown in scaling is compounded by the rapid rise in advanced technology nodes' wafer prices, making large on-die caches cost-ineffective. Then, the long cache datapaths resulting from the large area increase the cache latency and dynamic power, which diminishes the cache capacity instructions per clock (IPC) performance gains. Finally, while some markets demand the absolute highest performance, other markets can be more value focused, for which very large caches may be overkill.

As described in Section 12.1, chiplets can address some of the challenges above. By splitting the extra cache capacity into a separate die, cost can be lowered through yield enhancements associated with smaller dies, and the smaller dies can also avoid exceeding the reticle limit. In addition, the separate dies can enable product flexibility, allowing the extra cache die to be included in high-performance markets or removed for value conscious markets. However, 2D or 2.5D chiplets cannot address the latency and power penalties incurred from the long lateral datapaths, and side-by-side dies occupy valuable real estate in the package cavity, limiting package-level integration. Three-dimensional chiplets, on the other hand, mitigate the long lateral

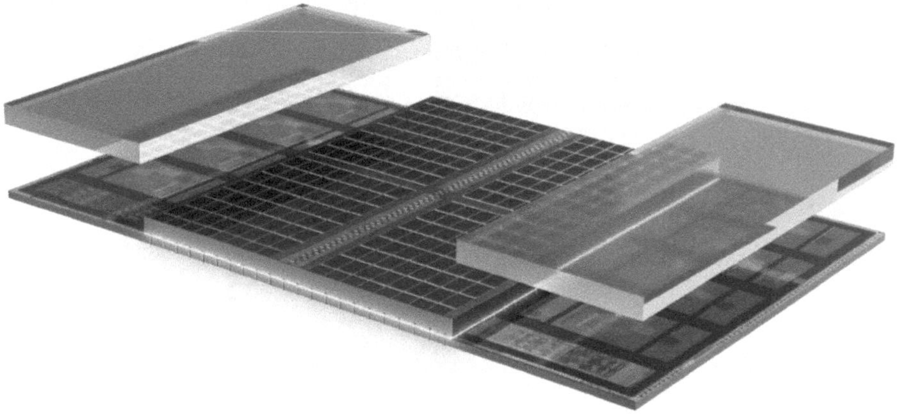

FIGURE 12.13 Schematic view of AMD's 3D V-Cache™ stacked die [11, 12].

datapath distances through their vertical interconnect, and reduce the module footprint to allow more transistors and more cores to be integrated into the package.

The AMD 3D V-Cache™ is the industry's first implementation of this 3D cache concept utilizing the HB technology [11]. As shown in Figure 12.13, there are three main components in the 3D V-Cache™ stacked die. On the bottom facing the substrate is the "Zen 3" CPU core complex die (CCD), which is identical to the CCD used in 2D "Zen3" products. Each CCD includes eight CPU cores, and the cores share a 32MB L3 cache. Stacked face-down on top of the CCD is the extended L3 die (L3D), placed over the central 32 MB L3 cache region of the CCD. As the name implies, the L3D serves as an extension to the CCD's L3 cache, increasing the total L3 cache capacity from 32 MB to 96 MB. And, by placing the low-power-density L3D, which does not require high-supply current, over the similarly low-power-density L3 cache region in the CCD, both the thermal and the power TSV overhead penalties can be mitigated. Finally, structural silicon dies, which are good thermal conductors, are placed on the edges of the CCD where the CPU cores reside to provide structural support for the thinned CCD, and to assist in heat dissipation for the high-performance, high-power-density CPU cores.

To enable CCD sharing between 2D and 3D products and to allow the CCD to interface with the same substrate through standard C4 bumps, the L3D uses a Face to Back (F2B) integration scheme to attach to the backside of the CCD. Figure 12.14 shows the topology of the TSMC-SoIC™ HB technology used by the 3D V-Cache™, in which the bond pad metal (BPM) serves as the die-to-die signal interface. BPM connects to the bottom die's circuitry through the TSV, and the top die's metal stack through the bond pad via (BPV).

The CoW stacking process is used for the AMD 3D V-Cache™, which supports different sized KGDs to be selected and stacked [12]. In one product configuration, as shown in Figure 12.15, eight CCDs are combined with an I/O die (IOD) to form a 64-core CPU. For the 3D stacked version on the right, support silicon is attached to allow the overall CCD module *z*-height to match 2D and 3D versions. As a result,

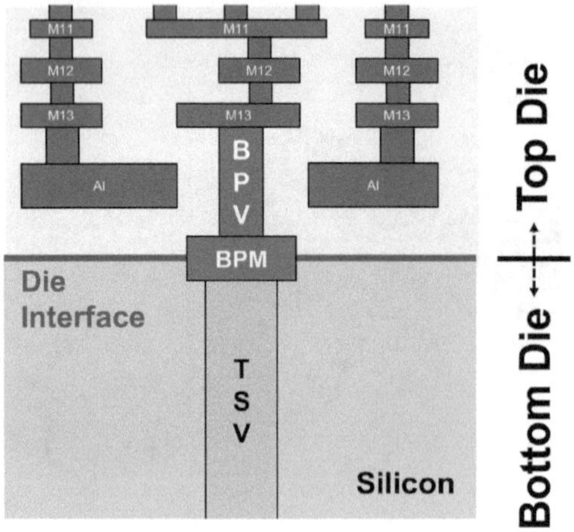

FIGURE 12.14 AMD 3D V-Cache™ HB topology [12].

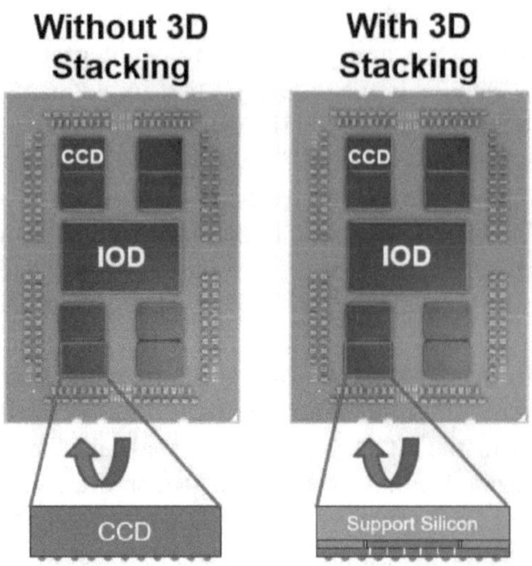

FIGURE 12.15 An MCM package with/without stacked die in a package [12].

the 2D and 3D versions of the products have identical interfaces and physical dimensions, allowing for AMD 3D V-Cache™'s seamless integration.

While the 3D V-Cache™ tripled the L3 cache capacity from 32 MB to 96 MB, only four CPU clock cycle delays are added to the overall L3 cache latency [1]. In addition, the L3D supports more than 2 TB/s of bandwidth to the CCD. The combination of low latency and high bandwidth highlights the advantage of HB and 3D. These advantages also translate to product performance. With the 3D V-Cache™, 15% average gaming performance improvement was achieved for AMD Ryzen™ desktop processors, and 66% faster RTL verification was enabled in AMD EPYC™ servers while running the Synopsys VCS® tool [1 & 12].

12.4.3 NAND Memory Applications

Data production and consumption increased significantly with strong growth of applications like IoT, AI, data centers, etc. These applications require high-speed, reliable, and low-power storage memories. To cater this huge storage demand, 3D NAND memory emerged as a great solution to address the density/physical limitations in planar memory cells by stacking up more memory layers in vertical direction and achieved higher bit growth density. The 3D NAND architecture consists of a memory array and CMOS periphery devices to command signals to and from the memory array. Memory array layer count increased with each generation without change in lithography tech node, resulting in increased memory density and a significant bit cost decrease. Samsung started with 32-layer 3D NAND, and currently 232-layer 3D NAND by Micron is in commercial use, as of early 2023. Increases in layer count with each generation leads to longer process/cycle times and also manufacturability/process challenges such as asymmetric etching profile and bow during high-aspect-ratio etching of memory array [13–17].

WoW HB appears as a great process solution to increase the memory capacity of 3D NAND. As shown in Figure 12.16, YMTC introduced 3D NAND using WoW HB, where high-performance CMOS peripheral circuits are integrated on one wafer, while NAND memory array cells are integrated on a separate wafer [18]. Both are bonded through metal interconnect vias. This approach has advantages in cycle time as the product development cycle for both wafers can be processed in parallel, and CMOS wafers can be processed with high temperature without the risk of impacting/degrading the NAND memory cell [15].

As in the typical bonding process, both CMOS and NAND memory wafer surfaces are prepared for bonding, followed by WoW bonding and annealing to establish metal connections between both wafers. Figure 12.17a shows a scanning electron microscopy (SEM) image cross section of YMTC's first-generation 64-layer 3D NAND manufactured using Cu–Cu HB of 1 μm pitch. The M3 copper layer from the NAND array wafer is connected to the M4 copper layer of the CMOS wafer, as shown in Figure 12.17b. With annealing, copper metals expand and merge into each other to create a strong copper-to-copper metal direct bond [15]. As shown in Figure 12.17c, the NAND array wafer M3 copper size is slightly smaller (~30 nm smaller on each side) than the M4 copper size from the CMOS wafer. This offset helps to provide more process margin for misalignment during WoW bonding.

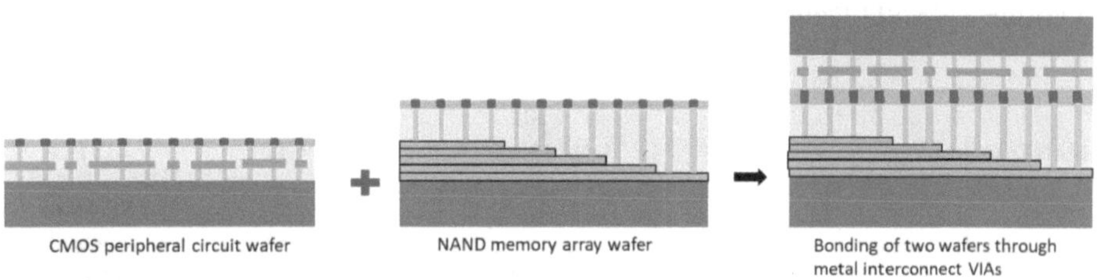

FIGURE 12.16 Schematic flow of YMTC's 3D NAND Xtacking® technology. (Adapted from Ref. [18].)

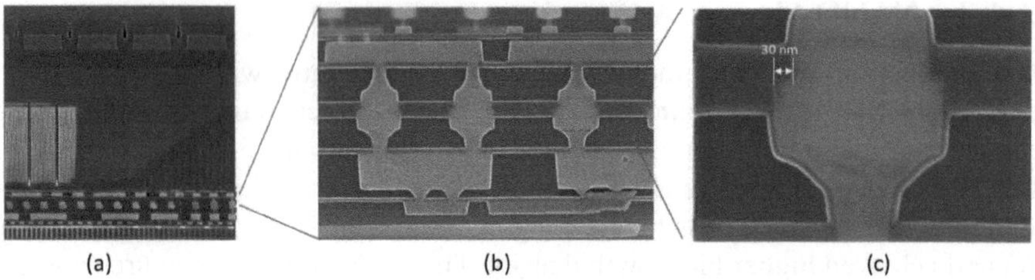

FIGURE 12.17 YMTC Xtacking® (a) 64-layer 3D NAND cross section, (b) Cu–Cu HB, and (c) Cu–Cu HB metal offset [15].

Comparing to 3D NAND architecture/wafers with CMOS under array and conventional periphery on side, hybrid bonded wafers produce denser memory dies that use less silicon real estate. As shown in Table 12.1, 128-layer hybrid-bonded NAND dies have approximately 8.5% smaller die size compared to CMOS under array NAND dies. With significant momentum of 64- and 128-layer 3D NAND using HB, YMTC introduced 232-layer 3D NAND using HB process. As shown in Figure 12.18, YMTC introduced new BSSC (backside source connect) technology to simplify the source connection process along with HB. Also significant improvement in cost and throughput was achieved by eliminating the interlayer between decks [20].

A few years ago, there were significant doubts around whether 3D NAND layers can exceed 200 due to the technical challenges around its enablement. Today, there are multiple SSD products with >200 layers that had overcome these technical challenges, including warpage and high-aspect-ratio etching. Now the next big question is: Will NAND ever reach >1,000 layers to support the exponential growth of higher speed, higher memory density, and low-power storage requirements? [20]. It is highly likely that HB may be a key contributor to reach the goal of >1,000 layers with lower memory cost. In addition to YMTC, more storage memory vendors are announcing roadmaps with HB. Recently, Kioxia and Western Digital announced their newest flash memory using HB [21]. Figure 12.19 illustrates a hypothetical future 3D NAND design, where each stack of memory is integrated on an individual wafer and those are hybrid bonded together to increase the layer count [15].

TABLE 12.1
Comparison of YMTC Hybrid Bonded 128-Layer NAND versus Storage Vendor's NAND Products [19]

NAND Technology	YMTC 128L Xtacking®	Samsung 128L V-NAND	Micron 128L CuA CTF	SK Hynix 128L 4D PUC
Device	Asgard AN4 1TB	Samsung EVO 870 1 TB	Crucial BX500 480 GB	SK Hynix Gold P31 1TB
Capacity per die	512 Gb	512 Gb	512 Gb	512 Gb
Die size	60.42 mm^2	74.09 mm^2	66.02 mm^2	63 mm^2

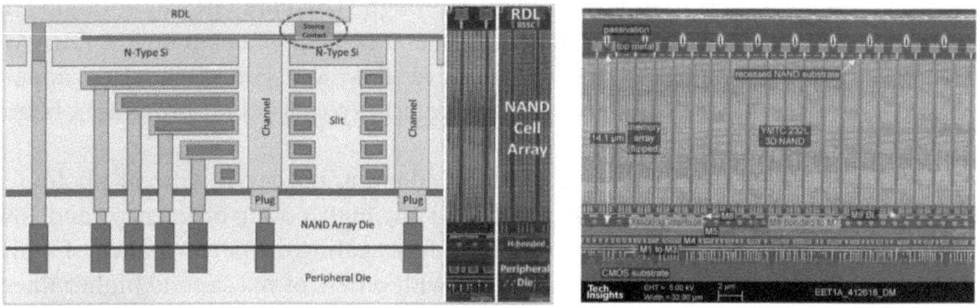

FIGURE 12.18 YMTC Xtacking®3.0 232-layer 3D NAND cross section [20].

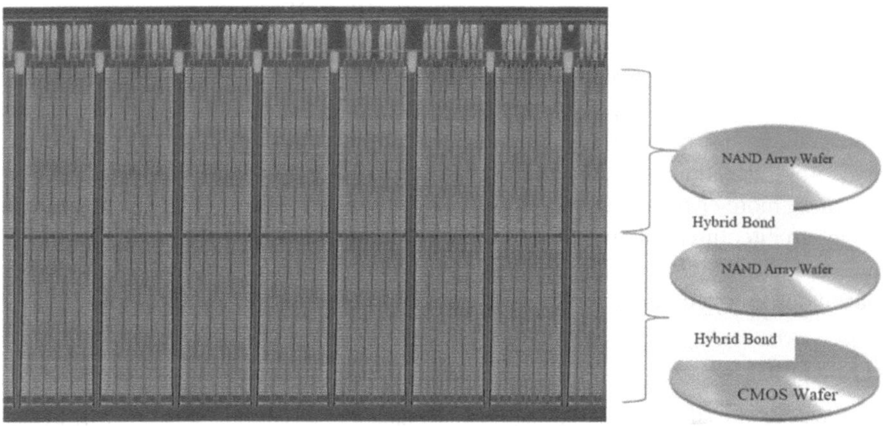

FIGURE 12.19 Toshiba-Kioxia 3D NAND cross section. Illustration of future 3D NAND stack using HB [15].

12.5 FUTURE APPLICATIONS

Beyond the early adoption and recent introduction of HB technology in the CIS market, 3D V-Cache™ and 3D NAND, many future applications of this technology are being actively explored to meet the never-ending need and requirement of further scaling in interconnection density and bandwidth, especially in the era of AI computing, which brings significant demand in both computing and memory capability.

High-bandwidth memory (HBM) density increases have been extremely sought after given the modern computing development, especially the AI computations that require more data and keep demand booming. However, the conventional 3D integration process using TSV and microbumps limited the stacking silicon thickness and interconnection density due to size and handling. Furthermore, microbump stacking and interposer connections inherited challenges such as large resistance, heavy capacitance, and a limited number of stacking levels. Contrarily, HB with WoW technology can increase the interconnect density from 10^3/cm^2 (microbump based) to 10^6/cm^2 (bumpless based) and provide a solution path for higher bandwidth and higher energy efficiency DRAM [22–24].

In addition to DRAM multilayer stacking through HB, more DRAM-to-logic HB applications are being investigated to address memory-bounded challenges. DRAM-to-logic direct interconnect through aggressive TSV and HB pitch also support DRAM scaling in another dimension, improve energy usage, and promote a better manufacturing and cost structure [25, 26].

Another exciting area for HB technology adaption resides in silicon photonics applications. Optical I/O, for its high bandwidth, high power efficiency, and low loss, is deemed as the inevitable solution for future scaling the performance of high-throughput SoC architectures. HB interfaces with orders of magnitude higher interconnect densities and Cu-to-Cu direct bonds can provide superior performance as the electrical–optical interconnects in the EIC/PIC chiplets integration [27].

The future of 3D stacking is a function of TSV pitch and can spawn many architectural innovations, including but not limited to IP-on-IP stacking; macro-on-macro stacking; IP folding/splitting, as well as circuit-level slicing. Three-dimensional stacking technology progression along with other advanced packaging techniques will enable beyond Moore's law scaling this decade and enable complex heterogeneous integration schemes not possible with monolithic designs. Design technology co-optimization (DTCO) and system technology co-optimization (STCO) are becoming more critical than ever to address the challenges around enabling more 3D chiplet architectures. Managing and mitigating thermal issues that result from high current densities, along with identifying efficient power delivery solutions, require continuous innovations in design, software, and integrated solutions.

REFERENCES

1. J. Wuu et al., "3D V-Cache™: The Implementation of a Hybrid-bonded 64 MB Stacked Cache for a 7 nm X86-64 CPU," ISSCC, 2022.
2. S. Sukegawa et al., "A 1/4-inch 8 Mpixel Back-illuminated Stacked CMOS Image Sensor," 2013 IEEE International Solid-State Circuits Conference Digest of Technical Papers, San Francisco, CA, 2013, pp. 484–485.
3. H. Wakabayashi et al., "A 1/2.3-inch 10.3 Mpixel 50 frame/s Back-illuminated CMOS Image Sensor," 2010 IEEE International Solid-State Circuits Conference (ISSCC), San Francisco, CA, 2010, pp. 410–411.
4. H. Rhodes et al., "The Mass Production of BSI CMOS Image Sensors," International Image Sensor Workshop, pp. 27–32, 2009.
5. Shin Iwabuchi, Y. Maruyama, Y. Ohgishi, M. Muramatsu, N. Karasawa, T. Hirayama, "A Back-illuminated High-sensitivity Small-pixel Color CMOS Image Sensor with Flexible Layout of Metal Wiring," 2006 IEEE International Solid-State Circuits Conference: Digest of Technical Papers, San Francisco, CA, 2006, pp. 1171–1178. doi: 10.1109/ISSCC.2006.1696162.
6. H. Tsugawa et al., "Pixel/DRAM/Logic 3-Layer Stacked CMOS Image Sensor Technology," 2017 IEEE International Electron Devices Meeting (IEDM), San Francisco, CA, 2017, pp. 3.2.1–3.2.4. doi: 10.1109/IEDM.2017.8268317.
7. T. Haruta et al., "4.6 A 1/2.3 inch 20 Mpixel 3-Layer Stacked CMOS Image Sensor with DRAM," 2017 IEEE International Solid-State Circuits Conference (ISSCC), San Francisco, CA, 2017, pp. 76–77. doi: 10.1109/ISSCC.2017.7870268.
8. Y. Kagawa et al., "An Advanced CuCu Hybrid Bonding for Novel Stacked CMOS Image Sensor," 2018 IEEE 2nd Electron Devices Technology and Manufacturing Conference (EDTM), Kobe, Japan, 2018, pp. 65–67. doi: 10.1109/EDTM.2018.8421453.

9. Y. Kagawa et al., "Development of Face-to-face and Face-to-back Ultra-fine Pitch Cu–Cu Hybrid Bonding," 2022 IEEE 72nd Electronic Components and Technology Conference (ECTC), San Diego, CA, 2022, pp. 306–311. doi: 10.1109/ECTC51906.2022.00057.
10. Y. Oike, "Evolution of Image Sensor Architectures with Stacked Device Technologies," in IEEE Transactions on Electron Devices, vol. 69, no. 6, pp. 2757–2765, 2022. doi: 10.1109/TED.2021.3097983.
11. L. Su, Computex Keynote, 2021.
12. R. Agarwal et al., "3D Packaging for Heterogeneous Integration," ECTC, 2022.
13. H. Kim, S.-J. Ahn, Y. G. Shin, K. Lee, E. Jung, "Evolution of NAND Flash Memory: From 2D to 3D as a Storage Market Leader," 2017 IEEE International Memory Workshop (IMW), Monterey, CA, 2017, pp. 1–4. doi: 10.1109/IMW.2017.7939081.
14. J. Lee, J. Jang, J. Lim, Y. G. Shin, K. Lee, E. Jung, "A New Ruler on the Storage Market: 3D-NAND Flash for High-density Memory and Its Technology Evolutions and Challenges on the Future," 2016 IEEE International Electron Devices Meeting (IEDM), San Francisco, CA, 2016, pp. 11.2.1–11.2.4. doi: 10.1109/IEDM.2016.7838394.
15. 3D NAND Memory Comparison, Yole SystemPlus, 2019. https://www.yolegroup.com/
16. Micron press release. https://investors.micron.com/news-releases/news-release-details/micron-ships-worlds-first-232-layer-nand-extends-technology
17. Z. Yang et al., "Asymmetric Etching Profile Control during High Aspect Ratio Plasma Etch," 2018 29th Annual SEMI Advanced Semiconductor Manufacturing Conference (ASMC), Saratoga Springs, NY, 2018, pp. 211–215. doi: 10.1109/ASMC.2018.8373146.
18. Yangtze Memory Technology Company (YMTC), "Crystal Stack Xtacking®: Technical Introduction," Available at: https://www.ymtc.com/ (accessed April 6, 2023).
19. https://www.3dincites.com/2021/12/iftle-504-ymtc-implements-hybrid-bonding-for-128-layer-3d-nand/
20. https://www.techinsights.com/blog/ymtc-leading-pioneer-3d-nand
21. https://www.kioxia.com/en-jp/business/news/2023/20230330-1.html
22. Y. C. Tsai, C. H. Lee, H. C. Chang, J. H. Liu, H. W. Hu, H. Ito, Y. S. Kim, T. Ohba, K.-N Chen, "Electrical Characteristics and Reliability of Wafer-on-Wafer (WoW) Bumpless Through-silicon Via," in IEEE Transactions on Electron Devices, vol. 68, pp. 3520–3525, 2021.
23. K. Sakui, T. Ohba, "High Bandwidth Memory (HBM) and High Bandwidth NAND (HBN) with the Bumpless TSV Technology," Proceedings of the International 3D Systems Integration Conference (3DIC), Sendai, Japan, October 8–10, 2019, pp. 1–4.
24. Y. S. Kim, S. Kodama, N. Maeda, K. Fujimoto, Y. Mizushima, A. Kawai, T. C. Hsu, P. Tzeng, T. K. Ku, T. Ohba. "Electrical Characteristics of Bumpless Interconnects for Through Silicon Via (TSV) and Wafer-on-Wafer (WoW) Integration," Proceedings of the International Conference on Electronics Packaging (ICEP), Hokkaido, Japan, April 20–22, 2016, pp. 74–78.
25. D. M. Niu, S. C. Li, Y. H. Wang, W. Han, Z. Zhang, Y. J. Guan, T. C. Guan, F. Sun, F. Xue, L. D. Duan et al., "184QPS/W 64 Mb/mm 23D Logic-to-DRAM Hybrid Bonding with Process-Near-Memory Engine for Recommendation System," Proceedings of the 2022 IEEE International Solid-State Circuits Conference (ISSCC), San Francisco, CA, February 20–26, 2022, pp. 1–3.
26. https://www.techinsights.com/blog/disruptive-event-china-breaks-through-restrictions-advanced-chiplet-strategy
27. H. Hsia et al., "Heterogeneous Integration of a Compact Universal Photonic Engine for Silicon Photonics Applications in HPC," 2021 IEEE ECTC, pp. 263–268.

Index

A

Activation energy, 135, 370, 375
Alignment accuracy, 71, 80, 81, 82, 83, 88, 89, 106, 122, 202, 204, 206–211, 218, 238, 256, 275
Autoclave, 93, 94, 161, 368, 369

B

Backside bonding layer fabrication, 63–65
Blade dicing, 77, 79
Bond energy, 58, 69, 70, 123, 124, 240
Bond pad formation, 184, 186, 188, 190, 195
Bond strength, 35, 55, 95, 99, 123–126, 205, 210, 211, 214, 229–231, 233, 259, 260, 340, 346, 347, 351, 371
Bonding layer fabrication, 59, 60, 63–66, 106
Bonding mechanism, 19, 20, 26, 56–58, 122, 381, 382, 384
Bonding temperature, 26, 33, 38, 39, 42, 44, 45, 121, 123, 195
Bonding void, 70, 72, 82, 83, 85, 106, 340, 371
Bump pitch, 6, 7, 14, 55, 131, 132

C

Chiplet architecture, 270, 278, 286, 288, 289, 292, 431, 432, 444
Chip-on-wafer (CoW), 299, 433, 434, 439
Coefficient of thermal expansion (CTE), 48, 59, 125, 226, 365
Computer modeling, 340, 347, 355, 363, 379
Coplanarity, 58, 154–156
Copper diffusion, 125, 141, 157, 345, 366, 368, 374
Copper grain, 154, 156, 157, 161, 162
Corrosion, 64, 77, 87, 92–94, 103, 169, 174, 179, 180, 362, 366, 367, 369
CTE mismatch, 9, 92, 93, 102, 125, 353, 387, 409
Cu oxide, 21, 22, 27–30, 35, 41
Cu recess, 64, 66, 68, 75, 92, 95, 97, 98, 106

D

Damascene, 56, 60, 75, 106, 126, 131, 140–143, 146, 148, 151, 152, 154–156, 159, 168, 173, 195
Debonding, 64, 87–89, 116–119, 122, 202, 211–219, 249, 397, 398, 400, 409
Die attach, 116, 121, 122, 240, 259
Die on tape, 79, 82, 85, 88, 89
Die to wafer, (D2W) 56, 121, 122, 216–219, 222, 369, 398
Dielectric layer, 58–61, 64, 78, 83, 93, 106, 204, 205, 229, 257, 315, 316, 318, 371, 380, 397, 421

E

Electrodeposition, 131–133, 141, 142, 145, 147, 151, 152, 155, 156, 161
Electromigration, 42, 61, 62, 92, 93, 172, 301, 340, 362

F

Failure mechanism, 361, 362, 365–368, 372, 379, 421
Failure mode, 255, 361, 362, 367, 368, 376, 377
Finite element, 101, 310, 316, 374, 384, 385, 388, 390, 395, 398, 400, 402, 404, 410, 413, 415
Flip chip, 2, 54, 55, 75, 79, 80, 102, 218, 222, 236, 239, 243, 287, 296, 362, 363
Fusion bond, 119, 201, 202, 204, 206, 208, 209, 217, 230, 240, 264

H

Heterogeneous integration, 1, 19, 107, 109–111, 119, 216, 219, 223, 228, 246, 270, 272, 280, 286, 288, 293, 300–302, 304, 365, 428, 444
Highly accelerated stress test (HAST), 40, 368, 372, 377, 379

I

Indirect alignment, 207, 208, 211
Interconnect density, 6, 7, 9, 12, 54, 105, 268, 290, 365, 430–433

J

Junction temperature, 315, 410

K

Known good die (KGD), 9, 75, 121, 216, 284, 248–250, 258, 284, 433, 434

M

Machine learning (ML), 54, 183, 272, 301
Mechanical stress, 212, 214, 275, 295, 296, 340, 365, 366–368, 374, 417, 421
Metrology, 95, 106, 127, 151, 159, 168, 170, 173, 183, 210, 215, 216, 231
Microbump, 54, 55, 82, 102–105, 107–109, 116, 187, 225–227, 265, 270, 274, 275, 283, 290, 301, 362–364, 410, 411, 413, 431–433
Multi-chip module (MCM), 430, 431, 440

O

Optical alignment, 205, 207–209

P

Parasitic capacitance, 105, 224, 411
Peeling stress, 381, 393, 397–400, 404, 406, 408
Pitch scaling, 55, 71, 89, 102, 106, 369
Plasma activation, 61, 62, 68, 69, 89, 95, 100, 122, 124, 204, 217, 230, 233, 237, 247–249, 252, 260
Plasma dicing, 77–79, 106, 127
Plasma treatment, 40–42, 58, 68, 123, 124, 127, 188, 204, 205, 230, 383
Plastic deformation, 44, 46, 343, 385
Polymer dielectric, 119, 122, 124–126, 404, 408
Prebonding, 23, 37, 48, 124, 194
Process modeling, 381, 382, 387, 398, 404

R

Redistribution layer (RDL), 6, 11, 60, 64, 83, 106, 119, 141, 142, 187, 256, 363, 364–366, 391, 430
Room temperature bonding, 26, 27, 47, 55, 123

S

Signal integrity, 14, 225, 268, 273, 277, 281, 282, 295–298, 380, 411
Silicon interposer, 6, 54, 430, 431
Silicon photonics, 299, 301, 444

T

Thermocompression, 202, 222, 265, 304, 404

W

Wafer-to-wafer bonding, 10, 186, 188, 192, 209, 216–218